Geomorphology and Reclamation of Disturbed Lands

Geomorphology and Reclamation of Disturbed Lands

Terrence J. Toy
Richard F. Hadley

Department of Geography
University of Denver
University Park
Denver, Colorado

ACADEMIC PRESS, INC.
Harcourt Brace Jovanovich, Publishers
Orlando San Diego New York Austin
Boston London Sydney Tokyo Toronto

ACADEMIC PRESS, INC.
Orlando, Florida 32887

United Kingdom Edition published by
ACADEMIC PRESS INC. (LONDON) LTD.
24–28 Oval Road, London NW1 7DX

Library of Congress Cataloging in Publication Data

Toy, Terrence J.
Geomorphology and reclamation of disturbed lands.

Includes index.
1. Geomorphology. 2. Reclamation. 3. Man—Influence on nature. I. Hadley, R. F. (Richard Frederick) Date . II. Title.
GB406.T69 1987 551.4 86-32041
ISBN 0–12–696960–4 (alk. paper)

PRINTED IN THE UNITED STATES OF AMERICA

87 88 89 90 9 8 7 6 5 4 3 2 1

To our wives, Linda and Gerry, for their unabated support and patience during this project, before and after.

Contents

5. The Drainage Basin: Geomorphic and Hydrologic Response

6. Lands Disturbed by Grazing

7. Lands Disturbed by Recreational Use

8. Lands Disturbed by Surface Mining of Coal

9. Lands Disturbed by Surface Mining of Uranium

10. Lands Disturbed by Construction Activities

11. Geomorphic Perspectives on the Design and Management of Disturbed Lands

Preface

The science of geomorphology concerns the character of the earth's land surface, the forms and processes responsible for its development and evolution. An understanding of geomorphic processes is essential to the comprehension of the consequences of human disturbance on environmental systems and the formulation of effective reclamation programs. Hence, it is the purpose of this book to demonstrate the linkages among geomorphic principles, environmental impacts caused by human activities, and the effectiveness of reclamation practices. While there are many descriptive accounts of environmental degradation resulting from various land uses, we emphasize the geomorphic processes responsible for such changes and the reasons why various reclamation practices are valuable in environmental management.

A lengthy list of human activities that degrade the natural environment could be compiled. It is impractical, however, to attempt to address all of these. Therefore, we choose to focus on a few activities that produce various levels of impact and present somewhat different challenges to geomorphologists, hydrologists, or reclamation specialists. Conspicuous by their absence are the topics of agriculture and anthropically accelerated coastal erosion; silviculture is treated briefly in the chapter concerning construction. We feel that agriculture is adequately discussed in numerous other books, and reiteration is unwarranted. Neither of us has direct field experience with coastal processes, and so it seemed prudent to concentrate on the processes with which we have been directly involved.

We decided that our discussion of selected types of disturbances should include consideration of the legal framework surrounding the activity wherever possible. This has become a very real operational constraint for some industries and their reclamation personnel. If the industries engaged in activities that disturb the environment and the regulatory agencies charged with protecting the environment can acquire a better understanding of the physical processes involved in reestablishing stable landscapes, many practical and political problems could be alleviated. We have attempted to clarify some of these issues in this book.

The text is intended for both an academic and industrial audience. In addition to a comprehension of the relationships among geomorphology, environmental impacts, and reclamation, the members of academe will, we hope, gain an appreciation of some economic and legal considerations that influence industrial decision-making and operations. This book is probably the most useful for advanced undergraduate and graduate students, together with their professors. In addition to an understanding of the characteristics and complexities of environmental systems, both natural and disturbed, the readers from industry will, we hope, discover that there is commonly a firm scientific foundation for most laws and regulations that govern their activities. Recognizing diverse earth-science backgrounds for each group, we attempt to simplify concepts whenever possible. To be sure, our effort to serve more than one master in our writing has proved to be a distinct challenge.

The subject matter is divided into three sections. The first deals with basic principles of geomorphology and, in addition to the Introduction, contains chapters on geomorphic concepts and processes, hillslopes, streams, and drainage basins (Chapters 1–5). The second section presents an analysis of several types of land disturbance and includes chapters on grazing, recreational uses, surface mining of coal, surface mining of uranium, and construction activities (Chapters 6–10). The third section consists of a single chapter (Chapter 11) that discusses geomorphic perspectives for the design and management of disturbed lands. Although this book is entirely a joint venture, some may be interested in the person responsible for individual chapters: Terrence Toy—Chapters 3, 8, 9, 10, 11, and part of 2; Richard Hadley—1, 4, 5, 6, 7, and part of 2.

Acknowledgments

A number of our colleagues graciously invested their time and wisdom toward the enrichment of this volume. We extend sincere appreciation to Athol Abrahams (Chapter 11), Genevieve Atwood (Chapter 8), Robert Churchill (Chapter 3), Gregg Lusby (Chapter 6), R. Dale Smith (Chapter 9), Des Walling (entire manuscript), and Wally Walters (Chapter 9). Should a residual aberration be discovered, the responsibility must be ours.

For permission to publish copyrighted material, we gratefully acknowledge the following publishers:

Chapter 2:

Figures 1 and 5	John Wiley & Sons, Schumm (1977)
Figure 2	Prentice-Hall International, Inc., Chorley and Kennedy (1971)
Figure 6	Annals, American Association of Geographers, Peltier (1950)
Figure 8	American Geophysical Union, Langbein and Schumm (1958)
Figure 9	American Journal of Science, Graf (1977)
Figures 10 and 12	Prentice-Hall, Inc., Steila (1976)
Figure 13	Kendall-Hunt Publishing Co., Branson *et al.* (1981)
Figure 14	W. H. Freeman and Co., Dunne and Leopold (1978)
Figure 17	Cambridge University Press, Carson and Kirkby (1972)
Figure 19	McGraw-Hill Book Co., Morisawa (1968)
Figure 20	W. H. Freeman and Co., Leopold, Wolman, and Miller (1964)

Chapter 3:

Figures 1b, 2a	Royal Tropical Institute, Ruhe and Walker (1968)
Figures 1c, 2b	Longman Group Ltd., Young (1977)
Figure 4	Houghton Mifflin Co., Ruhe (1975)
Figure 5	Zeitschrift für Geomorphologie, Dalrymple *et al.* (1968)
Figure 6	Geological Society of America, Toy (1977)

Chapter 4:

Figure 6	John Wiley & Sons, Schumm (1977)

Chapter 5:

Figure 3	Geological Society of America, Strahler (1952)
Figures 8 and 9	International Association of Hydrological Sciences, Walling and Kleo (1979)

Chapter 6:

Figure 1	John Wiley & Sons, Baver (1965)
Figure 2	Kendall-Hunt Publishing Co., Branson *et al.* (1981)
Figure 4	American Geophysical Union, Langbein and Schumm (1958)

Chapter 7:

Figure 2	Springer-Verlag, Inc., Webb (1983)

Chapter 8:

Figure 8	American Society of Agronomy, Crop Science Society of America, Soil Science Society of America, Gardner and Woolhiser (1978)

Chapter 10:

Figure 3	American Geophysical Union, Wolman and Schick (1967)
Figure 4	Charles E. Merrill Publishing Co., Keller (1985)
Figure 5	Geografisker Annaler, Wolman (1967)
Table 13	Van Nostrand Reinhold Co., Gray and Leiser (1982)

Complete references can be found at the end of each chapter. In addition, we feel that it is appropriate to recognize the essential research published by various government agencies in the United States. This information, which is in the public domain, was of critical importance in the preparation of this work.

We wish to convey our gratitude to the Department of Geography at the University of Denver (Chairman, Laurance C. Herold) for its assistance in the production of our book.

Finally, we would like to especially acknowledge the assistance, cooperation, and, foremost, patience of our administrative assistant and typist, Grace Bader, in the preparation of this text.

Geomorphology and Reclamation of Disturbed Lands

1

Introduction

Purpose of the Book

The study of landscapes, together with the physical and chemical processes responsible for the types of landforms and rates of formation, constitutes, in essence, the science of geomorphology. Coincident with the development of geomorphic principles in the past 100 years, there has been an ever-increasing percentage of the land surface of the earth subjected to disturbance by human activity. This disturbance is the consequence of numerous endeavors, such as agriculture, silviculture, construction of highways and pipelines, urbanization, exploitation of minerals and petroleum, recreational land uses, introduction of grazing herds, and a variety of other land uses. Agriculture and silviculture land use, by themselves, have impacted large areas of the earth for many centuries and have been the subject of extensive research and evaluation. Therefore, we will not attempt to reiterate the results of the voluminous literature in this book except in a peripheral way. Rather, we will consider primarily the effects of land disturbance by surface mining, recreational land uses, grazing, and the construction of highways and urban areas. We have chosen this group of disturbances because of the differences in areal extent, intensity, and duration. Each is an example of categories of disturbance to which there may be numerous members. These disturbances alter landforms, hydrologic regime, and erosional stability in varying degrees depending on the areal extent of the disturbance, the climatic regime, and the legal constraints pertaining to mitigation of impacts by reclamation of disturbed lands.

In order to evaluate the impacts of human activity on landscapes, and to develop possible mitigations for a wide variety of land disturbances, we must first understand the interrelated geomorphic and hydrologic processes that operate on hillslopes, stream channels, and drainage basins. Landforms result from physical and chemical weathering of unconsolidated surficial materials and rock, subaerial erosion, mass movement, sediment transport, and deposition. Each of these processes operates at a rate that is governed, directly or indirectly, by inputs of energy from solar radiation and precipitation (climate) and forces of gravity. Thus, if we equate environment with the composite of climate, vegetation, geology, and soils, the variety of naturally formed landforms and the processes that shaped them are greatly influenced by environment. This analogy, of course, is proposed with the recognition of the influence exerted by differences in rock type, soils, vegetation, and geologic structure.

It is apparent that when we are faced with the problems associated with land disturbance, people from many disciplines are involved: hydrologists, soils scientists, foresters, engineers, and planners, as well as geomorphologists. As pointed out by Ritter (1978), the geomorphologist probably understands the natural framework of geologic processes involved in causes and effects encountered in the surficial environment better than other groups of scientists or engineers. This statement reinforces our belief that there is a need for a book that illustrates the role of process geomorphology in developing solutions to the problems of land disturbance.

The Problem

The phenomenon of land disturbance is not new on the face of the earth. As noted by Box (1978), humans have always disturbed the land. The fact that the activities of early man did not alter the landscape appreciably may be attributed to the small population and the lack of refined tools. Nevertheless, it represents the beginning of the cumulative effects that we observe today. Accelerated erosion caused by human overuse of the land was confined to parts of Eurasia and was of limited areal extent and duration until about the sixteenth century (Butzer, 1974). Population increases and colonization of the remainder of the earth since that time have increased environmental deterioration until it is the serious problem encountered today worldwide.

Many investigators have estimated the impacts of human activity on erosion of upland hillslopes, transport of sediment, and deposition of sediment on lowlands and valley floors, in estuaries, and along coastlines. Evidence for these estimates of erosion and sedimentation has been gleaned from historical accounts and studies of the stratigraphy of alluvial deposits. There are some problems involved in accepting these estimates. Pre- and postdisturbance conditions are generally inferred, based on numerous assumptions, and are rarely accompanied by quantitative data based on measurements. However, the relative differences are acceptable in illustrating the impact of land abuse.

Hughes (1982) compiled data on the extent of erosion and sedimentation in the classic Greek and Roman periods and concluded that human activity, primarily deforestation and grazing, accounted for severe erosion problems. In another study of that same period, Judson (1968) estimated that erosion rates near Rome increased dramatically during the second century B.C. He states that the erosion rate increased from an average of 2 to 3 cm/1000 yr before the second century B.C. to 20 to 40 cm/1000 yr after that time; this represents an increase of from 700 to 2000%. Judson also attributes the sharp increase in erosion to the removal of trees from forested areas.

In more recent time, society has become more complex and land disturbances now reflect the impacts brought about by advanced technology in surface mining, highway construction, dam construction, and many other activities in addition to the continuing impacts of agriculture and silviculture. The areal extent of land disturbances generally has also increased with technological advances. For example, dam construction affects large areas upstream and downstream from a dam site (Cooke and Doornkamp, 1974). On the other hand, many human activities only influence a very limited area near the location of the disturbance.

In the conterminous United States alone (excluding Alaska and Hawaii) 57% (1.28 billion acres) of the land mass is devoted to agricultural use and 43% (981 million acres) is occupied by non-agricultural uses (Paone et al., 1978). The most significant nonagricultural land use is ungrazed forest land, which is followed by urban areas, recreation and wildlife habitats, transportation networks, and public installations (Paone et al., 1978). The types of landscape disturbance that we will consider affect a small percentage of the land area, but they are probably the most destructive in altering landforms and producing erosional and sedimentation problems. Here, it seems that area, intensity, duration, and the effects of both disturbance and reclamation on these components constitute the key elements.

Objectives

The objectives of this book can be summarized as follows: (1) to describe the fundamental geomorphic and hydrologic processes that control the development and evolution of landforms and that are primarily related to inputs from solar energy, hydrology, and sediment transport; (2) to illustrate the influence of climate, soil, and vegetation factors on geomorphic and hydrologic process rates; (3) to demonstrate the variabilities, in time and space, of land disturbances as they are related to geomorphic systems including hillslopes, stream channels, and drainage basins; (4) to describe the effects of land disturbances on geomorphic and hydrologic processes by using case studies that represent a variety of land uses and to propose possible mitigations through reclamation practices; and (5) to propose some principles of landscape design and management based upon established geomorphic principles that may lead to long-term stability and reduced maintenance costs on reclaimed or reconstructed landscapes.

What Constitutes Disturbances?

Landforms that comprise the earth's landscapes tend toward dynamic equilibrium if we assume that they have been developed during a long and continuous period. Variability of landscape features in space generally is controlled by rock and soil type, vegetation, climate, and geologic structure. If the surface is disturbed by human activity, such as surface mining, urbanization, or highway construction as examples, the vegetation is destroyed, soil properties are altered, and erosion is generally greatly accelerated. The geomorphic effect of such disturbances is the thesis of our book. It is recognized that many of these disturbances are minor and transitory and that the predisturbance landscape often is reclaimed to a productivity and form that may be more acceptable and conforms to a prior use plan.

Using surface mining of coal as an example of land disturbance, Toy (1984) visualizes a complete geomorphic investigation as being divided into three time periods: (1) predisturbance, (2) active disturbance, and (3) postdisturbance. During the predisturbance period, the landforms and geomorphic processes of an area are taken to represent natural conditions, although other human activities or land use may invalidate this assumption. To the extent that the surficial environment is undisturbed, there exists an approximate balance between the environmentally controlled geomorphic processes and surface form. Geomorphic data collected during this period constitute the "baseline" and provide a goal for eventual reclamation programs.

The active disturbance period is a time of maximum disequilibrium between surface form and geomorphic processes. Consequently, process rates may be vastly different from those measured during the predisturbance period as the systems operate to reestablish a balance between prevailing forces and resistances. Without effective management, on-site and off-site environmental impacts during this period of natural adjustment could be damaging to adjacent undisturbed lands.

The postdisturbance period follows the application of reclamation practices. Restoration of an entirely balanced condition between surface form and geomorphic processes is impossible; therefore, a state of disequilibrium will remain. However, surface features should reflect those of the predisturbance period that will tend to reduce disequilibrium. As the vegetation cover improves in density and vigor, process approaches a balance with form. Ideally, process rates should approach the "baseline" conditions of the predisturbance period. This succinct and simplified description summarizes the approach we will use in organizing our discussion of geomorphic systems and their response to surficial disturbances.

References

Box, T. M., 1978, The significance and responsibility of rehabilitating drastically disturbed lands: *in* Reclamation of drastically disturbed lands, Madison, Wisconsin, Am. Soc. of Agronomy, Crop Science Soc. of Am., Soil Sci. Soc. of Am., pp. 1–10.

Butzer, K. W., 1974, Accelerated soil erosion—a problem of man-land relationships: *in* Perspectives on environment, I. R. Manners and M. W. Mikesell, editors, Washington, D. C., Association of American Geographers, p. 57–78.

Cooke, R. U. and Doornkamp, J. C., 1974, Geomorphology in environmental management—an introduction: Oxford, Clarendon Press, 413 p.

Hughes, J. D., 1982, Deforestation, erosion, and forest management in ancient Greece and Rome: Journal of Forest History, v. 26, no. 2, pp. 60–75.

Judson, S., 1968, Erosion rates near Rome, Italy: Science, v. 160, pp. 1444–1446.

Paone, J., Struthers, P., and Johnson, W., 1978, Extent of disturbed lands and major reclamation problems in the United States: *in* Reclamation of drastically disturbed lands, Madison, Wisconsin, Am. Soc. of Agronomy, Crop Sci, Soc, Am., Soil Sci. Soc. Am., pp. 11–22.

Ritter, D. F., 1978, Process geomorphology: Dubuque, Iowa, Wm. C. Brown Co., Publishers, 603 pp.

Toy, T. J., 1984, Geomorphology of surface-mined lands in the western United States: *in* Developments and Applications of Geomorphology, J. E. Costa and P. J. Fleisher, editors, Berlin, Springer-Verlag, pp. 133–170.

2

Geomorphic Concepts and Processes: Application to Disturbed Lands

Introduction

Progress in evaluation of the effects of land disturbance and development of techniques to mitigate the impacts on landforms by reclamation practices require an understanding of the processes that shape the landscape. The complex nature of landforms, both in form and behavior, and the processes that produce them result from the interaction of several components at the surface of the earth: the atmosphere, hydrosphere, lithosphere, and biosphere. In order to comprehend the complexities of geomorphology, we must reduce the interrelated components of these spheres into their simplest parts (Ritter, 1978). In the past 40 years, the study of geomorphology has been directed largely to an understanding of processes and the quantification of responses. Strahler (1952) was one of the foremost leaders in this shift to quantification, and Hack (1960) renewed the concept of equilibrium.

Strahler (1952) proposed a dynamic approach to geomorphology. His approach was based on principles of physics and fluid dynamics, and his purpose was to enable the field investigator to view processes as manifestations of various types of shear stresses. Utilizing this perspective, his work focused on the complex interrelationships among the factors that are responsible for the shaping of landforms.

Strahler (1952) divides the shear stresses that affect earth materials into two major categories: gravitational and molecular. The gravitational stresses activate all downslope movements of matter, including mass movement and fluvial processes. Molecular stresses may act in any direction with respect to gravity and are those induced by temperature changes, absorption of water and dessication, and freeze–thaw cycles. A third category, chemical processes, is considered separately and includes processes of solution and acid reactions. Chemical processes do not produce shear stresses directly yet are important in landform development. Strahler viewed the dynamic approach as requiring analysis of geomorphic processes in terms of well-defined open systems that tend to achieve steady states of operation and are self-regulatory.

Forces and Resistance

With the move toward quantification in studies of landforms, we have become more aware of geomorphic processes and the mechanisms of their operation. Geomorphic processes operate when geomorphic work is done. Landforms are a reflection of a balance between driving forces in geomorphic systems and the resistance of earth materials upon which work is being done.

The driving forces that impart energy to earth materials for use in geomorphic work can be simply defined for our purposes into two groups: climate and gravity. Climate, which results from the distribution of solar energy in the atmosphere, may be reduced to temperature and precipitation patterns around the earth. The energy transmitted by solar radiation and raindrop impact control, to some degree, fluvial processes operating on the hillslopes of landscapes. The second major driving force is gravity. Gravity, together with the variables of climate, controls the fluvial processes of mass movement, overland flow, and streamflow, which will be discussed in detail in Chapters 3 and 4.

The resisting framework in geomorphology, as discussed by Ritter (1978), is composed of two major variables, lithology and structure. Rocks and surficial mantle materials of different chemical and mineralogic composition respond to weathering and erosion processes in different ways and rates. Geologic structure affects the topography (relief) due to differential erosion caused by varying resistance to erosion of rock units and deformation resulting from tectonism.

Magnitude and Frequency

A primary contrast between undisturbed and disturbed systems centers on the concept of magnitude and frequency of forces. For our purposes, we may simply express this concept in the form of two questions: (1) How much force is exerted by a particular process, and (2) How often does it operate? Together, magnitudes and frequencies of particular forces will determine the amount of work performed by those forces. Wolman and Miller (1960) concluded that the relative amount of work done during different events is not necessarily correlated with their importance in the shaping of landscapes, including hillslope forms and river patterns.

Although considerable attention has been focused on the role of extreme events by some geomorphologists, probably most still accept the conclusion of Wolman and Miller (1960) that a large part of geomorphic work is the consequence of events that generate a moderate amount of force and recur with moderate frequency. Their analysis was based primarily on examples of sediment transport in streams, and unfortunately, there is no well-documented counterpart related to hillslope processes.

Geomorphic Systems

The systems approach to geomorphology was amplified by Chorley (1962), who recognized the value of a general framework for geomorphic investigation based on a general systems theory. Chorley drew an analogy between open systems and drainage basins, slopes, and streams that are all maintained by continual fluxes of material and energy. Chorley and Kennedy (1971) expanded the concept further and described the drainage basin open system as an exchange of both mass and energy with their surroundings. The drainage basin receives energy or *input* from solar radiation, precipitation in the form of rainfall or snow, and relief of topography. The mass is represented by water, dissolved solids, and erosional debris. The *output* is heat, water, and erosional debris, both inorganic and organic material. They further explained the increasing complexity of natural systems by establishing three types of systems: morphologic systems, cascading

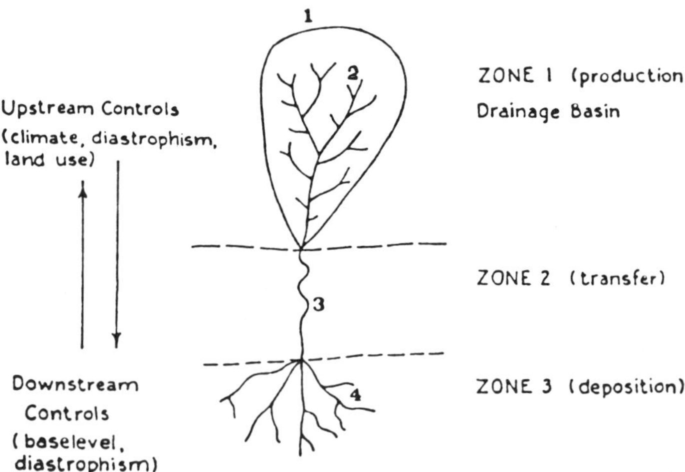

Figure 1. Diagram of an idealized fluvial system (Schumm, 1977. Copyright ©, reprinted by permission of John Wiley & Sons, Inc.)

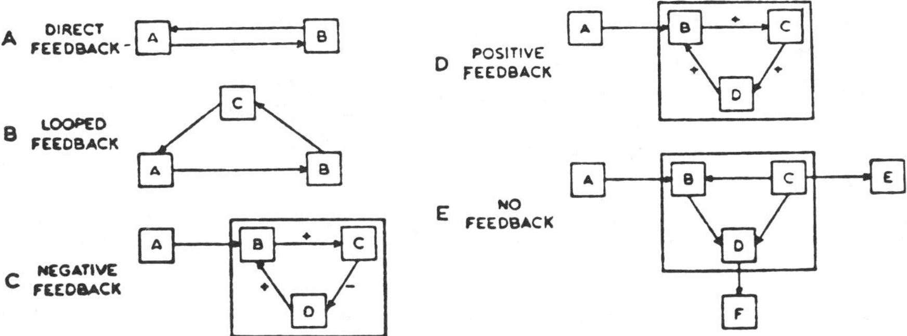

Figure 2. Symbolic diagrams illustrating the various types of feedback relationships (Chorley and Kennedy, 1971).

systems, and process–response systems. Schumm (1977), in describing the fluvial system, used the morphologic and cascading subsystems. He viewed the fluvial system as being divided into three zones: the drainage basin, the zone of transfer or transport, and the zone of deposition (Fig. 1). In each of these zones the landforms comprise the morphologic system and the cascading system is represented by the energy and matter flowing through each zone. Outputs from one subsystem become inputs to the adjacent downstream subsystem.

The process–response system as described by Chorley and Kennedy (1971) involves linkage between at least one morphologic system and one cascading system so that the process–response system demonstrates the manner in which *form* is related to *process*. Chorley and Kennedy (1971) cited an example by identifying relationships between processes and forms. When the infiltration cascade fills soil moisture storage, the result is overland flow, linking overland flow with the streamflow cascade. In this way, the morphologic systems of the slope and the stream channel become linked.

Feedback Mechanisms

Feedback in physical systems has been described by Chorley and Kennedy (1971) as the feeding back of part of the output of a system for another phase of operation, especially for self-regulation or control. Through this mechanism, the output is adjusted to input in natural geomorphic systems, and a measure of dynamic equilibrium is established within the system. These feedbacks may be either direct or looped (Fig. 2), involving one or more components of a system, and these feedbacks also may be either positive or negative. The most common mechanism in natural systems is negative feedback, whereby an externally generated variation produces a closed loop of change, which tends to stabilize the effect of the original change at a new equilibrium or a return to the former equilibrium state (Fig. 2c). A commonly used example of negative feedback in a geomorphic system is the graded stream, which transports all of the debris load delivered to it. If the external inputs of discharge and load remain constant, the graded stream neither erodes nor deposits.

Positive feedbacks occur when an external change in input is reinforced through a closed loop and changes increase in the same direction as the initial action (Chorley and Kennedy, 1971) (Fig. 2d). Positive feedbacks have an inherent self-destructive element and therefore can operate only over limited time periods. An example of a positive feedback loop exists where a decrease in infiltration capacity in the soil mantle on a hillslope causes an increase in surface runoff which, in turn, causes an increase in hillslope erosion (Fig. 3). Accelerated erosion will contine as the more permeable layers of soil are removed and

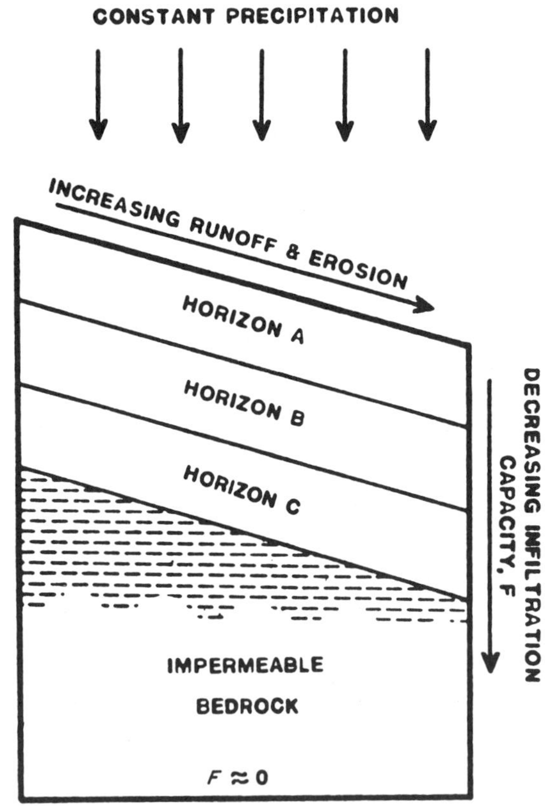

Figure 3. Hypothetical example of positive feedback mechanisms.

the infiltration capacity is further decreased. The positive feedback loop can only continue to operate as long as erosion progressively decreases the infiltration capacity (Chorley and Kennedy, 1971).

Ritter (1978) discusses the feedback mechanisms that are common in geomorphic systems, and he frames them in the context of resisting forces of lithology and geologic structure. These variables leave their mark on landforms as well as on the operation rates and types of geomorphic processes. Lithologic types influence runoff, erosion rates, debris loads in streams, and channel characteristics because of physical properties of sediment. This interrelationship is primarily a process–response phenomenon, but the adjustments that occur in channel geometry, longitudinal profile, and hillslope form with lithology differences also reflect the feedback mechanisms that are ubiquitous in geomorphic systems. For example, it was found that different lithologies produced widely different long-term sediment yields in the Cheyenne River basin of eastern Wyoming (Hadley and Schumm, 1961). Rock units that are predominantly shale have lower infiltration rates and consequently produce more runoff, and rock units that are predominantly sandstone have higher infiltration rates and runoff rates are generally lower.

The relatively simple concept of self-regulation or control by negative feedback is complicated by the existence of *geomorphic thresholds* and *complex response*. Complex response of geomorphic systems results from external changes in input that affect the operation of one or more processes in a system, and a threshold is a failure point that can produce drastic change in a geomorphic system in a short period.

The Concept of Thresholds

When considering the maintenance of an equilibrium state through the operation of negative feedback, the presence of thresholds presents problems because the passage of a system across a threshold may be irreversible (Chorley and Kennedy, 1971). *Geomorphic thresholds* were defined initially as the condition at which there is a significant landform change without a change of external controls such as base level, climate, and land use (Schumm, 1979). The definition has been expanded to include abrupt landform change resulting from progressive change of external controls. In a geomorphic system, such as a stream channel, a change in runoff characteristics may not always produce slow, continuous change. However, a threshold or failure point may be reached, and a short period of drastic change may occur.

Schumm (1973) further divided thresholds into *extrinsic thresholds* and *intrinsic thresholds*. An extrinsic threshold is triggered by changes in an external variable such as precipitation. The threshold exists within the system, but the abrupt change occurs under the influence of an external variable. An intrinsic threshold is activated when input is relatively constant, but a progressive change of the system produces instability and failure occurs. Schumm defined geomorphic thresholds as those that are developed within geomorphic systems, such as hillslopes, and produce changes with time. Field evidence supports the concept of geomorphic thresholds. Investigations of valleys of ephemeral streams in Wyoming showed that the occurrence of discontinuous gullies is related to the slope of the valley floor (Schumm and Hadley, 1957). The point where incipient gully erosion begins is generally on the steepest part of the valley floor (Fig. 4). Applying the concept of geomorphic thresholds to these observations, Schumm (1973) concluded that for a given region of relatively uniform geology, land use, and climate, a critical valley slope will exist beyond which the valley floor is unstable.

To further test this hypothesis, Patton and Schumm (1975) measured valley gradients in the Piceance Creek basin of northwestern Colorado where there is concern for environmental problems that may result from the mining of oil shale deposits. They made measurements of valley floor gradients in valleys that were occupied by discontinuous gullies and in valleys where no gullies existed. They also measured the drainage areas upstream from each gully. They developed a relation between drainage area and valley slope that shows the distribution of gullied and ungullied valleys (Fig. 5). This study defines, for a given drainage area, the valley slope where instability will occur. It should be noted that only drainage basins larger than about 4 sq miles (10.4 sq km) were used because of

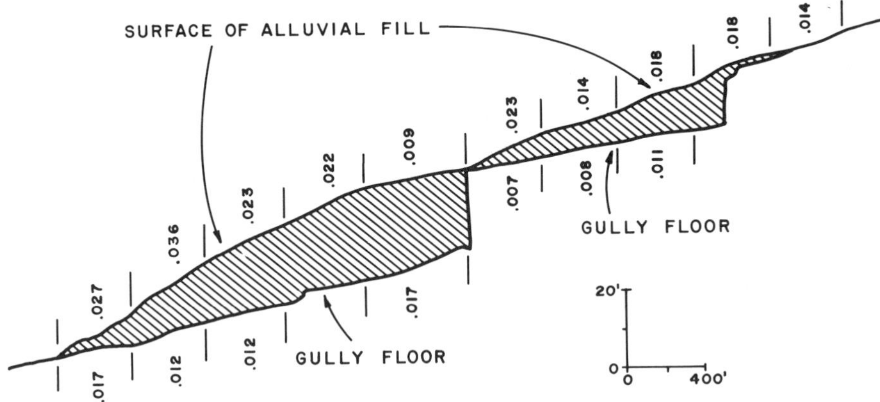

Figure 4. Profiles of discontinuous gullies, eastern Wyoming. Figures are gradients in feet per foot for each segment of profile (Schumm and Hadley, 1961).

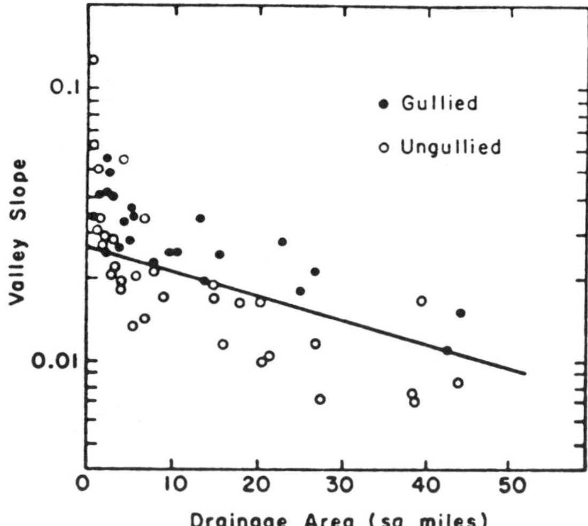

Figure 5. Relationship between valley slope and drainage area, Piceance Creek basin, western Colorado (Patton and Schumm, 1975. Copyright ©, reprinted by permission of John Wiley & Sons, Inc.)

variations in aspect, vegetation cover, and properties of alluvium that existed in the smaller basins. This is a good example of a threshold condition that may be used to predict the stability or instability of a geomorphic system.

Effects of Process

The concept that landforms are the products of geomorphic processes operating on earth materials through time is recognized by most geomorphologists. Expressed in terms of forces and resistances, the processes are dynamic, driving forces, while resistance is a property of the surficial materials upon which the forces impinge. Work is the product of this interaction between force and resistance.

The geomorphic processes operating on landscapes may be divided into two types: (1) *endogenetic* and (2) *exogenetic*. Endogenetic processes are generated within the earth's crust and mantle and include diastrophism and volcanism. These processes produce the primary topographic relief. Exogenetic processes are those that operate at or near the earth's surface. These processes include weathering, mass movement, erosion, and deposition, and they will vary with climatic conditions and operate at different rates depending on climate.

Effects of Geomorphic Scale

Geomorphic events that involve both processes and landforms occur in the real world at varying scales of time and space. In most studies of landscapes that have been disturbed by human activities, observations and data collection are confined to small-scale landforms and short time spans. This is primarily due to the small areal extent of most disturbances and the mitigating measures imposed by reclamation practices to restore equilibrium as soon as possible.

Time Scales

In an attempt to determine the significance of time spans in geomorphic interpretation of cause and effect, Schumm and Lichty (1963) considered that time in geomorphology can be divided into three classes: cyclic, graded, and steady time.

Cyclic time represents a long period of time, or as visualized by Schumm and Lichty (1963), it represents the span from the present back to the beginning of the erosion cycle. A landscape in cyclic time is continually changing as material is re-moved by erosion. Time is probably the most important variable of a cyclic time span.

Graded time is a much shorter time span than cyclic time and represents a period within cyclic time when a graded condition or dynamic equilibrium exists. Landforms are in a state of equilibrium with the processes acting on them. An entire system, for example, a drainage basin, that is losing mass by erosion cannot be in dynamic equilibrium. Therefore, graded time applies only to discrete components of the system, such as a hillslope or stream segment.

Steady time applies to very short time spans, a few months or years. A true steady state exists when the variables do not change with time and the landforms are considered to be time independent. Steady-state conditions apply to very small areas and short time spans. Schumm and Lichty (1963) showed that drainage basin variables vary in significance depending on the period of time being considered (Table 1). In the table, the major divisions are the three time spans previously described.

Spatial Scales

Scales of both time and space are closely related in geomorphic phenomena. At different scales of space, different processes become significant. For example, in a drainage basin, erosional processes on steep headwater hillslopes may produce large quantities of debris per unit area. Farther downstream where depositional features such as valley floors and floodplains are well developed, sediment yield per unit area is generally much less. This example also applies to the determination of downstream cumulative impacts of land disturbance where

Table 1 The Status of Drainage Basin Variables during Time Spans of Decreasing Duration[a]

Drainage basin variables	Status of variables during designated time spans		
	Cyclic	Graded	Steady
Time	Independent	Not relevant	Not relevant
Initial relief	Independent	Not relevant	Not relevant
Geology (lithology, structure)	Independent	Independent	Independent
Climate	Independent	Independent	Independent
Vegetation (type and density)	Dependent	Independent	Independent
Relief or volume of system above base level	Dependent	Independent	Independent
Hydrology (runoff and sediment yield per unit area within system)	Dependent	Independent	Independent
Drainage network morphology	Dependent	Dependent	Independent
Hillslope morphology	Dependent	Dependent	Independent
Hydrology (discharge of water and sediment from system)	Dependent	Dependent	Dependent

[a]Source: Schumm and Lichty (1963).

spatial scale must be considered in geomorphic interpretation. The geomorphologist must be aware of the spatial scale of the problem he or she is attempting to solve and how scale differences affect processes and response.

Effects of Climate

One of the most important factors responsible for the spatial and temporal variabilities of landforms is climate. If we assume that the components of a landscape that constitute a drainage basin represent an open system, then the two inputs of energy to the system attributable to climate are precipitation and solar radiation. As pointed out by Peltier (1950), climatic distinctions based solely on mean annual precipitation and temperature are obviously not refined; however, they do permit a generalized regionalization of the earth into nine morphogenetic regions based on the relative significance of geomorphic processes in relation to climate (Peltier, 1950). Peltier was not the only one to recognize the relationship of climate to landforms and process rates. The work of French geomorphologists (Tricart and Cailleaux, 1965) describes the basis for defining morphoclimatic zones, and Büdel (1948) proposed a system of morphogenetic zones. All of these proposals are simply a "reflection of the effect of climate on geomorphological processes over the surface of the earth" (Gregory and Walling, 1973, p. 11).

The landscapes of Peltier's nine morphogenetic regions as he visualizes them are a result of mechanical and chemical weathering processes, followed by the transport, erosion, and deposition of the weathering products. He concluded that chemical weathering increases with more precipitation and higher temperature because the increased vegetation cover in warm humid climates produces organic acids in soils (Peltier, 1950). Mechanical disintegration, according to Peltier, is largely produced by frost action in the presence of water and frequent temperature fluctuations about the freezing point (Peltier, 1950). Therefore, mechanical action is strongest in regions with mean annual temperatures in the range between −1° and −18°C (0° and 30°F) and mean annual precipitation in the range between 250 and 1000 mm (10 and 40 in.). Conversely, mechanical weathering decreases in the direction of warmer and drier climates, and in colder and drier areas (Peltier, 1950). The combined effects of mechanical and chemical weathering, as illustrated by Peltier (1950), are shown in Fig. 6.

Quantitative correlations of weathering, erosion, and transport with variations in climate have been limited by lack of data in many regions of the world. However, streamflow records from gauging stations in the United States have been used to demonstrate the relation between climate and streamflow (Langbein *et al.*, 1949). The family of curves (Fig. 7) illustrates that mean annual runoff increases as mean annual precipitation increases and that, with constant precipitation, runoff decreases as temperature increases. Langbein and Schumm (1958) developed a relationship between mean annual sediment yield and mean annual precipitation for drainage basins in the United States averaging about 3500 km. They adjusted the precipitation to an annual mean temperature of 50°F to obtain an "effective precipitation" based on the findings of Langbein *et al.* (1949). The resulting curve of Langbein and Schumm (1958) is shown in Fig. 8 and shows that at 50°F, maximum sediment yields may be expected where annual mean precipitation is about 305 mm (12 in.). As pointed out by Schumm

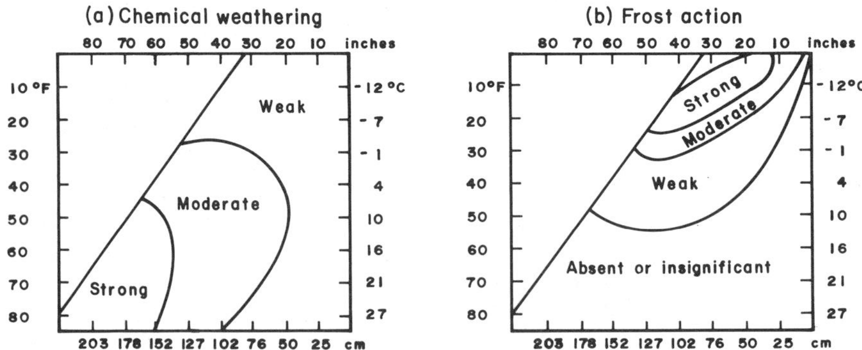

Figure 6. Relationship between temperature and precipitation and chemical and mechanical weathering (Peltier, 1950).

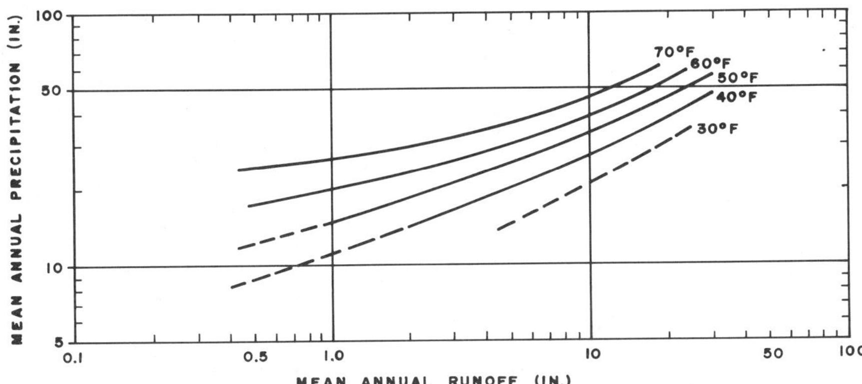

Figure 7. The effect of average temperature on the relationship between mean annual runoff and mean annual precipitation (Langbein *et al.*, 1949).

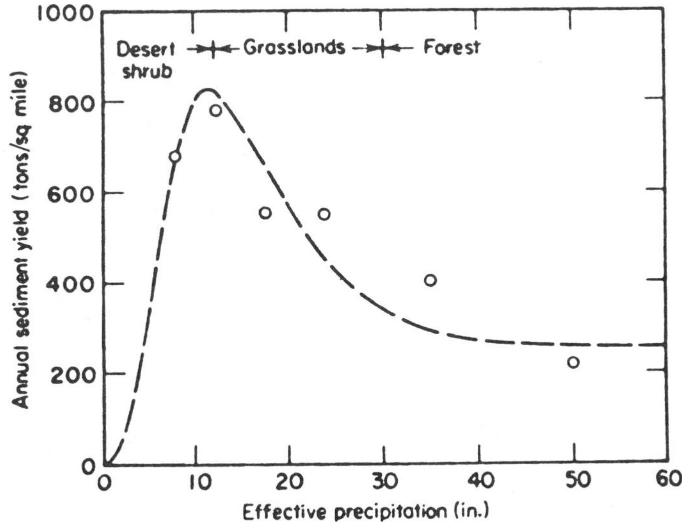

Figure 8. Variation of sediment yield with climate as based on data in the United States (Langbein and Schumm, 1958. © by the American Geophysical Union. Reprinted with permission.)

(1977), this relationship between annual precipitation and sediment yield is explained by the relationship of precipitation to vegetation growth. It must also be recognized that sediment yield is also greatly influenced by land use and seasonal variability of precipitation (Schumm, 1977). Therefore, although the Langbein–Schumm curve is a useful tool for general estimates of the magnitude of sediment yield, there are other factors such as relief and drainage area that may change the relationship.

In addition to the effects that climate may have on geomorphic and hydrologic processes under "natural" environments, it may also influence the rate of recovery after disturbance. For example, Graf (1977) considered the effects of human activities on the disruption of geomorphic systems. He assumed a steady state in fluvial systems, adjusted to climate and geology, prior to disturbance and a return to a "new" steady state during an intervening interval, a relaxation time (Fig. 9). The relaxation time will vary with climatic conditions in that geomorphic responses will be different for various materials in arid or humid climates (Graf, 1977). Graf also draws an analogy between the effects of climatic change and anthropogenic changes on landscapes. He states that both changes will alter fluvial systems by changing vegetation and land surfaces.

In a similar approach, Wolman and Gerson (1978) considered the absolute magnitude of climatic events and the absolute time intervals between events and concluded that they do not provide satisfactory measures of geomorphic effectiveness of events of various magnitudes and frequencies. In defining geomorphic effectiveness, Wolman and Gerson (1978, p. 190) distinguished be-

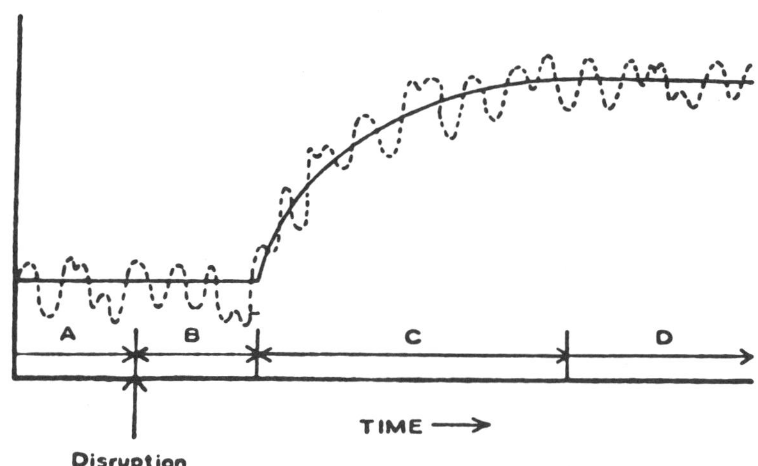

Figure 9. Graphic representation of the response of a geomorphic system to disruption (Graf, 1977). A, B, C, and D represent time periods.

tween work performed and sculpture of the landscape. Effectiveness is defined "as the ability of an event or combination of events to affect the shape or form of the landscape." Because the form of many landscape features is determined by both erosional and depositional processes, the three things that determine the effectiveness of a destructive event are (1) the force exerted, (2) the recurrence interval of the event, and (3) the magnitude of the restorative processes during the intervening period, or the "relaxation time" of Graf (1977). We may conclude from the studies of Graf (1977) and Wolman and Gerson (1978) that the rate of process operation or geomorphic effectiveness is dependent on climate to a great extent, and that both process rate and effectiveness will vary in both time and space. These concepts are important with regard to problems of reclamation of disturbed lands and will be considered further in the discussion of sediment/delivery ratios (see Chapter 5).

Soils and Hydrology

The soil mantle on the landscape plays an important role in the operation of both geomorphic and hydrologic processes that shape landforms. Soils serve as the host for plant life, the reservoir for soil moisture, and to a large extent it is the regulator for erosion processes such as raindrop impact, overland flow, and freeze and thaw. Soil is a combination of weathered earth materials, either residual or transported, and consists of a dynamic complex of organic and inorganic materials. Its ability to support life depends on its capacity to absorb, store, and transfer energy and water (National Resource Council [NRC], 1981).

The processes that are responsible for soil development are slow acting and generally are only measurable in terms of decades or centuries. Before a soil has been developed that is capable of supporting plant life, many alterations of the parent material have been made. Therefore, it may be assumed that any disturbance by human activity of a soil–vegetation complex that has been developed over a long period of time and that is in equilibrium with its environment will be difficult, if not impossible, to restore in a few years. It should be recognized, however, that many soils are developed on sites with detrimental characteristics such as poor drainage or steep gradients, and these are often highly erodible. The parent material because of salinity or nutrient deficiency may also be undesirable for the formation of soils in the case of residual soils. Therefore, the reclamation of disturbed soil–vegetation complexes does not necessarily imply that precise restoration of the predisturbance landscape and soil, if that were possible, is always the most desirable goal (NRC, 1981). Rather, the establishment of hydrologically and geomorphically stable landscapes is probably preferable.

Some of the physical and hydrologic characteristics of soils that are affected by disturbance are noteworthy here, as well as their relationship to hydrologic variables. These variables are porosity, permeability, bulk density, infiltration, moisture-holding capacity, and texture.

The part of an ideal loamy soil that is composed of pore spaces is approximately 50%. This pore space is occupied by air or water in varying amounts depending on climatic conditions and the length of time since the last precipitation event. The volume of a three-dimensional soil column that is made up of pore spaces is termed *porosity* and is expressed as a percentage. The weight of the same column of soil that has been oven dried to drive out the soil water divided by its volume is termed *bulk density* and is expressed by the following equation:

$$\text{Bulk density} = \frac{\text{weight of dried soil sample (g)}}{\text{volume of soil sample (cm}^3\text{)}}$$

The greater the bulk density, the less the pore space (Fig. 10). Both of these soil characteristics are important in determining infiltration rates and moisture-holding capacity of a soil, and they affect plant growth and erosion rates. The importance of these characteristics is discussed in later chapters. Also, the porosity and bulk density of a soil probably will be altered by disturbance and reclamation of a site, thus changing the hydrologic processes of infiltration, surface runoff, and erosion.

The process by which water enters the soil surface is termed *infiltration*. The

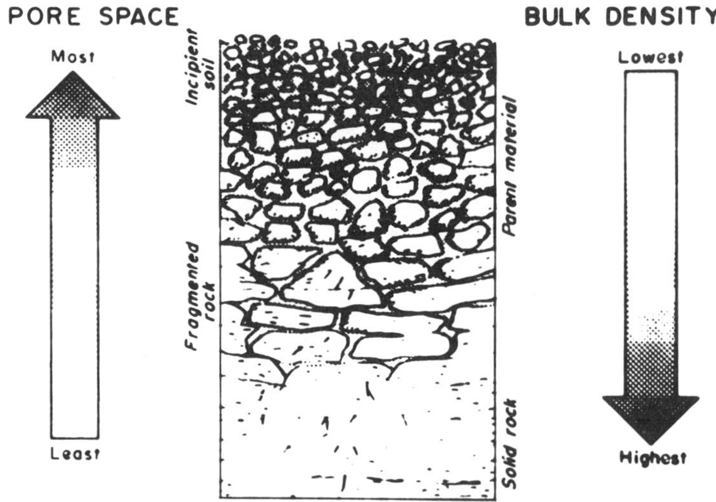

Figure 10. Pore space and bulk density vary inversely; in inorganic soil the total pore space increases while bulk density decreases (Steila, 1976. ©, Reprinted by permission of Prentice-Hall, Inc., Englewood Cliffs, New Jersey.)

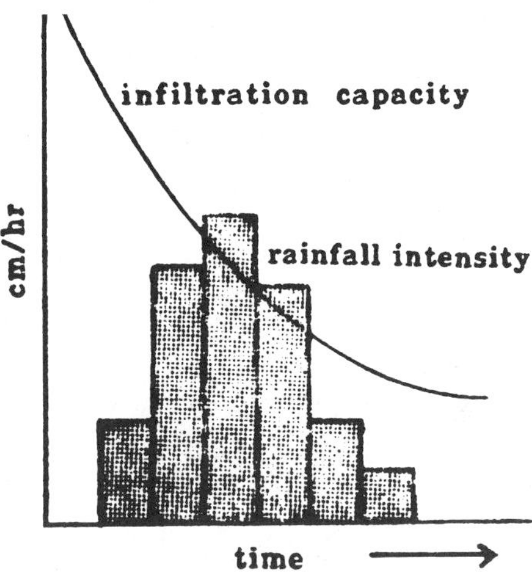

Figure 11. Infiltration capacity and precipitation intensity versus time. Runoff is produced when precipitation intensity exceeds infiltration capacity.

role of infiltration in the hydrologic response of a land surface was considered in detail by Horton (1933). The rate at which water can enter the soil surface is the *infiltration rate* and is expressed in centimeters per hour (cm/hr). There are several factors that influence the infiltration rate, and the most important for this discussion probably are (1) vegetal canopy cover, (2) litter and rock on the surface, (3) condition of the surface crust on the soil, (4) rainfall intensity or energy, (5) antecedent moisture in the soil, (6) soil texture, (7) bulk density, and (8) gradient of the land surface. All of these factors will affect infiltration rates of soils to some degree. In unsaturated soils, the maximum infiltration rate usually occurs at the beginning of a precipitation event. If the precipitation intensity (in mm/hr) does not exceed the infiltration capacity of the soil, the total water supplied is used to replenish soil moisture. After starting at a high rate, the infiltration rate decreases, and as the soil approaches saturation, it maintains a more or less constant minimum rate, approaching the percolation rate of the soil (Fig. 11). In cases where soils are wet at the beginning of a storm (antecedent moisture) or when precipitation intensity exceeds the capacity of the soil to accept it, water begins to accumulate on the land surface and eventually contributes to surface runoff. The shaded area within the rainfall histogram in Fig. 11 represents the excess rainfall that contributes to runoff.

All of the water that enters the soil is not available for plant growth. Some of it is rapidly evaporated to the atmosphere from the upper few centimeters of the soil. The remaining soil water that is stored in or moves through the soil reservoir can be classified as *hygroscopic*, *capillary*, or *gravitational* water. Hygroscopic water is the film of water held tenaciously by the soil particles and is not generally available for plant growth. Capillary water envelops the hygroscopic water and is not held as tightly to the soil particle. Gravitational water drains freely through the soil and usually occurs in the profile for only a short period after a precipitation event or application of irrigation water. After the gravitational water has drained through the profile, the soil moisture condition is of *field capacity*. Plants derive most of their moisture requirements from the capillary layer, and as the force required to remove water from the soil increases, plants begin to wilt. Here, the soil moisture condition is at the *wilting point*. Soil water that is available for plant growth can be described simply as the quantity of water between the field capacity and the wilting point of plants in a soil (Fig. 12). As expressed by Steila (1976), when fine particles are predominant in a soil the water-holding capacity increases but the proportion of available water to the total water held will decrease.

This is a very simplified discussion of the complex physical processes that are operative in the relationship between soils and water. However, it will serve as a background for the following discussion of erosion processes and the impacts of soil and vegetation disturbances on hydrology and erosion.

Erodibility of Surface Materials

Before we discuss processes of erosion and sediment yield in the context of how they are influenced by land disturbance, we should consider the variability of soil erodibility and the controlling factors. Middleton (1930) was a pioneer in the conduct of experiments with the purpose of defining an index of erodibility.

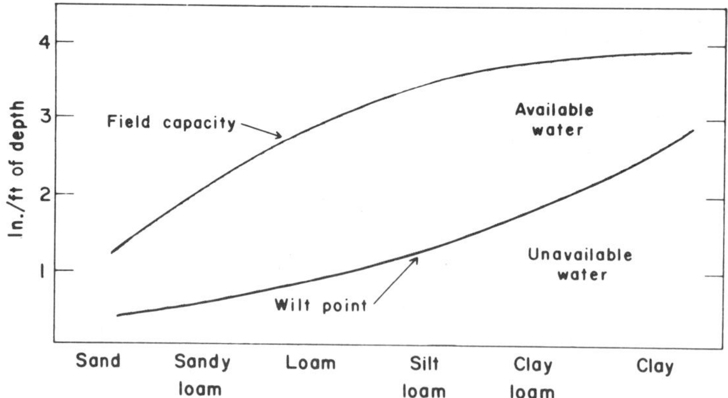

Figure 12. Diagram of general relationship between soil moisture availability and soil texture showing water holding capacities in different soil textures. (Steila, 1976. ©, Reprinted by permission of Prentice-Hall, Inc., Englewood Cliffs, New Jersey.)

His experiments were based on tests to determine the ease with which soils would disperse and transmit soil moisture. Bouyoucos (1935) treated the problem of dispersion by proposing the "clay ratio" as an index of erodibility. His ratio is the relationship of sand to silt and clay, as a measure of material binding. His work resulted in a simple determined index but does not reflect field observations. In addition to dispersion characteristics of soil, investigators found that soil aggregation and aggregate stability were related to erodibility.

In more recent work, Bryan (1968) considered the dominant factors that control inherent soil erodibility to be (1) rainfall, (2) topography, and (3) vegetative cover. Interestingly, these factors are not soil properties. However, even where these factors are held constant, variations in soil properties produce variations in soil loss. When the surface of the land is disturbed, the vegetation cover is generally destroyed, the topography may be altered, and soil structure is modified. Therefore, it seems likely that the erodibility will increase during the period between initial disturbance and reclamation (see Coal, Chapter 8).

Bryan (1968) concluded that none of the several indices of soil erodibility that have been proposed meet the requirements for such an index, which are simplicity of measurement, reliability, and universality of application. Bryan feels that the most profitable course of further study may be aggregate stability and distribution. Also, the experimental method that may be most significant in determining absolute erodibility is terminal velocity raindrop simulation rather than the gentle action of wet sieving. In a more recent paper, Bryan (1977) pointed out that most of the experimental work on erodibility has involved agricultural soils, where surfaces may be dominated by discrete particles because of tillage. However, predisturbance indices of erodibility are required for lands used for urbanization, surface mining, recreational use, and construction of highways that are generally dominated by undisturbed (coherent) soil bodies. In these cases, the only soil property of significance in the determination of erodibility may be shearing strength, which is determined by cohesion and by pore water pressure patterns (Bryan, 1977).

Determination of reliable indices of erodibility is difficult because of the interrelationships among the factors of topography, vegetation cover, rainfall, and slope length. It appears that erosion rates can be greatly reduced if slope lengths are shortened and some kind of protective mulch can be placed on the reclaimed soils until vegetation has become reestablished. Soil conditions will vary from place to place, therefore generalizations about erodibility are difficult to make.

Soil Erosion

Soil erosion is a normal process in landscape development by which surficial material is removed from the land surface by wind, water, or ice. We will

Table 2 Examples of Accelerated Erosion[a]

Cause	Effect	Setting	Reference
Agriculture			
Cultivation	Erosion rate: 444.5 tons/ha/yr	Small, intensely cultivated watersheds in western Iowa	Gottschalk and Brune (1950)
Grazing	Sediment yields: 19.7 tons/h/yr; averaged 154% of sediment yield from ungrazed watershed	Small watersheds in arid rangeland of Colorado plateau in western Colorado	Lusby (1979)
Timber harvesting	Sediment rate: 1.46 tons/ha/yr; up to fivefold increase in annual sediment rate in years following harvest	Small watershed near Newport, Oregon, 82% clear-cut	Beschta (1978)
Mining			
Underground	Estimated tractive force of 10-yr flood in stream increased from less than 1 dyne before settlement to more than 8 dyne after severe landscape alteration	Arroyo and gully development associated with gold and silver mining in Central City district, Colorado	Graf (1979)
Surface mining	Sediment yield: 2.38 tons/ha/yr[b]; sediment yield 11 times greater than undisturbed basin	Surface-mined basin (unreclaimed) near Sheridan, Wyoming	Ringen *et al.* (1979)
Construction			
Highway	Gross erosion from construction area averaged 338.5 tons/ha/yr; about 200 times that expected from grassland	Scott Run basin about 9.7 km above Potomac Estuary, Fairfax County, Virginia	Vice *et al.* (1969)

Urbanization	Construction site sediment yields ranged from 16.1 to 226 tons/ha/yr; cultivated fields averaged 6.2 tons/ha/yr	Small watersheds in Montgomery County, Maryland	Yorke and Herb (1976)
Channelization	Erosion of channel from cross-sectional area of 38 m^2 to 160–484 m^2 over 60 yr without evidence of stabilization	Blackwater River in Johnson County, Missouri	Emerson (1971)
Recreation			
Off-road	Sediment yield from used basin was 5.52 tons/ha/yr[c] while quantity of sediment delivered to reservoir below unused basin was not measurable by survey	Panoche Hills, western side of San Joaquin Valley, California, used for motorcycle hill climbs	Snyder *et al.* (1976)
Hiking	Gullying and soil profile truncation along hiking trails, soil analyses, and photographic evidence	Mountain hiking trails near Grövelsjön, Sweden	Bryan (1977)
Downhill skiing	Photographic evidence of soil erosion, deposition, and runoff turbidity (at time of study, no quantitative evidence of sedimentation rate)	Heavenly Valley ski resort	Seibert and Penneman (1975)

[a]Source: Toy (1982).

[b]For comparison: U.S. Soil Conservation Service estimate of soil loss limit to sustain crop production; *T* value = 11.21 tons/ha/yr (Logan, 1977). Rate of geologic erosion ranges between 0.224 and 2.242 tons/ha/yr (Smith and Stamey, 1965).

[c]Assuming 12,866 tons/cu hectometer.

consider erosion processes by water, or fluvial erosion, here. Wind erosion can be severe locally and is generally prevalent in arid and semiarid regions where the delicate balance of equilibrium in the naturally dry environment may be upset by destructive land use practices.

Under natural conditions, geologic or *normal* erosion is a normal process. *Accelerated* erosion, which is that erosion that occurs at a rate greater than normal for the prevailing environmental conditions of climate, soils, vegetation cover, and relief, is usually the result of human activity. Accelerated erosion has long been recognized as a serious problem with detrimental effects on soils. Table 2 provides some examples of accelerated erosion for purposes of illustrating the spectrum of human activities that contribute to the problem (Toy, 1982). Quantitative data for erosion and sediment yield rates in Table 2 are not intended to be average values; in fact, in some cases they probably represent values that approach the upper extreme.

The problems associated with accelerated erosion, with the concomitant possibility of legal ramifications, offer opportunities for pure and applied research. Some fundamental questions must be answered, such as: (1) What proportion of total erosion is directly attributable to human activities on the land? (2) What rate of erosion is tolerable within various pedogenic regimes? (3) What is the spatial and temporal applicability of our research results? The answers to questions (1) and (3) require a better understanding of the erosion–sediment transport systems and improved methods of measurement.

Question (2) assumes that tolerance limits are to be based on the maximum erosion rates permissible to allow sustained crop yields. However, other criteria may be more appropriate. Logan (1977) reviewed the basis for the tolerance values used by the Soil Conservation Service and suggested that more restrictive values may be required to meet regional water quality and sediment abatement needs in some areas.

The accelerated erosion data contained in Table 2 were collected, for the most part, in small watersheds over short periods of time. One might ask how long erosion might be expected to continue at this rate and to how large an area these rates apply. The National Academy of Sciences (1974) estimated that 76,279 ha (190,698 acres) of land will be disturbed by surface mining in nine western U.S. states from 1972 to 2000. This is 0.036% of the land area of these states and represents an impacted area of 2724 ha/yr (6810 acres/yr). Additionally, only part of any mining area is radically disturbed at any one time, and reclamation must proceed concurrently with the extraction process. In a discussion of the impact of timber harvesting in the eastern United States, Patric (1977) noted that the most serious erosion problems arise from construction of logging roads, but these comprise only 6–16% of the area logged.

Erosion, in general terms, is also related to vegetation or ground cover, which in turn is related to mean annual precipitation. In deserts where rainfall is very

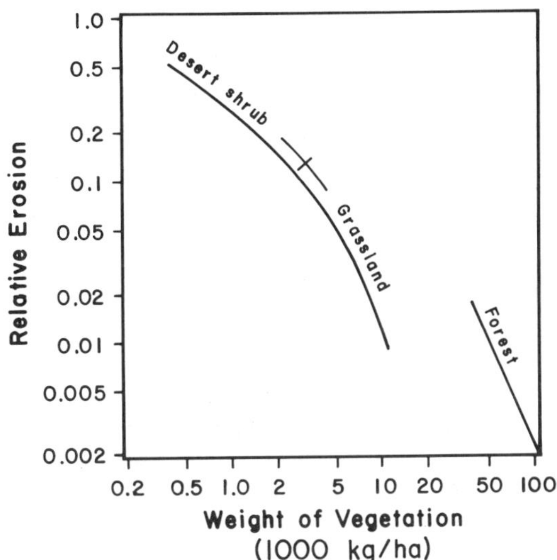

Figure 13. Relative erosion compared with vegetation weight per unit of area (Branson *et al.*, 1981).

low, there is also very little vegetation, and erosion rates from infrequent, high-intensity convective rain storms may be high. In semiarid regions where annual precipitation is 300–350 mm (11–13 in.), the vegetation cover is denser and the land surface is partially protected from fluvial erosion. As annual precipitation increases beyond about 350 mm (13 in.), the vegetation cover progresses from grasslands to forest cover and the erosion rates continue to decrease. In general, erosion increases with an increase in annual precipitation in natural environments and decreases with higher densities of vegetation cover, with the highest erosion rates occurring in semiarid regions. Figure 13 illustrates this relationship of relative erosion versus weight of vegetation per unit area; this is used as a surrogate for annual precipitation.

Movement of soil from one place to another on a slope requires that work be done. Work, by definition, is a force acting through a distance and is energy consuming. The fundamental forces in detachment, entrainment, and transportation of soil material are raindrop impact and overland flow of water.

Hillslope Processes

This discussion of hillslopes will focus on exogenetic processes, particularly erosion, because these are the most seriously affected by human-induced land disturbances and, except in local areas, accomplish most of the geomorphic work performed on hillslopes.

Force and Resistance on Hillslopes

All hillslopes are subjected to natural forces that are opposed by inherent resistances within and upon these landforms. All forces impinging upon hillslopes produce shear stresses within hillslope materials. Force requires energy, and the principal sources of energy for geomorphic systems can be ultimately traced to (1) gravity, (2) climate, and occasionally (3) heat generated or stored within the interior of the earth. Carson and Kirkby (1972) contend that mechanical and hydraulic forces move most hillslope debris because these operate at high efficiency, and of these forces, water flow is by far the most important both in total energy expended and in total debris transport achieved.

The principal sources of resistance on hillslopes are derived from (1) the structural strength of geologic materials, (2) the cohesiveness of soil mantles, and (3) vegetation cover. Collectively, these constitute the shear strengths of hillslopes which oppose the shear stresses. Mechanically, they tend to transform kinetic energy into heat of friction which will dissipate harmlessly into the surrounding environment.

Whenever forces exceed resistance, or shear stresses exceed shear strengths, the materials involved are deformed or translocated and work is performed. During these intervals, it may be said that processes are in operation. Ritter (1978) remarked that processes are the tools that forces apply against resistances. Embleton and Whalley (1979) suggested that a process may be considered as the work done by all the forces in a system. Hence, the examination of a geomorphologic process is largely the resolution and measurement of a set of interacting forces that obey certain mechanical laws. It should be noted that, here, the theory is much simpler than the practice. Carson and Kirkby (1972) define geomorphic process as a distinctive mechanism or group of mechanisms for the transportation of debris. By inference, this definition must include processes that prepare materials for transportation.

It is useful to reiterate that most geomorphic processes probably operate on a particular hillslope at one time or another. However, the frequency and magnitude of the force that they exert on resisting materials may vary considerably in time and space in response to concomitant changes in the geomorphic environment, especially climatic conditions. Although greatly simplified, Dunne and Leopold (1978) illustrate the concept in Fig. 14.

Likewise, resistance may vary considerably in time and space. Many times, the resistances of the geologic materials control the amount of work that takes place. Accordingly, the geomorphic processes on a hillslope may be "weather

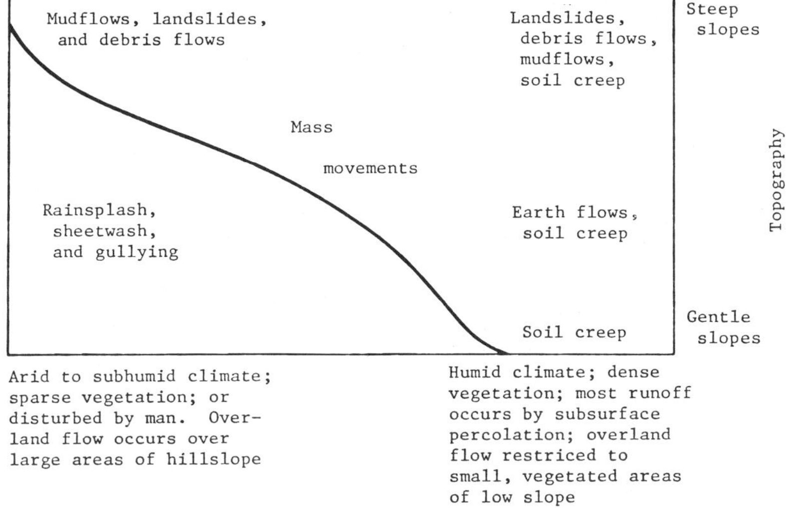

Figure 14. Conditions of climate, vegetation, land use, and topography under which various geomorphic processes dominate hillslope erosion, although they may not be restricted to the areas shown (Dunne and Leopold, 1978).

limited" or "transport limited" (Carson and Kirkby, 1972; Selby, 1982). In the first case, the weathering rate of hillslope materials is less than the transportation capacity of the operational geomorphic processes. Until weathering processes have reduced the resistances of these materials, the natural forces are unable to cause deformation or translocation. Thus, the actual rate of transportation can only equal the weathering rate. The absence of soil or weathered rock (regolith) on a hillslope surface suggests a weathering-limited circumstance.

In the second case, the weathering rate exceeds the transportation capacity of the prevailing processes. Soil and regolith accumulate on the hillslope surface to await potent transportation processes. Eventually, this mantle may itself insulate the underlying materials from some weathering processes. The presence of soil or regolith on a hillslope then suggests a transport-limited circumstance.

It is possible to envision a third possibility; a hillslope surface upon which there is a precise balance between weathering and transportation rates resulting in neither denudation nor accumulation of soil or regolith. These are usually called "transportation hillslopes" but should not be confused with transport-limited geomorphic processes. Pediments are often considered transportation hillslopes (Hadley, 1967). These are cut rock surfaces produced by mountain front retreat and are generally covered with a veneer of coarsely textured regolith. The material entering the pediment system from above equals the amount leaving the system below.

From earlier discussions of climate and geologic controls of hillslopes, it can be suggested that weather-limited hillslopes tend to occur more frequently in arid and semiarid regions while transport-limited hillslopes occur more frequently in humid regions.

A goal of exogenetic processes is to create a balance between forces and resistances. This goal cannot be achieved as long as parts of a landscape possess greater potential energy than other parts. Consequently, a fundamental objective of exogenetic processes must be the rendering of the earth's surface to a common level, known to geomorphologists as "baselevel." The ultimate baselevel is, of course, the datum plane of the world's oceans; however, for specific areas, such as enclosed mountain drainage basins and hillslopes, streams may grade to a local baselevel. The reduction of relief is accomplished by degradational hillslope processes, including mass movement and erosion, together with aggradational processes of deposition. Noteworthy in this regard is the comment by Thornes (1979) that change in form of the earth's surface indicates the operation of geomorphic processes, but the lack of observable change need not imply that no processes are operating. Indeed, the complete absence of geomorphic processes would certainly be a rarity, if it has ever happened. However, the rate of operation may be very small and imperceptible because the forces and resistances involved are small or because the ratio of force and resistance approaches unity.

From the foregoing discussion, it is evident that a basic understanding of geomorphic processes is critical to an understanding of hillslope form, development, and evolution. We shall be concerned here only with processes taking place on natural hillslopes while reserving discussion of those on disturbed hillslopes for a later chapter. Nevertheless, the principles are governed by physical and chemical laws in both situations, and as a result, it is usually only the frequency and magnitude of processes that differ.

Weathering

The rocks and minerals from which hillslopes are sculptured by geomorphic processes had their origin within the earth's crust under conditions of temperature and pressure typically much higher than characteristic of the surface. As landscape denudation created and destroyed hillslopes over millions of years, these rocks and minerals were brought progressively closer to the surface. At this interface of the atmosphere, hydrosphere, biosphere, and geosphere, these lithologies were subjected to low temperatures and pressures and to chemically active solutions of oxygen and water.

While the rock and minerals may have been at or near equilibrium with their environment of formation, their emergence at the surface produces or exacerbates a disequilibrium to which these materials respond. Birkeland (1974) defined "weathering" as the process of rock and mineral alteration to more stable forms under the variable conditions of moisture, temperature, and biological activity that prevail at the surface. The basic principles and the relationship between hillslopes and weathering processes shall be the focus here. Those desiring in-depth treatment are referred to Carroll (1970), Birkeland (1974), and Ollier (1975).

Two types of weathering are generally recognized: physical and chemical. As noted by Ritter (1978), the distinction between the two is real because the processes of physical weathering cause rock disintegration, involving no chemical reactions but simply generating small particles from larger ones. The processes of chemical weathering, conversely, cause the decomposition of minerals making up the affected rocks with the production of new minerals chemically more stable or nearer to equilibrium with the physio-chemical conditions of the earth's surface. Some workers (e.g., Selby, 1982) identify a third type of weathering resulting from biotic processes. However, the forces exerted on rocks and minerals by these processes have physical or chemical (organic as opposed to inorganic) origins; so the twofold categorization seems sufficient.

Carroll (1970) contends that chemical weathering takes place in two stages. The first stage is the production of rotten rock or "saprolites" upon which the second stage, soil formation, takes place. The first stage is identified as "geochemical" weathering and is mostly the inorganic alteration of solid rocks; however, in the second stage, identified as "pedochemical" weathering, the effects of vegetation, both living and dead, along with the effects of metabolism of microorganisms living in the geochemically altered rock material, are added to the continued inorganic processes. This combination produces a soil that is the culmination of weathering at that particular locality.

Although various weathering types and processes may be designated, it must be understood that many processes will usually operate together. It may be difficult to separate the effects of physical weathering from those of chemical weathering. Ritter (1978) observed that through this simultaneous operation of physical and chemical weathering, each may directly affect the character and rate of the other. For example, disintegration of larger rocks into smaller particles increases the total surface area exposed to chemical attack. Likewise, the expansion of minerals due to chemical decomposition may create sufficient internal stress to hasten the disintegration of the rock.

Table 3 contains brief descriptions of several physical weathering processes. The relative importance of each varies considerably from place to place. It should be noted that some controversy surrounds the effectiveness of temperature change in causing physical disintegration of rock. The argument in favor of the

Table 3 Physical Weathering Processes

Cause	Process	Effect
Stress relief	Unloading	Removal of confining and compressive pressures results in sheeting and exfoliation along dilation joints which develop in rocks.
Crystallization	Ice crystal growth (frost weathering) Salt crystal growth (salt weathering)	Growth of ice or salt crystals in crystals in joints, cracks, or pores exert lateral pressure resulting in granular disintegration of rocks, heaving, or cracking of soil.
Temperature change	Differential expansion of rock minerals; a.k.a., thermal expansion and contraction (insolation weathering)	Differences in expansion rates of minerals composing rocks in response to temperature changes (perhaps in presence of water) cause intra- and intergranular stresses resulting in granular disintegration of rocks.
Wetting and drying	Hydration of minerals Swelling of clays	Expansion causes stresses within rocks and soil resulting in spheroidal weathering, granular disintegration of some rocks and heaving or cracking of soils.
Plant growth	Root wedging	Enlargement of joints and cracks as a result of root, stem, or trunk growth of plants.
	Colloidal plucking (biotic weathering)	Soil colloids, perhaps produced by complex biochemical reactions loosen or extract small bits of rock from surfaces with which they come into contact.

process appeals to logic and has enjoyed the support of knowledgeable investigators (e.g., Ollier, 1975). However, laboratory experimentation has failed to prove its capability (Blackwelder, 1933; Griggs, 1936). Nevertheless, this process is generally included in discussions of physical weathering with assorted disclaimers or appeals.

Table 4 contains a list of common chemical weathering processes. Again, the relative importance varies geographically; however, it is frequently suggested that solution is the first phase of chemical weathering (Ollier, 1975). The formulas used as examples in Table 4 present a simplistic picture of the chemical environment within which weathering occurs. The products of one reaction may quickly become reagents in another. Exactly which reactions occur is substantially determined by the presence of free hydrogen (pH) and redox potential (Eh). Concurrently, some elements may be physically removed from the environment by soil water leaching.

Although complex in specifics, general consequences of the combined action of physical and chemical weathering processes have been identified by Brunsden (1979):

1. A movement toward a more stable state where mineral assemblages, formed under conditions of high pressure and temperature within the earth, adjust to the new environmental conditions, atmospheric pressure, and lower temperature of the earth's surface.
2. Irreversible changes from a massive to a clastic or plastic state as breakdown proceeds.
3. Changes in volume, density, grain size, surface area, permeability, consolidation, and strength.
4. The formation of new minerals, aggregates, and solutions.
5. Alteration of the resistance of original minerals.
6. The transfer, translocation, dispersion, aggregation, and concentration of minerals and salts.
7. The preparation of rock for subsequent erosion and transport by geomorphic processes.
8. The production of new land surfaces and deposits.

Table 4 Chemical Weathering Processes

Reaction	Process	Example
Solution	Minerals dissociate into their component ions in aqueous medium.	$CO_2 + H_2O \rightleftarrows H_2CO_2$ $CaCO_3 + H_2CO_3 \leftrightarrows Ca^{2+} + 2HCO_3^-$ (Limestone) (Carbonic acid)
Hydration	Absorption of water molecule into crystalline structure of mineral.	$2Fe_2O_3 + 3H_2O$ $2Fe_2O_3 \cdot 3H_2O$ (Hematite) (Limonite)
Hydrolysis	A chemical weathering process in which minerals are altered by chemically reacting with water and acids.	$K(AlSi_3O_8) + H_2O \leftrightarrows H(AlSi_3O_8) +$ KOH (Orthoclase) (Potassium Hydroxide)
Oxidation	Element within mineral loses electron to oxygen ion.	$4FeO + O_2 \rightarrow 2Fe_2\,O_3$ (Ferrous oxide) (Hematite)
Reduction	Element within mineral gains electron, often release of oxygen from compounds.	$2Fe_2O_3 \rightarrow 4FeO + O_2\uparrow$ (Hematite) (Ferrous oxide)
Carbonation	Reaction of carbonate or bicarbonate ions with minerals or rocks.	$2KAlSi_3O_8 + 2H_2O + CO_2 \rightarrow$ (Orthoclase) $Al_2Si_2O_5(OH)_4 + K_2CO_3 + 4\,SiO_2$ (Kaolinite) (Potassium carbonate)

Full consideration of weathering rates and influential environmental controls is beyond the scope of this discourse. However, brief discussion is useful in understanding the role of hillslopes. Climate and topography (sometimes expressed as *relief*) are major factors in determining the rate and depth of rock and mineral weathering. Climate has been regarded variously as a very important (Ollier, 1975; Brunsden, 1979), or the most important, control (Carroll, 1970; Cooke and Doornkamp, 1974) of weathering rates. This variable determines the temperature and moisture conditions under which the weathering processes occur. Concerning physical weathering, Brunsden (1979) observed that temperature, especially absolute values and seasonal and diurnal variations, control the frequency and magnitude of freeze–thaw, and partly control wetting, drying, and salt crystallization cycles. Concerning chemical weathering, a 10°C increase in temperature often doubles or triples the reaction rate. Toy *et al.* (1978) found a highly significant correlation between air temperature and shallow soil temperature.

Precipitation is the source of nearly all water in the weather environment, and as noted by Ollier (1975), water is the most important reactant in almost all forms of weathering. Concerning physical weathering, it is obvious that water is essential to ice crystal growth and the physical component of hydration. Additionally, it may be important in the process of thermal expansion and contraction. Concerning chemical weathering, Brunsden (1979) remarked that the intensity, frequency, and duration of precipitation–evaporation events are important in weathering studies because chemical reactions depend on a supply of water for many essential processes (hydration, hydrolysis, and solution) to do the following: leach and remove soluble constituents and thereby enhance further reactions; control the concentration of hydrogen ions, organic content, and oxidation–reduction conditions; and control the distribution of clay minerals and the zonation of the weathering mantle.

The dominance of climatic controls on weathering, both geochemical and pedochemical, has allowed identification of broad weathering regions based on climate parameters. Strakhov (1967) and Arkley (1967) provided valuable documentation of this relationship.

Topography affects weathering through its influence on soil moisture and erosion processes. In mountainous areas, there is a vertical or altitudinal zonation of soils and pedochemical processes in response to altitudinal zonation of climate. However, the topographic factor manifests itself markedly on local scales as well. The spatial distribution of soil moisture varies with hillslope form, including both profile and plan characteristics. As hillslope gradient increases, runoff increases at the expense of infiltration. Hence, it may be assumed that as gradient increases, soil moisture will usually decrease, other factors being constant. Concerning hillslope plan, the concave outward configuration, known as a valley head hillslope, causes the convergence and concentration of flow in the downslope direction. As a result, the soils of valley head hillslopes are often nearly saturated in humid climatic regions while nearby spur end hillslopes possess substantially drier soils. It may be inferred, therefore, that water is less likely to limit weathering on valley head hillslopes than on spur end hillslopes.

As will be described more completely in the following section, gradient is a major control of the erosion rate (Musgrave, 1947; Zingg, 1940; Wischmeier and Smith, 1965). In a study of the relationship between erosion and hillslope geometry, Hadley and Toy (1977) found that the greatest soil loss occurred in the straight segment with significantly less in the convex and concave elements. Ollier (1975) commented that the removal of material promotes more rapid weathering and renewed soil formation. Norton and Smith (1930) found an inverse relationship between gradient and the thickness of the A-horizon. Brunsden (1979) contends that weathering of primary minerals tends to increase downslope with a positive feedback mechanism in which more clay holds more moisture, which in turn enhances chemical processes.

The compass direction that a hillslope faces, known as hillslope aspect, along with its gradient, also influences soil moisture. Aspect and gradient determine the amount of direct solar beam irradiation received on hillslopes at a given latitude. This largely controls soil temperature and evapotranspiration rates. Brunsden (1979) remarked that these in turn partially control soil moisture variation, the distribution of organic matter, pH, depth to base saturation, and thus weathering reactions. Several investigators have documented the systematic variation of soil characteristics with hillslope aspect (Cooper, 1960; Finney *et al.*, 1962). As reported by Birkeland (1974) and Ollier (1975), the weathering and pedogenic environments differ considerably on north- and south-facing hillslopes. Some contrasts are presented in Table 5.

Similarly, several investigators have documented systematic variation of soil characteristics with position on hillslopes (Ruhe and Walker, 1968; Walker, 1966). As reported by Ruhe (1975) for soils in Iowa, as one progresses from an upper convex element (summit) to a lower concave element (backslope), the thickness of surficialsediment increases, gravel decreases, clay increases, and the geometric mean particle size decreases. Ruhe and Walker (1968) also pointed out that soil age may also vary with hillslope position and that this further complicates establishment and explanation of relationships between characteristics and hillslope position.

Mass Movement

The terms *mass movement, mass wasting,* and *hillslope failure* have been used to describe the movement of soil or rock materials downslope under the influence of gravity, but without the direct involvement of a transportation agent such as wind, water, or ice. If differentiation of the terms was considered useful, then it

Table 5 Comparison of Soil Characteristics on North- and South-facing Hillslopes[a]

South aspect	North aspect
Processes and Environment	
Higher evapotranspiration rates	Lower evapotranspiration rates
Higher soil temperatures	Lower soil temperatures
Lower soil moisture	Higher soil moisture
More freeze–thaw cycles in winter	Fewer freeze–thaw cycles in winter
More wetting and drying cycles in summer	Fewer wetting and drying cycles in summer
Lower vegetation cover	Greater vegetation cover
More xeric plant species	More meric plant species
Greater soil erosion	Less soil erosion
Soil Characteristics	
Shallower soils	Deeper soils
Redder soils	Darker soils
Less organic matter	Greater organic matter

[a]Adapted from data and conclusions compiled by Birkeland (1974), Ollier (1975), and Hadley (1962).

could be suggested that mass movement refers to the process while mass wasting refers to the consequences, namely, a change in landform with a reduction of relief and hillslope gradient in many cases. Hillslope failure then becomes a generic term including both of the others.

There have been numerous attempts to classify the various types of hillslope failure, and each has achieved a measure of success. None, however, have gained universal acceptance. A fundamental problem arises from the fact that a particular failure may be composed of several types of mass movement. For example, a debris avalanche may become a mudflow as it progresses downslope incorporating more water into the materials it transports. Further, the actual manner of movement may include sliding, flowing, and heaving. For a pure "slide," movement occurs along a well-defined surface; and the mass of material above this plane experiences no internal deformation. In a pure "flow," there is no sharply defined failure plane, but deformation takes place: throughout the mass of material in motion. With a pure "heave," the soil expands perpendicularly to the surface and subsequently contracts. This by itself does not produce lateral transport of materials, but provides the basic mechanism for the slow downslope migration of particles in response to gravity and may trigger more rapid events by reducing the cohesiveness of the earthen mass or increasing the internal stress beyond the threshold of material strength.

Derbyshire *et al.* (1979) redefined the heave manner of movement to include "subsidence" and called this group "vertical displacements." This is a useful inclusion because human activities, such as fluid extraction of oil, gas, and water as well as underground mining, have caused significant modification of landscapes through subsidence in some locales.

Nevertheless, Carson and Kirkby (1972) contend that actual mass movement processes are rarely, if ever, a pure slide, flow, or heave, but that all are a combination of the three, perhaps with some basal sliding, some internal shear or deformation, and some response to the triggering action of heaves.

Despite the aforementioned deficiencies inherent to all such classifications of hillslope failure, the scheme and nomenclature prepared by Varnes (1958, 1975)

Table 6 Factors Contributing to Mass Movement in Soils[a]

Types	Major mechanisms
Factors contributing to high shear stress	
Removal of lateral support	Stream or water erosion
	Subaerial weathering, wetting, drying, and frost action
	Slope steepness increased by mass movement
	Manmade quarries and pits
Overloading by	Weight of rain, snow, talus
	Fills, wastepiles, structures
Transitory stresses	Earthquakes
	Manmade vibrations
Removal of underlying support	Undercutting by running water
	Subaerial weathering, wetting, drying, and frost action
	Subterranean erosion (eluviation of fines or solution of salts)
	Mining activities
Lateral pressure	Water in interstices
	Freezing of water
	Swelling by hydration of clay
Factors contributing to low shear strength	
Composition and texture	Weak materials such as volcanic tuff and sedimentary clays
	Loosely packed materials
	Smooth grain shape
Physicochemical reactions	Cation (base) exchange
	Hydration of clay
	Drying of clays
Effects of porewater	Buoyancy effects
	Reduction of capillary tension
	Viscous drag of moving water on soil grains
Changes in structure	Spontaneous liquefaction

[a]From Selby (1970) (Varnes, 1958, original).

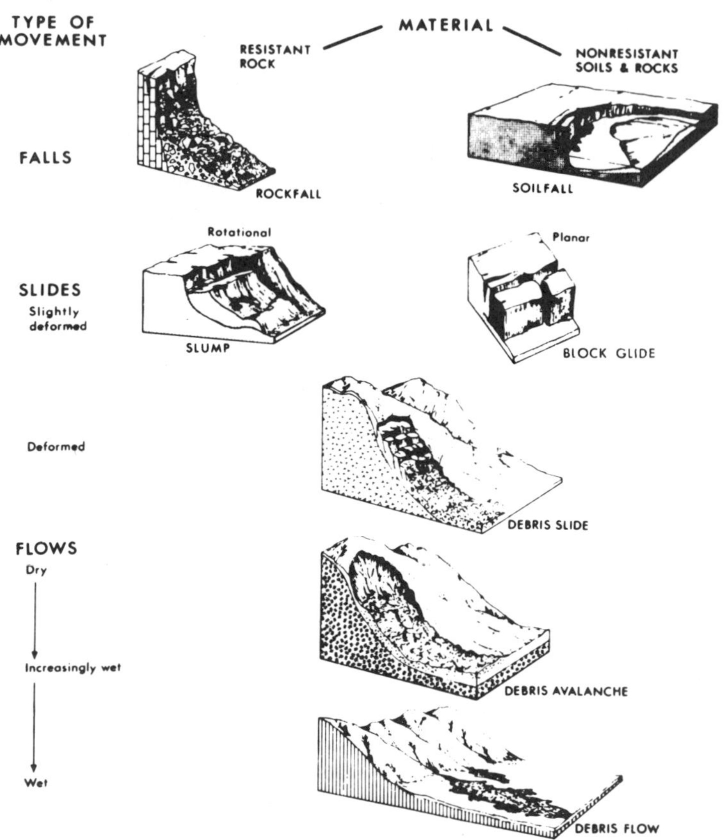

Figure 15. Types of mass wasting (Varnes, 1958).

is probably the most commonly used for descriptive and pedogogic purposes (Selby, 1982). Table 6 shows the Varnes (1975) system of classification. Here, the term *debris* refers to coarsely textured materials while *earth* refers to soil separates of sand, silt, and clay. Figure 15 illustrates various types of hillslope failures based upon the earlier (1958) but very similar Varnes system.

The examination of causes for hillslope failure is relatively simple in theory but very complex in practice. Various geomorphic processes driven by gravity produce shear stress within hillslope materials. Deformation, however, is resisted by the assemblage of material properties, collectively known as shear strength. Whenever shear stress exceeds shear strength, dislocation and movement of the involved materials will occur. This concept is often expressed as the safety factor (F_s) of hillslope stability, a ratio of shear strength to shear stress along a given plane at some depth within a hillslope:

$$F_s = \frac{\text{sum of resisting forces (shear strength)}}{\text{sum of driving forces (shear stress)}}$$

Values of the safety factor (F_s) greater than unity indicate stability while lesser values indicate instability. The plane within the hillslope with the lowest safety factor is the critical or failure surface, and its safety factor governs the overall stability of the hillslope (Gray and Leiser, 1982).

Although considerably simplified, the influence of hillslope gradient on particle or mass stability may be examined using the safety factor concept. The gravitational force acting on an inclined surface can be divided into may be examined using the Safety Factor concept. The gravitational force acting on an inclined surface can be divided into two parts: (1) the normal component, acting at right angles to the surface and tending to hold the material in place on the hillslope, and (2) the downslope component, acting parallel to the surface and tending to pull the material downward along the hillslope. Accordingly, the safety factor becomes:

$$F_s = \frac{\text{normal component}}{\text{downslope component}} = \frac{m \cos S}{m \sin S} = \frac{\cos S}{\sin S}$$

where: F_s is the safety factor, m the mass (weight) of material, S the hillslope gradient. Thus, as hillslope gradient increases, the safety factor decreases, other things being equal. Hillslope materials, then, are inherently less stable on steep hillslopes as compared to gentle hillslopes. As will be demonstrated, however, actual situations are far more complex because gravity is not the only operational variable.

From these basic principles, the next level of examination focuses upon the factors influencing shear stress, shear strength, and the interplay of the resulting forces. Even in theory, the mechanics of hillslope failure can become highly complicated as various elements in the pertinent equations are adjusted to reflect the specific conditions of particular environmental settings. A simple example serves to illustrate. The shear strength of a dry or freely draining, cohesionless material, such as sand, is given by:

$$S = \sigma \tan \theta$$

where S is the shear strength, σ the normal stress along shear plane, and θ the angle of internal friction. Many natural materials do possess some inherent cohesion (c) among particles, so it is necessary to include this factor to produce the following expression, commonly called Coulomb's Law:

$$S = c + \sigma \tan \theta$$

Soils containing clay-sized particles usually exhibit some cohesion. For large areas of the earth's land surface, soils are at least partially wet for all or part of the year. The pore water pressure (u) of soils must then be taken into account yielding the following equation:

$$S = c + (\sigma - u) \tan \theta$$

At this point, one could consider the geometry of the failure plane, whether it is rectilinear or curvilinear. Further elaboration of the mechanics of hillslope failure is beyond the scope of this discussion. However, excellent treatments of the topic are available (Selby, 1982; Gray and Leiser, 1982; Brunsden, 1979; Carson and Kirkby, 1972).

From the foregoing discussion, it is obvious that any factor that increases shear stress or reduces shear strength will tend to promote hillslope failure. Varnes (1958) compiled the list of such factors presented in Table 6. As noted by Wilson and Cooke (1980), however, there is seldomly only one cause of hillslope failure. Instability usually results from a sequence of events that ends with downhill movement of materials. Selby (1970) added that the processes that lead to the event might well have begun when the rock was formed or when soil development began, and subsequent movements, weathering, and erosion have all contributed to the potential instability of the hillslope. The last event that causes the mass to begin moving downhill may be quite trivial. Sowers and Sowers (1961) remarked that in most cases a number of causes exist simultaneously, so attempting to decide which one finally produced failure is not only difficult but also incorrect. Often the final factor is nothing more than a trigger that set in motion earth that was already on the verge of failure.

Nevertheless, some authors have offered generalizations based upon experience and research concerning the probable causes of hillslope failure. Gray and Leiser (1982) assert:

1. Removal of lateral support by either natural or human agencies is the most common of all factors leading to hillslope instability.
2. The addition of water to a hillslope may contribute to both an increase in stress and a decrease in strength. Water, in fact, has been implicated as either the controlling factor or a primary controlling factor in 95% of all landslides.

Dunne and Leopold (1978, p. 561) summarized the general controls of hillslope stability thusly:

> Landslides occur when the gravitational forces acting along the potential failure surface exceed the resistance of geologic materials to failure. Some processes influence the downslope disturbing force by affecting the weight and distribution of material on the hillside, or by altering the angle of slope. Other processes affect the shear strength of the material by changing the cohesion (c), by changing the frictional properties of the rock ($\tan \theta$), or by increasing the piezometric (pore) pressure at the failure surface.

Although most agree that water is an important and common control of hillslope failure, there has been some misconception concerning the actual function. In discussing the issue, Ritter (1978) observed that, in the past, the water effect was interpreted to be lubrication along the failure surface. However, water serves as an antilubricant when applied to some abundant minerals such as quartz (Terzaghi, 1950), and in humid regions soils usually contain more than enough water to cause lubrication, yet they also fail after rainstorms. He contends that the response of cohesion (c), normal stress (σ), and internal friction (θ) to wetting is significantly more important in the initiation of slippage than is lubrication. The rise of the water table or the piezometric surface, which accompanies extended periods of precipitation, may be the most common cause in sliding. As the water table rises, the pore pressure (u) increases within hillslope materials. This reduces the effective normal stress and hence shear strength.

Because of the importance of water in hillslope stability, it is permissible to infer seasonal variations in the safety factor (Brunsden, 1979). Seasonality of precipitation and evapotranspiration influence pore water pressures in soils, groundwater levels, and piezometric surface levels with the results described above.

Hillslope Erosion

The erosion process consists of the detachment, entrainment, and downslope transportation of surface materials. Water, wind, or ice may be agents of erosion and transportation media. Discussion here, however, shall focus on erosion by water and wind because these are the principal concerns in areas of land disturbance.

Erosion constitutes a form of geomorphic work that occurs whenever the driving forces impinging upon surface materials exceed the resisting forces encountered. The driving forces of impact, drag, and traction are generated by water falling on the hillslope or water and wind flowing across it. Resisting forces result from the frictional characteristics of the surface itself and the cohesive properties of the materials involved. Comprehensive descriptions of driving and resisting forces are provided by Carson (1971) and Kirkby and Morgan (1980).

Generally, water is the most effective agent of erosion. This fluid causes the translocation of surface materials in many ways. Meyer *et al.* (1972) identify four erosion subprocesses: (1) detachment by rainfall, (2) transport by rainfall, (3) detachment by runoff, and (4) transport by runoff.

Rainsplash. On the basis of hillslope hydrology and the driving forces produced by falling or flowing water, it is possible to designate four categories of hillslope erosion: (1) rainsplash, (2) sheet erosion, (3) rill erosion, and (4) gully erosion. The last category is actually a type of channel erosion, but will be included here to complete the continuum from hillslope interfluve to channel bed as contained in the land surface model of Dalrymple *et al.* (1968). Erosion by water on a hillslope usually begins with rainsplash at the start of a precipitation event. Meyer *et al.* (1976) remarked that several trillion raindrops annually bombard each hectare of land in the humid United States at impact velocities up to 9 m/sec (29.7 ft/sec). Wischmeier and Smith (1978) estimated that, for a common 30-min thunderstorm in the cornbelt states, the dead weight of water falling as billions of drops may well exceed 100 tons/acre (44 tons/ha) with an average velocity of nearly 20 miles/hr (32 km/hr) producing a total rainfall energy in excess of 2 million ft-lb/acre during this 30-min period. Stocking (1977) estimated that a large tropical storm of 30-min duration might deliver a dead weight of 350,000 kg/ha (312,000 lb/ac) of water with an approximate energy equivalency of 10 million J/ha.

This expenditure of energy on unconsolidated surface materials results in considerable geomorphic work. Ellison (1944) found through laboratory experimentation that stones 4 mm in diameter could be moved by rainsplash as far as 20 cm (8 in.), while smaller particles could be moved as far as 150 cm (60 in.). In a subsequent report (1947), he suggested that, for a nonvegetated soil surface, as much as 100 tons/acre (44 tons/ha) could be splashed into the air in a heavy storm. Robins and Neff (1963) stated:

> Computed total rainfall energy indicates the detachment potential to be in the order of four million foot-pounds of work per acre for a one-hour rain falling at a rate of two inches per hour. This is sufficient to lift the seven-inch topsoil layer two feet. High intensity rains on bare, unconsolidated soils in the field may set in motion as splash up to 600,000 pounds of soil material per acre.

Relationships between rainfall energy and rainfall intensity have been established by numerous investigators. Lal (1985) provided a noteworthy review of this research. One commonly cited equation is that of Wischmeier and Smith (1958):

$$y = 916 + 331 \log_{10} X$$

where y is the kinetic energy in foot-tons per acre,and X the rainfall intensity in inches per hour. Similarly, relationships exist between rainfall energy and rainsplash erosion. Again, Lal (1985) provided a useful review. Free (1960) reported that soil loss due to rainsplash alone from sand varies as kinetic energy to the 0.90 power, and from a silt loam soil as kinetic energy to the 1.46 power. Young and Wiersma (1973) discovered that decreasing rainfall impact energy by 89% without reducing rainfall intensity reduced soil loss by 90% or more.

Rainsplash contributes to the erosion process in at least five ways: (1) detachment of particles, (2) transportation of particles, (3) sealing of the soil surface, (4) splash creep of particles, and (5) generation of turbulence in overland flows. Mutchler and Young (1975) asserted that raindrop impact is the primary source of energy for detaching soil from any land area not protected by cover of some sort. Similar statements have been made by Young and Wiersma (1973), Foster and Meyer (1977), and Morgan (1978, 1979), among others. Quansah (1981) demonstrated that there are exponential relationships between soil detachment and kinetic energy for various soil types. Of course, the smaller separated particle is more easily entrained and transported than the larger mass from which it was derived.

The role of rainsplash in downslope sediment transportation is somewhat controversial as reported by Lal (1985). However, most opinion and evidence indicates that net downslope movement increases with hillslope gradient. On a horizontal surface, debris will be transported outward from the point of impact if the forces are great enough to move it and the directions of movement will be completely random (Carson and Kirkby, 1972, p. 189). However, for inclined hillslopes, segments, and elements, these authors contend:

> On a sloping surface, the pattern of movement becomes asymmetric and there is a net transport of debris down slope. The asymmetry is produced because (1) the down slope component of raindrop momentum acts directly to move debris and (2) the same set of splash trajectories will produce longer jumps when directed downhill rather than uphill. Both of these effects produce a net down slope transport, and outweigh any asymmetric tendency that may exist in the direction of rainfall.

Very similar descriptions have been presented by Ellison (1947) and Thornes (1979). Meyer *et al.* (1975) also submitted that the detached particles are splashed in all directions from the impact points, with net movement downslope.

The results of specific studies, nevertheless, have been conflicting, perhaps due to variability of research design. For example, Young and Wiersma (1973) concluded that for plots with 9% gradient the net movement of soil downslope by splash was minimal. Ellison (1947) submits that on a 10% slope, where there was a small amount of surface flow, considerably more than half of the particles that were splashed moved in downhill directions. Although the data source is unclear, Thornes (1979) remarks that on a 10% slope the downhill movement due to rainsplash is about three times that occurring in an uphill direction. Assuming that there is not a geomorphic threshold between 9 and 10%, it appears that there must be other influential factors involved that substantially affect the movement of particles by rainsplash. Again, in contrast, Lattanzi *et al.* (1974) found little effect of hillslope gradient on interrill erosion where rainsplash is the major cause of particle detachment.

Although some researchers have found little or variable relations between hillslope gradient and rainsplash transport (Morgan, 1978; Bryan, 1979), numerous others have found strong associations. Mosley (1973) showed that the percentage of sand splashed in a downslope direction increases rapidly to 95% where the sand surface is inclined at an angle of 25°, while 50% is splashed in each direction when the slope is horizontal. DePloey and Savat (1968) also reported a positive relationship with 75% of splashed material landing downslope for a gradient of 20° and 80% for a gradient of 30°. Citing this result, Young (1972) suggested that an approximate rule is that the proportion of all splashed particles that travel downslope is 50% plus the percentage slope. Finally, it has been concluded (Savat, 1981) that the work performed by splash (transport) is proportional to the sine of the slope angle to the 1.90 power. This means that splash transport is proportional to the sine of the slope angle to the 0.90 power.

The note concerning a small amount of surface flow in the statement by Ellison (1947) may be very significant. Palmer (1963) found that splash loss increases with water layer depth up to a threshold approximately equal to raindrop diameter and that it decreases thereafter. Mutchler and Young (1975) concluded that rainsplash will be highly erosive when the water depth is equal to about one-third the drop diameter and the underlying soil particles will be essentially protected from the erosive action of rainsplash when the water depth is equal to about three drop diameters. This may also account for the variability of research results, concerning erosion by rainsplash.

Rainsplash also tends to seal the soil surface. Compaction by direct raindrop impact and dispersion of soil aggregates appear to be the primary mechanisms of seal development. Gray and Leiser (1982) commented that increases of 15% in density in a 1-in. surface layer of soil have been attributed to compaction by raindrops. Several authors acknowledge the dispersion process with the inwash of fine-grained soil particles reducing porosity and permeability of the surface layer (e.g., Robins and Neff, 1963; Young, 1972; Ritter, 1978). Lal (1985) observed that raindrop impact can also close intraparticle pores by particle reorientation resulting in a change in soil bulk density and water transmission characteristics. Morgan (1979), however, citing the research of Farmer (1973) and Bryan (1969), suggested that the dispersion of clay-sized particles may be due to "slaking," the breakdown of soil aggregates by the compression of air ahead of a wetting front as rainfall starts to infiltrate a dry soil from the surface downward, rather than as a direct result of raindrop impact.

Regardless of the exact mechanism, the result is the same. McIntyre (1958) observed that the surface crust frequently consists of two parts, a very thin (about 0.1 mm), nonporous upper layer and a zone as much as five mm in depth that was attributed to the inwash of fine-grained particles. Edwards and Larson (1969), through laboratory experimentation using rainfall simulation and computer modeling, found that for silt loam soils, estimated 2-hr infiltration was reduced by as much as 50% by surface sealing. The reduction of soil infiltration capacity produces an increase in runoff and, hence, an increase in erosion by sheetwash. This is part of the annual cycle of erosion in western Colorado described by Schumm (1964).

Moeyersons and DePloey (1976) have identified another form of downslope particle movement caused by rainsplash which they call "splash creep." After fine-grained particles had been splashed from a sloping surface, the remaining lag material, composed of coarse-grained particles, creeped downslope under raindrop impact. Mosley (1973) discussed a similar process observed in both laboratory and field experiments; he concluded that in the Lusk, Wyoming badland, surficial creep due to raindrop impact is possibly the most important single agent of erosion. Carson and Kirkby (1972) also described the indirect method by which rainsplash may move large caliber surface stones by removing the fine-grained material on the downslope side causing eventual toppling of the stones.

Although Robins and Neff (1963) have remarked that the impact of rain onflowing water prevents establishment of fully turbulent flow which reduces erosive power, it seems that many authors hold a contrary opinion. Emmett (1978) commented:

> Over the length of slope for most overland flow, Reynolds numbers, a measure of fluid turbulence, normally remains in the regime considered as laminar flow, but the flow is not truly laminar because of the disturbance by falling raindrops and the influence of topographic irregularities. Such a disturbed flow is capable of eroding and transporting sediments.

Morgan (1979) noted that as the thickness of the surface water layer increases, so does splash erosion; this is believed to be due to the turbulence that impacting raindrops impart to the water. Lastly, Lal (1985) stated that the turbulence caused in the water by impacting raindrops increases the detaching capability. From the foregoing, it is evident that rainsplash functions in a variety of ways to cause the detachment, entrainment, and transportation of surface materials.

Sheet Erosion. Whenever rainfall intensity exceeds the infiltration capacity of surface materials, water begins to collect on the hillslope surface and fill microtopographic depressions. For a particular hillslope, there must be a finite volumetric capacity for depression storage (surface detention, surface retention) depending upon roughness, vegetation and litter cover, and perhaps season of the year, as it may correlate with the first two factors. As described above, this layer of ponded water itself may exacerbate the erosive action of rainsplash, here depending upon its depth in relation to raindrop diameter during a particular event.

Eventually, depression storage capacity is exhausted and the excess water begins to flow across the inclined hillslope surface. A plethora of terms with slight differences in connotations have been employed to describe the resulting occurrences. However, there are two basic processes in operation: (1) the hydrologic process of overland flow (runoff, surface flow, and sheet flow), and (2) the geomorphic process of sheet erosion (sheet wash, wash, sheetwash erosion, and slope wash) reflecting the work performed by rainsplash and overland flow. Historically, both processes were envisioned to function in "sheetlike" fashion with thin, perhaps monomolecular, layers of water cascading, sliding, or flowing downslope and, as a result, removing a rather uniform thickness of surface material from the hillslope. From this hypothesis, it can follow that overland flow would consist of laminar or "layered" movement of water molecules. But such shallow depths of laminar flow at moderate velocity could exert little force and produce minimal amounts of erosion by itself.

Emmett (1970, 1978) has shown that overland flow does not occur with uniform depth in most cases, nor is it laminar, as noted earlier:

> However, topographic irregularities that occur on natural slopes are sufficient to direct most runoff water into lateral concentrations of flow. These concentrations of flow weave anastomosing paths downslope and often give the appearance of flow in a wide, shallow-braided channel. (1978, p. 147)

In the earlier research, Emmett (1970) found flow concentration resulting from microtopographic irregularities of 3 mm in height, while Bryan (1979) observed flow concentration due to irregularity less than 1 mm in amplitude. Figure 16 shows the topography and flow pattern for one site included in Emmett's (1970) study.

Aside from surface and hydraulic properties, the uniformity of erosion on a hillslope often depends upon the time period considered. Morgan (1978) concluded that erosion did not take place uniformly across a hillslope during a single rainfall event. However, Carson and Kirkby (1972), Dunne and Leopold (1978), and Ritter (1978) remarked that shallow streams and threads of water shift their position back and forth across the hillslope and may, over a number of storms, remove a relatively uniform soil layer.

The force generated by overland flow that produces erosion when the resistances of the surface materials are overcome can be estimated thusly (Horton, 1945):

$$F = w\ d/12 \sin \alpha$$

where F is the eroding force (lbs/ft), w the weight of cubic foot of water (lbs), d the depth of flow (in.), and α is the gradient. Hence, the two principal factors governing erosive force are depth of flow and gradient. Assuming an ample supply of erodible material, constancy of material resistance, and hillslope gradient, the variability of erosion rate due to overland flow for a particular hillslope must be the result of variability of flow depth. Robins and Neff (1963) cited a cardinal principle in erosion control, namely, avoid high concentrations of flow.

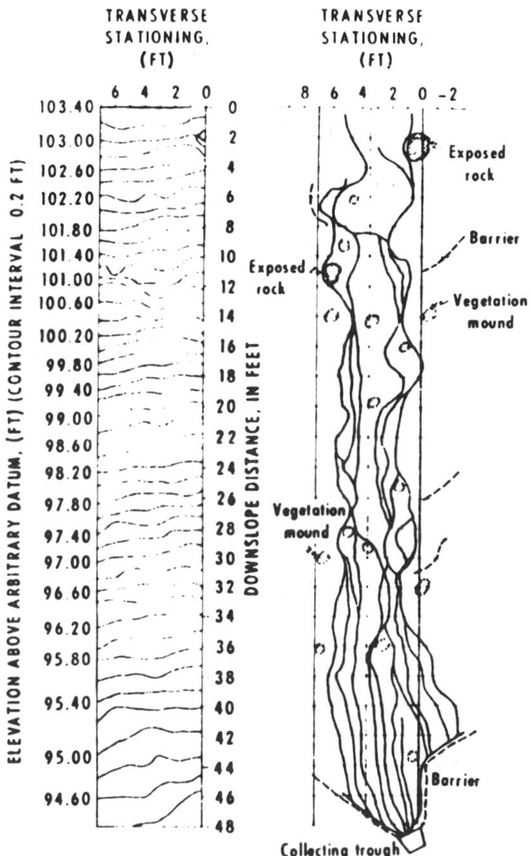

Figure 16. Topography (left) and flow patterns (right) for hillslope at Pole Creek, Wyoming, site 3 (Emmett, 1970).

It also follows that the depth of flow, erosive force, and erosion rate would be greater along the small channels depicted in Fig. 16 than for adjacent areas.

Three additional points regarding overland flow and sheet erosion are noteworthy. First, it is useful to clarify some common terminology. If the term *runoff* refers to all surface flow leaving a hillslope, then *overland flow* is one component of this total. Another will be discussed in the subsequent section concerning rilling. Further, *sheet flow* is a unique type of overland flow occurring only where the flow surface is very nearly featureless and where the flow is not mechanically disrupted by raindrop impact. Here it may be possible for water movement to be laminar, given appropriate hydraulic conditions. Foster and Meyer (1977) noted that such flow alone could transport only very small particles. Detachment and entrainment seem quite improbable.

Second, Emmett (1978) stated that overland flow is both unstable and spatially varied since it is supplied by rain and depleted by infiltration, neither of which is necessarily constant with respect to time and location.

Finally, it is important to remember that rainsplash and sheet erosion function in concert, except at the very beginning and following the rainfall event. Together, these two processes are much more effective in the detachment, entrainment, and transportation of materials than either taken separately, as demonstrated by Young and Wiersma (1973). While it is convenient to separate these geomorphic processes for pedagogic purposes, this overly simplifies the mechanism of erosion on hillslopes.

Rill Erosion. Where overland flow is concentrated due to microtopographic irregularities, or on disturbed sites, tillage, and grading marks, the depth and velocity of flow are substantially greater than for adjacent areas. In their laboratory study, Young and Wiersma (1973) found that depths of flow in the rills were on the order of 50 times greater than the average depths of overland flow in the areas between the rills, whereas flow velocities were about 10 times greater.

From the preceding equation, it is apparent, then, that rill flow possesses substantially more erosive force than overland flow. Foster and Meyer (1977) noted that flow in rills detaches soil particles when the flow's transport capacity is more than its sediment load and that the flow's shear stress exceeds the resistance of the soil to detachment. Erosive force of rill flow has been expressed as critical hillslope length (Horton, 1945), depth of flow (Selby, 1982), shear stress (Foster and Meyer, 1977), critical tractive force (Meyer *et al.* 1976), and critical discharge (Meyer, 1981). In most cases, these are closely related terms.

This locally intensified erosion produces distinct channels on the hillslope. Rills may be defined as a set of well-defined subparallel channels, a few to several centimeters in width and depth, which may be easily obliterated by frost-heaving of the soil, tillage, or grading operations. The development of a rill system on a hillslope is well described by Meyer *et al.* (1975, p. 179):

> Rill erosion begins when the eroding capacity of the flow at some point exceeds the ability of the soil particles to resist detachment by flow. Once rilling begins, the concentrated flow tends to enhance the detachment capabilities in that vicinity, and rilling progresses. Rill development often proceeds upslope as headcuts. Successive headcuts commonly exist along a rill, and the local erosion at each is very intense because of the erosive overfall condition. Where headcuts develop, they apparently contribute the major portion of the soil being lost from the rill. However, rill erosion may occur without major headcuts as a relatively uniform increase in the erosion rate with distance along the rill.

These authors also comment that the extent of rilling and the type of rill patterns that develop on sloping land depends on soil properties, slope steepness and other topographic characteristics, runoff rate, and tillage (for disturbed surfaces). Similarly, Carson and Kirkby (1972) suggest that rill demensions are controlled by the erodibility of the soil. Meyer *et al.* (1976) contend that the tendency for rills to meander decreases as the hillslope steepens.

The particle size distribution of sediments produced by rill erosion differs from that of sheet erosion. Foster and Meyer (1977) have found that the primary particle sizes of rill sediment are generally similar to those of the original soil mass and that aggregates from rill erosion tend to be larger than those carried from adjacent areas by rainsplash and overland flow because not much selective sorting occurs during rill erosion and the particles are not subject to repeated raindrop impact.

Once established, a rill system along a hillslope undergoes an evolution into an integrated drainage network which most agree follows a sequence discussed by Horton (1945). Because of especially effective erosion, a few rills on the hillslope will become more deeply incised into the surface than the others. These escape periodic destruction to become the longest and deepest channels, called "master rills." During particularly intense rainstorms, the flow in the smaller rills may overtop the divide between themselves, and the master rill, which due to greater incision, occupies a slightly lower position on the hillslope. This diversion of flow, in turn, causes the erosion of the divide. The breaking down of divides between adjacent rills and diverting the flow from the higher into the lower rill is called "micropiracy." As rill flow is turned toward the master rill and erodes along its course, a regrading of the surface takes place so that the initial subparallel rill system becomes an emergent dendritic drainage pattern. This process, which alters microtopography, is called "cross-grading." Carson and Kirkby (1972) noted that the cross-grading process clearly tends to form a permanent channel that cannot be destroyed between flows and shows one way in which hillslope processes can begin to form a river. Figure 17, adapted by Carson and Kirkby (1972) from Horton (1945), depicts the processes of micropiracy and cross-grading in operation on a hillslope through time. Cross-grading can be envisioned if one imagines small valleys on either side of each flow line.

It is necessary to acknowledge in this discussion that not all hillslopes develop rill systems. Carson and Kirkby (1972) observed that where there is a good vegetation cover, rills are less common than on bare soil because overland flow is less frequent and less intense so that sufficient erosive force to produce a rill is rare. In his study of hillslope hydrology and erosion, Emmett (1978) found:

> Sediment concentrations at New Fork River Site I were the highest observed at that time in the investigation and rilling was considered most likely to occur at this site. However after nearly

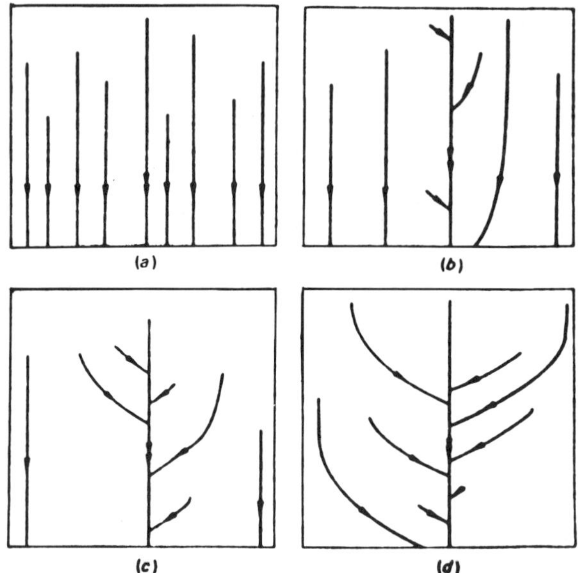

Figure 17. Hypothetical stages in the development of a master rill by cross grading as it enlarges its valley (Carson and Kirkby, 1972. ©, reprinted by permission of Cambridge Univ. Press).

> 10 hr of sprinkling at an intensity of 8.5 in./hr, no observable rills had formed. The sprinkling intensity was raised to 10.5 in./hr and continued for over 6 hr. Still no rills were formed by the increased runoff.

Evans (1980) suggested that whether rills or gullies form probably depends on soil factors as well as on the velocity and depth of flow. The concept of "rillability" of soils was examined by Young and Onstad (1976). They also noted differences in the particle size distribution of soils eroded from rills and those from adjacent areas. A soil rillability factor was devised based on the percentage of sand, percentage of clay, and organic matter content of the soil.

Although the geomorphic processes of rainsplash, sheet erosion, and rill erosion have been considered separately herein, the term *sheet erosion* has often been used to include all erosion that can be obliterated by tillage practices, thus including rather sizable rills (Meyer *et al.* 1976). As a consequence, caution must be exercised in discussions of hillslope erosion and erosion prediction models to ensure comprehension of terminology. The Soil Conservation Service (1977) has prepared guidelines for estimating the extent of rilling in the field.

Gully Erosion. With the coalescence of rills through the processes of micropiracy and cross-grading, flow is further concentrated during rainfall events, resulting in still greater depths and velocities. This produces greater tractive force and resultant stress across the materials comprising the wetted perimeter of the rill. The dimensions of the channel are enlarged by erosion until they surpass those included in our definition of a rill; a gully has now formed. Dunne and Leopold (1978) commented that gully erosion produces incisions ranging in size from a 1 ft in depth and width and a few feet long to several tens of feet deep, hundreds of feet wide, and miles long.

Concise definition of a gully is elusive because it is a transitional type of channel between a rill and a stream. Further, or perhaps as a result, there are various types of gullies. The Soil Conservation Service (1977) remarked that the term *gully* includes upland and valley side-wall gullies as well as incised channels within floodplains, commonly called valley trenches. Heede (1975) noted that gullies have been classified as continuous or discontinuous and these are strikingly different in appearance. Herein, we are concerned only with the so-called valley side-wall type which may be either continuous or discontinuous. Heede (1975, p. 156) described these thusly:

> Continuous gullies begin their downstream course with many small rills, while discontinuous gullies start with an abrupt headcut. The headcut of the discontinuous gully may be located at any position on the slope of a hillside, while a continuous gully always starts high up on the mountainside and continues its course down to the main valley floor. The discontinuous gully may intersect the surface of the slope at any point, and thus be terminated.

Later, he suggested that discontinuous gullies develop into continuous gullies; while this may be true for the valley side-wall type, some doubt persists concerning valley trenches in semiarid and arid regions. Despite considerable variety, gullies possess similar morphologic properties that provide some basis of definition. Hence, a gully may be defined as a permanent channel with vertical or nearly vertical banks, a few decimeters or more in width and depth, which carries ephemeral flow during and immediately following rainfall events.

Branson *et al.* (1981) observed that the factors contributing to gully formation are often ill-defined, although grazing, intense summer storms, and general climatic changes have been noted as possible contributors. Similarly, Selby (1982, p. 107) commented that:

> Gully erosion nearly always starts for one of two reasons: either there is an increase in the amount of flood runoff, or the flood runoff remains the same but the capacity of water courses to carry the flood water is reduced. The most common causes of increases in runoff or deterioration in channel stability are changes in vegetation cover—especially removal of trees, increases in the proportion of arable land in catchments, excessive burning of vegetation, or overgrazing—or a climatic change with accompanying variations in rainfall periodicity and intensity.

The influence of soil characteristics on gully formation has been documented. Gerdel (1937) concluded that gully erosion appeared to be associated with the prismatic type of soil aggregates that apparently erode as aggregate units, whereas sheet erosion was more common under conditions of good granular structure and was associated with sufficient dispersion of the aggregates to indicate a predominantly texturally separate type of erosion. DePloey (1974) identified three types of gully networks related to particular soil types: (1) axial gullying, composed of individual gullies with single headcuts that grow upslope due to surface erosion and occur in gravel deposits; (2) digitate gullying, involving several headcuts growing toward several tributary depressions and characteristic of clay loam soils; and (3) frontal gullying, resulting from piping along river banks and found on loamy sand soils with columnar structure.

Dunne and Leopold (1978) admitted that geomorphologists do not yet understand exactly how rills and gullies form and remain stable. Even the obvious may fail us. For example, Piest *et al.* (1975) concluded for a loess region in western Iowa that the erosive power of runoff plays a minor role in gully growth, except for its relation to transport of eroded bank material and possibly some plunge pool action. Gray and Leiser (1982) concluded that a whole array of processes can act individually or in concert to produce and expand gullies, including (1) waterfall (plunge pool) erosion at the gully head, (2) piping and spring sapping at the head and along thc sides, (3) erosion and scour along the length of the gully bottom, (4) raindrop splash and rilling on the sides, (5) freezing and thaw erosion of the gully sides, and (6) mass wasting of the gully sides and head.

Aside from the references contained above, useful discussions of gullying are provided by Leopold *et al.* (1964) and Emmett (1968). Mitchell and Bubenzer (1980) have compiled a useful summary of attempts to model the gullying process.

Limits to Erosion. The detachment, entrainment, and transportation of surface materials may be accomplished by rainsplash, sheet erosion, rill erosion, or gully erosion. For most situations, it is the combination that is responsible for hillslope denudation. Erosion rates may be limited by detachment rates or transport rates. Meyer *et al.* (1976) commented that only soil particles that have been detached from the soil mass will be eroded, no matter how great the transport capability (detachment limiting), and only soil particles that can be transported by the flow will be eroded, no matter how much is detached and available for transport (transport limiting). They also contend that erosion control practices that are designed to reduce whichever of these two subprocesses is limiting will usually be more effective than those designed to reduce the other subprocess that is in excess. Concerning limiting factors of erosion, Meyer *et al.* (1975, p. 179) remark:

> On upslope areas, where there is no evidence of significant flow concentration or rilling, raindrop detachment is satisfying the transport capacity of the interrill flow, or the detachment capacity of the flow is not sufficient to detach a significant amount of soil from the soil mass. Where rills develop, the detachment capacity of the flow exceeds the ability of the soil to resist detachment, and the transport capacity of the flow is unfilled by raindrop-detached particles. If

rills have alluvial fans, these are locations where the transport capacity was not sufficient to carry the sediment load, and deposition resulted.

Disturbed Land Erosion. The erosion processes described above are probably characteristic of disturbed areas. Dunne (1978) stated that Horton overland flow is most common in arid and semiarid lands or in humid areas where the original vegetation and soil structures have been destroyed. Among the types of disturbances cited as causing overland flow are (1) compaction of soils by animal, human, and vehicular traffic; and (2) vegetation removal due to grazing, burning, mining, or toxic smelter fumes. However, surface erosion processes are not the only ones capable of removing hillslope materials and creating depressions or channels. Subsurface flow may also detach or dissolve, entrain, and transport.

Erosion by Subsurface Flow. A substantial proportion of rainwater that infiltrates the soil flows laterally downslope within the porous medium; this hydrologic process is called "throughflow." Atkinson (1978) distinguishes two types of sub-surface flow: (1) matrix flow, wherein soil moisture moves through the inter-granular or small structural pores and voids of a soil; and (2) pipe flow, wherein water passes through larger subterranean voids or channels. In terms of hydraulic regime, matrix flow is generally considered as laminar while pipe flow is regarded as turbulent. Dunne (1978) and Kirkby (1971) suggested that Hortonian overland flow is perhaps of greater importance in arid and semiarid regions, whereas throughflow is of greater importance in humid regions. Nevertheless, the type of subsurface flow known as piping is common in arid and semiarid areas (Branson *et al.* 1981).

The amount of erosional work performed by subsurface flow depends upon type (matrix or pipe) and process (mechanical or chemical). It would seem apparent that the capacity for mechanical detachment and transport by laminar matrix flow would be minimal. However, transport in solution may be another matter entirely. Carson and Kirkby (1972) commented that chemical removal (by solution) is a major form of hillslope erosion in temperate and humid tropical areas, of the same order of magnitude as all forms of mechanical erosion combined. Except in special cases, this may be something of an exaggeration or merely due to the inefficiency of mechanical processes because of dense vegetation cover and erosion-resistant soils. In another study, Roose (1970) found that soilwater flow accounted for only about 1% of material removal from a hillslope in Senegal.

There is far less controversy concerning the mechanical erosive power of pipe flow. Carson and Kirkby (1972) observed that these pipes or tunnels may grow to several meters in diameter. One might only ask how prevalent they are in a particular region and, as a result, what their contribution is to total erosion because their magnitude and frequency is usually much less obvious in an area than that of rills or gullies.

Subsurface flow and erosion can have various surficial expressions. Percolines, or seepage lines, are zones of deep soil on a hillslope along which the downslope movement of moisture becomes concentrated (Bunting, 1961). A linear pattern of dark, dampened soil on the surface may reveal their existence and location. Jones (1976) suggested a relationship between percolines and pipes. Thornes (1980, p. 155) described the process by which pipes become gullies:

> Once a pipe is started it provides a new base for the local hydraulic gradient. The water flow lines are then concentrated towards the pipe and the rate of piping accelerates. Eventually the roof becomes unstable and may collapse leading to gully formation.

It is clear that subsurface flow and erosion can be a very important, even dominant, geomorphic process in areas where most of the rainfall infiltrates the soil surface and where surface processes are rendered relatively impotent by soil and vegetation resistances. Because land disturbance generally reduces infiltration capacities of soils and vegetation cover, making overland flow more likely, we must limit our discussion of subsurface processes. However, excellent treatments of the subject are available (Atkinson, 1978; Whipkey and Kirkby, 1978; Dunne, 1978; Carson and Kirkby, 1972).

The Rill–Interrill Concept. Much of the contemporary erosion research in the United States centers on the "rill–interrill concept" described in detail by Meyer *et al.* (1975, 1976) and Foster and Meyer (1975, 1977). These authors have noted that the separation of the erosion processes into these two source areas is conceptually useful, in mathematical simulations of erosion processes, and warranted because the detachment and transport processes on each of these areas are distinctly different. The concept assumes Hortonian overland flow for the particular hillslopes under examination. The four subprocesses discussed earlier still provide the basis of this approach.

The areas adjacent to rills are called "interrill areas." The work of Young and Wiersma (1973), together with that of the authors cited above, has demonstrated that the primary force initiating soil particle detachment is the energy of rainsplash rather than the energy of surface flow. Interrill flow usually traverses short distances at shallow depth and has very little detachment capacity in the absence of raindrop impact. However, according to this concept, most of the soil detached in the interrill areas is transported to rills by this flow with raindrop impact normally accounting for a small proportion of total soil transportation. This would depend, of course, on the frequency of rilling on the hillslope. The detachment and transport processes on the interrill areas occur essentially independent of rill erosion; although considerable excess interrill transport capacity may exist, it does not add to rill transport capacity to directly transport particles downslope. Again, Young and Wiersma (1973) found that 80–85% of the soil loss from interrill areas was transported to a rill before leaving the research plot.

Two processes are active within the rill itself: (1) detachment of soil materials, and (2) transport of both these materials and those delivered to the rill by interrill flow and rainsplash. The creation of the rill is evidence of the detachment capability of rill flow. Rill erosion is a function of flow depth, flow velocity, and the resistivity of the soil material upon which the resulting forces are exerted. Once rilling begins, it frequently progresses upslope as one or more headcuts across which detachment is intensified due to an overfall effect. It is believed that raindrop impact does not directly detach any particles below the flow line in the rills, but increases the detachment and transport capacity of the flow.

Most of the materials eroded from the hillslope are transported downslope by rill flow. However, it appears that sediment from the interrill areas has little effect on rill detachment and transport. This may be because sediments from interrill areas tend to be fine-grained due to the splash shattering of soil aggregates and the slight detaching and transporting capacity of interrill flow. These interrill sediments may be carried in suspension, while larger aggregate and sand-sized particles produced by rill erosion move as bedload. On hillslopes with cohesive soils, the transport capacity of rill flow is much greater than the detachment capability of the flow; hence, erosion is detachment limited. However, for noncohesive soils the sediment load may approach the transport capacity.

The relative contribution of rill and interrill erosion to the total varies with hydrologic and soil conditions along the length of the hillslope. Interrill erosion is relatively constant across and along the hillslope as long as soil and vegetation conditions remain constant. Rill erosion is generally absent near the crest of the hillslope but increases in the downslope direction. Thus, interrill erosion can vary from 100% on short slopes with negligible rilling to a small percentage of the total on extensively rilled hillslopes. Some soils are more susceptible to rilling than others, and this will influence the relative contribution. Figure 18 illustrates the difference in rillability and its effect on total erosion along a test hillslope. Two plots were composed of the same soil but were managed differently with the result that Plot R-3 rilled very rapidly and produced 1169 lb (526 kg) of soil erosion from 5 in. of simulated rainfall. Plot R-4 showed little rilling and produced only 346 lb (156 kg) of eroded soil. Runoff from the rilled R-3 plot was only 7% more than from the unrilled R-4 plot. This figure shows that both plots were about equally susceptible to interrill erosion; thus, the substantially greater total for the R-3 plot was due to its tendency to rill. The implications of this figure concerning hillslope length and erosion will be considered in a later section.

Wind Erosion. Surface materials may be detached, entrained, and transported by the action of wind. However, scientists have not agreed on the role of

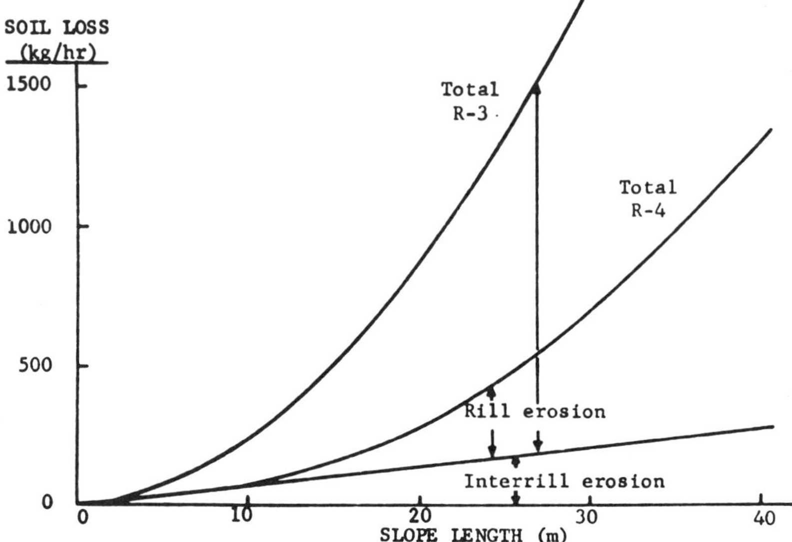

Figure 18. Relative contribution of rill and interrill erosion as affected by slope length for two conditions. R-3 and R-4 were equally susceptible to interrill erosion, but R-3 was much more susceptible to rilling (Meyer, *et al.*, 1976).

wind in geomorphology; irrefutable proof that wind is capable of significant geomorphic work is simply not available in many instances because definitive, quantitative data concerning eolian processes and features are woefully few. Israelsen *et al.* (1980) assert that in most areas of the United States the amount of erosion attributable to wind as opposed to that from water may be equal to or near zero.

But it is clear from the historical record of the "dust bowl" in the Great Plains of the United States, as well as this author's observation of nonrevegetated topsoil stockpiles perched atop ridges and filling valley head hillslopes near a divide, that wind erosion is capable of moving considerable volumes of material in relatively short time spans. Wilson and Cooke (1980) suggested that the human consequences of wind erosion are undoubtedly most serious in those agricultural areas that experience low, variable, and unpredictable rainfall, high temperatures and rates of evaporation, and high wind velocities, as is the case in semiarid areas, as well as some of the more humid regions that experience periodic drought. They offer an interesting table listing physical and economic effects of wind erosion (Table 7). Cooke and Doornkamp (1974, p. 51) described its effects on agricultural crops accordingly:

> Wind erosion may cause damage to crops by, for example, removing seeds, exposing plant roots, and blasting leaves. In addition, the soil is depleted of the more easily removed particles which include organic matter and smaller grains; such depletion, seriously and progressively reduce the soils water-retention properties and productivity.

Such consequences are equally detrimental to revegetation of disturbed lands.

In general terms, the basic principles and controls of water erosion may be applied to wind erosion (Gray and Leiser, 1982; Robins and Neff, 1963), recog-

Table 7 Some Physical and Economic Effects of Wind Erosion[a]

Physical effects	Economic consequences
Soil Damage (1) Fine material, including organic matter, may be removed by sorting, leaving a coarse lag. (2) Soil structures may be degraded. (3) Fertilizers and herbicides may be lost or redistributed.	Soil Damage (1,2,3) Long-term losses of fertility give lower returns per hectare. (3) Replacement costs of fertilizers and herbicides.
Crop Damage (1) The crop may be covered by deposited material. (2) Sandblasting may cut down plants or damage the foliage. (3) Seeds and seedlings may be blown away and deposited in hedges or other fields. (4) Fertilizer redistributed into large concentrations can be harmful. (5) Soil borne disease may be spread to other fields. (6) Rabbits and other pests may inhabit dunes and be trapped in hedges and feed on the crops.	Crop Damage (1–6) Yield losses give lower returns. (1–3) Replacement costs, and yield losses due to lost growing season. (5) Increased herbicide costs.
Other Damage (1) Soil is deposited in ditches, hedges, and on roads. (2) Fine material is deposited in houses, on washing, and on cars, etc. (3) Farm machinery, windscreens, etc., may be abraded, and machinery clogged. (4) Farm work may be held up by the unpleasant conditions during a blow.	Other Damage (1) Costs of removal and redistribution. (2,3) Cleaning costs. (4) Loss of working hours and hence productivity declines.

[a]From Wilson and Cooke (1980, p. 218).

nizing the substitution of air for water as the fluid medium involved. As described by Warren (1979), the lifting or "entrainment" of particles is a function of six forces: the driving forces of lift, shear, and ballistic impact from bouncing particles must overcome the resisting forces of gravity, friction, and cohesion. Shear along the surface is considered to be the major driving force of transportation. Wind velocity and turbulence combine to determine the magnitude of these driving forces. Although certainly a general figure, Gray and Leiser (1982) suggested that, for many soils, the wind velocity necessary to initiate particle movement is about 13 mph at a height of 1 ft above the surface. The velocity required to sustain transport, once started, is less than that required to initiate movement. Wind-driven sediment may be transported in three ways: (1) saltation, the bouncing of particles across the impact surface; (2) suspension; and (3) surface creep, the movement of large particles caused by the pushing action of saltating particles striking the larger from behind. Chepil (1945) found that 55–72% of particle movement was by saltation, 3–38% by suspension, and 7–25% by surface creep. The sediment load carried by wind decreases rapidly with distance from the surface. According to Chepil (1945), 62–97% of the total wind eroded soil is transported within 1 m of the surface.

The driving forces are opposed by resistances generated by soil and surface properties. Soil texture, moisture, and organic matter content combine to create aggregation and cohesion. Chepil and Woodruff (1963) found that a mixture of soil separates in the proportions 20–30% clay, 40–50% silt, and 20–40% sand offered the greatest resistance to wind erosion. Chepil (1956) found that erodibility of a soil decreased as the square of soil moisture increased. Likewise, Chepil (1955) concluded that erodibility decreased as organic matter content increased.

The surface properties that influence wind erosion are related to roughness and the disruption of air flow. Hence, energy is consumed through friction, reducing transport capacity. Wilson and Cooke (1980) distinguished five major groups of roughness elements: (1) vegetation, (2) clods and nonerodible fractions, (3) ridges, (4) field shelter belts, and (5) local changes in topography.

It is probably fair to infer that progress toward a complete understanding of wind erosion has lagged behind that of erosion by water. Wilson and Cooke (1980) stated:

> Wind-tunnel experiments have been concerned mostly with relatively simple situations in terms of erodibility and erosivity, and this has enabled the influence of various factors to be quantitatively assessed. In the field, however, the complexity of soil-erosion systems is much greater; and, in particular, erodibility and erosivity vary both spatially (and often over short distances) and temporarily.

Elsewhere, they remark that monitoring wind erosion is time consuming, expensive, and suitable equipment needs to be developed. Derbyshire *et al.* (1979)

affirmed that in the field there is considerable difficulty in adequately measuring (particularly over long periods) turbulence, strength, and direction of the wind at or near the ground level; this means that wind and sand flow patterns are generalized because of the lack of representative and reliable data. Actually, some of these same limitations apply to erosion by fluvial processes as well and may become crucial in both cases in the future as models are substituted for field data and experience.

Streams and Stream Channel Processes

In a drainage basin, hillslope erosion processes and stream channel erosion will yield sediment and dissolved solids that are transported by the rivers out of the basin system. Stream channels are simply the means of conveying streamflow and erosional products through the drainage network. The forces that shape the channel and determine its geometry are essentially those exerted by the streamflow. The load transported by stream is composed of a dissolved load carried in solution, a suspended load, and a bed load that is transported at or near the channel bed.

Channels also may be distinguished by the temporal distribution of flow in them. In arid and semiarid regions flow is ephemeral, occurring only in response to infrequent storms or snowmelt; they are dry most of the year. Perennial streams flow continuously and are supplied by surface runoff and groundwater. Intermittent streams flow seasonally and also are supplied by surface runoff and groundwater.

The Hydraulics of Streams

Flow of water in open channels is subject to two external forces: gravity and the resistance forces of friction. Gravity tends to accelerate the water downstream, and frictional forces within the water and between the water and the channel boundaries resists the downstream movement. The flow will accelerate downstream in a channel that is relatively steep and where the perimeter is smooth, and friction forces are low. However, in natural streams, as the flow velocity increases, the resistance to flow also increases. The resisting force exerted by the bed and banks of the channel is a shear stress (τ). The shear stress that can be generated on the bed of a river (τ_0) is a function of the specific weight of the fluid (γ), and both the hydraulic radius (R) and the channel slope (S) (see Fig. 19) as expressed in Eq. (1).

$$\tau_0 = \gamma RS \tag{1}$$

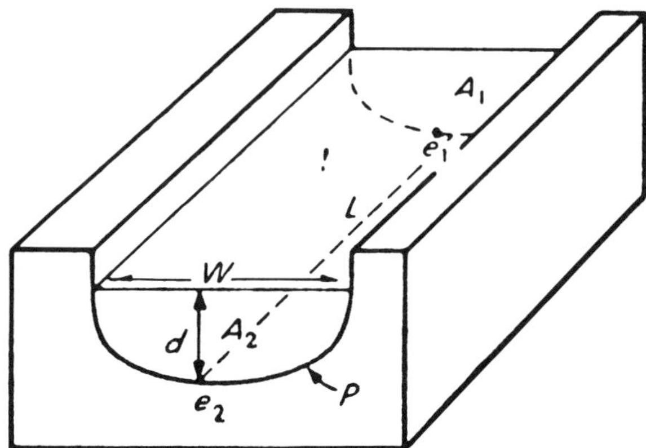

Figure 19. Stream channel morphometry. Stream width *W* is actual width of water in channel and *P* is the wetted perimeter. Cross-section *A* is the area of a transverse section of the river. Depth *d* is approximately equal to the hydraulic radius *R*. Stream gradient *S* is the drop in altitude between two points, e_1 and e_2, on the channel floor (Morisawa, 1968).

Minimal bed erosion occurs when channel slope (*S*) is low, and the wetted perimeter (*P*) is large when compared with the channel cross-sectional area. Therefore, the resisting force per unit area of channel perimeter is proportional to the product of the hydraulic radius and the channel slope.

The relationship between the forces of flow and resistance has been expressed by the Chezy equation and the Manning equation. The Chezy equation

$$V = C\sqrt{RS} \tag{2}$$

shows that mean velocity (*V*) is proportional to the square root of the product of hydraulic radius and slope. The Chezy coefficient (*C*) is a proportionality constant related to the resisting forces. Manning developed an empirical equation, using existing data, to estimate flow velocity

$$V = 1.49/nR^{2/3}S^{1/2} \tag{3}$$

The equation shows that mean velocity is directly proportional to the hydraulic radius and slope of the channel and inversely proportional to a roughness factor (*n*). The *n* factor is an estimated value based on judgment.

The Chezy equation and the Manning equation are similar; hydraulic radius *R* and channel slope *S* are directly proportional to mean flow velocity in both. The two equations can be related when

$$C = 1.49/nF^{1/6} \tag{4}$$

where *F* is the Froude number (explained below). Water, flowing in a channel, moves as laminar or turbulent flow. When velocity exceeds a critical value, the flow is transformed from laminar to turbulent. The Reynolds number ($R_e = \overline{V}d/v$), represents a measure of flow turbulence where $\overline{V}$ is the mean velocity, d the depth, and v the kinematic viscosity. High values of the Reynolds number indicate laminar flow. Flow in most natural streams is turbulent. Turbulent flow may be divided into two types: (1) tranquil and (2) rapid flow. These two types of turbulent flow can be determined from the Froude number:

$$F = \overline{V} / \sqrt{gD} \tag{5}$$

where $\overline{V}$ is the mean velocity, g the acceleration due to gravity, and D the depth of flow. When the Froude number (F) is less than one, the stream is in a tranquil flow regime. If the Froude number is greater than one, the stream is in a rapid flow regime, and a Froude number of one separates tranquil and rapid turbulent flow.

The velocity profile of turbulent flow becomes more uniform, or smoothed, with increasing values of the Reynolds number (R_e). Velocity theoretically is zero at the stream bed due to bed friction, and the maximum velocity occurs at some point below the water surface, usually about 0.2 of the total depth. Velocity also increases from the banks toward the center of the channel, and the maximum velocity may be at or close to the surface in the center of a symmetrical channel. The mean velocity of flow at a stream cross section occurs at a position of about 0.6 of the depth from the surface to the channel bed (Fig. 20). Turbulence and velocity are related to the erosion, transportation, and deposition that occurs in a stream. In other words, these processes constitute the geomorphic work that a stream can perform, and they represent the driving and resisting forces.

Stream discharge (Q) is the product of mean velocity and the cross-sectional area (A) at a channel section. It is expressed by the equation

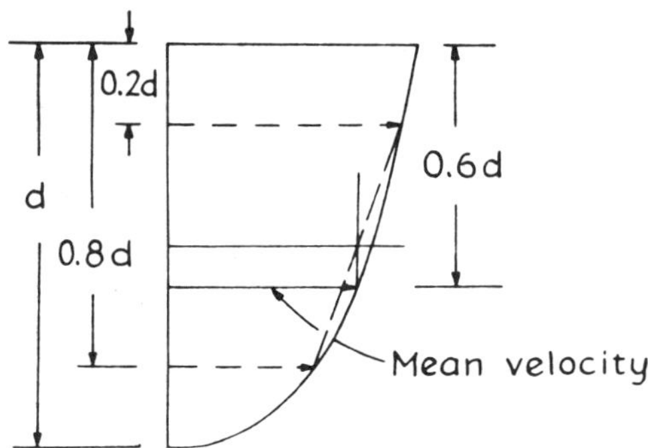

Figure 20. Relation of position of mean velocity to the distribution of velocity with depth (d) in a river (Leopold *et al.*, 1964).

$$Q = wd\overline{V} \qquad \text{or} \qquad Q = A\overline{V} \tag{6}$$

where A is the channel width (w) times depth (d). The interrelationships between the flow of water in channels and the transport of erosional debris will now be considered.

Transport of Sediment

Streams transport erosional debris as (1) suspended load, (2) bed load, and (3) dissolved load. The suspended load is composed of fine-grained silt and clay that is transported in the flow and suspended by turbulence. Bedload is transported at or near the bed of the stream by rolling or saltation (jumping) and is made up of the coarse particles of the total debris load. The dissolved load generally is not visible and is composed of chemical constituents derived from weathered rock and transported in solution.

Information on the dissolved loads of streams is useful in the evaluation of chemical denudation of river basins and in estimating the part of the load derived from chemical processes. Measurements of the variation of solute concentrations may be used as an indicator of human impact (Walling, 1984). Walling cites an example of an increase in salinity in streams of southern Australia resulting from the replacement of forest cover by croplands and pasture (Peck, 1976). Because much of the dissolved load in a stream is contributed by base flow from groundwater sources, concentrations are higher during low flow periods. In periods of high flow or floods when surface water is contributing most of the flow, concentrations are much lower. Table 8 (Walling, 1984) presents a summary of dissolved constituents in river waters of the world.

The separation of bed load and suspended load presents some problems. Suspended load is composed of silt and clay at low velocities and may move long distances downstream without being deposited. Coarser particles may become part of the suspended load temporarily when velocity and turbulence increase. The term *wash load* is used with reference to the fine fraction of the sediment load that consists of particles finer than those present on the stream bed. The suspended load moves through the channel at the velocity of flow while bed load is transported at a slower rate. The coarser particles found on the stream bed in abundance are termed the *bed material load*. Much of the bed material load is transported in suspension, so true bed load, although it is part of the bed material load, generally does not comprise more than about 10% of the sediment transported by a stream. There are exceptions to this generalization. Bed load may constitute as much as 70% of the total load in steep mountain torrents.

Most streams, at any discharge, can carry more fine-grained sediment than is supplied to them. Therefore, the relationship between suspended load and water discharge exhibits considerable scatter. This scatter is displayed in a logarithmic

Table 8 Mean Major Ion Composition (mg/liter) of River Waters of the World[a]

Region	HCO_3	SO_4	Cl	NO_3	Ca	Mg	Na	K	Fe	SiO_2	Total
North America	68	20	8	1	21	5	9	1.4	0.16	9	142
South America	31	4.8	4.9	0.7	7.2	1.5	4	2	1.4	11.9	69
Europe	95	24	6.9	3.7	31.1	5.6	5.4	1.7	0.8	7.5	182
Asia	79	8.4	8.7	0.7	18.4	5.6	9.3	—	0.01	11.7	142
Africa	43	13.5	12.1	0.8	12.5	3.8	11	—	1.3	23.2	121
Australia	31.6	2.6	10	0.05	3.9	2.7	2.9	1.4	0.3	3.9	59
World	58.4	11.2	7.8	1	15	4.1	6.3	2.3	0.67	13.1	120
World (% composition)	48.7	9.3	6.5	0.8	12.5	3.4	5.3	1.9	0.6	10.9	100

[a]Source: Walling (1984).

plot in Fig. 18, and may be caused by (1) inaccuracies of field and laboratory measurements, (2) dynamics of erosion and sediment yield in the basin, and (3) marked nonstationarity of basin response (Walling, 1977). These causes are difficult to isolate, but such factors as the lag of flood peaks in relation to maximum sediment concentration (Heidel, 1956) and hysteresis effects of sediment concentration with rising and falling stage are important (Walling, 1974). Large errors (±50%) may be associated with rating curve estimates of suspended sediment load, and they should only be used for the stream channels for which they were developed.

In contrast, whereas suspended sediment concentration is generally independent of flow conditions, a maximum bed load transport rate may be defined for a given discharge and sediment size (Richards, 1982). Richards attributes this phenomenon to the following factors: (1) the source for bed load is confined to the channel, (2) bed particle movement is brief and intermittent, (3) bed load is generally less than 10% of the total load, (4) exhaustion of supply is unlikely, and (5) bed load transport utilizes much of the available stream energy.

The force required to entrain a given particle on the stream bed is called the *critical tractive force* (τ_c). In a series of flume experiments using two sands having median diameters of 0.67 and 2.0 mm, Wolman and Brush (1961) determined velocities required to cause incipient motion. Discharge, area of cross section, water surface slope, and rate of particle movement were measured. The critical tractive force at which movement began was computed from the equation

$$\tau_c = \gamma R s \tag{7}$$

where γ the specific weight of the water, R is the hydraulic radius in feet, and s is the slope. The results of their experiments are illustrated in Fig. 21 along with results from work by Rubey (1938) and Bogardi and Yen (1938).

Rubey (1938) proposed the relation between velocity and critical tractive force, and he pointed out that for particles 5 mm and larger, velocity rather than the depth–slope product was the controlling factor. For particles less than 1.7 mm, the influence of velocity is reduced and movement is determined by the depth–slope product alone (Wolman and Brush, 1961). The data from Bogardi and Yen extend the relation to larger particles.

The sediment discharge of a stream, either as suspended load or bed load, will vary both spatially or temporally. Among the factors that determine the load are water discharge, velocity, channel geometry, roughness, channel slope, and the physical characteristics of the flow and of the sediment particles in the load. All of these variables are interrelated in a complex way. Over a long period of time, all these variables are dependent on the climate and lithology of the drainage basin.

Before we can discuss the morphology of stream channels and their hydraulic geometry, we must consider the effects of the variables that are responsible for

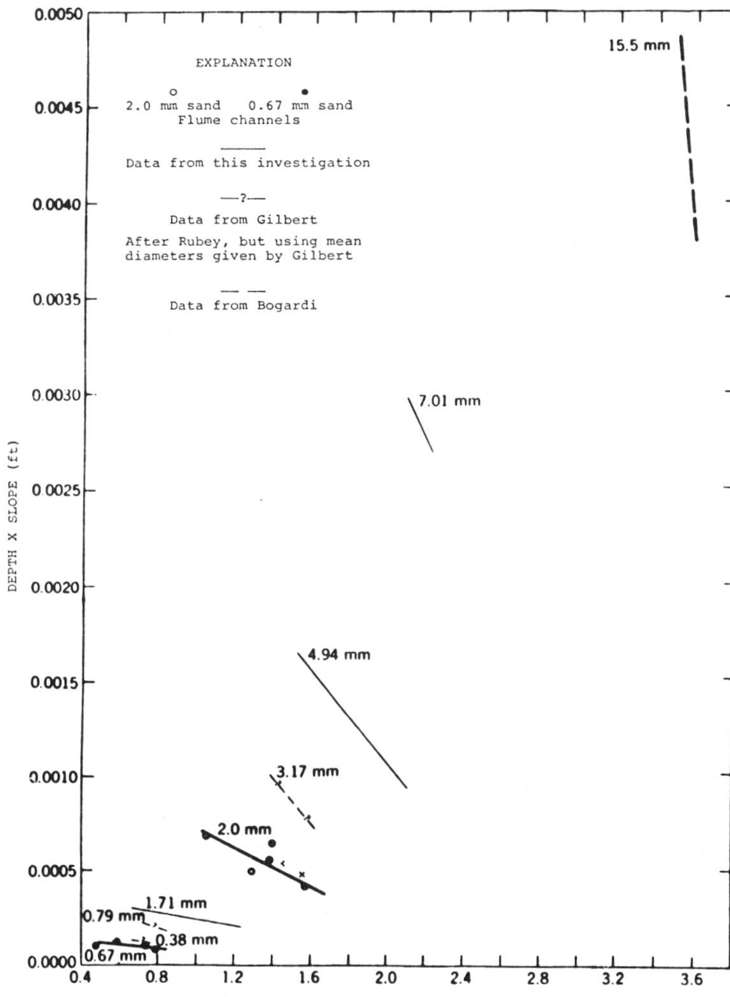

Figure 21. Conditions of incipient movement of particles of different sizes determined by depth times slope product and mean velocity (Wolman and Brush, 1961).

cause and effect relationships. Schumm (1977) approached this problem from the viewpoint that time is the most important independent variable of a cyclic time span. Schumm also divided the entire time span of landform evolution into three categories—cyclic, graded, and steady time. For example, a cyclic time span covers thousands of years, and a stream in this time span is an open system subjected to continual change with no constant relationships between independent and dependent variables. The graded time span refers to a shorter period (hundreds of years) when a steady-state equilibrium exists. The continual change of cyclic time may be progressing in the drainage basin, but the graded stream shows no progressive change. Schumm (1977) stated that such hydrologic variables as streamflow and sediment yield are important during graded time because they acquire statistical significance and define channel characteristics. The steady time span (defined by Schumm as one month or less) reflects a static equilibrium when instantaneous measurements of streamflow and sediment discharge can be observed at any point along the stream. Changes in streamflow and sediment discharge that occur during steady time may cause temporary changes in channel geometry that may be confusing to the geomorphologist who is interpreting processes.

As pointed out by Schumm (1977), it is important to differentiate between the time spans being considered in a study of landform evolution. When we are considering the sphere of interest of the geomorphologist, who is concerned with long-term river response, time is important. To the engineer, who is interested in the effects of human activities on hydrology and sediment discharge, time may not be important. For a detailed discussion of the variables that influence stream morphology and behavior, see Schumm (1977) and Chapter 5.

Soil Loss Prediction

We have considered the processes of erosion on upland slopes and the sources of sediment transported by streams from these areas. The next logical step is to quantify the soil loss from larger areas under a variety of land use practices. For many years soil conservationists have conducted experiments designed to estimate the soil loss from agricultural fields or plots. In 1915 the first experiments were begun to quantify the effects of land use on runoff and erosion. For an excellent summary of the evolution of experiments on the prediction of upland erosion see Kirkby and Morgan (1980). The pioneer work of the early investigators culminated in the development of the Universal Soil Loss Equation (USLE), the most widely used method of soil loss prediction in the United States. The equation is the product of soil erosion research in the U.S. Department of Agriculture under the leadership of W. H. Wischmeier.

Wischmeier and Smith (1965) stated that knowledge of the inherent erodibility of a soil is not sufficient for prediction of soil loss. The erodibility of a site is a function of several variables. These are (1) erosivity of the rainfall and runoff, (2) erodibility of the soil, (3) slope length, (4) slope steepness, (5) cover and agronomic management, and (6) supporting conservation practices. The USLE combines local values of these six factors to predict average annual soil loss from sheet (interrill) and rill erosion.

The equation is

$$A = RKLSCP \tag{8}$$

where A is the predicted average annual soil loss in tons per acre ($A \times 2.24 = 5$ tons/ha), R the rainfall and runoff erosivity factor, K the quantitative soil erodibility factor, LS the dimensionless factors for the effects of slope length and steepness, C the dimensionless factor that accounts for the effects of cover and management, and P the dimensionless factor that accounts for effects of supporting erosion control practices (contouring, strip cropping, and terraces).

Greater detail on the composition of and evaluation of factors L, S, C, and P and on the expected accuracy of USLE estimates is given in a paper by Wischmeier and Smith (1965).

The USLE was designed to be used by field technicians and is strictly an empirical tool that computes long-term average soil losses for specific combinations of physical and management conditions and rainfall characteristics. It does not predict specific year or specific storm losses (Wischmeier and Smith, 1965) nor does it compute watershed sediment yield. The gross erosion estimate predicted by the USLE must be adjusted with a factor called the sediment/delivery ratio which accounts for deposition in intervening areas between slopes and stream channels. The sediment/delivery ratio concept will be discussed in more detail in Chapter 5, (Drainage Basins).

The USLE has generally been limited until recently to use on croplands of the eastern United States. It is now potentially useful as a technique for estimating soil loss from lands of the western United States disturbed by mining and construction activities. Wischmeier's work in developing simplified relationships of soil characteristics for estimating the soil erodibility factor [K in Eq. (8)] was a major breakthrough. These relationships have removed many of the previous restrictions on use of the USLE.

There have been several attempts to develop models for estimating erosion and sediment yield from construction sites. Holberger and Truett (1976) (see Construction, Chapter 10) modified the Universal Soil Loss Equation through the addition of two sediment-loading functions. This modification will be discussed further in Chapter 10.

References

Arkley, R. J., 1967, Climates of some great soil groups of the Western United States: Soil Science, v. 103, no. 6, pp. 389–400.

Atkinson, T. C., 1978, Techniques for measuring subsurface flow on hillslopes: *in* Kirkby, M. J. (ed.) Hillslope Hydrology: New York, John Wiley and Sons, pp. 73–120.

Beschta, R. L., 1978, Long-term patterns of sediment production following road construction and logging in the Oregon Coast Range: Water Resources Research, Vol. 14, No. 6, pp. 1011–1016.

Birkeland, P. W., 1974, Pedology, Weathering, and Geomorphological Research: New York, Oxford University Press, 285 pp.

Blackwelder, E., 1933, The insolation hypothesis of rock weathering: American Journal of Science, v. 26, pp. 97–113.

Bogardi, I., and Yen, C. H., 1938, Traction of pebbles by flowing water: Unpublished thesis, Department of Mechanics and Hydraulics, State University, Iowa.

Bouyoucos, G. J., 1935, The Clay ratio as a criterion of susceptibility of soils to erosion: Journal of American Society of Agronomy, v. 27, pp. 738–741.

Branson, F. A., Gifford, G. F., Renard, K. G., and Hadley, R. F., 1981, Rangeland hydrology: Dubuque, Iowa, Kendall/Hunt Publishing Co., 340 pp.

Brunsden, D., 1979, Weathering: *in* Embleton, C. and Thornes, J. (eds.), Process in Geomorphology: New York, Halsted Press (John Wiley and Sons), pp. 73–129.

Bryan, Rorke B., 1968, The development, use and efficiency of indices of soil erodibility: Geoderma, v. 2, pp. 526.

Bryan, R. B., 1969, The relative erodibility of soils developed in the Peak District of Derbyshire: Geografiska Annaler, v. 51–A, pp. 145–159.

Bryan, Rorke B., 1977, Assessment of soil erodibility; new approaches and directions: *in* Erosion—Research Techniques, Erodibility, and Sediment Delivery, T. J. Toy, editor, Norwich, England, Geobooks, pp. 57–71.

Bryan, R. B., 1979, The influence of slope angle on soil entrainment by sheetwash and rainsplash: Earth Surface Processes and Landforms, v. 4, pp. 43–58.

Büdel, J., 1948, Das system der klimatischen Geomorphologie: Verhandlungen Deutscher Geographentag, v. 27, pp. 65–100.

Bunting, B. T., 1961, The role of seepage moisture in soil formation, slope development and stream initiation: American Journal of Science, v. 259, pp. 503–518.

Carroll, D., 1970, Rock Weathering, New York, Plenum Publishing Co., 203 pp.

Carson, M. A., 1971, The Mechanics of Erosion: London, Pion Press, 174 pp.

Carson, M. A. and Kirkby, M. J., 1972, Hillslope form and process: Cambridge, University Press, 475 pp.

Chepil, W. S., 1945, Dynamics of wind erosion: I. Nature of movement of soil by wind: Soil Science, v. 60, pp. 305–320.

Chepil, W. S., 1955, Factors that influence clod structure and erodibility of soil by wind, V. Organic matter at various stages of decomposition: Soil Science, v. 80, pp. 413–421.

Chepil, W. S., 1956, Influence of moisture on erodibility of soil by wind, Proceedings Soil Science Society of America, v. 20, pp. 288–292.

Chepil, W. S., and Woodruff, N. P., 1963, The physics of wind erosion and its control, Advances in Agronomy, v. 15, pp. 211–302.

Chorley, R. J., 1962, Geomorphology and general systems theory: U. S. Geological Survey Professional Paper 500–B, 10 pp.

Chorley, R. J., and Kennedy, B. A., 1971, Physical Geography—A Systems Approach: London, Prentice–Hall, 370 pp.

Cooke, R. U., and Doornkamp, J. C., 1974, Geomorphology in environmental management: London, Oxford University Press, 413 pp.

Cooper, A. W., 1960, An example of the role of microclimate in soil genesis: Soil Science, v. 90, pp. 109–120.

Dalrymple, J. B., Blong, R. J., and Conacher, A. J., 1968, A hypothetical nine–unit land surface model: Zeitschrift für Geomorphologie, v. 12, pp. 60–76.

DePloey, J., 1974, Mechanical properties of hillslopes and their relation to gullying in Central semi–arid Tunisia: Zeitschrift für Geomorphologie, v. 21, pp. 177–190.

DePloey, J., and Savat, J., 1968, Contribution a l'etude de l'erosion par le splash: Zeitschrift für Geomorphologie, v. 12, pp. 174–193.

Derbyshire, E., Gregory, K. J., and Hails, J. R., 1979, Geomorphological Processes: Boulder, Colorado, Westview Press, Inc., 312 pp.

Dunne, T., 1978, Field studies of hillslope flow processes: *in* Kirkby, M. J. (ed.), Hillslope hydrology, New York, John Wiley and Sons, pp. 227–293.

Dunne, T., and Leopold, L. B., 1978, Water in environmental planning: San Francisco, W. H. Freeman and Co., 818 pp.

Edwards, W. M., and Larson, W. E., 1969, Infiltration of water into soils as influenced by surface seal development: Transactions of American Society of Agricultural Engineers, v. 12, no. 4, pp. 463–465, 470.

Ellison, W. D., 1944, Studies of raindrop erosion. Agricultural Engineering, v. 25, pp. 131–6 and 181–182.

Ellison, W. D., 1947, Soil Erosion Studies—Part V, Soil transportation in the splash process: Agricultural Engineering, v. 28, pp. 349–351, 353.

Embleton, C., and Whalley, B., 1979, Energy, forces, resistances, and responses: *in* Embleton, C., and Thornes, J. (eds.), Process in Geomorphology: New York, Halsted Press (John Wiley and Sons), pp. 11–38.

Emerson, J. W., 1971, Channelization—a case study: Science, Vol. 173, pp. 325–326.

Emmett, W. W., 1968, Gully erosion: *in* Fairbridge, R. W. (ed.), The Encyclopedia of Geomorphology: Encyclopedia of Earth Sciences Series, Vol. III, New York, Reinhold Book Corp., pp. 517–519.

Emmett, W. W., 1970, The hydraulics of overland flow on hillslopes: U. S. Geological Survey, Professional Paper 662A, 68 pp.

Emmett, W. W., 1978, Overland flow: *in* Kirkby, M. J. (ed.), Hillslope Hydrology: New York, John Wiley and Sons, pp. 145–176.

Evans, R., 1980, Mechanics of water erosion and their spatial and temporal controls: an empirical viewpoint: *in* Kirkby, M. J., and Morgan, R. P. C. (eds.), Soil Erosion: New York, John Wiley and Sons, pp. 109–128.

Farmer, E. E., 1973, Relative detachability of soil particles by simulated rainfall: Soil Science Society of America, Proceedings, v. 37, pp. 629–633.

Finney, H. R., Holowaychuk, N., and Heddleson, M. R., 1962, The influence of micro climate on the morphology of certain soils of the Allegheny Plateau in Ohio: Soil Science Society of America, Proceedings, v. 26, pp. 287–292.

Foster, G. R., and Meyer L. D., 1975, Mathematical simulation of upland erosion by fundamental erosion mechanics: Present and Prospective Technology for Predict ing Sediment Yields and Sources, U.S.D.A., Agricultural Research Service, ARS–S–40, pp. 190–207.

Foster, G. R., and Meyer, L. D., 1977, Soil erosion and sedimentation by water—an overview: Soil Erosion and Sedimentation, Proceedings of National Symposium on Soil Erosion and Sedimentation by Water, American Society of Agricultural Engineers, 4–77, pp. 1–13.

Free, G. R., 1960, Erosion characteristics of rainfall: Agricultural Engineering, v. 41, pp. 447–449, 455.

Gerdel, R. W., 1937, Reciprocal relationships of texture, structure, and erosion on some residual soils: Soil Science Society of America, Proceedings, v. 2, pp. 537–545.

Gottschalk, L. C., and Brune, G. M., 1950, Sediment design criteria for the Missouri Basin loess hills: U.S. Department of Agriculture, Soil Conservation Service, Technical Paper 97, 21 pp.

Graf, W. L., 1977, The rate law in fluvial geomorphology: American Journal of Science, v. 277, pp. 178–191.

Graf, W. L., 1979, Mining and channel response: Association of American Geographers, v. 69, pp. 262–275.

Gray, D. H. and Leiser, A. T., 1982, Biotechnical slope protection and erosion control: New York, Van Nostrand Reinhold Company, 271 pp.

Gregory, K. J., and Walling, D. E., 1973, Drainage Basin Form and Process: New York, Halsted Press.

Griggs, D. T., 1936, The factor of fatigue in rock exfoliation: Journal of Geology, v. 44, pp. 781–796.

Hack, J. T., 1960, Interpretation of erosional topography in humid temperate regions: American Journal of Science (Bradley volume), v. 258–A, pp. 80–97.

Hadley, R. F., 1962, Some effects of microclimate on slope morphology and drainage basin development: U.S. Geological Survey Research, 1961, pp. B-32 to B-33.

Hadley, R. F., 1967, Pediments and pediment–forming processes: Journal of Geological Education, v. 15, pp. 83–89.

Hadley, R. F., and Schumm, S. A., 1961, Sediment sources and drainage basin characteristics in upper Cheyenne River basin: U.S. Geological Survey Water-Supply Paper 1531-B, pp. 137–198.

Hadley, R. F. and Toy, T. J., 1977, Relation of surficial erosion on hillslopes to profile geometry: Journal of Research, U. S. Geological Survey, v. 5, no. 4, pp. 487–490.

Heede, B. H., 1975, Stages of development of gullies in the West: Present and Prospective Technology for Predicting Sediment Yields and Sources, U.S.D.A., Agricultural Research Service, ARS–S–40, pp. 155–161.

Heidel, S. G., 1956, The progressive lag of sediment concentration with flood waves: Transactions of the American Geophysical Union, v. 37, pp. 56-66.

Holberger, R. L., and Truett, J. B., Sediment yield from construction sites: Proceedings of the Third Interagency Sedimentation Conference, PB-245-100, Water Resources Council, Washington, D.C., 1-47 to 1-48.

Horton, R. E., 1933, The role of infiltration in the hydrologic cycle: American Geophysical Union, Transactions, pp. 446–460.

Horton, R. E., 1945, Erosional development of streams and their drainage basins: Bulletin of Geological Society of America, v. 56, pp. 275–370.

Israelsen, C. E., Clyde, C. G., Fletcher, F. E., Israelsen, E. K., Haws, F.W., Packer, P. E., and Farmer, E. E., 1980, Erosion control during highway construction: Research Report, National Cooperative Highway Research Program Report 220, Transportation Research Board, National Research Council, 30 pp.

Jones, J. A. A., 1976, Soil piping and the subsurface initiation of stream channel networks: Ph.D. dissertation, Cambridge University.

Kirkby, M. J., 1971, Erosion by water on hillslopes: *in* Chorley, R. J. (ed.), Introduction to Fluvial Processes: London, Methuen and Co., Ltd., pp. 98–107.

Kirkby, M. J., and Morgan, R. P. C., (eds.), 1980, Soil Erosion: New York, John Wiley and Sons, 312 pp.

Lal, R., 1985, *in* Recent developments in erosion and sediment yield studies: Technical Documents in Hydrology, UNESCO, Paris, 127 pp.

Langbein, W. B., and Schumm, S. A., 1958, Yield of sediment in relationt o mean annual precipitation: Transactions of American Geophysical Union, v. 39, pp. 1076–1084.

Langbein, W. B., *et al.*, 1949, Annual runoff in the United States: U. S. Geological Survey Circular 52.

Lattanzi, A. R., Meyer, L. D., and Baumgardner, M. F., 1974, Influences of mulch rate and slope steepness on interrill erosion: Soil Science Society of America, Proceedings, v. 38, pp. 946–950.

Leopold, L. B., Wolman, M. G., and Miller, J. P., 1964, Fluvial Processes in Geomorphology: San Francisco, W. H. Freeman and Co., 522 pp.

Logan, T. J., 1977, Establishing soil loss and sediment yield limits for agricultural land: *in* Soil Erosion and Sedimentation, Proceedings of the National Symposium on Soil Erosion and Sedimenation by Water, St. Joseph, Michigan, American Society of Agricultural Engineers, pp. 59–68.

Lusby, G. C., 1979, Effects of grazing on runoff and sediment yield from desert rangeland at Badger Wash in western Colorado, 1953–73: U.S. Geological Survey Water-Supply Paper, 1532-I, 134 pp.

McIntyre, D. S., 1958, Permeability measurements of soil crusts formed by raindrop impact: Soil Science, v. 85, pp. 185–189.

Meyer, L. D., 1981, How rain intensity affects interrill erosion: Transactions of American Society of Agricultural Engineers, v. 24, pp. 1472–1475.

Meyer, L. D., Johnson, C. B., and Foster, G. R., 1972, Stone and woodchip mulches for erosion control on construction sites: Journal of Soil and Water Conservation, v. 27, pp. 264–269.

Meyer, L. D., De Coursey, D. G., and Romkens, M. J. M., 1976, Soil erosion concepts and misconceptions: Proceedings of the Third Federal Interagency Sedimentation Conference, PB–245–100, Water Resources Council, Washington, D. C., pp. 2–1 to 2–12.

Meyer, L. D., Foster, G. R., and Romkens, J. M., 1975, Sources of soil eroded by water from upland slopes: Present and Prospective Technology for Predicting Sediment Yields and Sources, U.S.D.A., Agricultural Research Service, ARS–S–40, pp. 177–189.

Middleton, H. E., 1930, Properties of soils which influence soil erosion: U. S. Department of Agriculture Technical Bulletin 178, pp. 1–16.

Mitchell, J. R., and Bubenzer, G. D., 1980, Soil loss estimation: *in* Kirkby, M. J., and Morgan, R. P. C. (eds.), Soil Erosion: New York, John Wiley and Sons, pp. 17–62.

Moeyersons, J., and DePloey, J., 1976, Quantitative determination of splash erosion simulated on unvegetated slopes: Zeitschrift fur Geomorphologie, v. 25, pp. 120–131.

Morgan, R. P. C., 1978, Field studies of rainsplash erosion: Earth Surface Processes, v. 3, pp. 259–299.

Morgan, R. P. C., 1979, Soil Erosion: New York, Longman Inc., 113 pp.

Morisawa, M., 1968, Streams: their dynamics and morphology: New York, McGraw-Hill, 175 pp.

Mosley, M. P., 1973, Rainsplash and the convexity of badland divides: Zeitschrift für Geomorphologie, v. 18, pp. 10–25.

Musgrave, G. W., 1947, The quantitative evaluation of factors in water erosion—a first approximation: Journal of Soil and Water Conservation, v. 2, pp. 133–138.

Mutchler, C. K., and Young, R. A., 1975, Soil detachment by raindrops: Present and Prospective Technology for Predicting Sediment Yields and Sources, U.S.D.A., Agricultural Research Service, ARS–S–40, pp. 113–117.

National Academy of Sciences, 1974, Rehabilitation potential of western coal lands: Cambridge, Mass., Ballinger Publishing Co., 198 pp.

National Resource Council, 1981, Surface Mining—Soil, Coal, and Society: Washington, D. C., National Academy Press, 233 pp.

Norton, E. A., and Smith, R. S., 1930, The influence of topography on soil profile character: Journal of American Society of Agronomy, v. 22, pp. 251–262.

Ollier, C. D., 1975, Weathering: London, Longman Group Ltd., 304 pp.

Palmer, R. S., 1963, The influence of a thin water layer on waterdrop impact forces: International Association of Scientific Hydrology Bulletin, v. 65, pp. 141–148.

Patric, J. H., 1977, Soil erosion and its control in eastern woodlands: Northern Logger and Timber Processor, v. 25, no. 11, pp. 431.

Patton, P. C., and Schumm, S. A., 1975, Gully erosion, northwestern Colorado—a threshold phenomenon: Geology, v. 3, pp. 88–90.

Peck, A. J., 1976, Interaction between vegetation and stream water quality in Australia: Proceedings of the Fifth Workshop of the United States/Australia Rangelands Panel, pp. 149–155.

Peltier, L., 1950, The geographical cycle in periglacial regions as it is related to climatic geomorphology: Annals, Association of American Geographers, v. 40, pp. 214–236.

Piest, R. F., Bradford, J. M., and Spomer, R. G., 1975, Mechanisms of erosion and sediment movement from gullies: Present and Prospective Technology for Pre dicting Sediment Yields and Sources, U.S.D.A., Agricultural Research Service, ARS–S–40, pp. 162–176.

Quansah, C., 1981, The effect of soil type, slope, rain intensity, and their interactions on splash detachment and transport: Journal of Soil Science, v. 32, pp. 215–224.

Richards, K. S., 1982, Rivers—Form and Process in Alluvial Channels: London, Methuen and Co., Ltd., publishers, 358 pp.

Ringen, B. H., Shown, L. M., Hadley, R. F., and Hinkley, T. K., 1979, Effect on sediment yield and water quality of a nonrehabilitated surface mine in north-central Wyoming: U.S. Geological Survey Water Resources Investigations 79–47, 23 pp.

Ritter, D. F., 1978, Process Geomorphology: Dubuque, Iowa, William C. Brown Co., 603 pp.

Robins, J. S., and Neff, E. L., 1963, Principles of the erosion process: Proceed ings of Forest Watershed Management Symposium, Oregon State University, pp. 127–139.

Roose, E. J., 1970, Importance relative de l'erosion, du drainage oblique et vertical dans la pedogenese actuelle d'un sol ferrallitique de moyenne Cote d'Ivoire, Cah. ORSTOM, ser. Pedol. 8, pp. 469–482.

Rubey, W. W., 1938, The forces required to move particles on a stream bed: U. S. Geological Survey Professional Paper 189–E, pp. 121–141.

Ruhe, R. V., 1975, Geomorphology: Geomorphic Processes and Surficial Geology: Boston, Houghton Mifflin Co., 246 pp.

Ruhe, R. V., and Walker, P. H., 1968, Hillslope models and soil formation: I, open systems: Transactions of Ninth International Congress of Soil Science, v. 4, pp. 551–560.

Savat, J. 1981, Work done by splash: Laboratory experiments: Earth Surface Processes and Landforms, v. 6, pp. 275–283.

Schumm, S. A., 1964, Seasonal variation of erosion rates and processes on hillslopes in western Colorado: Zeitschrift fr Geomorphologie, v. 5, pp. 215–238.

Schumm, S. A., 1973, Geomorphic thresholds and complex response of drainage systems: *in* Fluvial Geomorphology, Marie Morisawa, editor, Binghamton, S.U.N.Y. Publications in Geomorphology, pp. 299–310.

Schumm, S. A., 1977, The Fluvial System: New York, John Wiley and Sons, 338 pp.

Schumm, S. A., 1979, Geomorphic thresholds: the concept and its applications: Institute of British Geographers, Trans., v. 4, pp. 485–515.

Schumm, S. A., and Hadley, R. F., 1957, Arroyos and the semiarid cycle of erosion: American Journal of Science, v. 255, pp. 161–174.

Schumm, S. A., and Hadley, R. F., 1961, Progress in the application of landform analysis in studies of semiarid erosion: U.S. Geological Survey Circular 437, 14 pp.

Schumm, S. A. and Lichty, R. W., 1963, Channel widening and flood plain construction along Cimarron River in southwestern Kansas: U. S. Geological Survey Professional paper 352–D, pp. 71–88.

Schumm, S. A., Harvey, M. D., and Watson, C. C., 1984, Incised channels—morphology, dynamics, and control: Littleton, Colorado, Water Resources Publication, 200 pp.

Seibert, H., and Penneman, M., 1975, Revegetation and erosion control at Heavenly Valley: Lake Tahoe Research Seminar III, Proceedings, pp. 96–114.

Selby, M. J., 1970, Slopes and slope processes: Hamilton, New Zealand, New Zealand Geographical Society, 59 pp.

Selby, M. J., 1982, Hillslope Materials and Processes: Oxford, Oxford University Press, 264 pp.

Smith, R. M., and Stamey, W. L., 1965, Determining the range of tolerable erosion: Soil Service, v. 100, pp. 414–424.

Snyder, C. T., Frickel, D. G., Hadley, R. F., and Miller, R. F., 1976, Effects of off-road vehicle use on the hydrology and landscape of arid environments in central and southern California: U.S. Geological Survey Water Resources Investigations 76–99, 45 pp.

Soil Conservation Service, 1977, Procedure for determining rates of land damage, land depreciation and volume of sediment produced by gully erosion: *in* Food and Agriculture Organization of the United Nations, Guidelines for Watershed Management, Conservation Guide, pp. 125–141.

Sowers, G. B., and Sowers, G. F., 1961, Introductory soil mechanics and foundations: New York, MacMillan Co., 386 pp.

Steila, D., 1976, The geography of soils: Englewood Cliffs, N.J., Prentice–Hall, Inc., 222 pp.

Stocking, M. A., 1977, Rainfall energy in erosion: Research discussion, paper 13, Dept. of Geography, University of Edinburgh.

Strahler, A. N., 1952, Dynamic basis of geomorphology: Bulletin Geological Society of America, v. 63, pp. 923–938.

Strakhov, N. M., 1967, Principles of lithogenesis: Vol. I., Consultants Bureau, New York; London, Oliver and Boyd; Translation by Fitzsimmons, J. P., Tomklieff, S. I. and Hemingway, J. E. (eds.), 245 pp.

Terzaghi, K., 1950, Mechanism of landslides: Geological Society of America, Engineering Geology (Berkey) Volume, pp. 83–123.

Thornes, J., 1979, Introduction: *in* Embleton, C., and Thornes, J. (eds.), Process in Geomorphology: New York, Halsted Press (John Wiley and Sons), pp. 1–10.

Thornes, J., 1980, Erosional processes of running water and their spatial and temporal controls: A theoretical viewpoint: *in* Kirkby, M. J., and Morgan, R. P. C. (eds.), Soil Erosion: New York, John Wiley and Sons, pp. 129–182.

Toy, T. J., 1982, Accelerated erosion: process, problems, and prognosis: Geology, v. 10, pp. 524–529.

Toy, T. J., Kuhaida, A. J., and Munson, B. E., 1978, The prediction of mean monthly soil temperatures from mean monthly air temperatures: Soil Science, v. 126, no. 3, pp. 181–189.

Tricart, J., and Cailleaux, A., 1965, Introduction a la geomorphologie climatique: Paris.

Varnes, D. J., 1958, Landslide types and processes: *in* Eckel, E. B. (ed.), landslides and Engineering Practice: Highway Research Board, Special Report, v. 29, pp. 20–47.

Varnes, D. J., 1975, Slope movements in the Western United States: *in* Mass Wasting: Norwich, England, Geo Abstracts, pp. 1–17.

Vice, R. B., Guy, H. B., and Ferguson, G. E., 1969, Sediment movement in an area of suburban highway construction, Scott Run Basin, Fairfax County, Virginia, 1961–1964: U.S. Geological Survey, Water-Supply Paper 1591-E, 41 pp.

Walker, P. H., 1966, Postglacial environments in relation to landscape and soils: Iowa Agricultural Experiment Station Research Bulletin, 549, pp. 838–875.

Walling, D. E., 1974, Suspended–sediment and solute yields from a small catchment prior to urbanization: Institute of British Geographers Special Publication, no. 6, pp. 169–192.

Walling, D. E., 1977, Limitations of the rating curve technique for estimating suspended–sediment loads, with particular reference to British rivers: *in* Proceedings of the Paris Symposium, International Association of Hydrological Sciences Publication no. 122, pp. 34–48.

Walling, D. E., 1984, Dissolved loads and their measurement: *in* Erosion and Sediment Yield—Some methods of Measurement and Modelling, R. F. Hadley and D. E. Walling, editors, Norwich, England, Geobooks, pp. 111–117.

Warren, A., 1979, Aeolian processes: *in* Embleton, C., and Thornes, J. (eds.), Process in Geomorphology: New York, Halsted Press (John Wiley and Sons), pp. 325351.

Whipkey, R. Z., and Kirkby, M. J., 1978, Flow within the soil: *in* Kirkby, M. J. (ed.), Hillslope hydrology: New York, John Wiley and Sons, pp. 121–144.

Wilson, S. J., and Cooke, R. V., 1980, Wind erosion: *in* Kirkby, M. J., and Morgan, R. P. C., (eds.), Soil Erosion: New York, John Wiley and Sons, pp. 217–251.

Wischmeier, W. H., and Smith, D. D., 1958, Rainfall energy and its relationship to soil loss: Transactions of the American Geophysical Union, Vol. 39, No. 2, pp. 285–291.

Wischmeier, W. H., and Smith, D. D., 1965, Predicting rainfall–erosion losses from cropland east of the Rocky Mountains: U.S.D.A., Agricultural Handbook 282, 48 pp.

Wischmeier, W. H., and Smith, D. D., 1978, Predicting rainfall erosion losses: U.S.D.A., Agricultural Handbook 537, pp. 1-58.

Wolman, M. G., and Brush, L. M., 1961, Factors controlling the size and shape of stream channels in coarse noncohesive sands: U. S. Geological Survey Professional Paper 282–G, pp. 183–210.

Wolman, M. G., and Gerson, R., 1978, Relative scales of time and effectiveness of climate in watershed geomorphology: Earth Surface Processes, v. 3, pp. 189–208.

Wolman, M. G., and Miller, J. P., 1960, Magnitude and frequency of forces in geomorphic processes: Journal of Geology, v. 68, pp. 54–74.

Yorke, T. H., and Herb, W. J., 1976, Urban area sediment yield-effects of construction-site conditions and sediment-control methods: Proceedings of the Third Interagency Sediment Conference, PB-245-100, Water Resources Council, Wahington, D. C., pp. 2-52 to 2-64.

Young, A., 1972, Slopes: London, Longman Group, Ltd., 288 pp.

Young, R. A., and Onstad, C. A., 1976, Predicting particle–size composition of eroded soil: Transactions of American Society of Agricultural Engineers, v. 19, pp. 1071–1075.

Young, R. A., and Wiersma, J. L., 1973, The role of rainfall impact in soil detachment and transport: Water Resources Research, v. 9, no. 6, pp. 1629–1636.

Zingg, A. W., 1940, Degree and length of land slope as it affects soil loss in runoff: Agricultural Engineering, v. 21, no. 2, pp. 59–64.

3

Hillslopes: Form and Process

Introduction

Drainage basins are composed of two principal components: channels and hillslopes. At least areally, hillslopes are the dominant constituent. Young (1972) commented that most of the land surface of the earth is formed by valley hillslopes. Douglas (1977) concluded that hillslopes may be considered the cardinal relief feature, and the study of the form, significance, and development of hillslopes has long held a nodal position in geomorphology. However, Carson and Kirkby (1972, p. 1) observed

> Hillslopes have the distinction of being the commonest and, at the same time, the least studied, geomorphic features, especially in terms of the processes acting upon them. In some ways it is their very ubiquity which is responsible for this neglect because researchers have preferred to study unique or restricted features instead of facing the massive sampling problem which is involved in characterizing slopes. In addition, hillslopes present difficult research problems because their forms change either slowly or infrequently, particularly when compared to rivers.

It should be acknowledged that the concerted efforts of agricultural engineers, civil engineers, geographers, geologists, hydrologists, and soil scientists have made substantial strides in filling many of our knowledge gaps during the past two decades.

Additionally, hillslopes are the surfaces upon which mankind resides and engages in most economic activities. Dunne and Leopold (1978) remarked that hillslopes comprise the land to be managed, the sites upon which forestry, agriculture, urban construction, and other human activities must be carried on in harmony with natural processes. They also contend that the response of hillslopes to use controls the long-term productivity for cultivation or its long-term suitability for residential or other uses.

Spatial ubiquity and human necessity, then, provide a basis of our concern with hillslopes. In this chapter, we shall discuss the nature of both natural and disturbed hillslopes. First, it is appropriate to establish a foundation with a review of basic concepts and standard terminology.

Types of Hillslopes

The terms *slope* and *hillslope* are often used interchangeably in the literature. It seems preferable to employ the latter word in most cases to avoid any unintended confusion, through inference, with the simple form property of gradient. From the geomorphic perspective, a hillslope is the surface between the drainage divide and valley bottom. It is also an interface between the lithosphere, atmosphere, hydrosphere, and biosphere systems. It is a surface upon which geomorphic work is performed by geomorphic processes, such as weathering, erosion, and mass movement.

Even casual observation reveals the diversity of hillslope configurations from place to place. Ruhe and Walker (1968) noted that any hillslope may be defined geometrically in space by three components: (1) its inclination to the horizontal plane, which is the angle or gradient of the hillslope; (2) its distribution along the direction of gradient, or hillslope length; and (3) its distribution along the contour normal to the length, or hillslope width.

Components of Hillslope Form

For convenience of discussion or research design, hillslope form is often divided into two parts, the hillslope plan and the hillslope profile. The plan refers to linearity or curvilinearity of form along the horizontal dimension, or width, of a hillslope as would be depicted on aerial photographs or topographic maps (see Fig. 1). If a series of contour lines, representing a particular hillslope, is essentially linear, then the hillslope surface is rectilinear. Such a hillslope is called a valley side hillslope (side slope, valley side slope, or perhaps midslope). If the series of contour lines is curved, then there are two possibilities. The contour lines may be generally convex outward, with two adjacent valley side hillslopes attached and extending to higher elevations. Such a hillslope is called a spur end hillslope (nose, nose slope, or perhaps snout of a divide). Alternatively, the contours may be generally concave outward, with two adjacent valley side hillslopes attached and extending to lower elevations. Such a hillslope is called a valley head hillslope (hollow, head slope).

The hillslope profile refers to linearity or curvilinearity of form along the vertical dimension, or length, as represented in the spacing of contour lines on a topographic map. The hillslope profile is best taken along the down gradient line, orthogonal to the topographic contours, since this orientation represents the trajectory of fluids and particulate matter moving under the stress of gravity (Strahler, 1968). Hence, the hillslope profile follows the "true slope" in the same way that "true dip" follows the maximum gradient of geologic structures. If the series of contour lines, representing a particular hillslope, is evenly spaced along its length, then the hillslope possesses a straight profile. If the series of contour lines is not evenly spaced, then two possibilities again exist. An increasing distance between contour lines near the crest of the hillslope, at higher elevations, indicates that the hillslope possesses a convex profile. An increasing distance between contour lines near the base of the hillslope, at lower elevations, indicates that the hillslope possesses a concave profile.

Various terms have been applied to hillslope profiles and their parts in attempting to describe the form. Ruhe and Walker (1968) utilized the vocabulary depicted in Fig. 2A. Here the terms connote both geometry and position. For example, the shoulder is a convexly rounded component between the summit and the backslope. Expanding and somewhat quantifying the earlier framework of Savigear (1956), Young (1972) proposed the nomenclature shown in Figure 2B. Here, the hillslope profile is composed of slope units of two types: segments and elements. The segments are units through which the angle of inclination remains constant. The rectilinear, valley side hillslope, which appears straight in one dimension, is an example of a segment. Elements are curvilinear units of two types: convexities and concavities. This system is based solely upon profile geometry. However, adjectives may be added to provide location, such as "basal segment" or "basal element." Although less concise than the scheme offered by Ruhe and Walker (1968), the terminology proposed by Young (1972) will be used throughout this chapter because it is more popular among geomorphologists, and more importantly, it is more flexible in dealing with the irregularities of profile form.

Clearly, hillslope profiles may occur in a wide variety of forms depending upon the tectonic history and environmental setting in which they develop. Fault and fault-line scarps may produce virtually straight profiles. Active stream erosion at the base of a hillslope may produce an essentially convex profile. Deposition at the base of a hillslope may produce a concave profile. However, many writers note the prevalence of convex–concave or convex–straight–concave hillslope profiles (Bloom, 1978; Douglas, 1977; Wilson, 1968). Carson and Kirkby (1972) asserted that a large amount of evidence has been compiled to show that most, if not all, hillslopes possess rounded convex summits and are separated from stream channels at their bases by shallow concave elements. White (1966) remarked that hillslopes with prominent convex–concave elements and smooth regular profiles are commonly referred to as graded or equilibrium hillslopes. Figure 3 shows a hillslope with a convex element at the crest, a straight midslope segment, and a concave element at the base. In contrast, Leopold *et al.* (1964) submitted that there is no statistical evidence to support the contention that the convex–concave hillslope profile is indeed the most common.

Given the seemingly infinite combinations of tectonic history and environmental conditions which have operated jointly to produce landscapes worldwide, one can barely imagine the research design and sampling procedure that would be

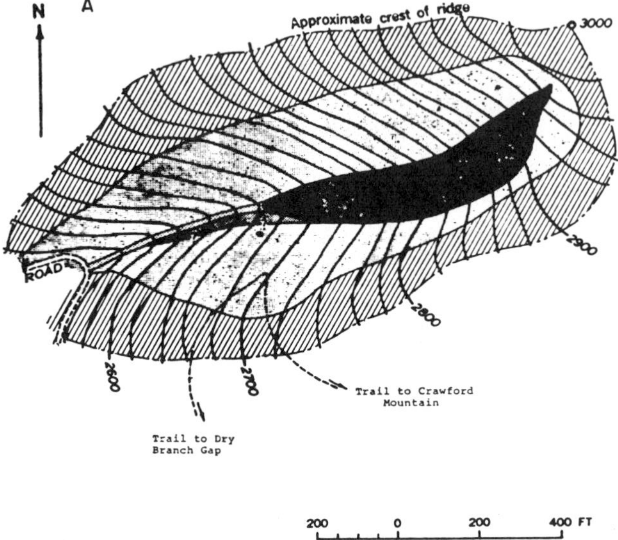

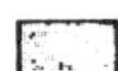

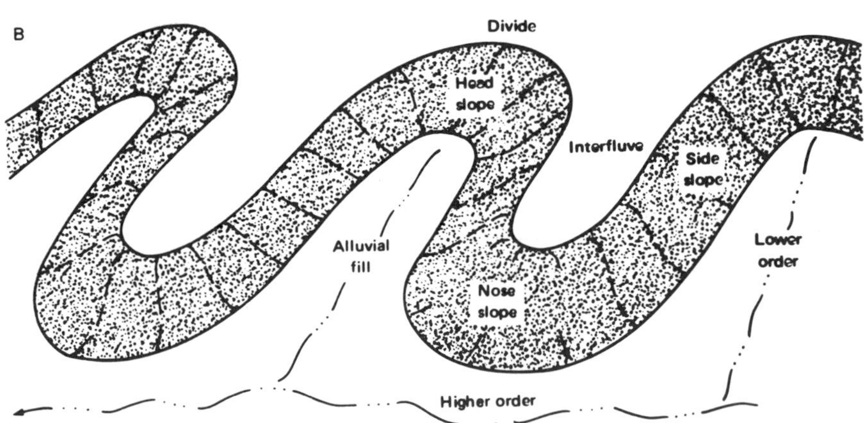

Figure 1. A. Terminology for hillslope form according to Hack and Goodlett (Hack and Goodlett, 1960). B. Terminology for hillslope form according to Ruhe and Walker (Ruhe and Walker, 1968). C. Terminology for hillslope form according to Young (Young, A., 1972, Slopes: London, Longman Group Ltd., 288 pp.).

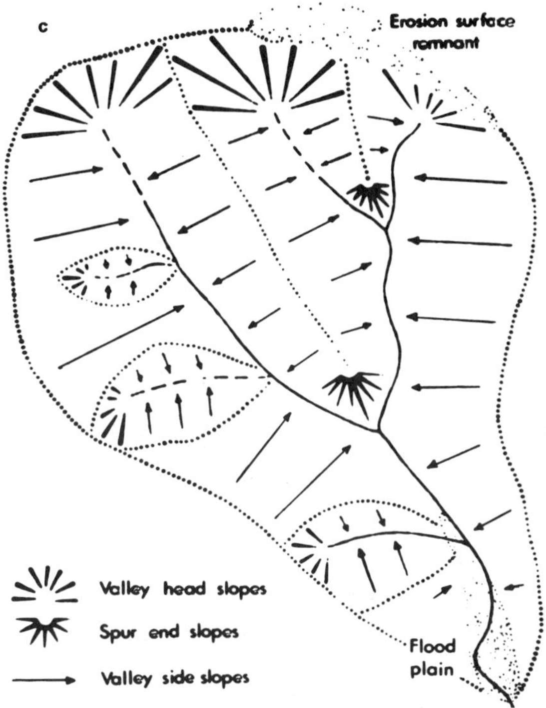

Figure 1 *(Continued)*

necessary to establish such statistical significance. Therefore, recourse must be given to the impressions gained through observation by generations of earth scientists, together with the available data from specific, albeit localized, studies.

While it is convenient to partition hillslopes into plan and profile components, they are actually three-dimensional landforms. It is not possible to fully comprehend the relationship between hillslope form and geomorphic processes by examining plans or profiles alone. Elaborating upon the prior work of Troeh (1965), a systematic examination of three-dimensional form was provided by Ruhe (1975). Both plan and profile may be either convex (V), concave (C), or linear (L). These combinations produce the nine basic slope geometries shown in Fig. 4. In the left column, hillslope plan is linear with hillslope profile changing from straight to convex to concave. In the top row, hillslope profile remains straight while the plan changes from straight to convex to concave. The remaining columns and rows illustrate other possible form combinations. Meyer and

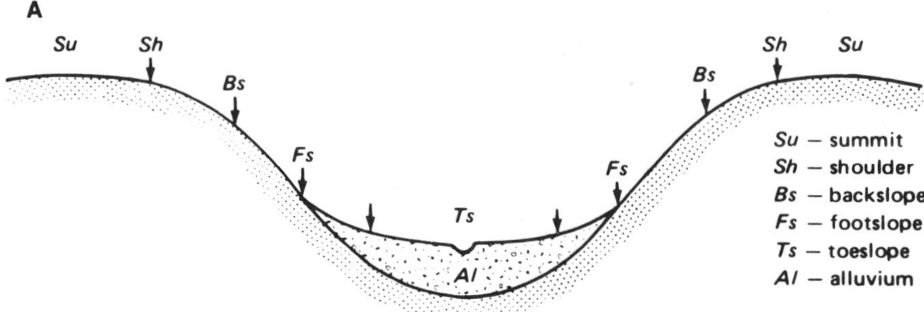

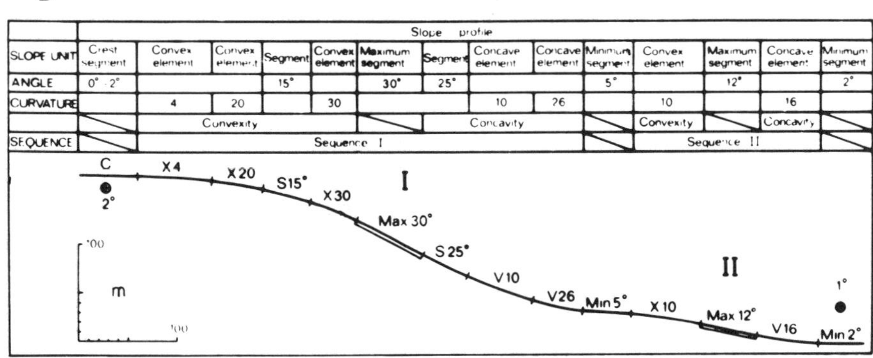

Figure 2. A. Hillslope profile terminology according to Ruhe and Walker (Ruhe and Walker, 1968). B. B. Hillslope profile terminology according to Young (Young, A., 1972, Slopes: London, Longman Group Ltd., 288 pp.) As follows (values of curvature in degrees/100 m): *slope unit,* a segment or an element; *segment,* a portion of a slope profile on which the angle remains approximately constant; *element,* a portion of a slope profile on which the curvature remains approximately constant; *convex element,* an element with a downslope increase in angle (i.e. with positive curvature); *concave element,* an element with a downslope decrease in angle (i.e. with negative curvature); *maximum segment,* a segment which is steeper in angle than the slope units above and below it: it may also form the lowest unit on a profile, having a gentler unit above it; *minimum segment,* a segment which is gentler in angle than the slope units above and below it; *crest segment,* a segment bounded by downward slopes in opposite directions; *basal segment,* a segment bounded by upward slopes in opposite directions; *irregular unit,* a portion of a slope profile within which there are frequent changes of both angle and curvature; *convexity,* all parts of a slope profile on which there is no decrease in angle downslope, but excluding maximum, minimum, and crest segments; *concavity,* all parts of a slope profile on which there is no increase in angle downslope, but excluding maximum, minimum, and crest segments; *profile sequence,* a portion of a slope profile consisting successively of a convexity, a maximum segment, and concavity.

Figure 3. Complex hillslope possessing convex, straight, and concave components.

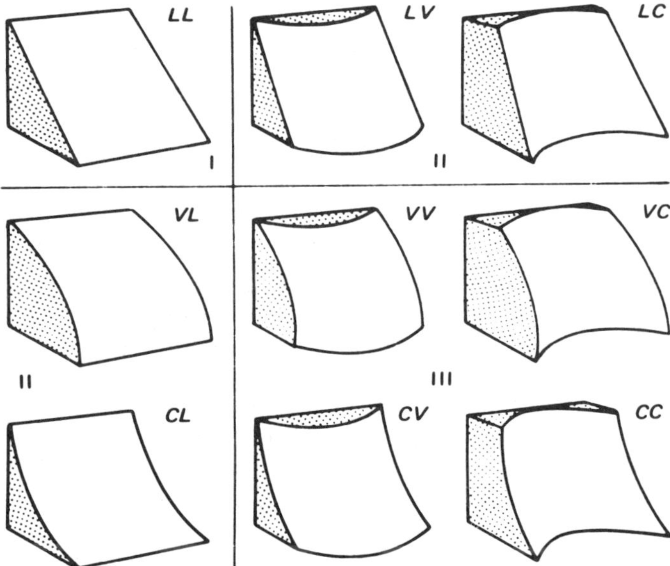

Figure 4. Three-dimensional hillslope forms (Ruhe, 1975).

Kramer (1968) examined the effect of profile form on runoff and erosion. This work will be discussed later.

Hillslope Form and Processes: Undisturbed Systems

Hillslope form refers to the three-dimensional configuration of the earth's terrestrial surface. Hillslope processes are the mechanisms by which work is performed upon this topography. The proposition that there must be a relationship between form and process is well rooted in the history of geomorphology. At the meetings of the Royal Society in Edinburgh on March 7 and April 4, 1785, a paper read by Dr. James Hutton contained the following passage.

> The operation of rains and torrents, modified by the hardness and tenacity of the rock, has worked the whole into its present form; has hollowed out the vallies, and gradually detached the mountains from the general mass, cutting down their sides into steep precipices at one place, and smoothing them to gentle declivities at another. [in Playfair, 1956, p. 114 (see Hutton, 1785)]

Nearly a century later, Gilbert (1877) endorsed the relationship between form and process with his explanations of topography throughout the Henry Mountains in Utah. Although the jargon changed somewhat during the next hundred years, the concept remained intact.

Thornes (1979) observed that a major task in geomorphology is to understand the relationships between form and process. He further noted that process–response models, where response denotes form or structure, are often used to describe and explain the relationships. Selby (1982) added that process–response models form the basis of most modern hillslope geomorphology.

Problems of Hillslope Research Concerning Form and Process

Nevertheless, the establishment of functional or even statistical relationships between form and process is an elusive task and a frustrating research direction in many instances. Among the theoretical and practical problems encountered are the following:

1. the "chicken or egg" argument (Thornes, 1979; Pitty, 1971; Wilson, 1968);
2. the slowness of process operation and form change (Pitty, 1971; Young, 1972);
3. the complexity of geomorphic processes (Pitty, 1971; Young, 1972);
4. the variability of processes through time and space (Young, 1972; Selby, 1982);

5. the difficulty and/or magnitude of representative sampling (Selby, 1982; Carson and Kirkby, 1972);
6. the generally unknown influence of human activities on process rates (Pitty, 1971; Cooke and Doornkamp, 1974);
7. the prospect of "equifinality" (Thornes and Brunsden, 1977; Young, 1972); and
8. the potential for physical danger encountered while observing or measuring some active processes (Pitty, 1971).

The "chicken or egg" argument centers around discussions of which came first, the form or the process. Essentially, this is a problem of identifying cause and effect, the dependent and independent variables. Thornes (1979) noted that just as forms are influenced by processes, so processes are influenced by form. Pitty (1971) concluded that in all situations there is the problem of deciding which came first; whether the process essentially preceded the form and thereby determined its characteristics, or conversely, whether the two co-exist in a closely interdependent state of "dynamic equilibrium." Today, many geomorphologists believe that, under natural conditions and over time periods of moderate to short duration, a "steady state" or "dynamic equilibrium" exists between form and process whereby changes in either produce changes in the other, sometimes through rather elaborate feedback mechanisms. R. R. Churchill (personal communication, 1985) suggests that over short time periods, form tends to govern process, while over very long periods, process governs form. In natural settings it is probably most often a change in process, perhaps due to the surpassing of intrinsic or extrinsic thresholds which causes a change in form. However, on drastically disturbed sites, forms may be entirely manmade and process rates are largely the consequence.

Carson and Kirkby (1972) remarked that hillslopes present difficult research problems because their forms change either slowly or infrequently, particularly when compared to rivers. From the practical viewpoint, Pitty (1971) observed that the slowness of process operation is especially troublesome in relation to the time span that one investigator can expect to devote to their study. It seems probable that the extended time frame necessary for meaningful studies of changes in hillslope form has had a negative impact on the availability of research funds.

It is likely that the rate of change in form is proportional to the magnitude of disequilibrium between form and process generated within a system. Under natural conditions, typical variabilities of processes, progressive reduction of relief, and alteration of surface material are likely to produce disequilibria of modest or minor magnitude in most cases. On drastically disturbed sites, the disequilibria may be much greater due to extensive modification of environmental conditions, and as a consequence, changes in hillslope form can occur much more rapidly.

As our knowledge of geomorphic processes grows, it becomes increasingly apparent that they operate in very complex ways. There are two aspects to this problem: (1) the manner in which processes affect hillslopes and (2) the manner in which hillslope processes interact among themselves. In both cases, our understanding is incomplete.

It is generally agreed that many geomorphic processes can operate on a given hillslope at one time or another. However, the magnitude and frequency of processes are variable in time and space. Temporal variabilities range from seasonality of processes (Schumm, 1964) to long-term changes, such as those since the Pleistocene Epoch. Pitty (1971) commented that even where a given form is seen to be the result of a known process, it may be difficult to ascertain how much the form owes to the present intensity of the process and how much to different conditions in the past. Many believe that landforms constitute palimpsests, where present form bears marks of past as well as present processes.

Spatial variabilities can also be examined on different scales. The loss of materials from a drainage basin indicates reduction in relief and, as a result, a reduction in the potential energy of hillslope systems contained therein. Likewise, spatial variabilities of process can be examined along a hillslope profile. Raindrop impact may be a dominant erosion process at the crest of a hillslope, while rill erosion is the dominant erosion process in the straight segment near midslope.

The mode and size of hillslope sampling poses additional obstacles in research. Selby (1982) remarked that the field or laboratory methods employed in hillslope analysis often involve the establishment of a statistical relationship between a hillslope change and measurements of soil or rock properties and a process such as rainfall; however, the primary difficulty in such methods is that of adequately sampling hillslopes, rock and soil types, and process types and rates. Concerning the mode of sampling, Young (1972) observed that in deciding the location of profiles, most studies have used purposive sampling, that is, deliberate selection of the line of profile. But purposive sampling has been criticized as subjective and therefore open to the selection, conscious or otherwise, of profiles that will demonstrate some preconceived hypothesis. He notes that controlled sampling is desirable if the profiles are analyzed statistically as a sample drawn from the total ground surface of the area. Unfortunately, practical difficulties of access may render random sampling impossible; even in these cases, the careful selection of lines judged to be representative of the landforms of the area can be regarded as a simple form of stratifed sampling and does not necessarily preclude statistical analysis.

Young (1972) then described procedures for random, systematic, and stratified sampling of hillslope profiles and acknowledged that each possesses additional problems. A sampling system based upon the extension of profiles from randomly selected points in a drainage basin tends to produce underrepresentation of land that is convex in plan and lies on the lower parts of the area, and overrepresentation of low-lying slopes that are concave in plan. Close to the interfluves, the reverse is true. These tendencies are the consequence of divergence and convergence of flow-lines downslope in areas that are convex and concave in hillslope plan. Questions of representativeness also arise with the other two sampling procedures, and attempts to adjust samples reintroduce the possibility of bias.

Sample size is another concern. Based upon an examination of hillslope angle in several mature regions, Strahler (1950, p. 685) proposed the "law of constancy of slopes" that states

> Within an area of essentially uniform lithology, soils, vegetation, climate, and stage of development, maximum slope angles tend to be normally distributed with low dispersion about a mean value determined by the combined factors of drainage density, relief, and slope-profile curvature.

Examining sample sizes from 20 to 200 readings, he found that a sample of 50 hillslopes is adequate to reveal the characteristics of the distribution in a homogeneous region of 6 to 12 first-order drainage basins, while a sample of 25 will closely fix the mean.

Research by others, however, presents conflicting results. Parsons (1982) examined 12 hillslope profile parameters in 7 first-order drainage basins developed within the nearly horizontal Upper Chalk of the South Downs in East Sussex, England. The results showed that if spatial randomness is assumed, the sample sizes required vary considerably from one parameter to another but are often very large. The basins developed on this highly permeable material, however, are much larger than those commonly encountered on disturbed lands and the measured hillslopes are very long by comparison. These conditions may account for some variability in the results. In contrast to Strahler's (1950) discovery of a normal distribution in hillslope angles, Speight (1971) found that frequency distributions of hillslope angle are often positively skewed, suggesting logarithmetic normality.

Regarding the development of mathematical hillslope models, Carson and Kirkby (1972, p. 9) commented

> However, the inherent difficulty in making large numbers of reliable measurements, especially of process rates, so limits sample sizes that there are severe practical problems in verifying or generating models. The best that can be done is to choose sample points which are contiguous and so minimize extraneous differences, but which at the same time retain a full range of internal variation due to the operation of the system—in other words to select sample points from within slope profiles or basins, which thus form suitable sample clusters as well as suitable spatial systems.

From the foregoing, a few recommendations for sampling of hillslopes seem permissible. First, the selected hillslopes should all be taken from a region homogeneous with respect to lithology, geologic structure, soils, vegetation, climate, and basin relief. Essentially, these hillslopes will have developed under similar environmental conditions. Samples from several discrete regions could, of course, be compared. Second, samples should be stratified by hillslope types, such as valley side hillslopes, spur end hillslopes, and valley head hillslopes. It seems preferable to avoid mixing the properties of various hillslope types in a single sample. Research may then focus upon the form and process of each component hillslope type but with the recognition that the landscape is composed of all types in various spatial proportions. Third, it appears that most standard sampling procedures are likely to produce sets of hillslopes that are impractically large or unrepresentative of the spatial distribution within a landscape. Subsequent "adjustment" of such samples may defeat the initial purposes of sampling, resulting merely in the delusion of randomness. Hence, judicious, purposive sampling, tempered by the above recommendations and field experience, may be the most realistic alternative. This probably would not be dissimilar to the selection of representative channel reaches for study. Finally, it would be prudent to test the normality of hillslope variables if they are to be used in statistical analyses based upon that assumption.

Through modern technology, human activities have proven remarkably effective in modifying hillslope form and processes (Toy, 1982). Pitty (1971) observed that in many areas there is difficulty, if interpretations of the past are to be made, in estimating the degree to which present-day erosion rates are due to artificially accelerated erosion since humans have swung the scales of the natural balance between the forces previously involved. Cooke and Doornkamp (1974) suggested that the data upon which the Langbein and Schumm (1958) curve was based, relating sediment yield to annual effective precipitation, was probably contaminated to some extent by human influences on environmental conditions in the areas of investigation. The question becomes, what are the forms and process rates under truly natural conditions? With the ubiquity of human presence in many parts of the world, where would one go to collect data for absolutely undisturbed sites? Although Thornes and Brunsden (1977) may have exaggerated by suggesting that all environments on the earth have now been substantially affected by humans, those engaged in the collection of predisturbance, baseline data must be aware of the possibility of human influences.

The "principal of equifinality" holds that, in any system, a given form can be produced in an infinite variety of ways (Von Bertalanffy, 1968). As applied to hillslopes, this means that different processes may produce similar forms. The concept is also known as "homologies" or "convergence" in landforms (Pitty, 1971). Thornes and Brunsden (1977) cited the example of a flat plain; it could be produced by marine abrasion, subaerial denudation, fluvial aggradation, or exhumation of a relict landform.

The implications of this prospect are serious. Young (1972) contended that because of equifinality, models can never be used to obtain process from the observed form. Selby (1982) remarked that because of equifinality, it is seldom possible to study the shape alone of the landform and so deduce its origin; it also implies, where detailed evidence is lacking, that it may be impossible to determine how a particular feature has evolved.

Detailed evidence and common sense are usually the keys. In their example, Thornes and Brunsden (1977) noted that supporting evidence, such as marine or fluvial gravels, may direct us to the correct formative process. Location of the plain in central Nebraska should probably eliminate the possibility of marine planation. Thus, an understanding of past and present geomorphic conditions in a particular area will lessen the problems associated with equifinality by reducing the reasonable choices.

Lastly, Pitty (1971) noted that while most geomorphic processes are theoretically observable, in practice it is difficult, if not dangerous, to do so; moreover, the physical risk tends inevitably to increase at times when mechanical processes are at their most active and in areas where biochemical activity is most vigorous. Mass movements, such as landslides, are probably the only perilous hillslope process. However, others are capable of destroying instrumentation or affecting its accuracy.

In light of all these obstacles, it is somewhat remarkable that any progress has been made in relating hillslope form and process. While "equifinality" presents difficulty in attributing particular processes to particular hillslope forms, it is possible to determine the form produced by certain processes. Perhaps the earliest suspicion of an association between form and process evolved from an understanding of the climatic influence on most subaerial geomorphic processes together with the readily observable contrasts between landscapes in dissimilar and distinct climatic regions. For example, the landscapes of desert regions can hardly be confused with those of arctic or alpine regions that experience glaciation. From this, the sub-discipline of climatic geomorphology emerged, with inductive logic filling the gaps between climatic extremes, while the body of research documentation develops. Tricart and Cailleux (1972), Derbyshire (1973), and Büdel (1982) provide detailed discussions for those wishing to pursue this topic.

Evidence of Form and Process Relationships

The generally accepted relationship between hillslope process and form rests upon several types of evidence that may be grouped into the following categories: (1) expert opinion, (2) climate–process–form inferences, (3) microclimate–process–form studies, and (4) specific process–form studies. Nearly all geomorphologists acknowledge the probable relationship between hillslope process

Table 1 Hillslope Forms Produced by Various Geomorphic Processes

Hillslope form	Geomorphic process	Proponents[a]
Convex element	Erosion/mass movement	
	Soil creep	Davis (1892); Gilbert (1909); Gotzinger (1907); Lehmann (1918); Penck (1924), Blanche (1942); Birot (1949); Baulig (1950); King (1953, 1957); Schumm and Hadley (1961); Schumm (1966); de Bethune (1967); Carson and Kirkby (1972)
	Rainwash (also sheetwash)	Maw (1866); Chamberlain and Salisbury (1904); Fenneman (1908); Lawson (1932); Baulig (1950); Horton (1945); Schumm (1956); King (1957); Schumm (1966)
	Rock weathering (including pressure release)	de la Noë and de Margerie (1888); Hicks (1893); Marr (1901); Gotzinger (1907); Lehmann (1918); Penck (1924); Blanche (1942); Birot (1949); Schumm (1966); Carson and Kirkby (1972)
	Rainsplash	Carson and Kirkby (1972); Mosley (1973)
	Basal erosion by stream (including incision)	Penck (1924); Schumm (1966)
	Overland flow along longitudinal axis of divide	Yair (1973)
	Deposition	
	Mantling of hillslope with windblown or volcanic debris	Schumm (1966)
Straight segment (inclined)	Erosion	
	Rainwash (also sheetwash)	Schumm (1956, 1966); Carson and Kirkby (1972)
	Basal erosion by stream (including incision)	Schumm (1966)
	Deposition	
	Talus, sand, volcanic debris at angle of repose	Schumm (1966)
Straight segment (vertical)	Mass movement	Schumm (1966)
	Basal erosion by stream (including incision)	Schumm (1966)

Table 1 *(Continued)*

Hillslope form	Geomorphic process	Proponents[a]
Concave element	Erosion	
	Rilling	Baulig (1950); Schumm (1966)
	Soil creep	Birot (1949); Schumm (1966)
	Piping	Schumm (1966)
	Basal erosion by stream (including incision)	Penck (1924)
	Solution	Carson and Kirkby (1972)
	Rainwash (also sheetwash)	Tylor (1875); de la Noë and de Margerie (1888); Davis (1892); Hicks (1893); Marr (1901); Gilbert (1909); Jeffreys (1918); Lake (1928); Meyerhoff (1940); Horton (1945); Birot (1949); King (1953, 1957); Holmes (1955); Carson and Kirkby (1972)
	Deposition	
	Talus accumulation at hill-slope base	Schumm (1966)
	Colluvium accumulation at hillslope base	Schumm (1966); Carson and Kirkby (1972)
	Alluvium accumulation at hillslope base	Lawson (1932); Schumm (1966); Carson and Kirkby (1972)

[a]Most references compiled from Cotton (1952), Schumm (1966), Wilson (1968), Carson and Kirkby (1972), and Young (1972).

and form, although there is often some uneasiness when the topic is discussed because of uncertainties about the causal mechanisms that generate this association. Such linkages between process and response are frequently elusive.

The positions of renowned geomorphologists with regard to the processes that produce particular forms are summarized by Cotton (1952), Schumm (1966), Wilson (1968), Carson and Kirkby (1972), and Young (1972). Table 1 further condenses this information. A large number of proponents do not necessarily add certainty to a specific process–form relationship. Indeed, it may indicate controversy. Often, the disagreement concerning what process produces a particular form seems to be the result of field experience in regions possessing dissimilar environmental conditions. However, this does not invalidate the conclusions drawn within a particular region. Bloom (1978) noted that lunar landscapes seem to lack the concavity of terrestrial foot slopes and suggests that this may be due to the total lack of water on the moon. Further, the concept of equifinality is well illustrated in this table; for example, a convex hillslope element may result from soil creep, rainwash, rock weathering, rainsplash, basal erosion by streams, overland flow along the longitudinal axis of a divide, or various types of deposition.

Dalrymple *et al.* (1968) also assessed the various perspectives concerning the process–form relationship in developing a nine unit model for valley side hillslopes as shown in Fig 5. Each unit is defined on the basis of two-dimensional geometry and the contemporary geomorphic process supposedly acting upon it. Selby (1982) observed that, in reality, it is unusual to find all nine units occurring on one hillslope profile. In addition, they do not necessarily occur in the order shown in the diagram and individual units may recur in a single profile. For example, the convex–concave hillslope might consist of units 1, 2, 3, 5, 6, and 7. A dominantly convex hillslope experiencing basal erosion by a stream might consist only of units 1, 2, 3, and 4. Young (1972) contended that this model is

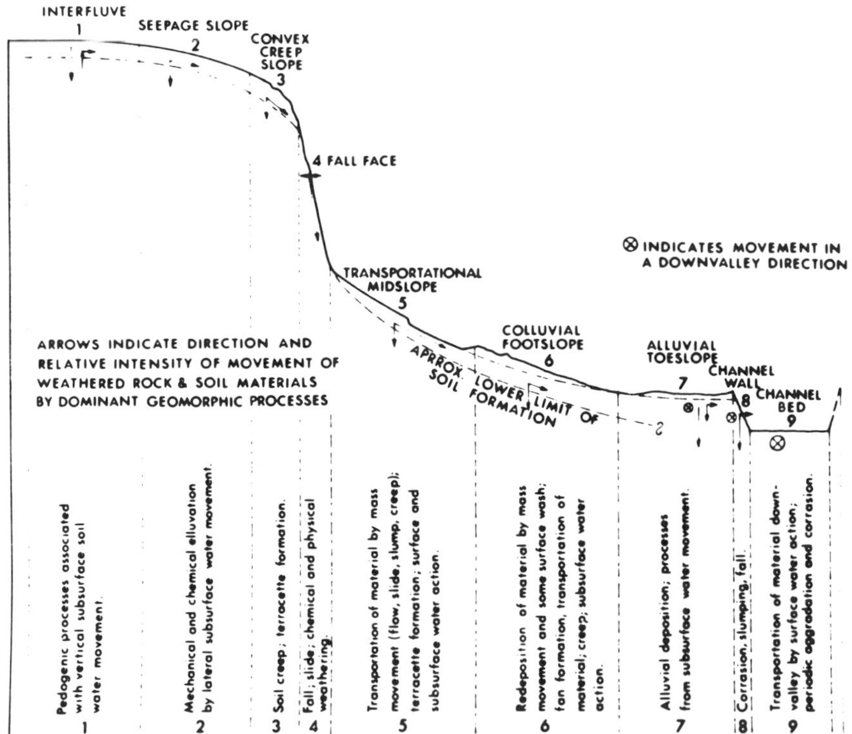

Figure 5. Hypothetical nine unit landsurface model showing proposed relationship between form and process (Dalyrymple *et al.*, 1968).

based upon a region with a humid temperate climate. However, Cooke and Doornkamp (1974) suggested that the model can be adapted to the character of most areas and that local details of bedrock and soils may be added to make it relate to a specific drainage basin. They also submit that the integrated view of form, materials, and processes implied by this approach is essential to the valid interpretation of river basin characteristics.

As discussed in Chapter 2, climate is a driving force in determining the suite of geomorphic processes that dominate in a particular locale. This, together with the relationships between hillslope process and form presented above, allows a logical connection between climate and form. Wilson (1968) remarked that most geomorphologists seem to believe that hillslope forms and processes (or process intensities) do vary with climate, despite the claims of King (1957) and others. Wilson (1968) also cited basic precepts of Davis (1930), namely, (1) hillslopes in humid climates are smooth, relatively gentle, mantled with soil and vegetation and consist of convex–concave profiles that decline with time; and (2) hillslopes in arid climates are rough, relatively steep, barren of soil and vegetation, and consist primarily of straight segments that retreat with time to produce pediments. Wilson (1968) then concluded that in humid temperate climates, detailed field studies, such as those by Hack and Goodlett (1960) and Schumm (1956), confirm many of the Davisian precepts.

Toy (1977) examined the relationship between hillslope form and climate at 28 sites extending along two traverses from the states of Kentucky to Nevada and Montana to New Mexico. Five hillslope parameters were compared with nine climate variables. It was found that 59% of the variation in the curvature of the convex element is explained by the climate variables, 43% of the variation in the slope of the straight segment is accounted for by climate, 37% of the variation in average hillslope gradient is explained by climate, and 26% of the variation in hillslope length ratio (field length/vertical distance from crest to base) is accounted for by climate. The remaining unexplained variation could be due to paleoclimatic conditions, other differences in geomorphic history of the regions, hillslope sampling error, or the use of weather station data collected in instrument shelters rather than surface climate data. Nevertheless, the statistical significance of these correlations helps to substantiate the relationships among climate, process, and hillslope form.

From this study, it is concluded that hillslopes in arid areas tend to be shorter, steeper, and have smaller radii of curvature of the convex element than those of humid areas. This is depicted diagrammatically in Fig. 6. It appears that some of the Davisian precepts may be statistically confirmed under controlled research designs.

Melton (1957) examined relationships among factors of climate, surface properties, and geomorphology for numerous drainage basins in the western United States. He concluded that the gradient of valley side hillslopes increases with

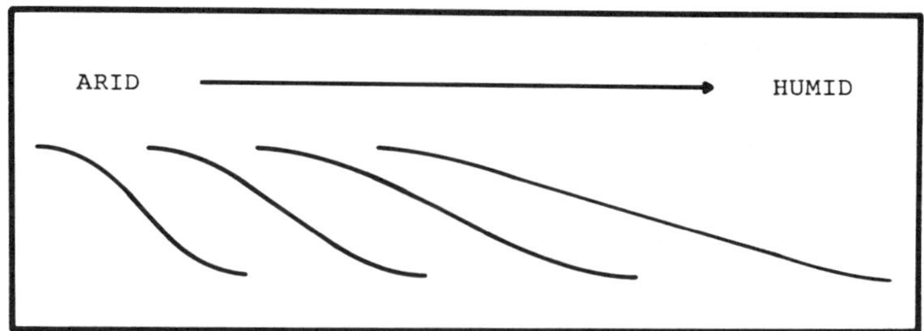

Figure 6. Conceptualized hillslope profiles showing the relationship between form and climate (Toy, 1977).

increases in the value of Thornthwaite's (1931) precipitation effectiveness index. Additionally, he discussed the complex manner in which climate affects basin characteristics, including hillslopes.

Dollfus (1964) examined hillslope form and climate in the central Andes Mountains. Here, climate zones result from altitudinal zonation and hillslope aspect or the direction faced by the hillslope with respect to the sun. It is concluded that these climate zones produce particular hillslope processes and forms.

The asymmetry of opposing valley side hillslopes is apparent in many places. Carson and Kirkby (1972) remarked that because this asymmetry appears to be especially common in west–east aligned valleys where differential insolation between opposite hillslopes is at a maximum, many geomorphologists believe that the influence of climate is not confined to latitudinal differences, but also exists within small areas. It is recognized that valley side asymmetry may be due to structural control (Young, 1972), asymmetric basal erosion by streams (Melton, 1960), or even the Coriolis effect (Von Bauer, 1860). Again, the multivariate system makes it difficult to ascribe a particular form to a single particular factor. Nevertheless, Carson and Kirkby (1972) submitted that, despite the importance of asymmetric lateral stream movement and structural influences on the intensity of denudation processes, many examples of valley side asymmetry clearly relate to contrasts in microclimatic conditions due to differences in hillslope aspect.

Douglas (1977) suggested that under dry conditions, low channel gradients favor the development of valley side asymmetry with the poleward-facing hillslopes becoming steeper as the channels and streams are moved against their bases due to the filling of debris from the equator-facing hillslopes that have more bare soil. He then suggests that such contrasts are less marked in humid regions and should be negligible in tropical regions, but that they are possibly significant in humid temperate regions where contrasts in insolation produce differences in frost action and snow retention. Thus, it would appear that microclimatic influences on hillslope form are, in part, dependent upon regional climatic conditions.

There are several studies that examine the relationship between hillslope aspect and morphology, including Hack and Goodlett (1960), Hadley (1962) Kennedy (1976), and Churchill (1981). Once again, the results are not always the same, due to differences in specific research situations. For example, Hadley (1962) found that north-facing hillslopes are generally steeper, less dissected, and support a more luxuriant vegetation growth than south-facing hillslopes within the Cheyenne River basin of east-central Wyoming. In contrast, Churchill (1981) found that north-facing hillslopes are less steep and more dissected by rilling than south hillslopes within the Big Badlands of South Dakota. He suggested that the relationship between aspect and hillslope inclination for the Badlands hillslopes is opposite that reported in a number of other cases largely because of the absence of protective vegetation cover. He also asserted that the important issue is not whether north-facing hillslopes are steeper than south-facing hillslopes or vice versa, but rather, that under controlled conditions a clear relationship can be demonstrated between aspect-induced topoclimatic (microclimatic) variation and hillslope morphology.

Specific process–form studies are of two basic types: (1) those that examine the form produced by particular processes and (2) those that examine the effect of form on particular processes. Some studies of the first type are contained in Table 1 (e.g., Mosley, 1973). Research of the second type tends to center on the effect of form on soil erosion.

Hillslope Form and Erosion

It is appropriate to elaborate upon the relationship between hillslope form and erosion rates because erosion is such an important consequence of land disturbance and because hillslope form is one factor that can be manipulated to some extent in many cases (surface mine reclamation and highway road construction). Under other circumstances, hillslope form is a "given"; however, an understanding of its function may help explain why erosion rates remain high in spite of attempts to manage this process through specific control practices and the reestablishment of vegetation cover. First, we shall consider the one-dimensional form components of hillslope gradient, length, aspect, profile, and plan shape; then we shall examine some three-dimensional form components.

The effects of hillslope gradient and length can be investigated at two scales: (1) the macrotopographic, field level and (2) the microtopographic, rill–interrill level. A fundamental assumption is Hortonian overland flow, with an increase in flow depth downslope. This appears reasonable for many disturbed lands. Thereby, an increase in hillslope gradient and length would produce an increase in flow velocity, an increase in tractive shear stress or force, and hence, a greater capacity for detachment, entrainment, and transport of soil particles by the flow.

Virtually all soil erosion or soil loss models include hillslope gradient and hillslope length variables in some form (Zingg, 1940; Musgrave, 1947; Wischmeier and Smith, 1965). Elwell (1984) listed five control variables as the major overriding factors determining soil losses and which constitute the "bricks" of model building; among these are hillslope gradient and length. Carson and Kirkby (1972) reviewed the role of these factors and concluded that power laws provide suitable empirical models for the variation of soil wash transport with topography.

While hillslope gradient and length factors are sometimes combined into a topographic factor when estimating soil erosion at the field level, it is useful to separate them in order to appreciate their individual contribution. Increases in hillslope gradient result in an increase in the downslope component of gravitational force acting on the water molecules engaged in overland and rill flow. This allows more rapid acceleration of the mass. Interestingly, the results of flume studies reported by Emmett (1978) indicate that flow depth decreases as gradient increases. While this should mean a decrease in flow transport capacity, it might also mean that soil detachment due to raindrop impact increases somewhat because the apparently critical "three drop diameter" depth (Mutchler and Young, 1975) is not reached until further downslope.

Additionally, soil aggregates may be inherently less stable on steeper hillslopes. A U.S. Environmental Protection Agency (1976, p. 29) report comments:

> As slope steepness increases, there is a corresponding rise in the velocity of the surface runoff, which in turn results in greater erosion. A doubling of the velocity of water produced by increasing the degree and length of the slope enables water to move soil particles 64 times larger, allows it to carry 32 times more soil material, and makes the erosive power, in total, four times greater.

Although the exact magnitude of increase could be debated, the principle is generally accepted.

Smith and Wischmeier (1957) found that soil loss was correlated with hillslope gradient according to the equation:

$$A = 0.43 + 0.30S + 0.043S^2 \quad (1)$$

where A is the soil loss and S the hillslope gradient. This and other studies report that polynomial functions best represent the relationship between hillslope gradient and soil loss. As presented in the review by Hadley *et al.* (1985), exponents range from 0.70 to 2.0 for various exponential relationships between these two variables.

The association between hillslope gradient and soil erosion has undergone close scrutiny in recent research and, again, produces some controversy. Under rather special conditions of a desert environment and changes in the texture of surface materials, Yair and Klein (1973) found an inverse relationship between erosion and gradient. Bryan (1979) found a highly variable relationship between erosion and gradient, although he admitted that these experiments were conducted on "extremely small sample plots."

Much of the controversy centers around the validity of the relationship on steep hillslopes. Most investigators probably accept an exponential relationship for hillslopes of low to moderate gradient, perhaps as high as 20%. However, considerable evidence is evolving that indicates that erosion rates decline for steep inclinations. Horton (1945) theorized a decrease in erosion for hillslopes exceeding gradients of 20° (36.4%) stating that the positive correlation is predicated on uniform turbulent flow and that this cannot occur on very steep hillslopes. Horton (1945) observed that this hypothesis corresponded well with the field observations of Renner (1936) wherein erosion increased with gradient up to approximately 40° (83.9%) and decreased thereafter to zero near a 90° gradient. Foster and Martin (1969) found that for soils with bulk densities of 80 to 90 lb/ft^3 (1282 to 1442 kg/m^3), soil erosion on short hillslopes would reach a peak and then decrease as the gradient approached 100%. Meyer *et al.* (1975) reported that erosion rates leveled off after about 20% in their experiments with various surface mulches.

Various reasons have been cited for this decline. The two basic categories are detachment limitation or transport limitation. Carson and Kirkby (1972) suggested that hillslopes able to stand at high gradients are composed of resistant materials. Meyer *et al.* (1975) inferred that the rate of soil detachment by rainfall changes slowly as hillslope gradient increases and that this largely governs the rate of soil loss; however, the transport capacity of interrill flow is believed to increase rapidly as the hillslope steepens. Hence, a detachment-limiting situation emerges. Bryan (1979) also suggested that the decline reflects a deficiency of erodible material due to heterogeneity or reduced depth of regolith.

As previously cited, Horton (1945) contended that uniform turbulent flow would not develop on very steep slopes. It would follow that the transport capacity of the flow would be limited, although Emmett (1978) showed that the flow may be "disturbed" by raindrop impact. Nevertheless, assuming the validity of the positive relationship between flow depth and transport and an inverse relationship between hillslope gradient and flow depth, it is conceivable that a transport-limiting situation could exist, at least for a time, on relatively noncohesive materials.

To add one further complication, a rarely cited report of the National Cooperative Highway Research program by Israelsen *et al.* (1980) concluded on the basis of numerous tests of hillslope plots up to 84% in gradient, that the universal soil loss equation (USLE) is valid for use on steep hillslopes. Their test plot dimensions of 19.5-ft long and 4-ft wide were much larger than some used in the research mentioned above. Erodible material was nonlimiting in these experiments.

Meyer *et al.* (1975, p. 183) described the role of hillslope gradient in rill–interrill erosion:

> The amounts of soil eroded from interrill areas and from rills vary quite differently with slope steepness. Interrill erosion will be appreciable at even small steepnesses but may increase relatively little for steeper slopes. In contrast, the influence of steepness on losses from rills may be very great as the slope steepens. The net effect of losses from interrill areas and from rilled areas may be expected to approximate the relationship used in the universal soil loss equation.

Upon first inspection, the notion that soil erosion increases with hillslope length, under the condition of Hortonian overland flow, seems quite acceptable. Runoff would accumulate in the downslope direction resulting in greater depth and velocity of flow; this, in turn, would probably favor rilling of the surface. However, this relationship requires additional assumptions as well. First, the hillslope must possess a relatively uniform or straight profile shape. Complications are introduced into the relation if the profile is convex–concave or simply concave. Second, rainfall intensity and surface infiltration capacity must continue to interact in such a way as to generate runoff throughout the entire hillslope length. Horton (1945) observed that rains of low intensity may produce erosion on scattered patches of soil as a result of local variations of infiltration capacity and resistivity and with little relationship to hillslope length. He also suggested that, other things being equal, runoff intensity in a given storm decreases somewhat as the hillslope length increases.

Nevertheless, where the aforementioned assumptions are valid, soil erosion can usually be expected to increase with hillslope length. Again, Zingg (1940), Musgrave (1947), and Wischmeier and Smith (1965), among others, included hillslope length in their soil loss models. In fact, Zingg (1940) found that length could be more important than gradient. Doubling the gradient of hillslope increased the total soil loss in runoff 2.61 times, whereas doubling the horizontal length of hillslope increased the total soil loss in runoff 3.03 times. Horton (1945) also concluded that the depth of erosion on a hillslope commonly increased with hillslope length. Kramer and Meyer (1969) and Laflen *et al.* (1978) observed increasing soil erosion with hillslope length.

Generally, the relationship between soil erosion and hillslope length is believed to be best expressed through the use of exponential equations. Depending on the specifics of research conditions, these exponents vary. Carson and Kirkby (1972, p. 210) stated

> The difference between these cultivated-field values and the exponents obtained for natural slopes may be due to the differences in soil and vegetation which occur under natural condi-

tions, but which are minimized by experimental procedures: for example, soils tend to become moister and vegetation somewhat denser downslope under natural conditions, so that the influence of slope length is reduced, leading to lower slope length exponents under natural conditions.

Wischmeier and Smith (1978) demonstrated the influence of hillslope gradient on the hillslope length factor used in the USLE. Herein, the length factor is given by the equation

$$L = (\lambda/72.6)^m \tag{2}$$

where L is the hillslope length factor, λ the field slope length, and m is 0.5 if slope $\geq$5%, 0.4 if slope $<$5% but $>$3%, 0.3 if slope $\leq$3% but $\geq$1%, and 0.2 if slope $<$1%.

Meyer *et al.* (1976) also described the role of hillslope length in rill–interrill erosion. Figure 7 shows that total erosion increases with hillslope length for both plots. However, the rate of increase is much greater for the rill-susceptible plot (R-3) than for the relatively rill-resistant plot (R-4), with interrill erosion approximately equal for each. Hence, the rate at which total erosion increases with hillslope length depends largely upon the propensity for rilling. Additionally, the relative contribution of rill and interrill erosion depends upon the hillslope length

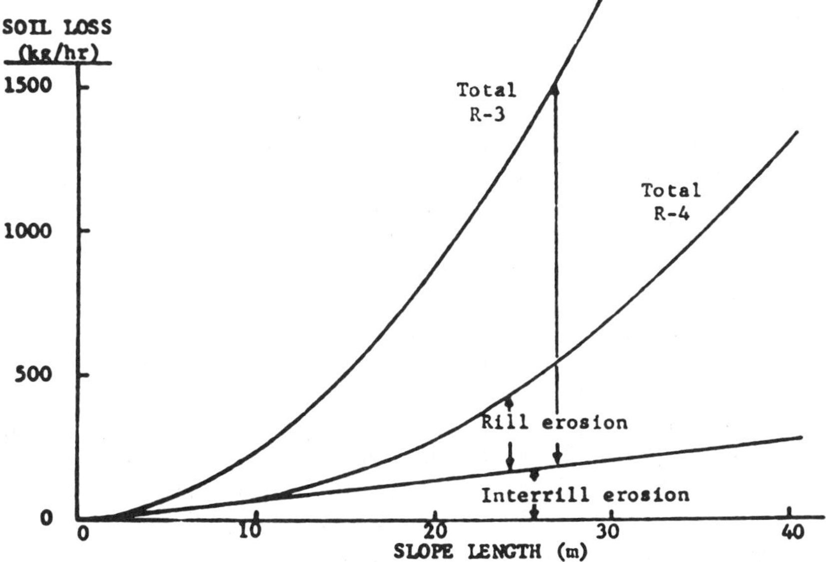

Figure 7. Relative contribution of rill and interrill erosion as influenced by hillslope length for two conditions (R-3 and R-4) are equally susceptible to interrill erosion, but R-3 was much more susceptible to rilling (Meyer *et al.*, 1976).

at which measurements are taken. All this suggests that the exponent for hillslope length in the USLE should increase as rill erosion increases. Because the amount of rill erosion relative to interrill erosion generally increases with both hillslope length and hillslope gradient, erosion will increase more with hillslope length on steep rather than gentle hillslopes. Thus, the larger exponent relating erosion and length, as presented earlier, is indeed appropriate. However, the fact that the data produced curves even when soil loss is plotted on a logarithmic scale suggests that an exponential relation may not adequately express the relationship at all. Other data contained in the same report support this conclusion.

Using their erosion data in a slightly different way, Meyer *et al.* (1975, pp. 180–181) provided added insight into the relationship between rill and interrill erosion on a rill-susceptible plot (R-3) as depicted in Fig. 8. They describe the relationship thusly:

> Nearly all the loss near the top of this plot was from the interrill portion. Thereafter, rill losses increased rapidly with distance until they equaled the cross-sectional area of loss for the interrill portion at about 25 ft downslope. By this distance, the erosion on the interrill portions was essentially constant.
>
> Although cross-sectional area of loss from rills and from interrill areas was equal at about 25 ft downslope, the cumulative volume of soil eroded from the rilled portions would not equal that from the interrill portion until about 45 ft downslope. In other words, the sediment load at about 45 ft would have been composed equally of soil originating from the rills and of soil from the interrill portions. Beyond that point, more than half the eroded soil would have been derived from rill portions.

Hillslope aspect influences erosion rates as well as hillslope gradients. In the western United States, Hadley (1962) found that south-facing hillslopes are subjected to greater surface erosion than their north-facing counterparts. Greater solar insolation on the south-facing hillslopes produces more xeric conditions resulting in sparse protective vegetation cover. Additionally, the drainage density, expressed as the total length of channels per unit drainage area (miles per square mile) averages 11.3 for south-facing sides of six drainage basins and only 5.2 for the north-facing sides. This is taken as evidence of higher erosion rates leading to greater dissection on the south-facing hillslopes.

For the eastern United States, Hack and Goodlett (1960) postulated that hillslope and channel erosion are more important processes on northwest- and southwest-facing hillslopes while soil creep is a more important process on northeast- and southeast-facing hillslopes. Drainage density appears higher for the hillslopes with a westerly aspect than those with an easterly aspect. Although there is some debate as to whether these differences in processes or process rates can account for valley side asymmetry, there seems to be agreement that hillslope aspect and attendant variation in solar insolation, resulting in variation of hillslope soil moisture and vegetation properties, does, in fact, produce differences in rates of soil erosion.

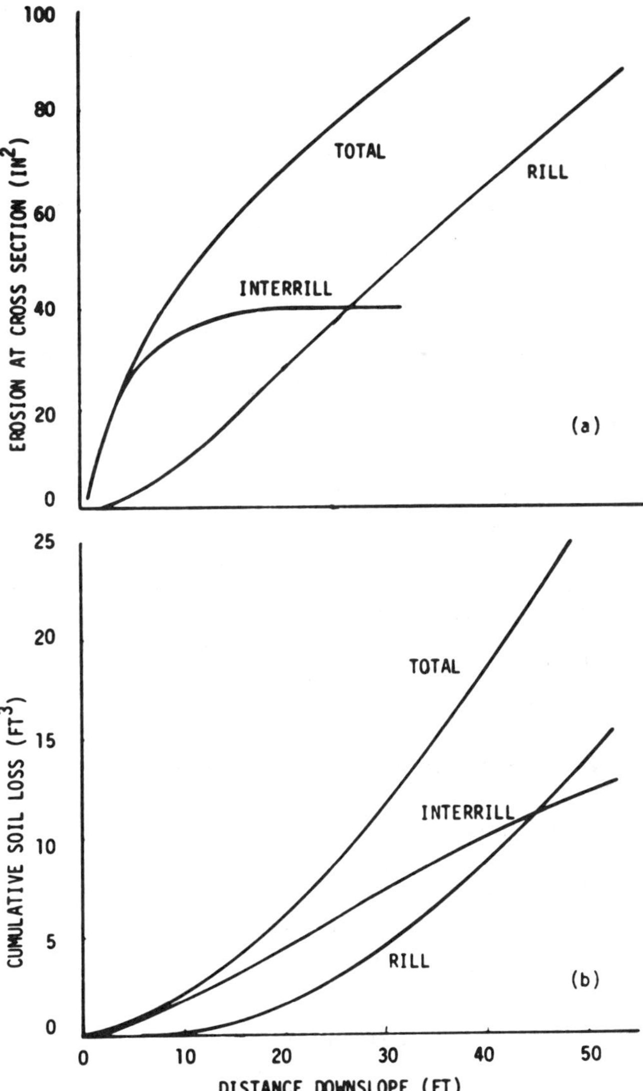

Figure 8. Eroded soil apportioned between rill and interrill sources for a rill-susceptible plot, R-3 (Meyer *et al.*, 1975).

Profile form also affects hillslope erosion. The profile depicts the nature of gradient changes over length. Following experimentation, Young and Mutchler (1969) concluded that erosion varies significantly from a given length with the same average gradient but with different configurations. Meyer and Kramer (1968) found that large differences in maximum sediment load, sediment carried beyond the toe of the hillslope, eroded depth, and subsequent hillslope profile occur for the different profile shapes, namely, uniform, concave, convex, and complex (convex–concave). Acknowledging these findings, Foster and Wischmeier (1974) devised a procedure for estimating soil loss from irregular hillslopes, based upon the USLE, by dividing the profile into segments or elements with essentially uniform gradients and soil properties and then summing the soil loss from these parts. Hadley *et al.* (1985) stated that it is justified to conclude that the nature of the hillslope (profile form) can be as much or even more important than hillslope gradient in determining erosion rates.

Meyer *et al.* (1975, p. 184) summarized the influence of profile form on soil erosion as shown in Fig. 9. They observed

> Erosion at successive points downslope increases gradually for uniform slopes. Losses for convex slopes are lesser near the top but increase very rapidly toward the end of the slope. For concave slopes, losses are greater near the top of the slope but decrease along the lower portions to the extent that deposition may occur. For complex slopes with their upper half convex and lower half concave, sediment load increases to a point approaching the toe of the slopes, and thereafter deposition may occur. The magnitude of the erosion and the location where deposition begins are functions of several factors including the relative susceptibility of the soil to rill and interrill erosion, the curvature of the slope, and the runoff rate.

The mathematical simulation approach of Li *et al.* (1976) yielded very similar results. In light of earlier discussion, it is somewhat disconcerting that their model assumes the effects of rainsplash to be negligible; however, it may be acceptable to assume that their effects are relatively uniform across the hillslope in some cases. The congruency between their results and those of field experimentation seems to support such an assumption.

Meyer and Romkens (1976, pp. 2–71) described the reasons for the tendencies contained in Fig. 9:

> A convex slope is more erodible than a uniform slope, because it is steepest near the toe where runoff is greatest. A uniform slope will yield more sediment than a concave one, because the concave slope is steepest where the flow is least and because some of the sediment eroded from the upper portions of the concave slope may deposit as it flattens near the toe. . . . A complex slope that is convex along its upper portion and concave along its lower portion will generally yield less sediment than a uniform slope. A flat section at the toe of a slope will also reduce sediment yield.

Of course, other research produces differences in the relative amount of erosion from various profile forms because of dissimilarities in research design and environmental conditions. Nevertheless, a brief review of these results is useful

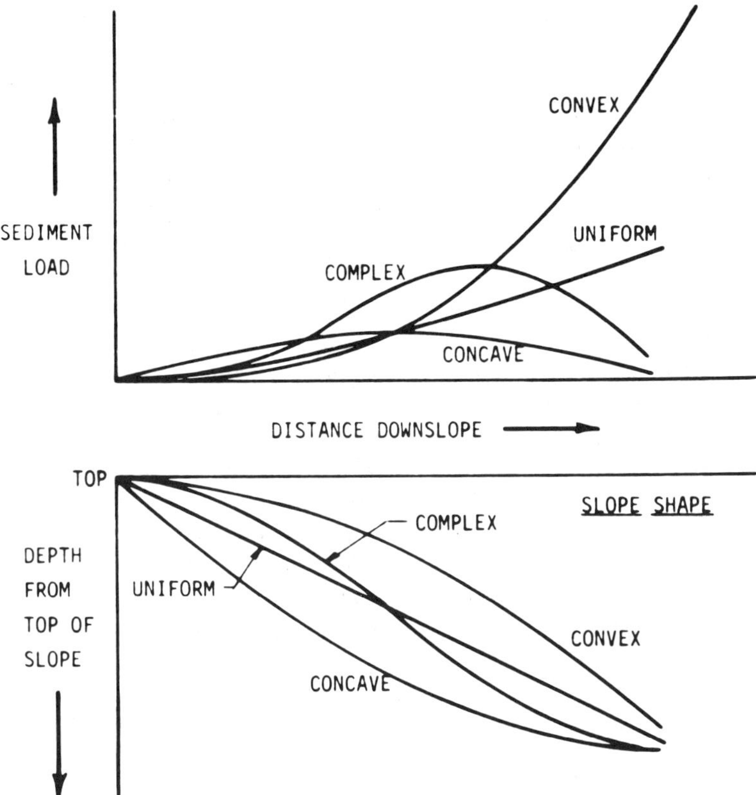

Figure 9. Influence of hillslope shape on sediment load and erosion rate; steepness of sediment load curve indicates rate of erosion and negative slope indicates deposition (Meyer *et al.*, 1975).

to illustrate the possible range of values which might occur in the field. Young and Mutchler (1969) used rainfall simulation on convex, uniform or straight, and concave profiles under vegetation cover of corn, oats, and cultivated fallow in the early stages of development. Average soil loss, in tons per acre (tonnes per hectare), for these three covers was 4.8 (10.8), 12.8 (28.7), and 13.7 (30.7) for concave, uniform, and convex profiles, respectively. Based upon estimating equations of soil loss, Meyer and Kramer (1968) found that the maximum depths during the first erosion period for the 5% slope range from 0.019 ft (5.8 mm) near the top of the concave hillslope to 0.028 ft (8.5 mm) at the end of the uniform hillslope, to 0.044 ft (13.4 mm) just above midway for the complex hillslope, to 0.129 ft (39.3 mm) at the end of the convex hillslope. At 10% the depths are greater; 0.035 (10.7 mm), 0.051 (15.5 mm), 0.080 (24.4 mm), and 0.218 (66.4 mm) ft, respectively, but in the same order. Hadley and Toy (1977) used rainfall simulation on natural hillslopes in western Colorado with a sparse vegetation cover consisting mainly of salt desert shrubs. Erosion from the convex, straight, and concave segments of these complex, convex–concave hillslopes averaged 0.0215 ft (6.6 mm), 0.0538 ft (16.4 mm), and 0.0277 ft (8.4 mm), respectively. Li *et al.* (1976) found from their simulation analysis that erosion rates on convex hillslopes are nearly five times greater than those on a uniform hillslope, whereas the erosion rates on concave hillslopes are much less than those on a uniform hillslope.

A couple of additional findings are noteworthy. Hadley and Toy (1977) observed that either erosion and deposition could occur within the basal concave element of convex–straight–concave hillslopes. Although the top of the convex element of such hillslopes is usually considered a zone of minimum erosion or within Horton's "belt of no erosion," Yair (1973) found that in arid regions there could be sheet erosion and rilling along the longitudinal axis of the divide.

Finally, Meyer and Kramer (1968) examined changes in profile form due to erosion based upon erosion equations. Fifty periods of erosion changed the profile of the convex profile the most and the concave profile the least. The uniform profile developed into a concave profile, and the initially complex profile also developed a concave profile except at the very upper part of the hillslope. After 200 periods of erosion, all hillslope profiles tended toward concave profiles. These trends are very similar for both the Zingg and Wischmeier erosion equations.

In another series of tests, "critical values" were incorporated into these erosion equations. Critical values are limits of gradient and length below which erosion is not appreciable. Erosion equations with critical values predict the "belt of no erosion" near the top of the hillslope as described by Horton (1945). When critical values are included, all original hillslope profiles except the concave soon develop complex profiles having inflection points somewhere along the upper half of the hillslope, similar to those commonly encountered in the field. As erosion progresses further, however, these complex profiles become dominantly concave.

It is interesting to consider whether this sequence, with the dominantly concave profile as a terminus to evolution, is actually characteristic of natural or disturbed hillslopes. It must be remembered that these models are based upon assumptions of uniform soil throughout the hillslope, accurate predictions of sediment loads based solely upon gradient and length, and only one geomorphic process, namely, surface erosion, shaping the hillslopes.

The plan of a hillslope refers to its linearity or curvilinearity along the horizontal dimension or width. Three categories of plan forms are valley side, spur end, and valley head hillslopes. When viewed from the two-dimensional perspective, the valley side hillslope is composed of an infinite number of adjacent uniform or straight profiles, the spur end hillslope is composed of an infinite number of adjacent convex profiles, and the valley head hillslope is composed of an infinite number of adjacent concave hillslopes. As such, the effects of these are adequately discussed above.

However, when viewed from the three-dimensional perspective, hillslope plan forms have added significance concerning the process of erosion. Figure 4 illustrates nine basic slope geometries (Ruhe, 1975) constructed from combinations of convex, concave, and straight profiles and plans. The most interesting of these are the valley head hillslopes (hollows), consisting of concave profile and concave plan, and the spur end hillslope (nose), consisting of convex profile and convex plan.

Unfortunately, the relationship between these three-dimensional hillslope forms and soil erosion rests largely upon principles of hillslope hydrology and hillslope erosion, together with deductive logic, rather than actual experimental or field data. The hydrologic argument contends that flow lines, which follow the direction of the true slope, tend to converge on valley head hillslopes so that there are, other things being equal, greater flows of water and sediment passing successive points downslope than on straight, valley side hillslopes. Conversely, on spur-end hillslopes, the flow lines diverge and flows of water and sediment tend to be less than on corresponding valley side hillslopes (Carson and Kirkby, 1972).

Indeed, Hack and Goodlett (1960) found that valley head hillslopes are moist areas and that the moisture in the ground increases toward the stream. Vegetation characteristics generally reflected this concentration of moisture. Further, the soil mantle is thicker in the valley head hillslope. They submitted that for this landform, the amount of water passing over the surface at any point is proportional to a quantity considerably greater than dictated by hillslope length alone.

In the same study, the spur end hillslopes were found to be the driest part of the valley. The amount of runoff crossing any place during a rainstorm is proportional to the radius of curvature of the hillslope contour.

Hack and Goodlett (1960) also suggested that runoff probably does not begin in all parts of a first-order valley simultaneously. Runoff may begin on spur end hillslopes, accelerating the saturation of the materials on the valley side hillslopes, and runoff from both areas may accelerate the saturation of soil materials in the hollows. Rainfall sufficient to saturate the soil mantle on the spur end and valley side hillslopes and cause saturated overland flow may be insufficient to saturate the thicker soil mantle in the valley head hillslope.

In contrast, Carson and Kirkby (1972) contended that the higher basal soil moisture contents and faster increase in soil moisture contents during storms produce saturation of the soil more frequently in valley head hillslopes than elsewhere so that overland flow is also more frequent. They contended that by the convergence of flow lines, more sediment tends to be produced in the valley head hillslopes than on the spur end hillslopes that experience divergence of flow lines. They concluded that where the increased water flow is able to transport more than the additional sediment supplied, the valley head hillslope will become a locus of erosion so that incipient hollows tend to grow larger and become a stable feature of the landscape. Conversely, where water flows are not large enough to carry additional sediment, incipient hollows tend to fill in and the landscape tends to remain very regular in outline.

From the foregoing, it is clear that the actual effect of three-dimensional hillslope form depends upon several site-specific characteristics. Relatively droughty spur end hillslopes, with thin soil and vegetation covers, could produce large amounts of runoff and sediment if subjected to high-intensity rainfall, unless erosion is detachment limited. Likewise, relatively humid valley head hillslopes with thick soil and vegetation covers could produce modest amounts of surface runoff and sediment if subjected to low-intensity rainfall; here erosion would be both detachment and transport limited. Subsurface flow and material translocation, however, would be relatively more important in these valley head hillslopes. As usual, rainfall, soil, and vegetation properties, togcther with hillslope form, combine to determine the actual erosion rate in a particular locale.

Influences of Geologic Materials on Hillslope Form

While the foregoing presents a strong case for a relationship between hillslope process and form, it must be clearly understood that geomorphic processes are not the sole determinant of hillslope form under natural conditions. Gregory (1978) concisely expressed the multivariate control of form in the "geomorphological equation" whereby:

$$F = f(PM)\ dt \tag{3}$$

where the landform F is the function (f) of processes (P) and materials (M) through time (dt). This is a symbolic representation of Strahler's (1950) "law of constancy of slopes." Similarly, it has been suggested by Ritter (1978) that landforms may be analyzed in terms of balances between the driving forces, resulting from climate and gravity, and the resistances offered by the geologic framework of the earth's surface. The two principal components of this framework are geologic structure and lithology, and it can be observed that either may dominate in the development of hillslope form. For this reason, any attempt to examine the relationship between process and form should attempt to hold constant the geologic influences. The study by Toy (1977) examined hillslopes that were produced on horizontal neritic shales.

As with attempts to isolate the role of climate and process, Wilson (1968) observed that a great many investigations regarding the influence of geologic structure and lithology have failed to yield general conclusions. Again, the problem is that geology acts in concert with other factors to produce complex landscapes. He also remarked that there is general agreement that arid climates produce erosional hillslopes with relatively little soil and vegetation, with the result that factors of structure and lithology are of more importance in controlling the landscape than in humid climates.

Young (1972) contended that, as a generalization, structure has a greater influence on hillslope form than lithology; specifically, the effects arising from the superposition of strata of differing lithology are greater than those caused by each individual rock type. Ritter (1978) noted that structural influence is readily apparent only when the rocks and climate involved are conducive to differential weathering and erosion. Nevertheless, the strike and dip hillslopes associated with domes, basins, anticlines, and synclines illustrate the significance of geologic control in numerous places.

Lithology may also affect hillslope form. There is general agreement that hillslopes eroded into weak rocks have gentler angles than those cut into hard resistant rocks. Selby (1982) showed a strong correlation ($r = 0.88$) between rock mass strength and hillslope angle for hillslopes measured in Antarctica and New Zealand. Wilson (1968), however, observed that, while resistant lithologies produce steeper hillslopes than those developed on weaker rock, highly cohesive weak materials such as clays may also yield steep hillslopes.

Complications arise from the fact that in any given climate each rock type will respond to the processes of weathering and erosion in a different manner and at a different rate (Ritter, 1978). The classic example, of course, is limestone, which often forms a caprock and scarp in arid climates because of its resistance to the geomorphic processes prevalent in those regions, while producing karst topography with sinkholes and caves in humid regions because of its susceptibility to solution processes.

Numerous references are available that describe geologic controls in detail (Selby, 1982; Scheidegger, 1964). However, this will not be emphasized in the subsequent section because on slightly or moderately disturbed lands the geologic framework may be considered as a "given." On drastically disturbed lands, such as those surface mined, the geologic framework is essentially homogenized by the resource extraction process. Hence, changes in geomorphic processes become paramount in importance.

Hillslope Form and Processes: Disturbed Systems

From the foregoing, it is apparent that hillslopes are highly complex landscape features, subjected to a variety of geomorphic processes and assuming a variety of two-dimensional and three-dimensional forms. Additionally, they are a part of a complex process–response system. The work performed by geomorphic processes on a particular hillslope may impact adjacent areas through the deposition of colluvium, materials deposited by mass movement phenomena, or alluvium, materials deposited by fluvial erosion processes. A change in the volume of water or sediment reaching streams at the base of a hillslope will produce an adjustment of its channel geometry and gradient. These responses may be transmitted back against or into the hillslope through gullying and rilling. It should be understood, therefore, that hillslopes are important, not only because of their spatial ubiquity and their utility in human activities, but because of their relationship to the drainage network in an integrated system. Hence, stream channels can be no more stable than the surrounding hillslopes, and the converse is true as well.

It is worthwhile to review a few basic geomorphic principles as a preface to the discussion of process and form on disturbed hillslopes.

1. It is generally accepted that, in nature, geomorphic processes strive to create and maintain a balance or "equilibrium" between form and process.
2. The operation of geomorphic processes results in forces being applied against the resistances of surface materials, shear stresses against shear strengths.
3. When stresses exceed strengths, materials are deformed and/or transported and work has been accomplished.
4. The characteristics and operation of forces and resistances, stresses and strengths, are governed by physical and chemical laws.
5. Most geomorphic processes probably operate on a particular hillslope at one time or another, although the magnitude and frequency of their operation will vary for particular processes and geomorphic environments (i.e., climate, geology, soil, vegetation, and relief).
6. The work performed by geomorphic processes may be weather or detachment limited, transport limited, or conceivably, both.

These same principles are equally applicable to disturbed hillslopes. There is still the general quest for dynamic equilibrium, still the interplay of forces and resistances, still the performance of work, and still the governing physical and chemical laws. Hence, the discussion of hillslope processes in undisturbed systems presented in Chapter 2 provides a firm foundation for the consideration of disturbed systems.

Geomorphic Consequences of Land Disturbances on Hillslopes

With a knowledge of basic geomorphic principles, it is now possible to understand the consequences of land disturbance for hillslope systems. Essentially, a disequilibrium is produced by altering the balance between force and resistance. The more severe types of disturbance, such as surface mining, highway construction, and residential development, also alter the balance between form and process through grading operations.

The effects of land disturbance on hillslopes may be examined in terms of force and resistance. Table 2 shows the impact of various types of disturbances on site conditions and, as a result, geomorphic processes. Perusal of this table reveals that most disturbance types increase the force exerted on or within hillslope materials while concurrently reducing resistance. It becomes clear why these alterations of the natural setting produce such significant changes in the rates of geomorphic processes. It should be acknowledged that the disturbances listed in Table 2 do not always produce the geomorphic response cited. Careful planning, coupled with an understanding of geomorphic processes and their controls, allow the mitigation of many adverse effects. However, the consequences included are by no means infrequent or unusual. Additionally, when many of these geomorphic responses do occur, there are remedial or reclamation practices which can correct them, or at least reduce their severity.

The geomorphic effects of hillslope disturbance can be demonstrated conceptually using a modified version of the diagram presented in Chapter 2, illustrating changes in valley floor stability through time. In this case, time is still depicted on the abscissa, and now the force resulting from geomorphic processes is depicted on the ordinate, as shown in Fig. 10. Line 1 represents decreasing resistance (increasing instability), and line 2 represents the failure threshold which must be exceeded in order for work to be performed. The vertical lines extending above line 1 represent the variations in force impinging upon the hillslope. The increase in force due to disturbance may be envisioned as lengthening the vertical lines extending above line 1. This, in turn, is likely to cause line A to be shifted to the left somewhat, indicating that the threshold would be achieved sooner.

The reduction of resistance may be envisioned in two ways. First, it may have the effect of lowering the threshold line 2. As such, line A would again be shifted to the left, indicating a sooner response to force. Second, the reduction of resistance may have the effect of increasing the gradient of line 1.

Collectively, these effects suggest that a threshold could be achieved by forces of lesser magnitude as line 2 is lowered or line 1 is steepened. Frequency distributions of various natural phenomena, such as stream flooding, show that events producing small force are quite common while events producing great force are quite rare. Thus, it would follow that external events capable of generating sufficient force to surpass a threshold should occur with greater frequency as a result of land disturbance. For example, prior to disturbance, a rainfall of 0.70 in./hr (18 mm/hr) could have been completely absorbed into the surface, producing no runoff erosion. However, after disturbance, this same intensity of rainfall may exceed the new infiltration capacity, producing runoff in sufficient

Table 2 Geomorphic Effects of Land Disturbance on Hillslopes

Change in force or resistance	Type of disturbance	Geomorphic effects
Increase in force	Removal or reduction in vegetation cover (grazing, logging, tillage, hiking, mining, traffic, residential development)	Reduction of soil infiltration capacity causing increase in depth and velocity of surface runoff
		Rainsplash impact on soil surface
	Compaction of soil (grazing, hiking, vehicular traffic)	Reduction of infiltration capacity of soil causing increase in depth and velocity of surface runoff
	Topsoil stripping (mining, residential development)	Reduction of infiltration capacity when organic matter content of soil is reduced causing increase in depth and velocity of surface runoff
		Reduction of infiltration capacity when soil texture is altered toward percentage of silt and clay-sized separates causing increase in depth and velocity of surface runoff
	Surface grading (highway construction, residential development, mining)	Increase in hillslope angle causing increase in surface runoff velocity
		Increase in angle causing increase in downslope component of gravitational force for soil and hillslope mass
		Increase in hillslope length causing greater depth and velocity of runoff toward base of hillslope
		Increase in hillslope height resulting in greater shear stress within soil and hillslope mass
		Increase in hillslope convexity resulting in acceleration of surface runoff
		Creation of valley head hillslopes which concentrate flow resulting in greater depth and velocity of flow
Reduction of resistance	Removal or reduction of vegetation cover (same activities as above)	Decrease in interception of raindrop energy
		Decrease in frictional surface opposing surface runoff
		Decrease in binding of soil particles and aggregates
		Increase in erodibility of soil by reduction of organic matter content

Table 2 *(Continued)*

Change in force or resistance	Type of disturbance	Geomorphic effects
		Decrease in surface resistance to wind flow across soil surface
	Compaction of soil (same activities as above)	Decrease in microtopographic resistance to water and wind flow as a result of surface smoothing causing higher velocities of both near the surface
		Increase in soil erodibility by destruction of soil aggregates or structure
	Topsoil stripping (same activities as above)	Increase in soil erodibility by destruction of soil aggregates or structure
		Increase in soil erodibility through decrease in organic matter content if stockpiled for long periods
	Surface grading (same activities as above)	Increase in soil erodibility through destruction of soil aggregates or structure
		Decrease in microtopographic resistance to water and wind flow as a result of final finegrading, causing higher velocities of both near the surface
		Removal of basal support for hillslope through removal of concave basal element, encouraging mass-movement
		Destruction of resistant geologic strata

depth and velocity to detach and transport soil particles. Perhaps it would take a rainfall intensity of only 0.40 in./hr (10 mm/hr) to produce enough runoff to exceed the threshold of surface resistance causing erosion; these happen more often than those of 0.70 in./hr (18 mm/hr).

Thus, in concept and in practice the effects of hillslope disturbance on geomorphic processes can be examined in terms of force and resistance, magnitude and frequency. Generally, the force exerted by geomorphic processes on hillslopes are increased while the resistance of the surface to that force is decreased. The magnitude of force exerted by geomorphic processes is increased while the force required to produce work, by overcoming resistance, is reduced. The

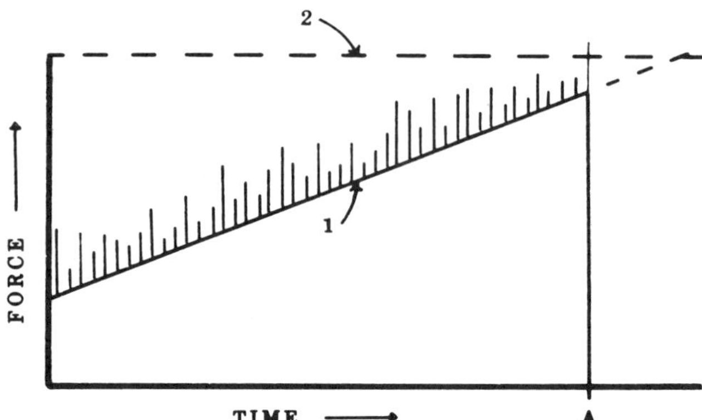

Figure 10. The effect of hillslope disturbance viewed conceptually in terms of force and resistance.

frequency with which processes generate sufficient force to overcome resistance increases, largely because lesser force is required and events generating force in smaller magnitudes recur with greater frequency. This summarizes the essence of geomorphic response to hillslope disturbance.

Geomorphic Consequences of Reclamation on Hillslopes

There are reclamation practices designed to ameliorate the impact of hillslope disturbance. Table 3 provides a summary of common procedures and their geomorphic effect. These techniques could also be examined in terms of their influence on force and resistance.

It is necessary to prevent runoff from upland areas from crossing the disturbed or newly reclaimed site. The velocity and depth of these flows are usually highly erosive to the unprotected surface. Water diversion above and adjacent to the reclaimed site is essential.

In cases of severe land disturbance, such as highway cut construction, surface mining, and sometimes residential development, the reclamation process usually begins with grading operations that reconfigure the surface, creating new hillslopes and hillslope forms. It is clear from the preceding discussion that the preferred profile form would be concave or convex–concave and that the three-dimensional spur end shape likely would be advantageous.

Usually, there is a trade-off between hillslope gradient and length. As angle decreases, length increases. This does not necessarily have to occur because these longer hillslopes could be divided into two or more in order to reduce hillslope length as well. For some industries, maximum hillslope gradients are specified by legislation, regulation, or guidelines. However, hillslope length is not controlled. This seems to be an oversight, given Zingg's (1940) findings, wherein doubling hillslope length had a greater effect on soil loss than doubling hillslope gradient. Other investigators have substantiated the importance of this hillslope property. The rule of thumb should be to minimize both gradient and length to the extent feasible. Further discussion of hillslope design will be included in subsequent chapters concerning surface mining of coal, surface mining of uranium, and geomorphic perspectives on the design and management of disturbed lands.

Following grading, stockpiled topsoil is commonly spread upon the hillslope. Packer and Aldon (1978) listed three benefits of this practice: (1) it provides fertility not usually encountered in raw spoils, (2) it furnishes a source for renewed microbiological activity to improve soil-building processes, and (3) it has better infiltration and soil stability characteristics. It is the latter advantage that is important from our present viewpoint because it reduces the volume and velocity of surface runoff generated and hence the erosive force while placing somewhat more resistant (less erodible) material at the surface.

The heavy equipment traffic on the hillslope during grading and topsoiling compacts the surface. Various surface manipulations are employed to relieve this problem, ranging from deep ripping which cuts through the topsoil and subsoil to a depth of 2 ft (0.6 m) or more, to discing of the upper few inches of the topsoil. These practices, taken together, serve to bind the topsoil and subsoil somewhat, increase the infiltration capacity, and cause ponding of runoff, thus again reducing flow velocities and erosive force.

Surface mulches and soil stabilizers may be applied in the next phase of reclamation. These are temporary erosion control measures intended to protect the soil surface until the establishment of vegetation cover. Mulches encourage infiltration, impede runoff, and reduce raindrop impact energy and, as a result, surface sealing.

Soil stabilizers may be sprayed directly onto the soil surface or used together with a mulch. These serve to bind the soil particles for a time and thereby reduce their erodibility. These treatments tend to be expensive, however, and are usually reserved for special problem areas.

Successful revegetation is a key to satisfactory long-term reclamation. Revegetation is the principal mechanism for substantially increasing surface resistance to erosive forces. Generally, the new vegetation cover should consist of perennial, maintenance-free species in sufficient density and diversity to effectively minimize soil erosion. It should be emphasized that accelerated soil erosion is inevitable without effective vegetation cover on the hillslope. Sheet erosion will lead to rilling and probably gullying.

Finally, the reclaimed land must be managed to minimize grazing damage by either domestic animals or wildlife. Emerging vegetation can be virtually

Table 3 Geomorphic Effects of Hillslope Reclamation Practices

Change in force or resistance	Reclamation practices	Geomorphic effects
Reduction in force	Surface mulching: straw, hay, stubble, paper products, wood products (Meyer *et al.*, 1972; Meyer and Romkens, 1976; Plass, 1978; Kay, 1978)	Interception of raindrop impact protecting soil surface Increases infiltration of water reducing depth and velocity of runoff
	Revegetation (Vogel, 1981; Long, 1978)	Increases infiltration causing decrease in depth and velocity of runoff Interception of raindrop impact protecting soil surface
	Water diversion (Glover *et al.*, 1978)	Reduces runoff crossing disturbed site by diverting it around
	Topsoiling (Schuman and Power, 1981; Power *et al.*, 1981)	Increases infiltration causing decrease in depth and velocity of runoff
	Surface manipulation: gouging, listering, dozer basins, scarification, tracking (Glover *et al.*, 1978; Dollhopf *et al.*, 1977; Verma and Thames, 1978)	Increases infiltration causing decrease in depth and velocity of runoff
	Surface grading (Glover *et al.*, 1978; Verma and Thames, 1978; Wright *et al.*, 1978; Meyer and Romkens, 1976)	Reduction in hillslope angle causing decrease in downslope component of gravitational force within soil and hillslope mass Reduction in hillslope angle causing decrease in surface runoff velocity Reduction of hillslope height causing decrease in shear stress within soil and hillslope mass Increase in hillslope concavity to reduce surface runoff velocity Creation of spur-end hillslopes to disperse surface runoff
Increase in resistance	Surface mulching (same materials and references as above)	Increase in raindrop energy interception Increase in frictional surface opposing surface runoff If anchored, reduces wind velocity by increasing surface friction Traps sediment on-site
	Revegetation (same references as above)	Increase in raindrop energy interception

Table 3 (*Continued*)

Change in force or resistance	Reclamation practices	Geomorphic effects
		Increase in frictional surface opposing surface runoff Increase in frictional surface opposing wind flow across surface Decrease in soil erodibility by addition of organic matter Increase in binding of soil by rootlets
	Surface manipulation (same techniques and references as above)	Increase in frictional surface opposing surface runoff Increase in frictional surface opposing wind flow Traps sediment on-site
	Soil stabilizers or tackifiers: asphalt, terratack, polyvinyl acetate, styrene butadiene, Portland cement (Kay, 1978; Israelson *et al.*, 1980)	Reduces soil erodibility through chemical binding or cementing effect May be used with mulch to produce a cohesive protective cover
	Surface grading	Rough grading increases frictional surface opposing surface runoff

stripped from the hillslope by rodents, rabbits, livestock, or big game animals in a short time unless specific action is taken to control their impacts. Additionally, larger animal species will compact the soil during their transgress of the reclaimed area.

The reclamation practices contained in Table 3, and briefly described above, collectively reduce the potential for geomorphic work on a hillslope by reducing force and increasing resistance. Consequently, only processes and events capable of generating considerable force to overcome the elevated threshold of resistance are capable of performing work. Because such events tend to recur less frequently, the total amount of work will also remain less.

In terms of Fig. 10, reclamation practices that reduce force can be envisioned as shortening the vertical lines extending above line 1. Those practices that increase resistance have the effect of elevating line 2. Successful reclamation may also reduce the gradient of line 1 to some extent, but some resistance factors that existed under natural conditions, such as those associated with geologic strata, are impossible to reproduce.

From the geomorphic perspective, the goal of reclamation is the reestablishment of a balance or dynamic equilibrium between process and form, force and resistance. However, the original site conditions can never be entirely replicated by reclamation practices. The development of a balance requires time during which an effective vegetation cover and root network are generated, soil structure and profile properties develop, and hillslope and channel characteristics come into adjustment.

Conclusions

Under natural conditions, hillslopes assume a wide variety of shapes from crest to base. These forms are generally taken to represent a balanced or dynamic equilibrium state between the forces produced by geomorphic processes and the resistances provided by hillslope materials (including vegetation). Geomorphic work takes place at a modest, nearly imperceptible rate whenever extrinsic or intrinsic thresholds are overcome by episodic or accumulative processes and forces. In most cases, surface erosion is the most important process due to its ubiquity. However, in some circumstances, mass movements are the primary manifestation of site instability.

Land disturbances create disequilibrium between force and resistance, process and form. Geomorphic work is accelerated toward a reestablishment of the balance. It is probably accurate to suggest that the rate of acceleration is proportional to the extent of disequilibrium. The products of this work may be transported off-site and result in another disturbance, creating other disequilibria in other systems, and engendering other geomorphic responses.

Hillslope reclamation attempts to recreate, as nearly as possible, the dynamic equilibrium state. In fact, it merely reduces the extent of disequilibrium because certain hillslope conditions, such as vegetation and soil properties, must self-generate through time. Nevertheless, the amount of geomorphic work performed by geomorphic processes will be reduced and, hence, the potential for off-site impacts.

Geomorphologists understand a great deal about hillslopes and their response to various geomorphic processes. This provides a foundation for evaluating the consequences of land disturbance on hillslopes. In candor, there remains much to be discovered. When examined in detail, the complexity of the hillslope system, which is a true interface between the atmosphere, hydrosphere, biosphere, and lithosphere, offers a real challenge.

In many ways, the disturbed hillslope system constitutes an intriguing geomorphic laboratory. The acceleration of geomorphic work and change in form can be observed and measured in time spans shorter than would be necessary to detect significant development on natural hillslope systems. Although this results in some research efficiency, one must reflect upon the extent to which the findings may be extrapolated to natural hillslope systems. Would a weathering-limited erosion process on a disturbed hillslope have been a transport-limited process under natural conditions, or vice versa?

References

Baulig, H., 1950, Essais de geomorphologie: Paris, pp. 125–147.

Birot, P., 1949, Essai sur quelques problemes de morphologie generale: Lisbon, pp. 17–33.

Blanche, J., 1942, Des versants aux rivieres: Rev. Geogr. Alp., 30, pp. 1–50.

Bloom, A. L., 1978, Geomorphology: Englewood Cliffs, New Jersey, Prentice-Hall, 510 pp.

Bryan, R. B., 1979, The influence of slope angle on soil entrainment by sheetwash and rainsplash: Earth Surface Processes and Landforms, v. 4, pp. 43–58.

Büdel, J., 1982, Climatic Geomorphology: Princeton, New Jersey, Princeton University Press, 443 pp.

Carson, M. A., and Kirkby, M. J., 1972, Hillslope form and process: Cambridge, University Press, 475 pp.

Chamberlain, T. C., and Salisbury, R. D., 1904, Geologic processes and their results: Geology, v. 1, New York, Henry Holt and Co.

Churchill, R. R., 1981, Aspect-related differences in badlands slope morphology: Annals, Association of American Geographers, v. 71, pp. 374–388.

Cooke, R. U., and Doornkamp, J. C., 1974, Geomorphology in environmental management: London, Oxford University Press, 413 pp.

Cotton, C. A., 1952, The erosional grading of convex and concave slopes: Geographical Journal, v. 118, pp. 197–204.

Dalrymple, J. B., Blong, R. J., and Conacher, A. J., 1968, A hypothetical nine-unit land surface model: Zeitschrift für Geomorphologie, v. 12, pp. 60–76.

Davis, W. M., 1892, The convex profile of badland divides: Science, v. 20, pp. 245.

Davis, W. M., 1930, Rock floors in arid and in humid climates: Journal of Geology, v. 38, pp. 1–27, 136–158.

de Bethune, P., 1967, Sur la developpement de la convexite sommitale des versants: Slopes Commission Report, v. 5, pp. 89–100.

de La Noë, G., and de Margerie, E., 1888, Les formes du Terrain: Paris, 205 pp.

Derbyshire, E. (ed.), 1973, Climatic Geomorphology: MacMillan Press, Ltd., 296 pp.

Dollhopf, D. J., Jensen, I. B., and Hodder, R. L., 1977, Effects of surface configuration in water pollution control on semiarid mined lands: Montana Agricultural Experiment Station, Montana State University, Research Report 114, 179 pp.

Dollfus, O., 1964, L'influence de l'exposition dans le modelé des versants des andes centrales Peruviennes: Zeitschrift für Geomorphologie, v. 5, pp. 131–135.

Douglas, I., 1977, Humid landforms: Cambridge, MIT Press, 288 pp.

Dunne, T., and Leopold, L. B., 1978, Water in environmental planning: San Francisco, W. H. Freeman and Co., 818 pp.

Elwell, H. A., 1984, Soil loss estimation: a modeling technique: *in* Hadley, R. F., and Walling, D. E. (eds.), Erosion and Sediment Yield, Norwich, England, Geo Books, pp. 15–36.

Emmett, W. W., 1978, Overland flow: *in* Kirkby, M. J. (ed.), Hillslope Hydrology, New York, John Wiley and Sons, pp. 145–176.

Fenneman, N. M., 1908, Some features of erosion by unconcentrated wash: Journal of Geology, v. 16, pp. 746–754.

Foster, G. R., and Wischmeier, W. H., 1974, Evaluating irregular slopes for soil loss prediction: Transactions, American Society of Agricultural Engineers, v. 7, pp. 305–309.

Foster, R. L., and Martin, G. L., 1969, Effect of unit weight and slope on erosion: Journal of Irrigation and Drainage Division, American Society of Civil Engineers, Proceedings, v. 95, no. IR–4, pp. 551–561.

Gilbert, G. K., 1877, Report on the geology of the Henry Mountains: U. S. Department of Interior, Washington, 160 pp.

Gilbert, G. K., 1909, The convexity of hilltops: Journal of Geology, v. 17, pp. 344–350.

Glover, F., Augustine, M., and Clar, M., 1978, Grading and shaping for erosion control and rapid vegetative establishment in humid regions: *in* Schaller, F. W., and Sutton, P., Reclamation of Drastically Disturbed Lands: Madison, Wisconsin, American Society of Agronomy, Crop Science Society of America, Soil Science Society of America, pp. 271–283.

Gotzinger, G., 1907, Beitrage zur Entstehung der Bergrüekenformen: Geogr. Abhl., v. 9.

Gregory, K. J., 1978, A physical geography equation: National Geographer, v. 12, pp. 13–141.

Hack, J. T. and Goodlett, J. C., 1960, Geomorphology and forest ecology of a mountain region in the central Appalachians: U. S. Geological Survey, Professional Paper, 347, 66 pp.

Hadley, R. F., 1962, Some effects of microclimate on slope morphology and drainage basin development: U. S. Geological Survey, Geological Survey Research, 1961, pp. B–32 to B–33.

Hadley, R. F., and Toy, T. J., 1977, Relation of surficial erosion on hillslopes to profile geometry: Journal of Research, U. S. Geological Survey, v. 5, no. 4, pp. 487–490.

Hadley, R. F., Lal, R., Onstad, C. A., Walling, D. E., and Yair, A., 1985, Recent developments in erosion and sediment yield studies: Technical Documents in Hydrology, UNESCO, Paris, 127 pp.

Hicks, L. E., 1893, Some elements of land sculpture: Bulletin of Geological Society of America, v. 4, pp. 133–146.

Holmes, C. D., 1955, Geomorphic development in humid and arid regions: A synthesis: American Journal of Science, v. 253, pp. 377–390.

Horton, R. E., 1945, Erosional development of streams and their drainage basins: Bulletin of Geological Society of America, v. 56, pp. 275–370.

Hutton, J., 1785, Phenomena common to stratified and unstratified bodies: *in* Playfair, J., 1956, Illustrations of the Huttonian Theory of the Earth: New York: Dover Publications Inc., pp. 97–140. (facsimile reprint)

Israelsen, C. E., Clyde, C. G., Fletcher, F. E., Israelsen, E. K., Haws, F. W., Packer, P. E., and Farmer, E. E., 1980, Erosion control during highway construction: Research Report, National Cooperative Highway Research Program Report 220, Transportation Research Board, National Research Council, 30 pp.

Jeffreys, H., 1918, Problems of denudation: Phil. Mag. v. 36, pp. 179–190.

Kay, B. L., 1978, Mulch and chemical stabilizers for land reclamation in dry regions: *in* Schaller, F. W., and Sutton, P., Reclamation of Drastically Disturbed Lands: Madison, Wisconsin, American Society of Agronomy, Crop Science Society of America, Soil Science Society of America, pp. 467–483.

Kennedy, B. A., 1976, Valley-side slopes and climate: *in* Derbyshire, E. (ed.), Geomorphology and Climate, London, John Wiley and Sons, pp. 171–201.

King, L. C., 1953, Canons of landscape evolution: Bulletin, Geological Society of America, v. 64, pp. 721–752.

King, L. C., 1957, The uniformitarian nature of hillslopes: Transactions of Edinburgh Geological Society, v. 17, pp. 81–102.

Kramer, L. A., and Meyer, L. D., 1969, Small amount of surface mulch reduces surface erosion and runoff velocity: Transactions, American Society of Agricultural Engineers, v. 12, pp. 638–641, 645.

Laflen, J. M., Baker, J. L., Hartwig, R. O., Buchele, W. F., Johnson, H. P., 1978, Soil and water loss from conservation tillage system: Transactions, American Society of Agricultural Engineers, v. 21, pp. 881–885.

Lake, P., 1928, On hill slopes: Geology Magazine, v. 65, pp. 108–116.

Langbein, W. B., and Schumm, S. A., 1958, Yield of sediment in relation to mean annual precipitation: Transactions of American Geophysical Union, v. 39, pp. 1076–1084.

Lawson, A. C., 1932, Rain-wash erosion in humid regions: Bulletin of Geological Society of America, v. 43, pp. 703–724.

Lehmann, O., 1918, Die talbildung durch schuttgerinne, *in* A. Penck Festband, Englehorns, Stuttgart, pp. 48–65.

Leopold, L. B., Wolman, M. G., and Miller, J. P., 1964, Fluvial Processes in Geomorphology: San Francisco, W. H. Freeman and Co., 522 pp.

Li, R. M., Simons, D. B., and Carder, D. R., 1976, Mathematical modeling of overland flow for soil erosion: National Soil Erosion Conference, Proceedings, Purdue University, pp. 210—216.

Long, S. G., 1978, Characteristics of plants used in Western reclamation: Fort Collins, Colorado, Environmental Research and Technology, Inc., 146 pp.

Marr, J. E., 1901, The origin of moels and their subsequent dissection, Journal of Geography, v. 17, pp. 63–68.

Maw, G., 1866, Notes on the comparative structure of surfaces produced by subaerial and marine denudation: Geographical Magazine, v. 3, pp. 439–451.

Melton, M. A., 1957, An analysis of the relations among elements of climate, surface properties, and geomorphology: Office of Naval Research, Technical Report No. 11, Project NR 389–042, 102 pp.

Melton, M. A., 1960, Intravalley variation in slope angles related to microclimate and erosional environment: Bulletin of Geological Society of America, v. 71, pp. 133–144.

Resources Council, Washington, D. C., pp. 2–1 to 2–12.

Meyer, L. D., and Kramer, L. A., 1968, Relation between landslope shape and soil erosion: Transactions of American Society of Agricultural Engineers, v. 11, pp. 1–14.

Meyer, L. D., and Romkens, M. J. M., 1976, Erosion and sediment control on reshaped land: Proceedings of the Third Federal Inter-Agency Sedimentation Conference, PB–245–100, Water Resources Council, Washington, D. C., pp. 2–65 to 2–76.

Meyer, L. D., Johnson, C. B., and Foster, G. R., 1972, Stone and woodchip mulches for erosion control on construction sites: Journal of Soil and Water Conservation, v. 27, pp. 264–269.

Meyer, L. D., Foster, G. R., and Romkens, M. J. M., 1975, Sources of soil eroded by water from upland slopes: Present and Prospective Technology for Predicting Sediment Yields and Sources, U.S.D.A., Agricultural Research Service, ARS-S–40, pp. 177–189.

Meyer, L. D., De Coursey, D. G., and Romkens, M. J. M., 1976, Soil erosion concepts and misconceptions: Proceedings of the Third Federal Interagency Sedimentation Conference, PB–245–100, Water Resources Council, Washington, D. C., pp. 2-1 to 2-12.

Meyerhoff, H. A., 1940, Migration of erosional surfaces: Annals of Association of American Geographers, v. 30, pp. 247–254.

Mosley, M. P., 1973, Rainsplash and the convexity of badland divides: Zeitschrift für Geomorphologie, v. 18, pp. 10–25.

Musgrave, G. W., 1947, The quantitative evaluation of factors in water erosion— a first approximation: Journal of Soil and Water Conservation, v. 2, pp. 133–138.

Mutchler, C. K., and Young, R. A., 1975, Soil detachment by raindrops: Present and Prospective Technology for Predicting Sediment Yields and Sources, U.S.D.A., Agricultural Research Service, ARS-S–40, pp. 113–117.

Packer, P. E., and Aldon, E. F., 1978, Revegetation techniques for dry regions: *in* Schaller, F. W., and Sutton, P. (eds.), Reclamation of Drastically Disturbed Lands: Madison, Wisconsin, American Society of Agronomy, Crop Science Society of America, Soil Science Society of America, pp. 425–450.

Parsons, A. J., 1982, Slope profile variability in first-order drainage basins: Earth Surface Processes, v. 7, pp. 71–78.

Penck, W., 1924, Die morphologische analyse. Ein kapitel der Physikalischen Geologie, Engelhorn, Stuttgart; English translation by H. Czech and K. C. Boswell, Morphological Analysis of Landforms: London: MacMillan Co., 429 pp.

Pitty, A. F., 1971, Introduction to Geomorphology: London, Methuen and Co., Ltd., 526 pp.

Plass, W. T., 1978, Use of mulches and soil stabilizers for land reclamation in the Eastern United States: *in* Schaller, F. W., and Sutton, P. (eds.), Reclamation of Drastically Disturbed Lands: Madison, Wisconsin, American Society of Agronomy, Crop Science Society of America, Soil Science Society of America, pp. 329–337.

Power, J. F., Sandoval, F. M., Ries, R. E., and Merrill, S. D., 1981, Effects of topsoil and subsoil thickness on soil water content and crop production on a disturbed soil: Soil Science Society of America Journal, v. 45, no. 1, pp. 124–129.

Renner, F. G., 1936, Conditions influencing erosion on the Boise River watershed: U.S.D.A., Technical Bulletin 528, 32 pp.

Ritter, D. F., 1978, Process Geomorphology: Dubuque, Iowa, William C. Brown Co., 603 pp.

Ruhe, R. V., 1975, Geomorphology: Geomorphic Processes and Surficial Geology: Boston, Houghton Mifflin Co., 246 pp.

Ruhe, R. V., and Walker, P. H., 1968, Hillslope models and soil formation: I, open systems: Transactions of Ninth International Congress of Soil Science, v. 4, pp. 551–560.

Savigear, R. A. G., 1956, Technique and terminology in the investigation of slope forms: Slopes Commission Report, v. 1, pp. 66–75.

Scheidegger, A. E., 1964, Lithologic variations in slope development theory: U.S. Geological Survey, Circular 485, 8 pp.

Schuman, G. E., and Power, J. F., 1981, Topsoil management on mined lands: Journal of Soil and Water Conservation, v. 36, pp. 77–78.

Schumm, S. A., 1956, The role of creep and rainwash on the retreat of badland slopes: American Journal of Science, v. 254, pp. 693–706.

Schumm, S. A., 1964, Seasonal variation of erosion rates and processes on hillslopes in western Colorado: Zeitschrift für Geomorphologie, v. 5, pp. 215–238.

Schumm, S. A., 1966, The development and evolution of hillslopes: Journal of Geological Education, v. 14, no. 3, pp. 98–104.

Schumm, S. A., and Hadley, R. F., 1961, Progress in the application of landform analysis in studies of semiarid erosion: U.S. Geological Survey Circular 437, 14 pp.

Selby, M. J., 1982, Hillslope Materials and Processes: Oxford, Oxford University Press, 264 pp.

Smith, D. D., and Wischmeier, W. H., 1957, Factors affecting sheet and rill erosion: Transactions, American Geophysical Union, v. 38, pp. 889–896.

Speight, J. G., 1971, Log-normality of slope distributions: Zeitschrift für Geomorphologie, v. 15, pp. 290–311.

Strahler, A. N., 1950, Equilibrium theory of erosional slopes approached by frequency distribution analysis: American Journal of Science, v. 248, pp. 673–696.

Strahler, A. N., 1968, Slope Analysis: *in* Fairbridge, R. W. (ed.), The Encyclopedia of Geomorphology: Encyclopedia of Earth Sciences Series, Vol. III, New York, Reinhold Book Corp., pp. 998–1002.

Thornes, J., 1979, Introduction: *in* Embleton, C., and Thornes, J. (eds.), Process in Geomorphology: New York, Halsted Press (John Wiley and Sons), pp. 1–10.

Thornes, J., and Brunsden, D., 1977, Geomorphology and time: London, Methuen and Co. Ltd., 208 pp.

Thornthwaite, C. W., 1931, The climates of North America according to a new classification: Geographical Review, v. 21, pp. 633–655.

Toy, T. J., 1977, Hillslope form and climate: Bulletin of the Geological Society of America, v. 88, pp. 16–22.

Toy, T. J., 1982, Accelerated erosion: process, problems, and prognosis: Geology, v. 10, pp. 524–529.

Tricart, J., and Cailleux, A., 1972, Introduction to Climatic Geomorphology: London, Longman Group, Ltd., 295 pp.

Troeh, F. R., 1965, Landform equations fitted to contour maps: American Journal of Science, v. 263, pp. 616–627.

Tylor, A., 1875, Action of denuding agencies, Geology Magazine, v. 22, pp. 433–73.

U.S. Environmental Protection Agency, 1976, Erosion and sediment control: surface mining in the Eastern United States, planning, technology transfer, EPA–625/3–76–006, 91 pp.

Verma, T. R., and Thames, J. L., 1978, Grading and shaping for erosion control and vegetative establishment in dry regions: *in* Schaller, F. W., and Sutton, P. (eds.), Reclamation of Drastically Disturbed Lands: Madison, Wisconsin, American Society of Agronomy, Crop Science Society of America, Soil Science Society of America, pp. 399–409.

Vogel, W. G., 1981, A guide for revegetating coal minespoils in the Eastern United States, U.S.D.A., Forest Service, General Technical Report NE–68, 190 pp.

Von Bauer, K. E., 1860, Über ein allgemeines Gesetzin der Gestaltung der Flussbetten: Bulletin of the St. Petersburg Imperial Academy of Science, vol. 2, pp. 1–49, 218–250, and 353–382.

Von Bertalanffy, L., 1968, General systems theory: foundations, development, applications: New York: George Braziller Co., 289 pp.

White, J. F., 1966, Convex-concave landslopes: a geometrical study: Ohio Journal of Science, v. 66, no.6, pp. 592–608.

Wilson, L., 1968, Slopes: *in* Fairbridge, R. W. (ed.), The Encyclopedia of Geomorphology: Encyclopedia of Earth Sciences Series, Vol. III, New York, Reinhold Book Corp., pp. 1002–1020.

Wischmeier, W. H., and Smith, D. D., 1965, Predicting rainfall-erosion losses from cropland east of the Rocky Mountains: U.S.D.A., Agricultural Handbook 282, 48 pp.

Wischmeier, W. H., and Smith, D. D., 1978, Predicting rainfall erosion losses: U.S.D.A., Agricultural Handbook 537, pp. 1–58.

Wright, D. L., Perry, H. D., and Blaser, R. E., 1978, Persistent low maintenance vegetation for erosion control and aesthetics in highway corridors: *in* Schaller, F. W., and Sutton, P. (eds.), Reclamation of Drastically Disturbed Lands, Madison, Wisconsin, American Society of Agronomy, Crop Science Society of America, Soil Science Society of America, pp. 553–583.

Yair, A., 1973, Theoretical considerations on the evolution of convex hillslopes: Zeitschrift für Geomorphologie, v. 18, pp. 1–9.

Yair, A., and Klein, M., 1973, The influence of surface properties on flow and erosion processes on debris-covered slopes in an arid area: Catena, v. 1, pp. 1–18.

Young, A., 1972, Slopes: London, Longman Group Ltd., 288 pp.

Young, R. A., and Mutchler, C. K., 1969, Effect of slope shape on erosion and runoff: Transactions, American Society of Agricultural Engineers, v. 12, pp. 231–233, 239.

Zingg, A. W., 1940, Degree and length of land slope as it affects soil loss in runoff: Agricultural Engineering, v. 21, no. 2, pp. 59–64.

4

Streams and Stream Channels: Form and Process

Introduction

Stream channels transport the water, particulate sediment, and dissolved solids generated by weathering and erosion on the hillslopes of the tributary drainage basin. Channels are dynamic and react to changes in the quantity of water and sediment delivered to them.

The processes of open channel hydraulics and sediment transport have been discussed in Chapter 2. The question of how a channel is formed by the water and sediment flowing in it, or channel morphology, is the subject of this chapter.

Types of Channels

There are three general types of stream channels based on the flow that they carry: perennial, ephemeral, and intermittent. Perennial streams are characteristic of humid regions and carry some flow at all times. Ephemeral streams flow only in response to summer thunderstorms or spring snowmelt and are dry most of the year. Intermittent streams flow seasonally and generally in separated reaches where ground water intersects the stream bed.

Channel Form and Process: Undisturbed System

Hydraulic Geometry

At any stream cross section, the width and depth of the flow in the channel and the velocity change with the amount of water flowing in the cross section. The relationship between discharge and these variables has been termed the *hydraulic geometry*. This differs from channel geometry, which is the bankfull width and mean depth of a channel cross section. Because discharge varies widely at any given cross section throughout the year, it is important to understand the concept of flow frequency.

The cumulative frequency of a given daily discharge is expressed as the percentage of time that the discharge rate is equalled or exceeded. Discharge rates for small streams generally will be less than for large rivers. Therefore, the small stream flowing only 5 cfs and the large river flowing 500 cfs may both be experiencing flow of the same frequency. Discharge varies not only at a given cross section on a stream but also in a downstream direction. This concept is illustrated in Fig. 1. Comparison of stations along a stream are made with the assumption that all are experiencing equal discharge frequency (Leopold and Maddock, 1953), while at a given cross section, different discharges have different frequencies.

Leopold and Maddock (1953) developed the hydraulic geometry relationships for several stream channels in the United States using streamflow records from gauging stations operated by the U.S. Geological Survey, relating discharge (Q)

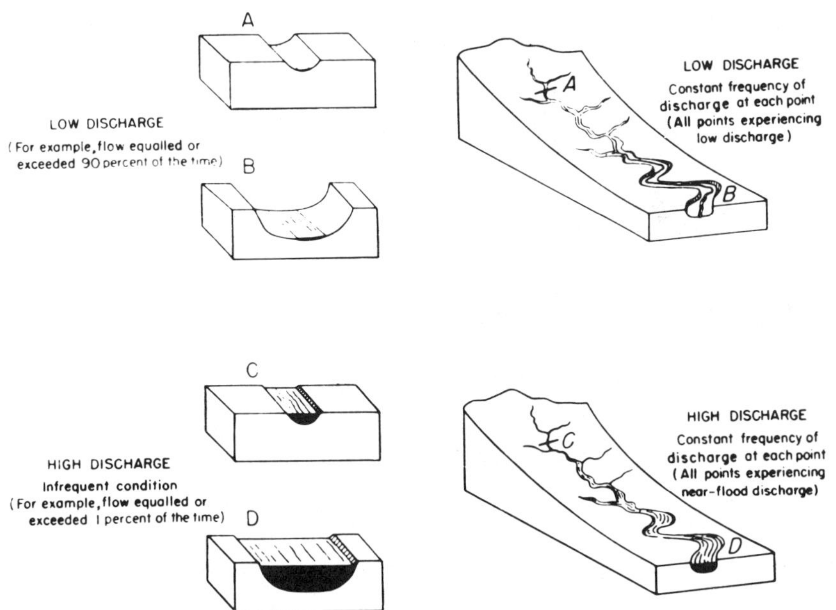

Figure 1. Comparison of different rates of discharge at a given river cross section and at points downstream (Leopold and Maddock, 1953).

variations to channel dimensions of width (w) and depth (d), and flow velocity ($\bar{v}$) at a station. Their equations are

$$w = aQ^b \tag{1}$$

$$d = cQ^f \tag{2}$$

$$\bar{v} = kQ^m \tag{3}$$

where Q is the discharge and w, d, and $\bar{v}$ represent the water surface width, mean depth, and mean velocity, respectively. The characters b, f, m, a, c, and k represent numerical constants, since $wd\bar{v} = Q$, $b + f + m = 1$, and $ack = 1$. The relationships of width, depth, and velocity to discharge are in the form of simple power functions as defined by Leopold and Maddock (1953) (Fig. 2).

Leopold and Maddock made the same kind of hydraulic geometry analysis for several rivers using data from gauging stations in a downstream direction. The

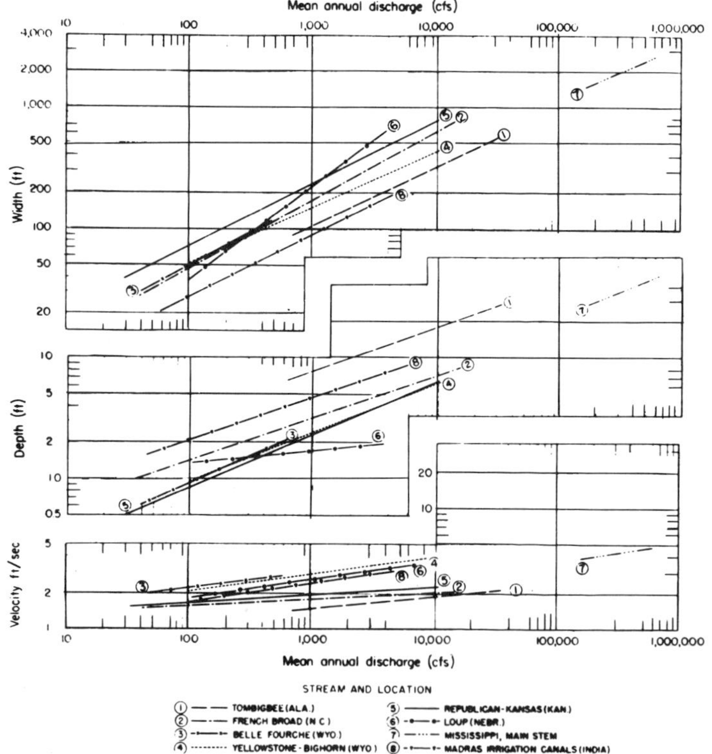

Figure 2. Width, depth, and velocity in relation to mean annual discharge as discharge increases downstream in various river systems (Leopold and Maddock, 1953).

channel dimensions and mean velocity of flow were plotted against mean annual discharge. They found that width, depth, and mean velocity increase with mean annual discharge, although the lines fitted to the data show considerable scatter. Also, average values for the river basin exponents are $b = 0.5, f = 0.4$, and $m = 0.1$. Width increases faster than depth or velocity as discharge increases. It should be noted that for some streams in arid and semiarid areas, discharge decreases in a downstream direction because of flow losses in ephemeral stream channels.

Suspended Sediment and Discharge

The suspended load of most streams usually increases with increasing discharge, and it usually increases more rapidly than water discharge. The relationship between suspended load and water discharge (sediment-rating curve) is commonly shown as a plot on logarithmic coordinates (see Fig. 3). These plots usually exhibit considerable scatter because the suspended sediment load of a river is generally a noncapacity load (Walling, 1977). Rainfall, runoff, and debris load from a drainage basin can vary widely from storm to storm. From the work of Leopold and Maddock (1953), it was approximated that the relationship of suspended sediment to the discharge of the sediment–water mixture at a stream cross section for most of the range of discharge may be represented by the equation

$$L = pQ^j \tag{4}$$

where L is the suspended load in tons per day and p and j are numerical constants. Because the sediment load usually increases faster than discharge at a station, the slope of the curve (j) is greater than unity. According to Leopold and Maddock (1953), values of j lie in the range from 2.0 to 3.0, and most of the sediment is derived from processes of sheet, rill, and gully erosion from hillslopes in the drainage basin, as discussed in Chapter 2.

Studies of suspended sediment discharge in a downstream direction indicate that sediment yield decreases with increasing drainage area. That is, the unit sediment yield (in tonnes per square kilometers) decreases downstream because of the diversity of topography and the storage of eroded material at the base of hillslopes, on floodplains, and in channels. This stored sediment may be remobilized during subsequent storms.

Sediment and the Longitudinal Profile

The longitudinal profile of a stream channel is, in essence, the gradient or slope along its length. Most longitudinal profiles, especially in humid regions

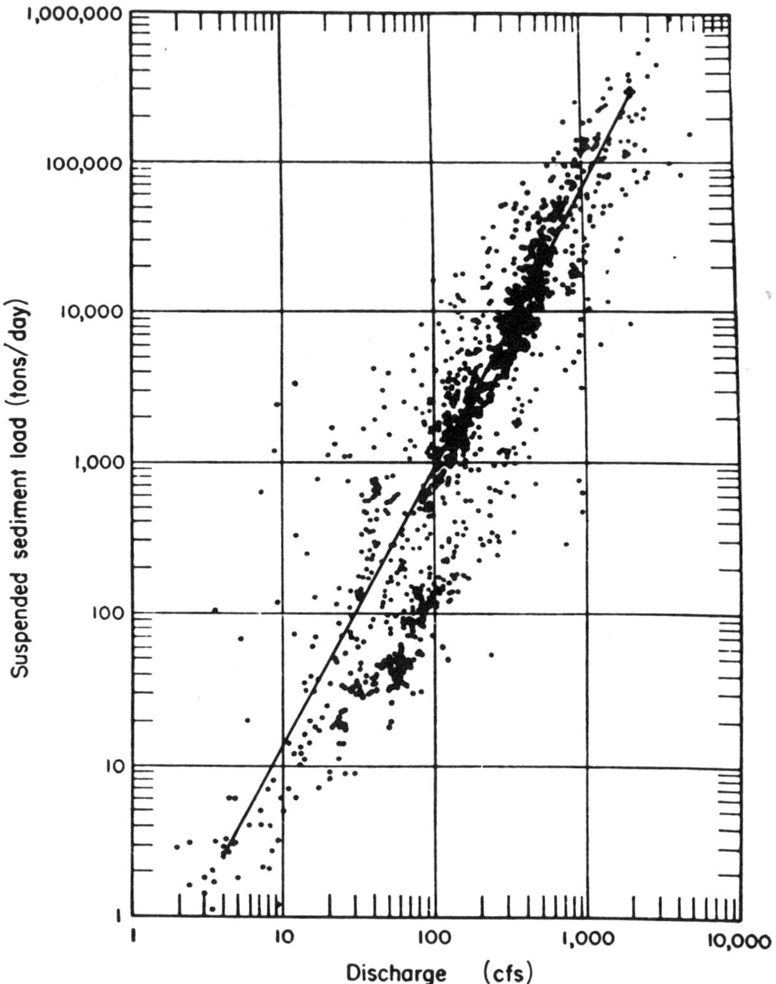

Figure 3. Relation of suspended sediment load to discharge, Powder River at Arvada, Wyoming (Leopold and Maddock, 1953).

and of perennial streams, have a shape that is concave upward, but not necessarily without local irregularities. The sediment load that is transported by the stream affects both the channel geometry and the longitudinal profile. Lane (1955) concluded that a channel can be kept in dynamic equilibrium if changes in sediment load and sediment size are balanced by offsetting changes in water discharge and stream gradient. Lane expressed these relationships with the following equation:

$$Q_s d_{50} = QS \tag{5}$$

where Q_s is the bed material load, d_{50} the median particle size, and S the stream channel gradient. Hack (1957), using drainage area as a surrogate for discharge, found that the slope of a stream at any point in the channel is approximately proportional to the 0.6 power of the ratio of the median size of bed material (M) and the drainage area upstream (A) from the sampling location. Hack expressed this relation by the equation:

$$S = 18 \left(\frac{M}{A} \right)^{0.6} \tag{6}$$

The slope of a stream channel, as indicated by Eq. (6), is influenced by both the flow characteristics and geology, or sediment size and rock type. Channels tend to adjust their slopes, or gradients, depending on the flows and sediment loads delivered to them for transport. This process involves aggradation, degradation,and changes in channel patterns.

Channel Patterns

In addition to changes that occur in stream channel cross section and slope, channels also have characteristic patterns in a downstream direction when observed in plain view. These patterns have been divided into three groups: braided, meandering, and straight. The most easily recognized are braided and meandering; long straight reaches are not common in natural streams. All of these patterns are a reflection of the water discharge, sediment load that the channel is transporting, channel slope, and cross section.

As mentioned above, straight channels over long reaches of a valley are not common. The thalweg, or deepest part of the channel, generally will wander from bank to bank even where the channel is relatively straight as the channel migrates laterally (Leopold *et al.*, 1964). Depositional bars accumulate opposite points of maximum depth. Flow and depositional patterns in straight channels are similar to meandering channels.

Braided streams are divided into separate channels, separated by bars and islands in low flow. In flood flows, these bars and islands are often submerged. Braided reaches of a stream are generally steeper, shallower, and wider than adjacent unbraided reaches (Ritter, 1978). The term *braiding* has been used interchangeably with *anastomosing* to describe multiple channel patterns. Anastomosing channels simply are those channels that divide and rejoin, but as Schumm (1977) pointed out, anastomosing channels are distinct single channels that are stable in contrast to the braided stream.

The meandering channel pattern exhibited by streams is the most common. Because nearly all streams tend to develop curves or loops, it has been suggested

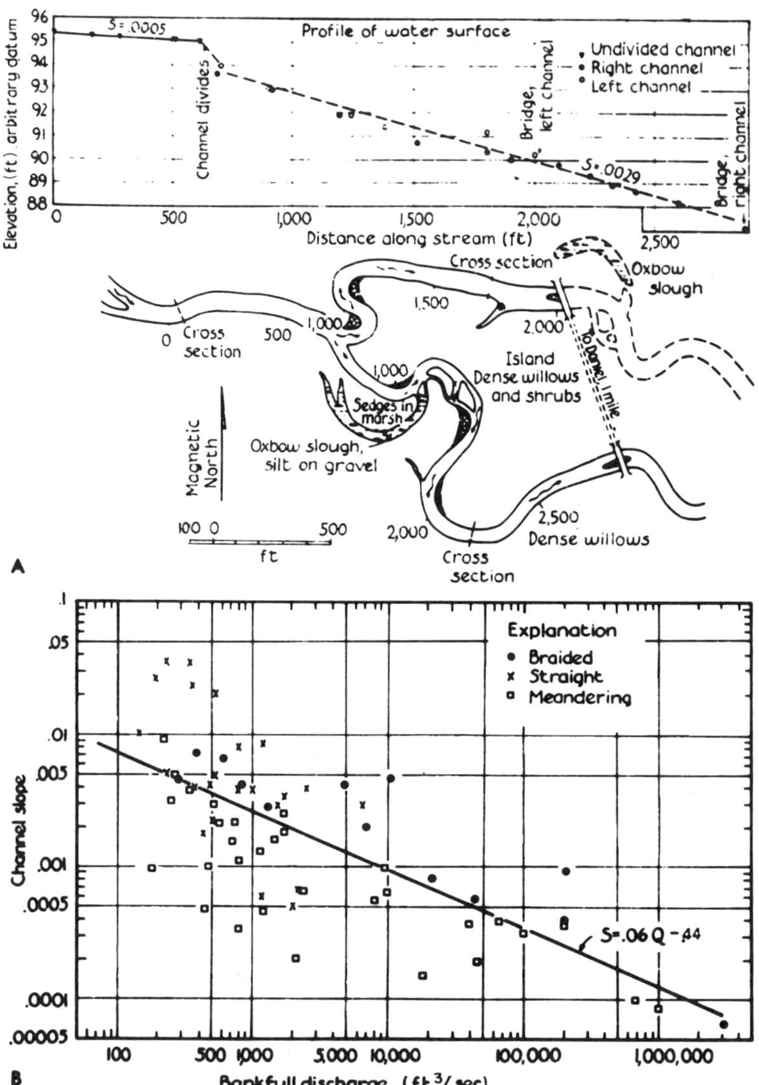

Figure 4. Relation of discharge to slope in braided and nonbraided rivers. A. Plan view and profile of a channel that divides around an island, showing that individual channels are steeper than the single one. B. Relationship of discharge to slope. Data from meandering and braided channels are separated by the line (Leopold and Wolman, 1957).

that meandering be used only for channels that display curves of considerable symmetry (Leopold *et al.*, 1964). The term that describes the degree of wandering of a channel is *sinuosity*, which may be defined as the ratio of channel length to valley length. The arbitrary division point between channels that are meandering and those that are braided or straight is a sinuosity of 1.5 (Leopold *et al.*, 1964). The relationship of channel slope to discharge for meandering, braided, and straight channels for the Green River near Daniel, Wyoming, as determined by Leopold *et al.*, is shown in Fig. 4.

Table 1 Classification of Alluvial Channels[a]

Mode of sediment transport and type of channel	Channel sediment (*M*) (%)	Bedload (% of total load)	Channel stability		
			Stable (graded stream)	Depositing (excess load)	Eroding (deficiency of load)
Suspended load	>20	<3	Stable suspended load channel. Width/depth ratio <10; sinuosity usually >2.0; gradient, relatively gentle.	Depositing suspended load channel. Major deposition on banks cause narrowing of channel; initial streambed deposition minor.	Eroding suspended load channel. Streambed erosion predominant: initial channel widening minor.
Mixed load	5–20	3–11	Stable mixed load channel. Width/depth ratio >10, <40; sinuosity usually <2.0, >1.3; gradient, moderate.	Depositing mixed load channel. Initial major deposition on banks followed by streambed deposition.	Eroding mixed load channel. Initial streambed erosion followed by channel widening.
Bed load	<5	>11	Stable bed load channel. Width/depth ratio >40; sinuosity usually <1.3; gradient, relatively steep.	Depositing bed load channel. Streambed deposition and island formation.	Eroding bed load channel. Little streambed erosion; channel widening predominant.

[a]Source: Schumm (1977).

Schumm (1977) developed a classification of channels that focuses on the factors that determine stream morphology. He separated all channels into two groups: bedrock controlled and alluvial channels. The basis for Schumm's classification is the freedom of the channel to adjust its shape and gradient. Bedrock channels are essentially dependent on the rock types that make up their bed and banks, while alluvial channels are able to adjust their shape, pattern, and gradient.

Alluvial river channels have been classified by Schumm into three classes according to the type of sediment load they transport: bed load, mixed load, and suspended load (Table 1). Within this classification, a continuum of channel patterns exists that can be recognized as braided, meandering, and straight. The three classes of stream channels, as defined by Schumm (1977), are further expanded to include two major types of stream systems (Fig. 5). These types are single channel and multiple channel streams. The classification of alluvial channels (Schumm, 1977) is based on channel stability and the predominant mode of sediment transport. Discharge is not used as a basis for classification, and M is the percentage of silt and clay in the channel perimeter. The classification is

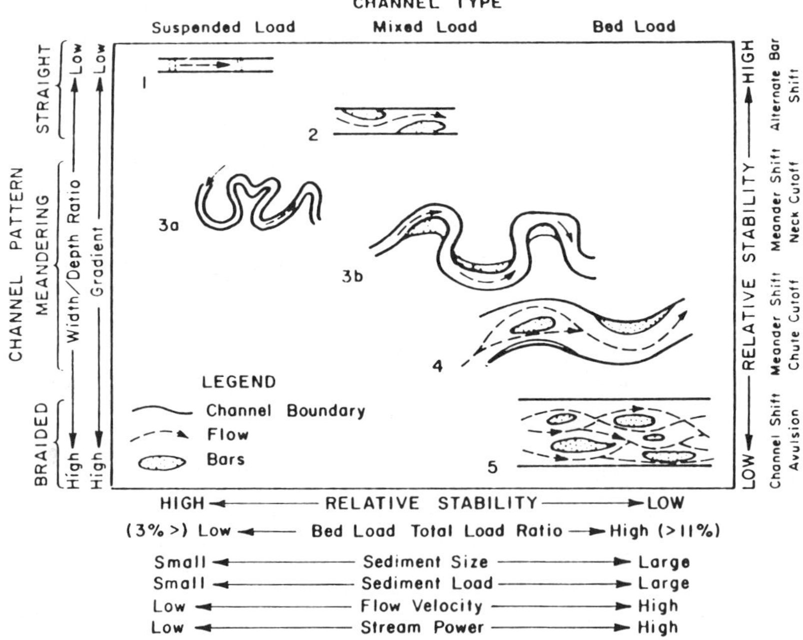

Figure 5. Channel classification showing relative stability and types of hazards encountered with each pattern (Nelson *et al.*, 1983).

applicable to channel segments rather than river basins, but adds a significant dimension not found in descriptions of channel patterns based primarily on channel geometry.

Disturbed Stream Channels

Entrenched Channels

The disturbance of stream channels in equilibrium with their environment will introduce external changes that cannot be predicted or explained by traditional approaches. This has led to the concept of geomorphic thresholds which seeks to explain abrupt changes in variables such as slope or velocity. These changes in morphology may be caused by changes in baselevel, land use, or climate. The first two are common changes that occur with human activity (Schumm, 1979). With a change in baselevel or land use, stream channels may become entrenched as they adjust themselves and establish a new equilibrium slope. The entrenched channels may vary in size from small rills on hillslopes or road cuts to valley floor gullies and entrenched streams, both ephemeral and perennial, according to the classification of Schumm *et al.* (1984).

The causes of channel entrenchment have been summarized by Schumm *et al.* (1984), and all of these causes can be related to human activity although they can also occur under natural conditions. The first cause they discuss is decreased erosional resistance, which occurs with decreased vegetation cover or decreased permeability and cohesion. The second cause is increased erosional forces resulting from constriction or concentration of flow by roads, ditches, or dikes. Steepening of gradients also increases energy slopes of flowing which can result from undercutting hillslopes, channelization of streams by cutting off meanders, and deposition of sediment.

Channel entrenchment will have effects both on the disturbed site as well as off-site, both upstream and downstream. On-site effects are incision as the channel adjusts to the new base level (Schumm *et al.*, 1984). The channel entrenchment will move both upstream and downstream. The sediment that is produced by the rejuvenation of the channel systems will cause downstream aggradation. Therefore, erosion and deposition are taking place at different locations in the system at the same time (Fig. 6).

Channel Adjustments

Stream channels respond to changes in discharge and sediment load by adjusting their morphology. These changes may be caused by either natural phenome-

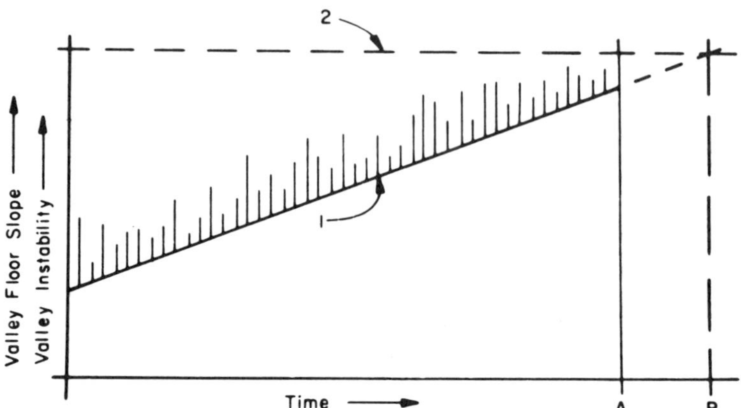

Figure 6. Changing valley floor stability through time. Line 1 shows increasing instability of valley floor as it approaches failure threshold (line 2) (Schumm, 1977. Copyright ©, reprinted by permission of John Wiley & Sons, Inc.)

na or related to human activity. Some of the natural changes are climate, tectonic activity, and baselevel fluctuations. A multitude of human activities, some of which we consider in this book, include land use that modifies vegetation cover, increases sediment loads, and alters topography, as well as river regulation by dams that trap sediment load and reduce flood peaks. All of these factors will cause a channel response to the imposed changes in external variables of discharge and sediment load.

Schumm (1977) developed a series of qualitative equations to describe the stream adjustments that may occur when discharge and sediment load are altered. The basis for Schumm's framework of equations is summarized here.

The effects of changing discharge and sediment load on channel morphology are expressed in the following equations. A plus or minus sign indicates how an increase or decrease in discharge (Q) or sediment load (Q_s) affects the morphologic variables:

$$Q+ \simeq b+,\ d+,\ \lambda+,\ S- \tag{7}$$

$$Q- \simeq b-,\ d-,\ \lambda-,\ S+ \tag{8}$$

$$Q_s^+ \simeq b+,\ d-,\ \lambda+,\ S+,\ P- \tag{9}$$

$$Q_s^- \simeq b-,\ d+,\ \lambda-,\ S-,\ P+ \tag{10}$$

where Q is the discharge, Q_s the sediment load, b the channel width, d the channel depth, λ the meander wavelength, S the channel slope, and P the sinuosity (Schumm, 1977). These equations illustrate how channel adjustment occurs with changes in discharge and sediment load.

Engineers and hydrologists have generally considered the relationships of water yield to changes in land use, and geomorphologists have examined changes in sediment yield with changing land use (Park, 1981). Park cited examples of studies that have been conducted to measure the effects of various land use practices on sediment yield directly and the historical approach where sediment deposition in reservoirs is used in basins experiencing changes in land use over long periods of time. Experiments in drainage basins also are often used where changes in land use are imposed under carefully controlled conditions. All of these approaches yield valuable results on the nature, magnitude, and distribution of changes in the fluvial system that are influenced by land use changes.

References

Hack, J. T., 1957, Studies of longitudinal stream profiles in Virginia and Maryland: U. S. Geological Survey Professional Paper 294-B, pp. 45–97.

Lane, E. W., 1955, Design of stable channels: American Society of Civil Engineers Transactions, v. 120, pp. 1234–1279.

Leopold, L. B., and Maddock, T., Jr., 1953, The hydraulic geometry of stream chan nels and some physiographic implications: U. S. Geological Survey Professional Paper 252, 57 pp.

Leopold, L. B., and Wolman, M. G., 1957, River channel patterns: braided, meandering, and straight: Geological Survey Professional Paper 282-B, pp. 39–85.

Leopold, L. B., Wolman, M. G., and Miller, J. P., 1964, Fluvial processes in geomorphology: San Francisco, W. H. Freeman and Co., publishers, 522 pp.

Nelson, J. D., Volpe, R. L., Wardwell, R. E., Schumm, S. A., and Staub, W. P., 1983, Design considerations for long-term stabilization of uranium mill tailings impoundments: NUREG-CR-3397 (ORNL-5979), Nuclear Regulatory Commission, 163 pp.

Park, C., 1981, Man, river systems, and environmental impacts: Progress in Physical Geography, v. 5, no. 1, 31 pp.

Ritter, D. F., 1978, Process Geomorphology: Dubuque, Iowa, William C. Brown Co., 603 pp.

Schumm, S. A., 1977, The Fluvial System: New York, John Wiley and Sons.

Schumm, S. A., 1979, Geomorphic thresholds: the concept and its applications: Inst. British Geographers, Trans., v. 4, p. 485–515.

Schumm, S. A., Harvey, M. D., and Watson, C. C., 1984, Incised channels—morphology, dynamics, and control: Littleton, Colorado, Water Resources Publications, 200 pp.

Walling, D. E., 1977, Limitations of the rating curve technique for estimating suspended-sediment loads, with particular reference to British rivers: *in* Proceedings of the Paris Symposium, International Association of Hydrological Sciences Publication no. 122, pp. 34–48.

5

The Drainage Basin: Geomorphic and Hydrologic Response

Introduction

Using the systems approach as described in Chapter 2, the drainage basin can be described as a process–response system, and as such, it is an organized segment of the landscape with geometric features that are interrelated (Chorley and Kennedy, 1971). The drainage basin provides a convenient and functional unit for the description of landforms and the measurement of inputs and outputs of energy and water. The inputs take the form of thermal energy from the sun and kinetic energy from precipitation, while the outputs are essentially water, sediment, and dissolved solids (Strahler, 1964), all of which are in balance. This adjustment of rivers and their tributaries, which make up drainage basins of all sizes, was recognized by Playfair (1802) and was amplified and expanded by Horton (1945) and Strahler (1964). The drainage basin can be divided into its geometric components in order to describe the morphometry of the system. The components of the drainage basin may be divided into linear, areal, and relief elements. Also, the basin is the natural integrator of such variables as precipitation, runoff, erosion, and sediment discharge as they relate to input and output in an open hydrologic system.

Drainage Basin Morphometry

Linear Drainage Basin Characteristics

The stream network in a drainage basin was examined quantitatively by Horton (1945) in terms of stream orders, stream numbers, stream lengths, bifurcation ratio, and length of overland flow. The relationships among these variables were expressed by Horton (1945) in mathematical equations that follow the laws of geometric series. Horton termed these relationships the *laws of drainage composition*, and they form the basis for much of the work in quantitative geomorphology done since his classic paper was published. (Some common relationships are summarized in Table 1.)

Stream ordering as defined by Horton (1945) states that fingertip or unbranched tributaries are to be designated as first-order streams. Streams that receive first-order tributaries, but these only, are second-order streams; third-order streams receive second- or first- and second-order streams, and so on until

Table 1 Some Common Relationships among Drainage Basin Variables[a]

Drainage basin variables	Nature of relationship: Direct	Nature of relationship: Inverse
Stream order versus:		
Stream number		X
Stream length	X	
Basin area	X	
Stream gradient		X
Stream discharge	X	
Drainage density versus:		
Mean annual runoff	X	
Mean annual flood ($Q_{2.33}$)	X	
Baseflow		X
P-E index (Thornthwaite)		X
Percentage of base surface	X	
Stream frequency	X	
Length of overland flow		X
Drainage area		
Stream length	X	
Stream discharge	X	
Sediment yield per unit area		X
Sediment delivery ratio		X
Total sediment yield	X	
Relief ratio versus:		
Sediment yield per unit area	X	
Bifurcation ratio		X
Other		
Infiltration capacity versus sediment yield		X
Precipitation variable versus sediment yield	complex (see Chap. 4)	

[a]Compiled from numerous sources; specific references can be found in Ritter (1978) and Chorley *et al.* (1984) as well as in this chapter.

the mouth of the basin is reached (Fig. 1A). When the order of the main stream is determined, it is projected headward through the basin along the stream that appears to be the main channel, which involves some subjective judgment.

Strahler (1952) proposed a modification of Horton's ordering system which focused on stream segments instead of entire streams. All streams having no tributaries are designated first-order streams. Where two first-order streams join, the resulting stream segment is a second-order stream; two second-order streams join to produce a third-order stream and so on downstream. In Strahler's ordering system, a stream segment may be joined by a channel of lower order without increasing the order. For example, a third-order stream may be joined by any number of first-order tributaries (Fig. 1B). The fact that all fingertip tributaries

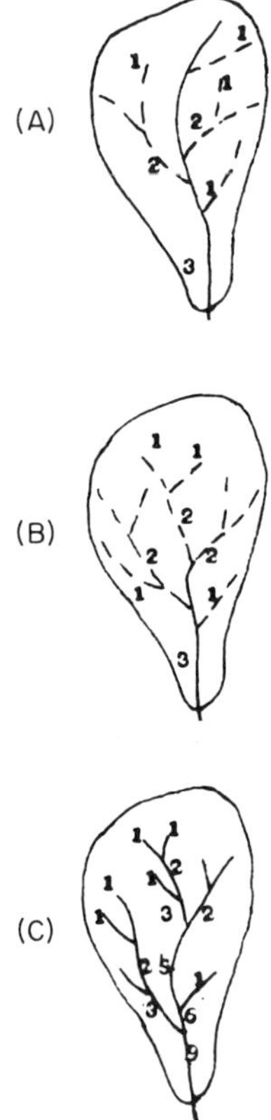

Figure 1. Three methods of determining stream orders in a drainage basin: A. Horton (1945); B. Strahler (1952); C. Shreve (1966).

(order 1) apparently are not accounted for in Strahler's method by an increase in order at every junction has been compensated for by the stream network analysis method of Shreve (1966) shown in Fig. 1C. Shreve considered streams as links in the drainage network, dividing the network at each junction, with the magnitude of each link reflecting the sum of the first-order tributaries upstream. More recently, Abrahams (1984) reviewed the development of methods of studying channel networks since 1966 when Shreve introduced his random topology model. The reader is encouraged to refer to the Abrahams paper for more detail. The method proposed by Strahler (1952), however, is most often used in morphometric analysis.

The stream-ordering method permits a quantitative expression of morphometry when comparing drainage basins. It should be noted, however, that investigators must indicate the method used to determine stream order and the scale of maps or aerial photographs used. Determination of stream orders in a basin can produce substantially different results depending on map scale. Most topographic maps do not show the detail, especially of lower order streams that can be seen on aerial photographs, and neither maps nor photographs will be as accurate as plane table mapping in the field.

The law of stream numbers expresses the relationship between the number of stream segments of a given order and the stream order for a particular basin (Horton, 1945). The number of streams of different orders in a basin approximates an inverse geometric series; as order increases, the number of streams decreases geometrically, in which the first term is unity and the ratio is the *bifurcation ratio* (R_b). The bifurcation ratio is simply the ratio of the number of streams of any given order to the number of streams in the next lower order, given by the ratios N_1/N_2, N_2/N_3 . . . N_5/N_6.

Horton's law of stream lengths states that the average lengths of streams of each order in a basin approximate a direct geometric series in which the first term is the average length of first-order streams.

The laws of stream network composition proposed by Horton (1945) demonstrate the adjustment of streams and their drainage basins in a quantitative way. Since his classic paper was published, many hydrologists and geomorphologists have utilized morphometric analysis in field studies with a variety of objectives ranging from a better understanding of geomorphic and hydrologic relationships to specific applied environmental problems.

An example of the application of Horton's laws of stream network composition to a field study is shown in Fig. 2 (Brush, 1961). The relationship of stream order to stream numbers, stream lengths, and drainage area plot as straight lines on semilogarithmic paper, that is, as simple geometric relationships. The orderliness of change from one order to another is especially noteworthy and reflects the adjustment in the drainage network.

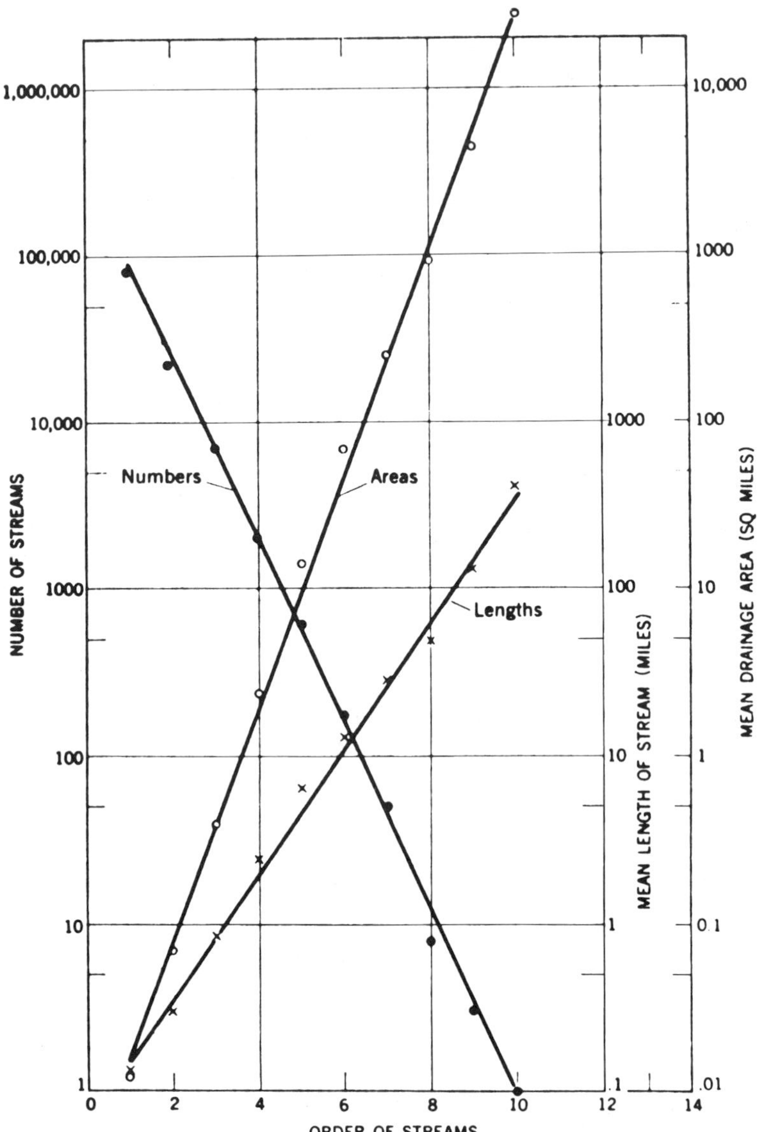

Figure 2. Relationship of number of streams, mean lengths of streams, and mean areas of drainage basins to stream order, for the Susquehanna River basin, using Horton's system of orders (Brush, 1961).

Areal Drainage Basin Characteristics

There are areal measures of drainage basin characteristics that may be related to both streamflow and sediment yield. These characteristics include (1) drainage area, (2) stream frequency, (3) basin shape, and (4) drainage density.

The drainage area (A), as measured on a horizontal plane, represents the area within the drainage divides and can be computed for a given order or for the total basin. The drainage area is important as the hydrological integrator for runoff and sediment yield.

Stream frequency (F) is defined as the number of stream segments of all orders per unit area; $F = \Sigma n_u / A$.

Basin shape, or geometry, has been expressed in several ways, but it was determined by Morisawa (1959) that only two, *circularity ratio* (R_c) and *elongation ratio* (R_e), showed a significant correlation with the rainfall/runoff ratio for 25 drainage basins on the Appalachian Plateau and, hence, appear to be of great geomorphic significance.

Miller (1953) defined basin circularity as the ratio of basin area (A_b) to the area of a circle having the same perimeter (A_c), thus the relationship for basin circularity (C) can be expressed as

$$C = \frac{A_b}{A_c}$$

Miller's rationale for this expression of basin shape is that a circle provides maximum area with a minimum perimeter, while an oblong basin will have a large perimeter but a smaller drainage area. Basin shape is related to the hydrograph characteristics of flood flows. The lag time, or time for concentration of flow from tributaries to the main channel, is less in ovoid or circular basins than it is in long, narrow basins.

Schumm (1956) expressed basin shape as the *elongation ratio* (R_e). It is the ratio of the diameter of a circle with the same area as the basin (A_c) and the maximum length of the basin (L_m) from the mouth to the headwater divide along the main channel. The elongation ratio can be expressed by the equation

$$R_e = \frac{A_c}{L_m}$$

Although shape does have some influence on the hydrologic response of a basin, such as streamflow characteristics and sediment yield, other characteristics considered here are probably more important (Strahler, 1964).

The drainage density (D_d), as defined by Horton (1932), is the total length of streams (ΣL) within a basin divided by the drainage area (A), or kilometers of channel per square kilometer (km/km^2). It may be expressed by the equation

$$D_d = \frac{\sum L}{A}$$

Drainage density is an expression of dissection of a basin by streams and is related to other characteristics such as rock and soil type, vegetation, climate, and infiltration. In the Cheyenne River basin of eastern Wyoming, drainage densities measured in small basins (<12 km^2) ranged from 3.3 km/km^2 on sandstone and sandy soils with high infiltration rates to 9.9 km/km^2 on shale and clay soils with low infiltration rates (Hadley and Schumm, 1961).

The greater the degree of dissection in a drainage basin, the shorter the length of hillslopes, which is the basis for Horton's length of overland flow. As defined by Horton (1945), the length of overland flow (l_o) is the distance that water flows over the ground surface before it becomes concentrated in a stream channel. The average length of overland flow (l_o) is generally half the distance between stream channels. Thus, it is approximately equal to the reciprocal of twice the drainage density and can be expressed by the equation

$$l_o \approx \frac{1}{2D_d}$$

Relief Characteristics of the Drainage Basin

In addition to linear and areal characteristics of a drainage basin, the relief or slope characteristics also are important. Gradient may be expressed as the slope of a stream channel in a short reach, or it can be averaged over the length of the channel profile. Another aspect of gradient is the angle of valley side slopes. Relief, or topographic differences in elevation between the low and high points in a drainage basin, is a simple measure of basin slope that has proven to be useful in studies of morphometry. It was expressed by Schumm (1956) as *relief ratio* (R_r) and described by the equation:

$$R_r = \frac{R}{L}$$

where R_r is the relief ratio, R the difference in elevation between the mouth of a basin and the headwater divide, and L the maximum length of the basin, measured in the same units as R along a line essentially parallel to the principal channel. Relief ratio is an expression of drainage basin relief in a two-dimensional form, elevation and distance. Relief ratio correlates better with hydrologic characteristics of a basin if relief is determined so that abnormally high, isolated points on the drainage divide are eliminated. Also, basins with more or less uniform slopes result in significant relationships between relief ratio and hydrologic variables, such as runoff and sediment yield.

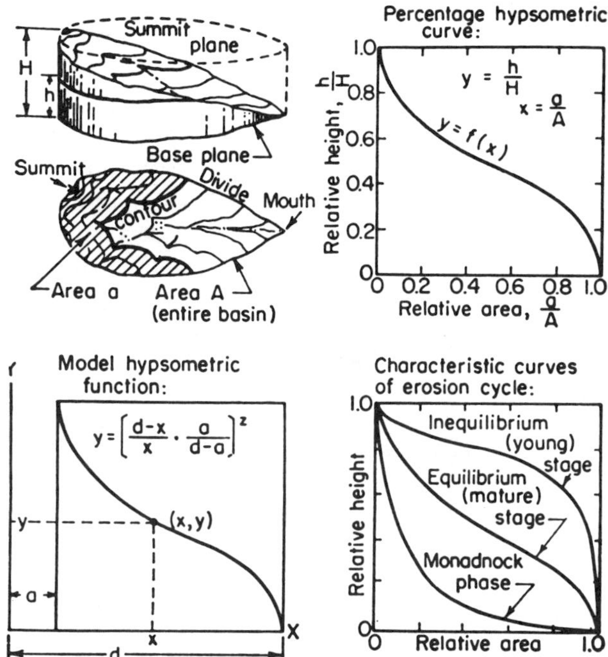

Figure 3. Hypsometric analysis of a drainage basin (Strahler, 1952).

Another method of studying relief is through the use of the hyposometric curve of Strahler (1957), which relates elevation and basin area, a three-dimensional view. As shown in Fig. 3, the drainage basin has vertical sides and is bounded by a basal horizontal plane passing through the basin mouth and a parallel summit plane passing through the divide. Hypsometric analysis determines how much of the drainage basin is located between cross sections or horizontal slices bounded by specific elevations or contour lines. The relative height (y) is the ratio of the height (h) of a specific contour above the horizontal base plane to the total relief (H). The relative area (x) equals the ratio a/A. In the ratio, a is the area of the basin above the specific contour, and A is the total basin area. The *hypsometric curve* shown in Fig. 3 represents the distribution of mass in the basin above the base plane datum. The *hypsometric integral* (f) expresses the volume of the basin, as a percentage, that is unconsumed by erosion. In natural basins underlain by essentially homogeneous material, values for the hypsometric integral are commonly in the range from 40 to 80%. Higher values indicate that much of the original basin remains uneroded. Hypsometric analysis of drainage basins is a valuable tool in estimating valley floor and floodplain storage and other variables related to the distribution of elevation in a basin.

Relationship of Basin Morphometry to Hydrologic Characteristics

As stated so aptly by Ritter (1978), the input to a drainage basin of water and energy helps to determine the morphometry, and the basin morphometry controls, to some degree, the hydrologic output by streams. Climate, vegetation, and hydrology are highly interdependent, which makes the separation of cause and effect difficult. For example, sediment yield is dependent on the gross erosion in the drainage basin, which is related to the vegetation cover, soil erodibility, precipitation, relief, and the transport efficiency of the channel network. Streamflow is also dependent on such hydrologic variables as precipitation and infiltration characteristics of soils and on the morphometric variables of basin size, basin shape, hillslope gradients, and relief.

In a study of 99 small drainage basins in the Cheyenne River basin of eastern Wyoming and adjacent areas in South Dakota and Nebraska, sediment yield and runoff were related to basin morphometry, vegetation cover, rock/soil characteristics, and infiltration characteristics (Hadley and Schumm, 1961). In this study, the measurements of drainage density and relief ratio were the geomorphic characteristics of small drainage basins that correlated best with runoff and sediment yield rates.

Drainage density, or texture of the topography, is controlled by the rock/soil type, vegetation, and climate. Basins that are underlain by shale generally have high densities (9.9 km/km^2), and basins underlain by sandstone have very low drainage densities (3.3 km/km^2).

In Fig. 4, the texture expressed as drainage density is plotted against mean annual runoff for 13 small basins. Basins underlain by Pierre Shale with drainage densities of about 9.9 km/km^2 yield a mean annual runoff of about 28,000 m^3/km^2. The basins underlain by sandstone of the Lance and Wasatch formations have average drainage densities of about 2.5 km/km^2 and yield mean annual runoff of about 1428 m^3/km^2. Further, the basins with the highest drainage densities have the lowest infiltration rates, and infiltration rates are inversely related to runoff.

Drainage density is also related to sediment yield from the small basins in the Cheyenne River basin. The drainage density is controlled primarily by lithology and infiltration rates. Table 2 shows the progressive increase in mean annual rates of sediment yield as lithology varies from sandstone to fine shale.

Schumm (1955) measured relief ratio in small basins in Arizona, New Mexico, and Utah where sediment yields had been determined (Hains *et al.*, 1952; King and Mace, 1953). Data from these studies, together with those from the Cheyenne River basin, were used to develop broad relationships between lithology and long-term sediment yield for the semiarid western United States (Fig. 5) (Hadley and Schumm, 1961).

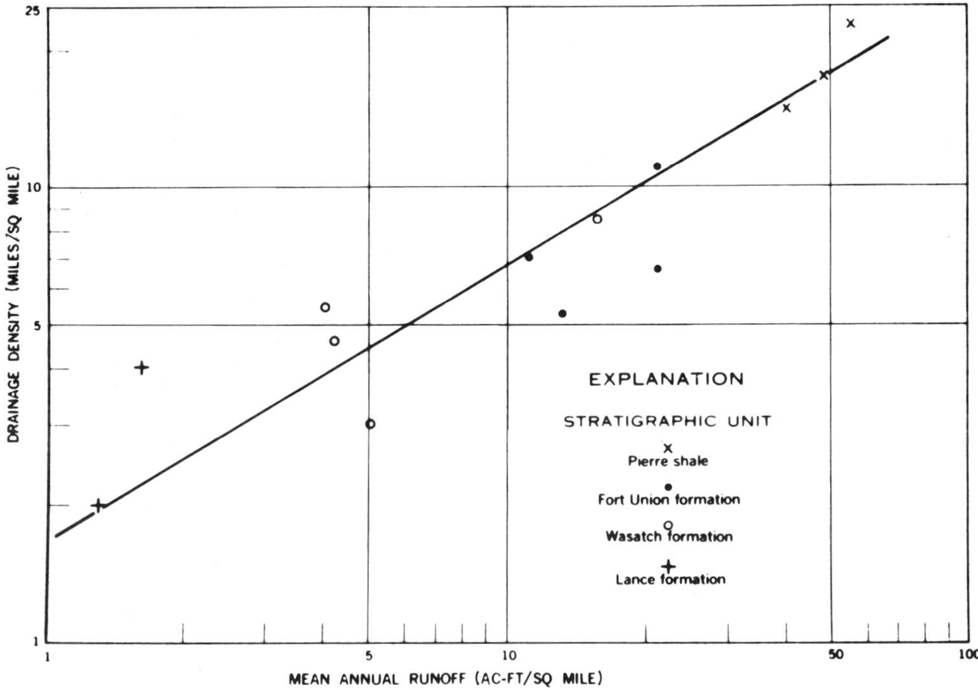

Figure 4. Relationship of mean annual runoff to drainage density for small basins, eastern Wyoming (Hadley and Schumm, 1961).

Table 2 Drainage Density and Mean Annual Sediment Yield Related to Lithology[a]

		Drainage density		Mean annual sediment yield	
Lithologic unit	Rock type	(mile/sq mile)	(km/km^2)	(ac-ft/sq mile)	(m^3/km^2)
Wasatch Formation	Sandstone	5.4	3.4	0.13	61.9
Lance Formation	Silty sandstone	7.1	4.4	0.5	238.1
Fort Union Formation	Sandy shale	11.4	7.0	1.3	619.0
Pierre Shale	Shale	16.1	10.0	1.4	666.7
White River Group	Shale	258.0	160.0	1.8	857.2

[a]Source: Hadley and Schumm (1961).

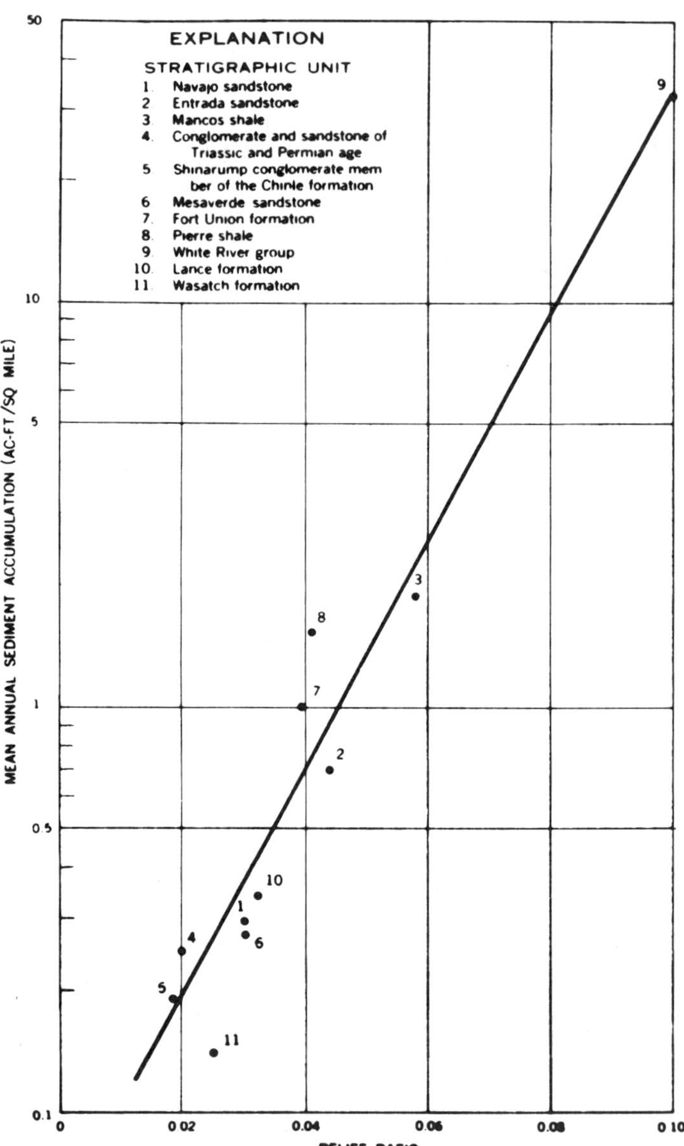

Figure 5. Relationship of mean annual sediment yield to relief ratio for small basins on various lithologies in the western United States (Hadley and Schumm, 1961).

In areas of similar climate, homogeneous lithology, and vegetation cover, the simple measurement of relief ratio may be useful in estimating sediment yield and might be applied to reclaimed land. Figure 6 illustrates the relationship between sediment yield and relief ratio for small basins (<12 km^2) on uniform lithology in eastern Wyoming. This removes the variability of lithology and improves the correlation. It should be noted, however, that the relief ratio is not a good measure of hydrologic response or sediment yield rates in basins that have an uneven distribution of land mass, for example, a basin with a hypsometric integral less than 20%.

The studies in the Cheyenne River basin (Hadley and Schumm, 1961) indicate

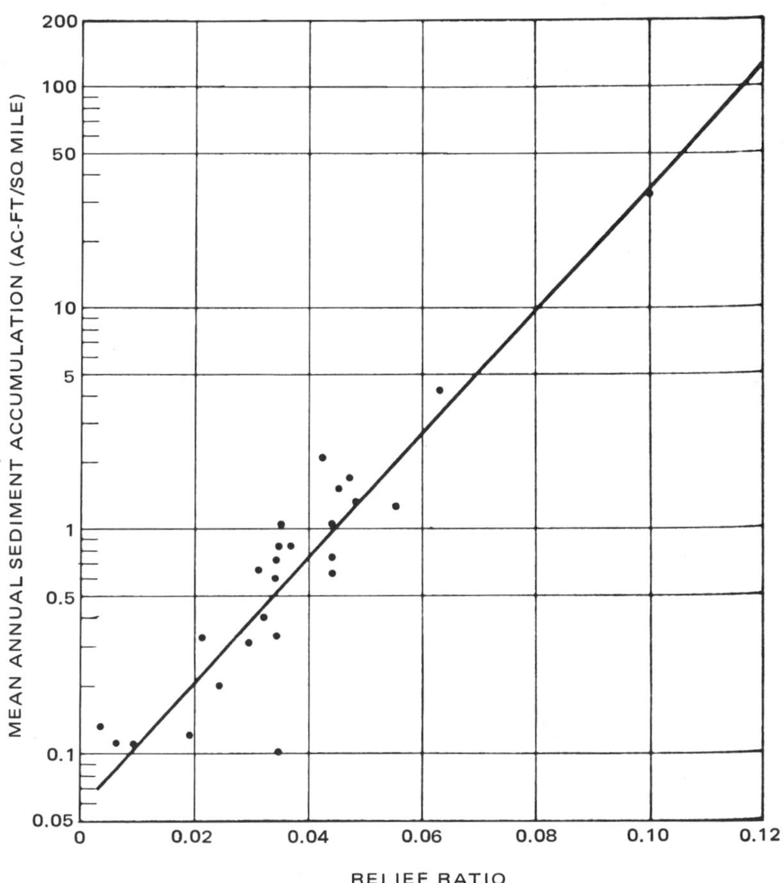

Figure 6. Relationship of mean annual sediment yield to drainage density for small basins on the same lithology, eastern Wyoming (Hadley and Schumm, 1961).

that relationships exist between basin morphometric properties and runoff and sediment yield. The relationship between lithology and drainage density and relief ratio has been associated with infiltration rates and lithology. Lithology, vegetation, soils, climate, and hydrology are variables that are so interrelated that it is difficult to isolate the influence of any one variable on drainage basin response.

In a study of the relationships among elements of climate, surface properties, and geomorphology, Melton (1957) used the Thornthwaite precipitation-effectiveness index, or P-E index, as a variable in correlations with several morphometric factors. The P-E index is a measure of the availability of moisture to vegetation and depends on the amount and distribution of both precipitation and evaporation. The infiltration capacity of an area is determined by the P-E index of the region, the surface cover, and the soil type. In equilibrium conditions, the drainage density will attain a value dependent on these variables (Melton, 1957). Figure 7 illustrates the relationship between P-E index and drainage density that Melton developed for areas in Western United States.

Sediment Yield and Climate

As discussed in previous sections, climate has significant control on drainage basin morphometry and hydrology. The Langbein–Schumm curve (1958), as

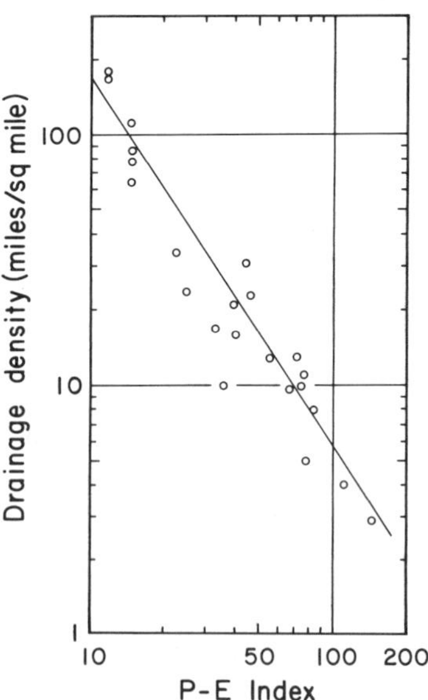

Figure 7. The control of drainage density exercised by Thornthwaite's precipitation-effectiveness (P-E) index (Melton, 1957).

demonstrated in Chapter 2, indicates that sediment yields from basins are generally higher in arid and semiarid regions because of the combined effects of low mean annual rainfall and insufficient vegetation cover to protect the land surface.

Since the work of Langbein and Schumm (1958), there have been many contributions to the literature regarding the relationship between sediment yield and climate. Several of these studies have developed relationships among mean annual sediment yields, mean annual precipitation, and mean annual runoff. An excellent comparative summary of the results of studies by Langbein and Schumm (1958), Dendy and Bolton (1976), Tabuteau (1960), Fournier (1960), Douglas (1967), and Wilson (1969) has been published by Walling and Kleo (1979). The graphical illustrations derived from the foregoing studies are illustrated in Fig. 8.

Langbein and Schumm (1958) correlated mean annual sediment yield and mean annual effective precipitation. Effective precipitation is estimated from mean annual runoff. Known values of annual runoff were converted graphically to effective precipitation at a temperature of 10°C. Effective precipitation is defined as that precipitation which produces the known amount of runoff. This limits the application of the curve in Figure 8Ai to semiarid and subhumid climates. Langbein and Schumm used sediment yield data from two sources: (1) small drainage basins with reservoirs and (2) sediment-gauging stations. The drainage basins averaged about 80 km^2 in area, and the sediment stations provided data from basins averaging about 3900 km^2. The curve in Figure 8Ai is based on data from the small basins with reservoirs. They found that the maximum sediment yield occurs at about an annual effective precipitation of 300 mm. There are some limitations to use of the Langbein–Schumm curve in estimating sediment yield rates from values of annual precipitation. The curve implies that sediment yield remains constant for values of effective precipitation greater than about 1000 mm. Data from other regions of the world suggest that this may not be valid.

Dendy and Bolton (1976) developed a relationship between mean annual sediment yield and mean annual runoff from data on reservoir sedimentation at more than 500 locations in the United States. As pointed out by Walling and Kleo (1979), the curve is drawn through group-averaged data. The Dendy–Bolton curve also indicates a continuous decrease in sediment yield from about 50 to 1200 mm mean annual runoff (Fig. 8Aii).

Walling and Kleo (1979) further reviewed other attempts to develop similar relationships based on data from several regions of the world, and they noted that there is a tendency for sediment yields to increase when mean annual precipitation exceeds 1000 mm and mean annual runoff exceeds 500 mm. The work of Fournier (1960) and Wilson (1969) are examples of attempts using worldwide data. Wilson's curve (Fig. 8Biv) shows one peak at about 750 mm and a second peak at 1750 mm of mean annual precipitation. Much of Wilson's data that contributed to the first peak was obtained from small basins developed on loess

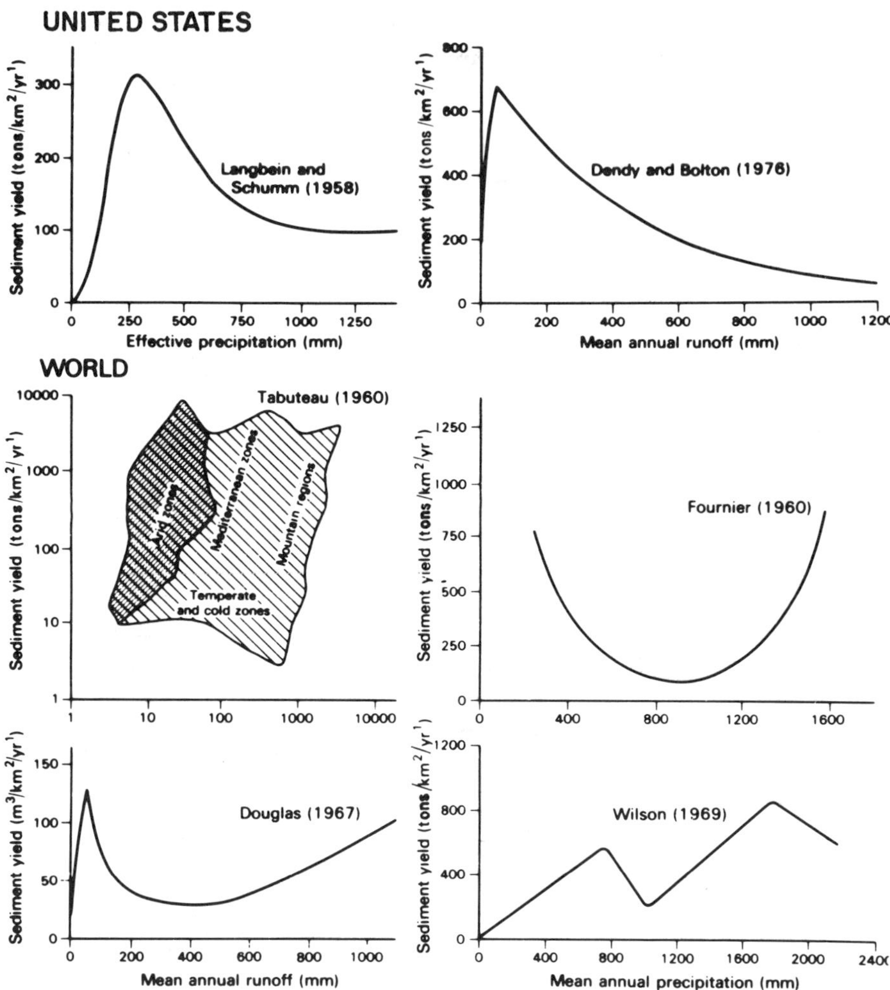

Figure 8. Comparative studies of runoff and precipitation versus sediment yield (Walling and Kleo, 1979).

that was subjected to intense cultivation. The second peak represents data from tropical and Mediterranean climates. For some of the data from tropical climates in basins over 25,000 km^2, the peak at 1750 mm of precipitation is absent. These results probably indicate that land use, soil properties, and drainage basin size will affect the magnitude of the sediment yield peak. Fournier (1960) found a second rise in sediment yield on the limb of his parabolic curve between 1200 mm and 1600 mm (Fig. 8Bii). These data represent monsoonal climates in India and China. Douglas (1967) developed a relationship using sediment yield and runoff for Asian river basins that is similar in shape to the Langbein–Schumm curve for lower values of mean annual runoff and somewhat similar to Fournier for higher values.

Walling and Kleo (1979), using sediment yield data from a worldwide data base representing 1246 sediment stations, attempted a more definitive analysis of the relationship between sediment yield and mean annual precipitation and runoff. Individual values of sediment yield did not produce any meaningful relationship when plotted against mean annual precipitation. Walling and Kleo concluded that other factors such as relief, soil characteristics, seasonality of precipitation, and the magnitude of human activity are important as controls of the magnitude of sediment yield.

In an attempt to develop a clearer relationship between the variables, Walling and Kleo plotted group-averaged sediment yield data versus mean annual precipitation and runoff for basins less than 10,000 km^2 in area (Fig. 9). The resultant curve (Fig. 9A) is much different than those in Fig. 8, but it does have

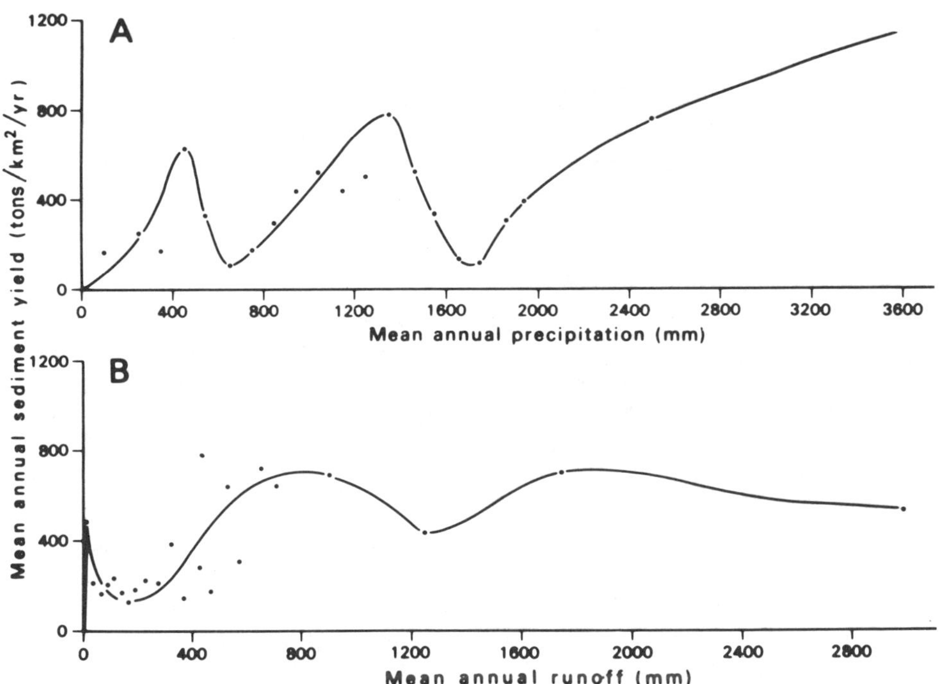

Figure 9. Sediment yield versus precipitation (A) and runoff (B) on a worldwide basis (Walling and Kleo, 1979).

some similarities, such as the first peak at about 450 mm which is comparable to the Langbein–Schumm peak and the double peaks of Wilson (1969), although they are at different precipitation values. The second curve (Fig. 9B) for the relationship between sediment yield and mean annual runoff is similar to the Douglas curve of Fig. 8Biii. These curves (Walling and Kleo, 1979) may not be any more definitive than the others discussed, but they do show that maximum sediment yields are not associated with arid and semiarid climatic regions.

The data analysis presented by Walling and Kleo (1979) suggested that rainfall or runoff energy is more significant in controlling the magnitude of sediment yield than the protection afforded by a denser vegetation cover in humid regions. They also pointed out that the true relationship between sediment yield and precipitation or runoff may be obscured by changes in natural vegetation cover associated with human activity.

Sediment Yield and Drainage Area

The relationships found to exist between topographic and climatic characteristics and sediment yield cannot be directly extrapolated from small drainage basins to larger drainage basins or vice versa. There is an inverse relationship between sediment yield per unit area and drainage area. The decrease in sediment yield per unit area with increasing drainage area may be due to one or more of the following: (1) absorption of storm flow in the channels of ephemeral streams resulting in the deposition of sediment load; (2) diversity of topography and landforms in larger drainage basins, including decline in hillslope and stream gradients in a downstream direction, thus providing sites for deposition of colluvium at the base of hillslopes and storage of alluvium on valley floors; and (3) the limited areal extent of convective thunderstorms that produce runoff on only a small part of a drainage basin. This relationship of mean annual sediment yield and drainage area is illustrated in Fig. 10 from data for 99 small basins in the Cheyenne River basin of eastern Wyoming (Hadley and Schumm, 1961), ranging in size from less than 0.13 km^2 to about 2.6 km^2. Sediment yield from the smallest basins (<0.13 km^2) is approximately equal to the gross erosion in the basins because of the steep hillslopes and lack of valley floor and floodplain development in headwater areas. In the larger basins, sediment storage increases.

These differences between on-site erosion and sediment yield at some measuring point downstream involve processes of both erosion and sediment transport and have been termed *sediment delivery*. The concept of a *sediment delivery ratio,* which is defined as the ratio of sediment delivered at the mouth of a drainage basin to gross erosion within the basin (both in metric tons per square kilometer per year), was described by Roehl (1962) and others in the U.S. Department of Agriculture. Sediment delivery ratio can be expressed by the simple equation

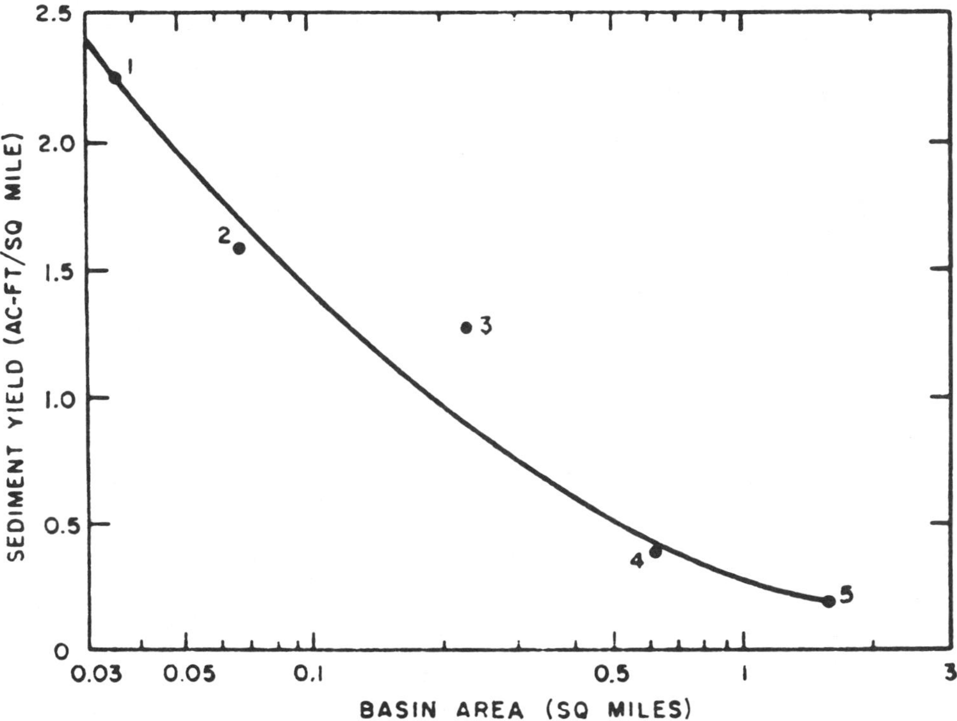

Figure 10. Relationship of sediment yield and drainage area for 99 small basins in eastern Wyoming (Hadley and Schumm, 1961).

$$D = y/T$$

where D is the sediment delivery ratio ($100D$ as a percentage), y the sediment yield at the downstream measuring location, and T the gross erosion, which includes gully, channel, and rill–interrill erosion upstream from the measuring location.

The graph in Fig. 11 illustrates the relationship of sediment delivery ratio versus drainage area for basins in central and southeastern United States (Roehl, 1962). As pointed out by Boyce (1975), if we accept the inverse relationship between sediment yield rates and drainage area, then the decrease in sediment yield rates with increase in drainage area cannot be explained as an upland hillslope phenomenon. The measurement of sediment yield is determined in the stream system, which includes channels, floodplains, lakes, and reservoirs; therefore, most of the deposited sediment must be on the floodplain, valley floor, or trapped in reservoirs. Boyce suggested that the differences between erosion and sediment yield (sediment delivery) cannot be fully explained by assuming continual floodplain aggradation, and he proposed that sediment delivery curves

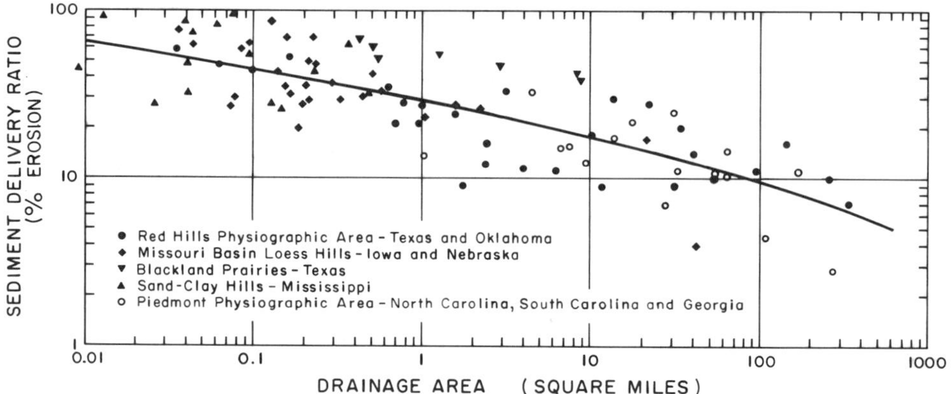

Figure 11. Relationship of sediment delivery ratio and drainage area for basins in central and southeastern United States (Roehl, 1962).

be based on average land slope. This approach recognizes the decrease in average land slope as drainage area increases.

There is a need for an improved understanding of the processes involved in the delivery of erosional debris through the drainage basin. The linkage between erosional processes on upland hillslopes and fluvial processes in floodplains, valley floors, and stream channels is poorly understood. The environmental problems associated with land use often include remobilization of sediment on disturbed valley floors that represents perturbations in the sediment delivery process.

References

Abrahams, A. D., 1984, Channel networks: A geomorphological perspective: Water Resources Research, vol. 20, no. 2, p. 161–188.

Boyce, R. C., 1975, Sediment routing with sediment-delivery ratios: *in* Present and Prospective Technology for Predicting Sediment Yields and Sources, Agricultural Research Service, U. S. Department of Agriculture, ARS-S–40, pp. 61–65.

Brush, L. M., 1961, Drainage basins, channels, and flow characteristics of selected streams in central Pennsylvania: U. S. Geological Survey Professional paper 282-F, pp. 145–181.

Chorley, R. J., and Kennedy, B. A., 1971, Physical Geography: A Systems Approach: London, Prentice-Hall, 370 pp.

Chorley, R. J., Schumm, S. A., and Sugden, D. E., 1984, Geomorphology, London, Metheun & Co., 605 pp.

Dendy, F. E., and Bolton, G. C., 1976, Sediment yield-runoff-drainage area relation ships in the United States: Journal of Soil and Water Conservation, v. 32, pp. 264–266.

Douglas, I., 1967, Man, vegetation, and the sediment yield of rivers: Nature, v. 215, pp. 925–928.

Fournier, F., 1960, Climat et erosion—la relation entre l'erosion du sol par l'eau et les precipitations atmospheriques: Presses Universitaires de France, Paris.

Hadley, R. F., and Schumm, S. A., 1961, Sediment sources and drainage basin charac teristics in upper Cheyenne River basin: U. S. Geological Survey Water-Supply Paper 1531-B, pp. 137–198.

Hains, C. F., Van Sickle, D. M., and Peterson, H. V., 1952, Sedimentation rates in small reservoirs in the Little Colorado River basin: U. S. Geological Survey Water-Supply Paper 1110-D, pp. 129–155.

Horton, R. E., 1932, Drainage basin characteristics: American Geophysical Union Trans., pp. 305–361.

Horton, R. E., 1945, Erosional development of streams and their drainage basins— hydrophysical approach to quantitative morphology: Bulletin, Geological Society of America, v. 56, pp. 275–370.

King, N. J., and Mace, M. M., 1953, Sedimentation in small reservoirs on the San Rafael Swell, Utah: U. S. Geological Survey Circular 256, 21 pp.

Langbein, W. B., and Schumm, S. A., 1958, Yield of sediment in relation to mean an nual precipitation: Trans. American Geophysical Union, v. 39, pp. 1076–1084.

Melton, M. A., 1957, Analysis of the relations among elements of climate, surface properties, and geomorphology: Project NR 389–042, Tech. Report 11, Columbia University, Department of Geology, ONR, Geography Branch, New York.

Miller, V. C., 1953, A quantitative geomorphic study of drainage basin characteris tics in the Clinch Mountain area, Virginia and Tennessee: Columbia University, Department of Geology, Tech. Report No. 3, Contract N6 ONR, 30 pp.

Morisawa, M. E., 1959, Relation of quantitative geomorphology to streamflow in representative watersheds of the Appalachian Plateau province, Columbia Univer sity, Department of Geology, ONR, Geography Branch, Project NR 389–042, Tech. Report 20, pp. 11–13.

Playfair, J., 1802, Illustrations of the Huttonian theory of the earth: Edinburgh, William Creech.

Ritter, D. F., 1978, Process Geomorphology: Dubuque, Iowa, Wm. C. Brown, publisher, 602 pp.

Roehl, J. W., 1962, Sediment source areas, delivery ratios, and influencing morpho logical factors: International Association of Hydrological Sciences, Pro ceedings of Bari Symposium, Publ. No. 59, pp. 202–213.

Schumm, S. A., 1955, The relation of drainage basin relief to sediment loss: Publ. no. 36, International Association of Hydrological Sciences, General Assembly of Rome, v. 1, pp. 216–219.

Schumm, S. A., 1956, Evolution of drainage systems and slopes in badlands at Perth Amboy, New Jersey, Bull. Geological Society of America, v. 67, pp. 597–646.

Shreve, R. L., 1966, Statistical law of stream numbers: Journal of Geology, v. 74, pp. 17–37.

Strahler, A. N., 1952, Hypsometric (area-altitude) analysis of erosional topography: Geological Society of America Bulletin, v. 63, pp.

Strahler, A. N., 1957, Quantitative analysis of watershed geomorphology: American Geophysical Union Transactions, v. 38, pp. 913–920.

Strahler, A. N., 1964, Quantitative geomorphology of drainage basins and channel networks: *in* Handbook of Applied Hydrology, V. T. Chow, editor, New York, McGraw-Hill, pp. 4–39 to 4–76.

Tabuteau, M. M., 1960, Etude graphique pour les consequences hydro-erosives du climat mediterrani: Assoc. Geogr. Francais Bulletin, v. 295, no. 5, pp. 130- 142.

Walling, D. E., and Kleo, A. H. A., 1979, Sediment yields of rivers in areas of low precipitation, International Association of Hydrological Sciences Publ. No. 128, pp. 479–493.

Wilson, Lee, 1969, Les relations entre les processes geomorphologiques et le climat modern comme methode de paleoclimatologie: Revue Geogr. Phys. Geol. Dynamique, ser. 2, 11, pp. 303–314.

6

Lands Disturbed by Grazing

Introduction

Lands that are primarily used for grazing have been classified worldwide as rangelands. Rangelands comprise about 40% of the earth's land surface and are found in many different climatic zones; however, 80% of the rangelands are located in arid and semiarid climatic regions (Branson *et al.*, 1981). The problems associated with arid and semiarid grazing lands, such as deterioration of vegetation cover by overgrazing, poor areal distribution of grazing herds, and soil compaction by trampling are exacerbated by hydrologic characteristics of low rainfall, high evapotranspiration potential, and low water yield. Because these problems are generally more severe in water-deficient regions, we will restrict our discussion to the grazing lands of the western United States.

Arid and semiarid lands of the intermontane valleys and basins in the western United States are generally vulnerable to erosion because of such factors as low precipitation, immature soils, steep hillslopes, and a sparse vegetation cover (Peterson and Hadley, 1960). Disturbance of the natural balance among these factors may alter the landscape stability resulting in excessive hillslope erosion and trenching of alluvial deposits on valley floors.

The grazing lands of the western United States that are considered here comprise the lands at lower altitudes interspersed among the mountain ranges. Precipitation on these lands is generally too low to permit the cultivation of crops without irrigation, but it is sufficient to produce a vegetation cover. Some of these lands produce runoff, but this is usually due to infrequent, high-intensity thunderstorms in the summer or prolonged periods of snowmelt in the spring. Most of the time they are water deficient, and all available moisture is used by the sparse vegetation cover or returned to the atmosphere. The only economic use for most rangeland is the pasturing of livestock (Calef, 1960).

Historical accounts indicate that the first settlers of the West encountered wide valleys generally with dense stands of grass and shrubs and occupied by shallow ephemeral or intermittent stream channels. The vegetation was sub-irrigated by shallow groundwater tables, or it was flood irrigated by overbank flows. Undoubtedly, there were occasional gullies and highly eroded uplands, but they were not common. As the population increased rapidly in the mid–1800s, the grazing pressures also increased, and livestock numbers doubled and redoubled. The increase in grazing of arid and semiarid rangelands in the western United States in the late 1800s was accompanied by a period of accelerated erosion that produced deep valley trenches (arroyos), lower water tables, and deterioration of the vegetation cover. The question that conservationists have raised is whether the widespread erosion was caused by overgrazing, climatic variations, or infrequent intense storms. Perhaps, a combination of all of these factors is responsible for the erosion observed.

There are many hydrologists, geomorphologists, and conservationists who attribute the cycle of erosion that began in the Southwest in about 1880 to land abuse by overgrazing (Antevs, 1952; Bailey, 1941; Bennett, 1939). The public lands were subjected to unrestricted grazing from the time of settlement in the nineteenth century until the Taylor Grazing Act was enacted in 1934. It is also probably true that large numbers of livestock on these lands led to general depletion of forage, trampling and compaction of soils, and trailing. All of these factors may have contributed to increased runoff and erosion, loss of protective plant cover, and reduced infiltration capacity. Whether or not overgrazing can be held responsible for widespread and contemporaneous erosion throughout the Southwest is still a question that must be answered (Branson, 1975).

Assessment of Impact

Analysis of suspended sediment discharge records at the Grand Canyon gauging station on the Colorado River in the heart of the arid and semiarid rangelands of the western United States lends support to the theory that land use may indeed affect erosion. Hadley (1974) examined the sediment records for two periods at the gauging station (1926–1941 and 1942–1960) and found that the suspended-sediment loads for the latter period were only 50% of the earlier period. These data combined with studies of livestock use on the Colorado Plateau rangelands (Peterson and Hadley, 1960) also showed the relationship between grazing and erosion. Livestock numbers have been greatly reduced since about 1940 in the Colorado River basin, and erosion control practices have been initiated in many critically eroding areas. The regulation of grazing by the Taylor Grazing Act of 1934, which is discussed in a later section, is largely responsible for the change in land use. Experimental studies by the U.S. Geological Survey in western Colorado indicate that grazing control alone can reduce runoff by as much as 30% and sediment yield by about 40% (Lusby *et al.*, 1971).

Some investigators felt that a shift in climate may have been responsible for the epicycle of erosion that began in the arid Southwest about 1880. Kirk Bryan (1925, 1941) was the foremost proponent of this theory. He felt that overgrazing was the "trigger" for the erosion, but the shift in climate was the primary cause. Bryan (1941) developed his argument by describing the evidence in many stream valley alluvial deposits that indicate cycles of erosion and deposition that occurred before human occupancy. Climatic changes, however, probably were not uniform from one river basin to another throughout a region as large as the western United States. Therefore, erosional and depositional features that remain as relics of past epicycles are not always easily correlated. Hadley (1960) concluded that alternating epicycles of erosion and deposition in the valleys of the West are related to postglacial climatic changes, but it does not seem necessary to expect alluvial fills and terraces to be contemporaneous from valley to valley.

There is evidence that valley trenching in the Southwest may occur as a result of intense storms that recur infrequently resulting in severe erosion over long reaches of a stream valley. Thornthwaite *et al.*, (1942) proposed this theory in a study of alluvial valleys in northeastern Arizona. Most proponents of this theory, however, relate the erosion resulting from infrequent storms to a threshold condition in stream valleys caused by drought or overgrazing.

Data from studies undertaken by investigators based on each of the three theories described seem to have one thing in common. Valley trenching on rangelands in the western United States is a common phenomenon and is the result of geomorphic processes that operate in an arid and semiarid environment. The initiation of the epicycles of erosion and deposition that have occurred probably cannot be attributed to any single cause. Overgrazing, drought, or infrequent storms all interact to cause perturbations in the transport of sediment through the drainage basin.

The foregoing discussion points out that widespread erosion occurred prior to human occupancy of the western United States that may have been caused by several factors. The present erosion problems, although still reflecting processes common to a water-deficient environment, are largely the result of land misuse.

We have described the general characteristics of rangelands in arid and semiarid regions; these are low annual precipitation, poorly developed thin soils, steep hillslopes, and sparse vegetation cover. All of these factors are closely related to the impacts caused by grazing animals. Smith and Wischmeier (1962) concluded that four interrelated factors have been considered as basic determinants of fluvial erosion. These are (1) climate; (2) soil characteristics, such as resistance to erosion, infiltration capacity, and transmission rates; (3) length and

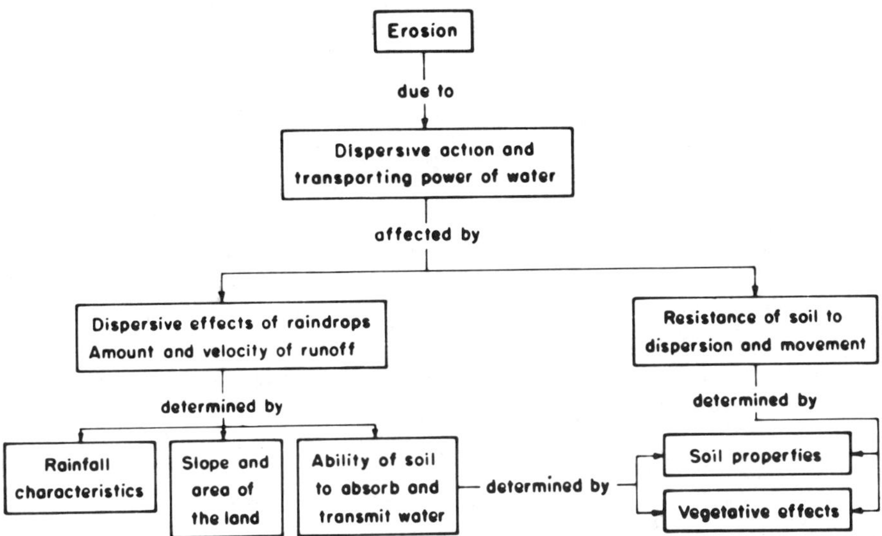

Figure 1. Factors affecting soil erosion by water (Baver, 1965).

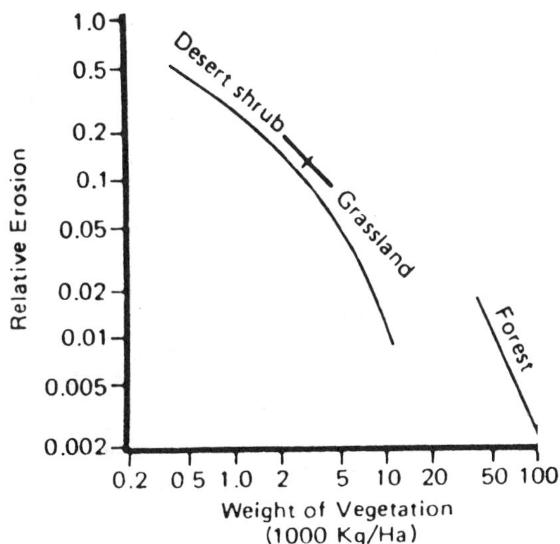

Figure 2. Relative erosion compared with vegetation weight per unit of area (Branson *et al.*, 1981).

gradient of hillslopes; and (4) vegetation cover. The diagram in Fig. 1 by Baver (1965) shows these factors and their interrelationships.

The introduction of grazing herds in the nineteenth century with no regulation on numbers of animals, seasonal or areal distribution of herds, or provisions for adequate stock water caused several environmental impacts. These are (1) soil compaction by trampling and trailing, (2) depletion of the vegetation cover, and (3) reduction of infiltration capacity of the soil. The depletion of the vegetation cover exposes more bare soil surface to raindrop impact, and when the infiltration capacity of the soil is reduced by compaction, runoff and erosion increase. Vegetation densities are generally low on rangelands, and erosion rates are normally higher than those occurring on lands receiving more precipitation. The relationship between relative erosion and weight of vegetation per unit area, used as a surrogate for density, is shown in Fig. 2.

Areal Extent of Disturbed Grazing Lands

There are approximately 305 million hectares (ha) of land in the 11 western states, of which approximately 60 million ha are administered by the Bureau of Land Management as grazing districts. Because these lands are located in a vast region about 2000 × 1200 km in area, stretching from Canada to Mexico and from the Great Plains to California, Washington, and Oregon, environmental conditions are quite varied. Therefore, the problems associated with land misuse are not concentrated in any particular part of the region. Rather, they are related to soils, vegetation cover, precipitation amount and distribution, and history of past use in critical areas.

Intensity of Land Disturbance

The intensity of erosion on grazing lands probably can be placed in proper perspective if we compare the accounts of the condition of much of the land at the turn of the twentieth century with conditions at the present time. Many of the valleys that were trenched by arroyos as much as 15 m deep lost most of the vegetation cover on the valley floors, and intense grazing depleted the hillslopes of their vegetation. The amount of sediment transported from the hillslopes and the valley alluvium is reflected in the high suspended sediment loads in the Colorado River and its tributaries prior to 1941 (Hadley, 1974). The reduction of sediment loads since about 1941 in these same rivers indicates that regulated land use and properly designed erosion control practices may reverse the consequences of the accelerated erosion that began in the last two decades of the nineteenth century.

Duration of Grazing Disturbance

Observations have shown that erosion caused by poor grazing practices will be prolonged and accelerated by continued misuse. Likewise, the impacts of land misuse can be reversed by regulation of grazing and effective erosion control. In the absence of erosion control practices, natural geomorphic processes would ultimately restore eroded valleys to equilibrium. These processes may be assisted by structures and treatments that utilize the limited amount of water more efficiently.

Legal Constraints

The free common use of the federal lands in the western United States for grazing was brought to an end by the passage of the Taylor Grazing Act in 1934. The act substituted a system of leasing these lands for grazing. The three objectives of the Taylor Grazing Act are (1) to stop injury to the public grazing lands by preventing overgrazing; (2) to provide for their orderly use, improvement, and development; and (3) to stabilize the livestock industry dependent upon the public range.

The Taylor Grazing Act is administered by the Bureau of Land Management, and the public grazing lands are divided into districts. Allotments are leased to ranchers based on the forage available. With the establishment of grazing regulation, quantitative studies also were initiated to determine the effects of grazing. The results of these studies have been useful in managing the grazing lands.

Mitigation of Grazing Impacts

Of the several methods that have been attempted to control erosion on grazing lands, the regulation of grazing is the simplest. This method requires a grazing system that carefully manages livestock numbers and excludes grazing on pastures during part of each year in order to prevent soil compaction and protect vegetation.

The effects of this type of grazing management on runoff and sediment yield was studied in four pairs of small drainage basins in Badger Wash basin, western

Colorado, during a 20-yr period (1953–1973) (Lusby, 1979a). In each pair of basins, one basin was grazed, using four different systems of livestock management, and the other basin remained ungrazed. This study is notable for the length of time that data were collected on the effects of the land-use practices.

Badger Wash is in an area of intricately dissected terrain underlain predominantly by poorly developed shale soils with some sand. Vegetation cover is sparse and is composed of salt desert shrubs. The climate is arid to semiarid; the average annual precipitation is about 230 mm (9.0 in.) based on a nearby precipitation station with 48 yr of record. According to statements by early settlers, the area supported a much better vegetation cover prior to 1890 when large numbers of livestock were introduced from Texas (Lusby, 1979a).

The results of the 20-yr grazing study show that complete grazing exclusion resulted in a reduction of runoff of about 20% during the period 1953–1965, and an additional 20% during 1966–1973. During the same periods, sediment yield was reduced by 35 and 28%,respectively, for a total of 63%. Changes in the season of grazing on the utilized areas from November 15–May 15 to November 15–February 15 each year reduced runoff and sediment yield about 29%. The same changes in use, except that grazing was permitted only every other year, resulted in a reduction of runoff and sediment yield of about 20% (Lusby, 1979a). Figure 3 shows a mass diagram of runoff in one pair of the basins at Badger Wash for the period 1953–1966 (Lusby *et al.*, 1971). It is apparent from this graph that the exclusion of grazing from basin 2-B at the beginning of the study (1953) had almost an immediate effect on runoff. The values of annual runoff began to diverge from the line of equality in 1954. The apparent reason is that the fine shaly soils are loosened and expanded by frost action in the winter months, and if they are not compacted by livestock trampling and raindrop impact, much of the precipitation is absorbed by the soils. All basins are subject to raindrop impact, but the additional compaction of trampling only occurs on the grazed basins.

Effects of Structures

In sharp contrast to the livestock management system applied to Badger Wash basin, erosion control by structural treatments is far more complex and costly. This type of conservation program was applied to the Willow Creek basin, northeastern Montana. The basin has a drainage area of 1422 km^2 (547 sq miles) underlain mostly by dark marine shales that weather to heavy-textured residual soils. The generally sparse vegetation is typical of the northern Great Plains consisting primarily of grasses and shrubs. Some of the upland slopes have dense stands of grass where soils are sandier. The average annual precipitation is about 305 mm (12 in.). The topography is characterized by rolling uplands with low relief, and stream valleys have relatively wide, flat floors with low gradients.

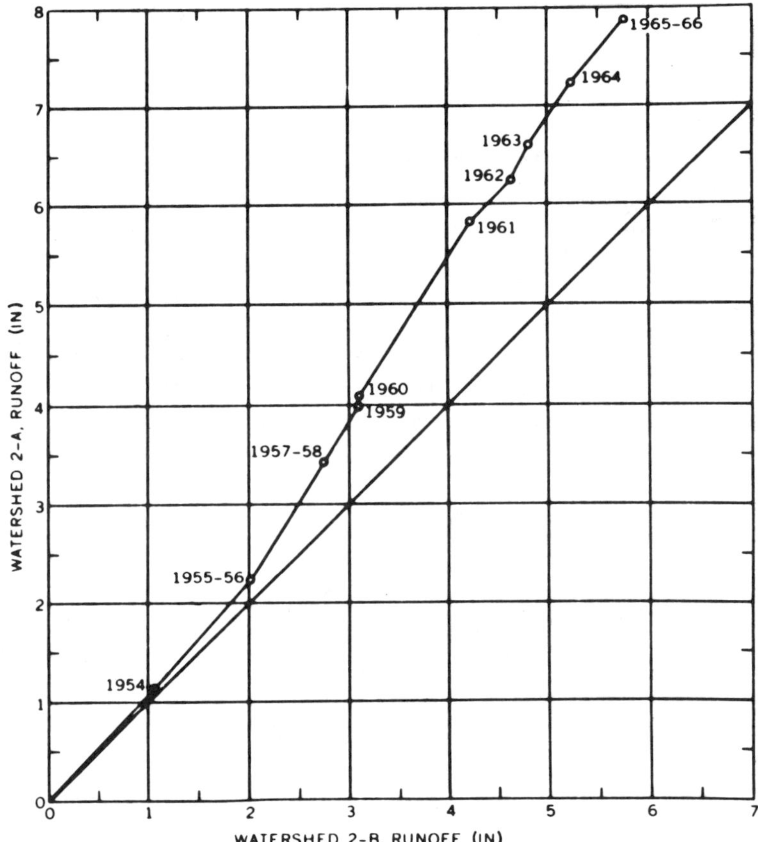

Figure 3. Mass diagram of runoff for pairs of basins, 2-A (grazed) and 2-B (ungrazed) in Badger Wash basin, Colorado, 1953–1966 (Lusby *et al.*, 1971).

The Willow Creek basin is used entirely for grazing and has been heavily used in the past. Many of the valleys have been trenched by gullies, and erosion is locally severe on upland slopes. During the 1950s, the Bureau of Land Management began an extensive conservation program in the basin with the objective of controlling runoff and sediment yield and improving the vegetation cover. The method of control was structural, including detention and retention reservoirs, water-spreading dikes, and contour furrows. The number of reservoirs in 1954 was 57 with a combined capacity of 883 hm^3 (7163 ac-ft), and in 1967 the number of reservoirs had increased to 190 with a combined capacity of about 5914 hm^3 (47,963 ac-ft) (Frickel, 1972). The reservoirs detain runoff and sediment from approximately 75% of the Willow Creek basin. It should be emphasized, however, that as sediment accumulates in these reservoirs, the cumulative storage is continually decreasing and regular maintenance of the spillways and conservation pools is necessary.

Analysis of runoff and sediment yield data collected at Willow Creek during the period 1954–1968 (Frickel, 1972) indicates that the conservation structures have reduced peak discharges but that they do not reduce the volume by a major amount (about 18%). It was also estimated that the average annual sediment yield in the Willow Creek basin is about 92 hm^3 (750 ac-ft). The suspended sediment discharge at the gauging station near the mouth of the basin for 11 yr (1954–1964) is about 61 hm^3 (500 ac-ft). Suspended-sediment concentrations at the gauging station were reduced about 55% during the 11-yr period. The valley trench still exists in Willow Creek valley because much of the sediment derived from upland sources is being deposited in reservoirs, water spreading systems, and contour furrows. These conservation structures, meanwhile, are losing about 1.6% of their capacity annually (Frickel, 1972).

Effects of Vegetation Cover

We have discussed the importance of vegetation cover in controlling erosion on grazing lands. If vegetation density is expressed as canopy cover, then the reciprocal of density is percentage of bare ground exposed to rainfall. Many drainage basins in the semiarid western United States have a plant cover that is predominantly sagebrush. Because of the competition of the individual shrubs for the limited soil moisture that is available, there generally are large interspaces between plants that are barren. This bare ground is highly susceptible to erosion because of the lack of a protective plant cover.

The U.S. Geological Survey conducted a 9-yr study (1965–1973) on four small drainage basins in western Colorado near Wolcott in an area named Boco Mountain (Lusby, 1979b). The dominant vegetation in the study area is sagebrush, which averages 0.6 m (1.98 ft) in height, and the shrub canopy varies from 45 to 55% of the drainage area. Two of the four basins were undisturbed in the study, and the sagebrush cover was plowed out in the other two basins. Wheatgrass was planted in the plowed basins. The objective of the study was to determine the changes that might occur in runoff and sediment yield with this vegetation conversion.

Measurements indicated that conversion to grass caused a reduction in runoff from summer rainstorms of about 75% and runoff from spring snowmelt increased about 12%. Annual runoff from the grass basins decreased about 20% (Lusby, 1979b). Sediment yield from the seeded basins was reduced by about 80%.

Percentage of bare soil on the Boco Mountain basins was related to the ratio of

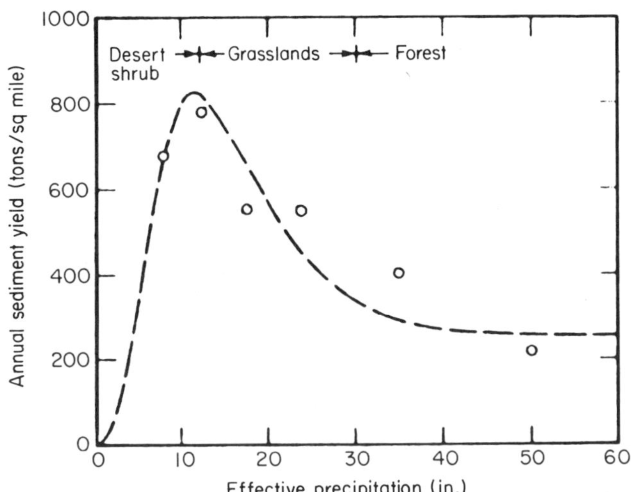

Figure 4. Variation of sediment yield with climate as based on data in the United States (Langbein and Schumm, 1958).

sediment yield to runoff. The regression of this ratio and percentage of bare soil indicated that a reduction of 38% in the area of bare soil that resulted from the vegetation change from sagebrush to grass effected a decrease of 73% in sediment concentration (Lusby, 1979b).

Schumm (1969), following an earlier study by Langbein and Schumm (1958), noted that sediment yield rates in the United States indicate that erosion reaches a peak in semiarid regions where mean annual precipitation is about 300 mm (12 in.). As precipitation increases, the sediment yields decrease primarily due to the influence of vegetation cover. If vegetation cover is destroyed in humid regions, then the sediment yield rates continue to increase (see Fig. 4). Schumm (1969) concluded that human potential for increasing erosion rates is reduced in arid regions where vegetation cover is sparse and natural erosion rates are high. However, Lusby *et al.*'s study (1971) indicated that the potential for reducing erosion is great when mean annual precipitation approaches 300 mm (12 in.) and vegetation cover can be modified to reduce the bare soil exposed to raindrop impact.

References

Antevs, E., 1952, Arroyo-cutting and filling: Jour. Geology, v. 60, pp. 375–385.

Bailey, R. W., 1941, Land erosion—normal and accelerated—in the semiarid West: Am. Geophy. Union Trans., v. 22, pt. 2, pp. 240–261.

Baver, L. D., 1965, Soil Physics: New York, John Wiley and Sons, Inc.

Bennett, H. H., 1939, Soil Conservation: New York, McGraw-Hill, 993 pp.

Branson, F. A., 1975, Natural and modified plant communities as related to runoff and sediment yields: *in* Ecological Studies 10, Coupling of Land and Water Systems, A. D. Hasler, editor, New York, Springer-Verlag, Inc., pp. 157–172.

Branson, F. A., *et al.*, 1981, Rangeland Hydrology: Dubuque, Iowa, Kendall-Hunt Publishing Co., 340 pp.

Bryan, K., 1925, Date of channel trenching (arroyo cutting) in the arid Southwest: Science, new series, v. 62, no. 1607, pp. 338–344.

Bryan, K., 1941, Pre-Columbian agriculture in the Southwest, as conditioned by periods of alluviation: Assoc. Am. Geographers Annals, v. 31, no. 4, pp. 219–242.

Calef, W., 1960, Private grazing and public lands: Chicago, The University of Chicago Press, 292 pp.

Frickel, D. G., 1972, Hydrology and effects of conservation structures, Willow Creek basin, Valley County, Montana, 1954–68: U. S. Geological Survey Water-Supply Paper 1532-G, G–1 to G–34.

Hadley, R. F., 1960, Recent sedimentation and erosional history of Fivemile Creek, Fremont County, Wyoming: U. S. Geological Survey Professional paper 352-A, pp. 1–16.

Hadley, R. F., 1974, Sediment yield and land use in Southwest United States: Proceedings of the IAHS Paris Symposium, IAHS Publication No. 113, pp. 96–98.

Langbein, W. B., and Schumm, S. A., 1958, Yield of sediment in relation to mean annual precipitation: Am. Geophys. Union Trans., v. 39, pp. 1076–1084.

Lusby, G. C., 1979a, Effects of grazing on runoff and sediment yield from desert rangeland at Badger Wash in western Colorado, 1953–73: U. S. Geological Survey Water-Supply Paper 1532-I, pp. I1 to I34.

Lusby, G. C., 1979b, Effects of converting sagebrush cover to grass on the hydrology of small watersheds at Boco Mountain, Colorado: U. S. Geological Survey Water-Supply Paper 1532-J, pp. J1 to J36.

Lusby, G. C., *et al.*, 1971, Effects of grazing on the hydrology and biology of the Badger Wash basin in western Colorado, 1953–66: U. S. Geological Survey Water-Supply Paper 1532-D, pp. D1 to D90.

Peterson, H. V., and Hadley, R. F., 1960, Effectiveness of erosion abatement practices on semiarid rangelands in Western United States: Int'l. Assoc. Sci. Hydrol., v. 53, pp. 182–191.

Schumm, S. A., 1969, A geomorphic approach to erosion control in semiarid regions: Trans. Am. Soc. Agricult. Engineers, pp. 60–68.

Smith, D. D., and Wischmeier, W. H., 1962, Rainfall erosion: Adv. Agronomy, v. 14, pp. 109–148.

Thornthwaite, C. W., Sharpe, C. F. S., and Dosch, E. F., 1942, Climate and accelerated erosion in the arid and semiarid Southwest, with special reference to the Polacca Wash drainage basin, Arizona: U. S. Department of Agriculture Tech. Bull. 808, 134 pp.

7

Lands Disturbed by Recreational Use

Introduction

The demand for lands dedicated to outdoor recreational uses has increased greatly since World War II, beginning in the 1950s. The impacts on ecosystems caused by the pressures of recreational uses have been serious and varied depending on the type of activity. Recreation has been classified into three categories by Clawson and Knetsch (1966): (1) user oriented, (2) intermediate, and (3) resource oriented. User-oriented areas generally are located in urban areas and include parks and playgrounds. Resource-oriented areas generally are larger and are located some distance from cities in remote open areas, such as rangelands, forests, and national parks. We are primarily concerned here with resource-oriented areas and the activities of hiking and off-road vehicle (ORV) use.

Because of the unique nature of land disturbance caused by recreational use, the structure of this chapter will deviate from the order used in the chapters on construction, coal, and uranium. First, the amount and kind of data are very limited for recreational activities in comparison with the other types of land use considered in this book. Also, the areal extent and duration of the disturbance vary considerably from one location to another. The inclusion of recreational activities in the book is justified because the intensity of impacts, caused primarily by ORVs, results in severe alteration to the landscape and the hydrology and geomorphology of the affected areas, and thus has been at the focus of public attention.

Nature of Disturbance

The principal kinds of impacts that are associated with these activities are related to hillslope processes, soils, and vegetation cover. Trailing by animals, hiking, and off-road vehicle use tend to damage or destroy the vegetation cover and compact soils. The increase in bare-soil areas and decrease in infiltration rates will increase the runoff, surface erosion, and sediment yield rates on disturbed sites and downstream in off-site areas.

The evaluation of the impacts that provides a basis for regulation of the acceptable intensity of these recreation activities can be accomplished only with quantitative data gathered in the field in studies of geomorphic and hydrologic response to disturbance. Unfortunately, there are a limited number of quantitative studies in the technical literature except for such activities as hiking and off-road vehicle use.

Degradation of Mountain Trails

Since about 1970, there has been a great increase in the number of people who are using recreation areas in the mountains, forests, national parks, and wilderness areas of North America and western Europe. The most popular means of travel by those who are exploring these remote areas are hiking and horseback riding (McQuaid-Cook, 1978). The increased use of these areas will cause damage to soils and vegetation primarily by compaction of soils on relatively level sites and gully erosion on steep slopes coupled with destruction or damage to the vegetation cover. McQuaid-Cook (1978) summarized the physical processes that result from pedestrian hiking. Soil permeability also is greatly decreased along with significant reduction of vegetation due to the lack of air and soil-water movement to the plant roots in the dense soil. The reduced infiltration capacity increases overland flow and soil erosion.

Soil compaction will vary with type of soil, soil moisture content, frequency of travel on a trail, and the distribution of force exerted by the feet of the humans or animals traversing the trail. Shod horses loosen soil more than unshod horses; similarly, hikers with lugged-soled boots compact the soil and destroy vegetation more than hikers with soft-soled shoes (McQuaid-Cook, 1978).

In a study of impacts of hiking on mountain trails in northern Sweden, Bryan (1977) compared the influence of soil properties to trail erosion and landscape degradation. All of the areas he investigated are above treeline in the Swedish mountains. Where intensity of trail use is uniform, he found that damage is controlled by soil properties and profile characteristics. The soil properties that are of particular significance are abundance of rocks, texture homogeneity, iron pan morphology, aggregate stability, and organic content of the soil.

The Grövelsjön district, the site of Bryan's study, is a high plateau with some mountain peaks rising to 1200 m (3960 ft). The surface is dominated by glacial features, and the soils are generally thin and stony, with the exception of peat-filled depressions and some sandy deltaic deposits (Bryan, 1977). Six trails, representing a variety of terrain and intensity of use, were selected for detailed study. Because Bryan (1977) felt that trail degradation reflects not only surface erosion but alteration of the complete profile, his study methods examined the profile characteristics at 33 sites. Pits were dug in the trails and at nearby undisturbed locations.

Bryan (1977) concluded that the hiking trails at Grövelsjön have been severely degraded with intensity of use being an important contributing factor. The damage can occur at any location, but at lower intensities of use, the degree of damage is closely related to soil, vegetation, and topography.

Topography as a factor in trail erosion is related to whether or not the trail parallels contours. Where trails follow the fall line of the hillslope, damage is severe from fluvial erosion regardless of the gradient. Where trails follow the hillslope contour, damage is minimal unless water is diverted down the slope by incision. Erosion is also increased where trails cross the slope at an oblique angle, crossing contours. Bryan was unable to determine any critical threshold slope gradient for initiation of erosion.

The initiation of erosion is the result of the downslope component of gravitational forces, which determine flow velocity, and the resistance afforded by vegetation and soil properties. Where vegetation remains intact, erosion is minimal. Vulnerablity to erosion is highest when the soil is saturated. Bryan stated that vegetation cover is completely broken down by an intensity of use that approaches 800 to 1000 trail users. If the intensity of use continues throughout the short growing season, the damage is cumulative from year to year, even on trails that experience very little use.

Where vegetation has been completely destroyed, the resistance to erosion is controlled by the soil properties. As stated earlier, the soils at Grövelsjön are generally thin and stony. The highest erosion rates, however, occur on the fine-grained, homogeneous deltaic sands. The soil profiles that are stony or which include an iron pan offer some resistance to incision, but once a threshold of resistance is exceeded, these soils also erode rapidly. Organic-rich peat soils also are subject to severe erosion if they remain wet throughout the summer (Bryan, 1977).

In a similar study conducted in the southern Rocky Mountains of Colorado, Summer (1982) determined indexes of soil erodibility for alpine tundra in Rocky Mountain National Park. Field experiments were conducted using a portable rainfall simulator to estimate an erodibility index (grams of detached soil per unit area) at 71 sites representing eight soil types. The physical properties of the soils that were measured by Summer (1982) included texture, water-absorption capacity, organic carbon, and aggregation. The erodibility indexes obtained using the rainfall simulator were divided into three classes as follows:

1. High erodibility: 1,038 g/m^2
2. Moderate erodibility: 555 g/m^2
3. Low erodibility: 247 g/m^2

These values of soil erodibility provide the best estimates available for predicting relative magnitudes of potential soil loss under barren alpine conditions (Summer, 1982).

When compared with field erodibility indexes, results of simple and multiple regressions show that 29% of the variance in erodibility is explained by the measured variables listed earlier. The strongest predictor of field erodibility is soil aggregation ($r = 0.42$). The unexplained variability (71%) may be due to unmeasured soil properties, nonlinearity in the data, random processes, bias, and experimental errors (Summer, 1982). It is clear that determination of a soil erodibility index that reflects accurately all of the variables involved in a system so complex is very difficult. The laboratory parameters used by Summer (1982) should not be used as predictive tools; rather, the field determination of detachability with the rainfall simulator may be identified as the relative erodibility index until further research is done.

The impact on alpine soils by hikers is strongest in areas near Trail Ridge road where most of the trails occur. Also, the gentle and uniform topography on the summit surfaces eliminates the influence of slope variability.

Effects of Off-Road Vehicles

Probably no single recreational activity has caused as much intensive damage to the environment in localized areas as the use of off-road vehicles (ORV), primarily motorcycles and dune buggies. This is especially true in desert environments of California and other western states. The impact of these vehicles on desert ecosystems is easily observed in areas of concentrated ORV use, and even remote parts of the desert have not been spared entirely. In a study conducted by the Arid Lands Committee of the American Association for the Advancement of Science (1974), it was estimated that, in the southern California desert alone, visitor days increased from 4.8 million in 1968 to 13 million in 1973. Off-road vehicles can cause damage to all lands, but especially arid lands where precipitation is low and recovery of the ground cover will be slow. Ironically, as late as 1971, the recreation activities projected through the year 2000 by the U.S. Bureau of Outdoor Recreation do not include ORV use (Kockelman, 1983).

At about the same time that the recreational use of the deserts in the western United States increased (1967–1971), the number of motorcycles registered in the United States also increased dramatically. The following table summarizes this increase (Kockelman, 1983).

Registered motorcycles	Years
0.5 million	1960s
5 million	1972
8 million	1976
12 million	1983

In 1978, 4.7 million motorcycles were used for off-road recreational activities according to an estimate by the Motorcycle Industry Council (1980). The figures in the table do not include dune buggies or four-wheel drive ORVs. It should also be pointed out that many trail bikes that are used for hill climbing are not required to be registered in all states.

As the number of vehicles in use increased, more questions were being asked by the government agencies that administer much of the desert land and by conservation groups about the long-term impacts of ORVs on the environment. No ready answers were available, however, because of the lack of quantitative data.

Since about 1970, studies have been initiated to evaluate the effects of ORVs on the environment, and the federal government has become involved in attempts to control the problem. In 1972 President Nixon issued Executive Order 11644 to control the use of ORVs on federal public lands, and in 1977 President Carter issued Executive Order 11989 to provide for immediate closure of areas or trails to ORVs causing adverse effects. We will summarize some of the results of studies that have been undertaken since about 1970 that have quantified the impacts on the environment attributable to ORVs, primarily in arid regions in the western United States.

Compaction of Soils by ORVs

As discussed in this chapter in the section on the impact on soils by hiking, the surface loading of soils produces stresses that compress the soil and increases soil bulk density. Compaction may be characterized by changes in bulk density, porosity, or void ratio (ratio of void volume to solid volume) (Webb, 1983).

Studies of soil compaction induced by motorcycle use have been made in California deserts (Snyder *et al.*, 1976; Webb, 1983). Webb's study was located on the flat desert floor of the western Mojave Desert, California, near Fremont Peak, and consisted of penetrometer measurements on four trails representing 1,

10, 100, and 200 motorcycle passes. In contrast, the study by Snyder *et al.* (1976) was located in a steep upland valley on the western edge of the Central Valley of California (Panoche Hills) near Los Banos (Fig. 1). Bulk density measurements were made of soil samples taken at intervals to a depth of 1.2 m in hill-climbing trails; the trails are on hillslopes with gradients of about 65%. Bulk densities were also determined for samples from sites adjacent to the trails that were assumed to be virtually undisturbed. It was not possible to determine the number of motorcycle passes that had been made on the Panoche Hills trails because they had been in use for at least 3 yr before the study began.

In the study by Webb (1983), the moisture content of the soil at the time of compaction measurements was 6.2% and appeared to be typical for the region in the interval between spring rains. The motorcycle trails were noticeable even with only one pass. As the number of passes increased, a definite indentation was formed on the desert floor with berms on each side of the trail about 10 to 30 mm higher than the trail. Annual vegetation remained in the 1-pass trails, but most of the annuals had been destroyed after 10 passes (Webb, 1983). Figure 2 shows the relationship between the number of passes and the soil bulk density. The bulk density in the 0–60 mm depth increased as a logarithmic function of the number of passes and may be expressed by the following equation:

Figure 1. Motorcycle trails on steep valley hillslopes in Panoche Hills, near Los Banos, central California.

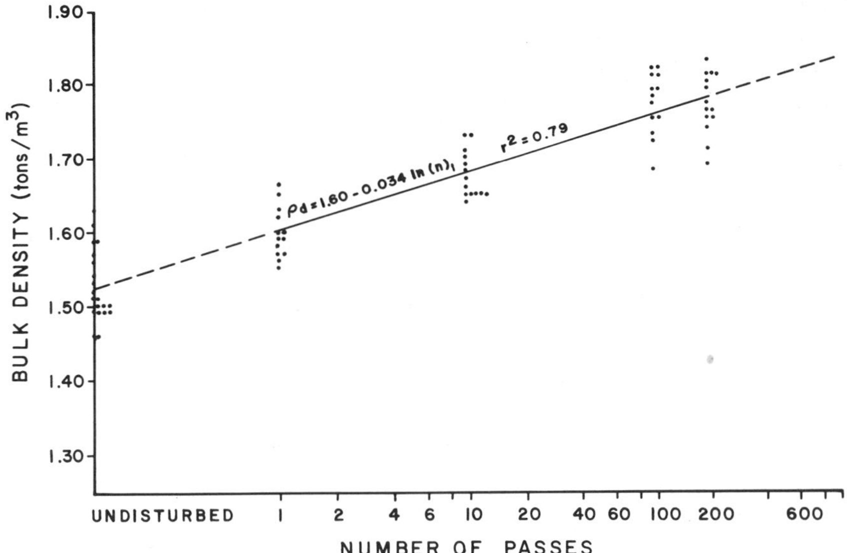

Figure 2. The effects of repeated motorcycle passes on the bulk density in the upper 0–6 cm of loamy sand at the Fremont Peak study site (Webb, 1983).

$$\rho_d = 1.60 + 0.034 \ln(n) \qquad r^2 = 0.79 \tag{1}$$

where ρ_d is the dry bulk density, (in metric tons per cubic meter), n the number of passes, and r^2 the coefficient of determination.

In the Panoche Hills study (Snyder *et al.*, 1976), bulk density was measured using gravimetric samples of known volumes. Figure 3 shows the bulk density of the surface soils in the trails to be 37% greater than for soils of the undisturbed areas. This difference decreases somewhat erratically to less than 12% at a depth of 1.2 m, but it is, nonetheless, evidence of compaction to a depth of over 1 m. We might expect a decrease in the bulk density of soils through time after motorcycle use stopped, but a comparison of bulk densities of the first and last samplings (1971–1974) showed no such trend. This may be due to the limited precipitation and frost action at the sites.

Alteration of Vegetation Cover

In the Panoche Hills study (Snyder *et al.*, 1976), it was evident that extreme alteration of the vegetation cover resulted from ORV use. The trails were completely denuded and deeply rilled. Even the intertrail areas were damaged, making the contrast with undisturbed areas adjacent to the study site even more pronounced. The entire study area was composed of a single drainage basin that was divided into two parts by a small earthfill dam. The downstream part of the study area had never been used for motorcycle hill climbing, and the upstream part of the study area had been heavily used by motorcycles prior to 1971. A summary of the ground cover measurements in both areas is given in Table 1.

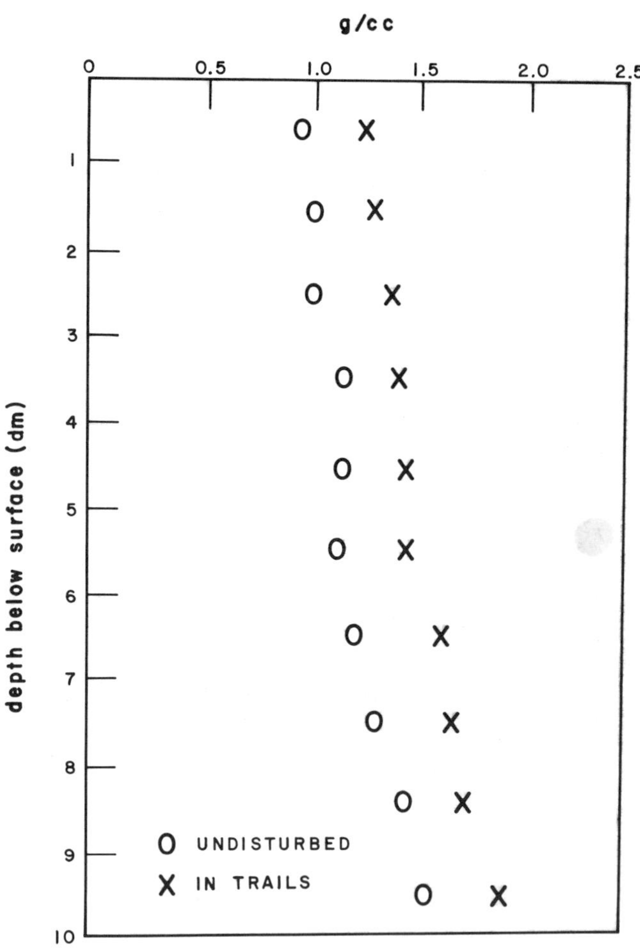

Figure 3. Variations of average bulk density with depth for two types of land use, Panoche Hills, central California (Snyder *et al.*, 1976).

Table 1 Percentage of Ground Cover in Panoche Hills Basins, 1971–1975[a]

	East-facing slope				West-facing slope					
	In trails		Beside trails		In trails		Beside trails		Area average[b]	
Disturbed by ORV use	1971	1975	1971	1975	1971	1975	1971	1975	1971	1975
Woody vegetation										
Mormon Tea	T[c]	T	3	2	0	0	0	3	3	3
Rabbitbrush	T	6	3	3	0	0	0	T	1	3
Herbaceous vegetation	27	30	71	70	13	42	44	52	37	58
Total vegetation	27	36	77	75	13	42	44	55	42	65
Bare soil	72	64	23	25	83	53	44	42	55	32
Rock	1	T	T	T	4	5	2	3	3	3

	East-facing slope		West-facing slope		Area average[b]	
Undisturbed	1971	1975	1971	1975	1971	1975
Woody vegetation						
Mormon Tea	4	4	1	2	3	2
Rabbitbrush	2	1	T	1	3	6
Herbaceous vegetation	71	77	57	57	60	67
Total vegetation	77	82	58	60	66	75
Bare soil	22	18	41	39	31	23
Rock	1	0	1	1	3	2

[a] Source: Snyder *et al.* (1976).

[b] These values obtained from random transects crossing whole slope and are not averages of the other values in this table.

[c] T, Trace (less than 1%).

Runoff and Fluvial Erosion

Analysis of these data with regard to the hydrologic effects of ORV use reveal why runoff and sediment yield are so much higher on the disturbed area. At the beginning of the study in 1971, the average percentage of bare soil in the disturbed area was 55% and in the undisturbed area it was only 31%.

As previously noted, the volume of runoff and sediment yield is related to the percentage of ground cover or its inverse percentage of bare ground. In the Panoche Hills study in California (Snyder *et al.*, 1976), both runoff and sediment yield were measured in small reservoirs at the downstream end of the disturbed and undisturbed areas. The differences in runoff from the two areas is striking. Ten runoff events occurred (1971–1975) in the disturbed area, while only five events occurred in the undisturbed area. The total unit runoff (volume per unit area) from the disturbed was nearly eight times more than from the undisturbed area. These differences are attributed to lack of vegetation cover and soil compaction in the motorcycle trails.

In studies on ORV-used areas in the Mojave Desert, California, Hinckley *et al.* (1983) found that soil compaction by ORVs decreased infiltration rates and increased the volume and duration of runoff. For example, low-intensity storms of 4.0 mm/hr produced runoff from heavily compacted soils, while observations on undisturbed soils indicated that no runoff was generated unless precipitation intensities exceeded 40 mm/hr.

In order to document the effects of ORVs on the landscape, almost all quantitative studies of ORV-used areas include monitoring of erosion rates using a variety of field techniques. The most common methods are erosion transects on hillslopes that are surveyed periodically to determine surface erosion along a line, and erosion pins where point erosion is measured periodically by determining differences in pin exposure. Both of these methods are often difficult to replicate with a high degree of accuracy using standard survey methods.

Sediment yield rates, or the volume of sediment derived from erosion in a drainage basin and delivered to a measurement point (either a gauging station or reservoir) in a specified time period, differ from erosion rates. Sediment yield per unit area usually decreases as drainage area increases because eroded material is deposited and stored at the base of hillslopes, on floodplains and in channels enroute to the measurement point. Therefore, for very small drainage basins (<0.4 km^2), the sediment yield approaches the erosion rate from the contributing area.

In the Panoche Hills study, erosion was measured on hillslopes using surveyed transects, and sediment yields were determined by measuring the sediment accumulation in reservoirs on both an ORV use area and an undisturbed area. The surveys of erosion transects did not produce results that could be related to the sediment yields in the reservoirs (Snyder *et al.,* 1976). Apparently the short length of the study (1971–1975) and the low precipitation during the study period did not produce erosional changes that could consistently be detected with the measurement techniques used.

The sediment yield records from the two reservoirs at Panoche Hills, however, reflect the relative differences in erosion between undisturbed and ORV-used areas. From September 1973 to April 1975, seven storms delivered 857 m^3/km^2 (1.80 ac ft/sq mile) of sediment to the reservoir on the ORV use area. In the same period, only four storms produced runoff on the undisturbed area, and the sediment yield was so small that it could not be measured with the survey methods used. The total runoff reaching the reservoir was only 16 m^3 (0.013 ac-ft) during the period 1973–1975.

In another study of the effects of motorcycle ORV use located in Dove Spring

Figure 4. View of motorcycle trail in Dove Springs area, Mojave Desert, California, showing granular material loosened by vehicles (Snyder *et al.*, 1976).

Canyon, Kern County, California, Snyder *et al.* (1976) measured surface erosion on trails using erosion transects. The transects were surveyed during the 5-yr period 1971–1975. The valley of Dove Spring Canyon is underlain by a weathered granite lithosol (Fig. 4). The friable surface material breaks down quite easily under mechanical impact leaving a coarse, sandy detritus. This coarse detrital mantle greatly reduces runoff by absorbing most of the precipitation falling on it. In high-intensity storms, the loess material is eroded leaving exposed a rilled unweathered surface (Fig. 5). The vegetation is sparse and consists primarily of desert shrubs. Average annual precipitation at Red Rock State Park, located 2 miles east of the Dove Spring Canyon study site, was 138.2 mm (5.4 in.) for the 3-yr period 1973–1975.

Figure 6 shows some typical results of erosion transect surveys. The most significant changes were found in transects 3 and 4, located on the upper part of a heavily used hillslope. Transect 3 shows erosion of 0.15 to 0.23 m (0.5 to 0.75 ft) of surface material from the trail areas between 1970 and 1975. The eroded material probably was removed by a combination of mechanical action of the motorcycles along with the downslope component of gravity forces which also tends to move the soil material in a downslope direction. The Dove Spring

Figure 5. Erosion on motorcycle trail shown in Fig. 4 after loosened material had been removed by overland flow. Note the deep rilling of the underlying material (Snyder *et al.*, 1976).

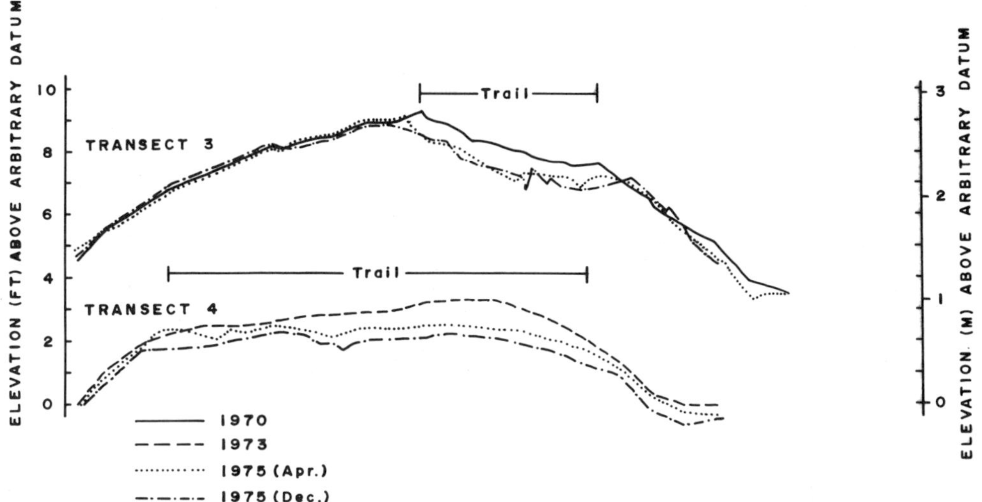

Figure 6. Typical erosion transects on motorcycle trails, Dove Springs, Mojave Desert, California (Snyder *et al.*, 1976).

Canyon area is still being used for ORV hill climbing. Erosion rates, resulting from both sheet erosion and mass movement, as high as 0.1 m/yr, represent severe erosion and are probably the result of heavy use, noncohesive soils, and sparse vegetation cover.

Reclamation of ORV Use Areas

It is evident that ORVs have a damaging impact on the soil, plant cover, and hydrologic processes. The severity of impacts is related to the intensity of use, and in arid environments, the reclamation of damaged areas is constrained by the availability of soil moisture.

The impact of motorcycle and dune buggy use on plant cover was evident in the ORV use areas at Panoche Hills and Dove Spring Canyon. There has been a marked recovery of plant cover, from 20 to 39%, in the trails at Panoche Hills with a concomitant reduction in percentage of bare soil since access was restricted in 1971. However, even after 5 yr of nonuse (1976), the trails in the previously heavily used area still had 58% bare soil compared with 23% in the undisturbed area. It has not been determined how much time will be required for these areas to attain the equilibrium they had before disturbance.

Compaction of soils and reduction of infiltration seem to be the most serious effects of ORV use. These factors deprive vegetation of moisture needed for growth. The damage may be irreversible unless mechanical scarification can be applied to improve moisture penetration.

References

American Association for the Advancement of Science, 1974, Committee on Arid Lands, Off-road vehicle use: Science, v. 184 (4135), pp. 500–501.

Bryan, Rorke B., 1977, The influence of soil properties on degradtion of mountain hiking trails at Grövelsjön: Geografiska Annaler, v. 1–2, Ser. A., pp. 49–65.

Clawson, M., and Knetsch, J., 1966, Economics of outdoor recreation: Baltimore, Johns Hopkins Press.

Hinckley, B. S., Iverson, R. M., and Hallet, B., 1983, Accelerated water erosion in ORV-use areas: *in* Environmental Effects of Off-Road Vehicles, R. H. Webb and H. G. Wilshire, editors, New York, Springer-Verlag, pp. 81–94.

Kockelman, W. J., 1983, The wonders and fragility of arid lands, *in* Environmental Effects of Off-Road Vehicles, R. H. Webb and H. G. Wilshire, editors, New York, Springer-Verlag, pp. 1–10.

McQuaid-Cook, J., 1978, Effects of hikers and horses on mountain trails: Journal of Environmental Management, v. 6, pp. 209–212.

Motorcycle Industry Council, 1980; The recreational trailbike planner: Newport Beach, California, v. 1, no. 9.

Snyder, C. T., Frickel, D. G., Hadley, R. F., and Miller, R. F., 1976, Effects of off-road vehicle use on the hydrology and landscape of arid environments in central and southern California: U. S. Geological Survey Water-Resources Investigations 76–99, 45 pp.

Summer, Rebecca M., 1982, Field and laboratory studies on alpine soil erodibility, southern Rocky Mountains, Colorado: Earth Surface Processes and Landforms, v. 7, pp. 253–266.

Webb, R. H., 1983, Compaction of desert soils by off-road vehicles: *in* Environmental Effects of Off-Road Vehicles, R. H. Webb and H. G. Wilshire, editors, New York, Springer-Verlag, pp. 51–79.

8

Lands Disturbed by Surface Mining of Coal

Introduction

In the United States, land disturbances associated with the surface mining of coal resources have attracted greater public concern in the past two decades than disturbances caused by any other human activity. Curtis (1981) commented that it is indeed unfortunate that surface mining—one of the most efficient methods of mining—is also the most destructive of other natural resources. The societal anxiety directed toward this industry is manifested in a plethora of legislation and regulation with which the mine operator must contend.

The intent of these controls includes (1) maximum resource recovery, (2) minimum environmental impact, and (3) land reclamation subsequent to the disturbance. It is the latter two objectives that create the interface between government, industry, and the science of geomorphology. To minimize environmental impact, it is necessary to minimize the areal extent, intensity, and duration of the environmental disequilibria produced by the extraction process. Similarly, effective reclamation reduces the severity of disequilibria and provides essential conditions for the natural regeneration of a steady state that requires a minimum of geomorphic work and produces landform and landscape adjustment.

The purpose of this chapter is to provide a basic understanding of the relationships among surface mining, reclamation, and geomorphic principles. Toward this end, the following topics will be addressed: (1) an assessment of land disturbance caused by surface mining of coal; (2) an examination of the specific nature of this disturbance, including consideration of spatial aspects of the disturbance, cumulative impacts, baseline data requirements, and geomorphic perspectives regarding land disturbance; (3) discussion of the legal constraints placed upon the mining industry; and (4) discussion of some common reclamation practices employed by the industry.

Surface Mining Processes

First, however, it is useful to briefly review the surface mining process because there often exists a measure of confusion regarding the terminology and procedures. For example, strip mining is only one type of surface mining, and there are, in turn, various types of strip mining. Table 1 contains a summary of surface mining techniques with commentary concerning geomorphic processes

Table 1 Surface Mining Techniques[a]

Technique	Uses	Comments
Open pit	Quarries for limestone, sandstone, marble, granite, sand and gravel, iron, copper, infrequently for coal in western United States	Insufficient spoil to refill pit, excavation remains as depression after reclamation attempts
Dredging	Placer mining of gold, sand, and gravel from stream beds	Potential for flooding and downstream sedimentation, gravelly materials difficult to reclaim
Hydraulic mining	Extensively used for gold mining in past, limited use now	Downstream sedimentation often a major problem; gravelly materials difficult to reclaim
Strip mining		
Contour stripping	Used extensively in hilly terrain, such as Appalachia, for coal extraction, some in West and Midwest	Runoff control above cut essential, severe erosion and landslides if overburden cast downslope from cut
Mountain top removal	Used commonly in hilly terrain, such as Appalachia, for coal extraction, some in West	Adaptation of area mining to contour mining, permanent change of topography, produces large plateau
Area stripping	Used in flat to gently undulating terrain, such as West and Midwest United States, for extraction of coal	Runoff and sediment often controlled internally, large tracts of land disturbed but only small amount need be unreclaimed at any given time, final cut and highwall pose reclamation problem
Auger mining	Actually underground technique often used with contour mining	Land subsidence is potential surface expression and problem, also disruption of subsurface flows and water pollution

[a]Compiled from Doyle (1976), Maneval (1979), National Research Council (1981), and Ramani and Grim (1978).

A

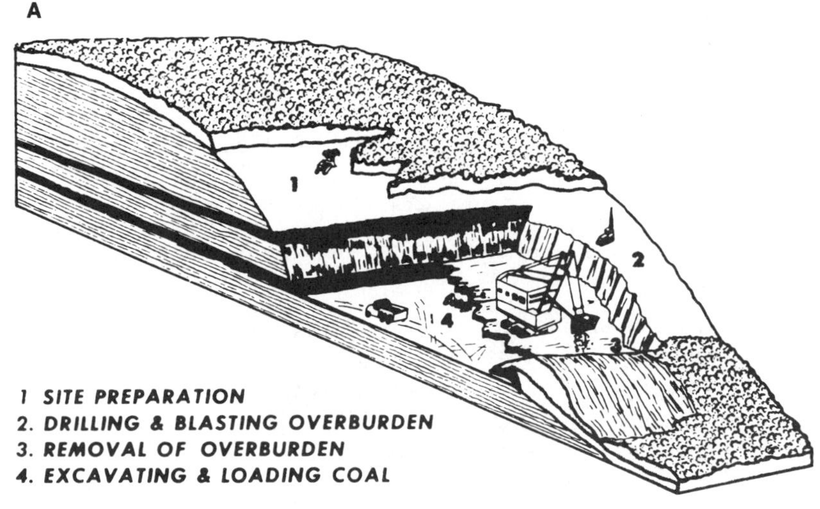

B

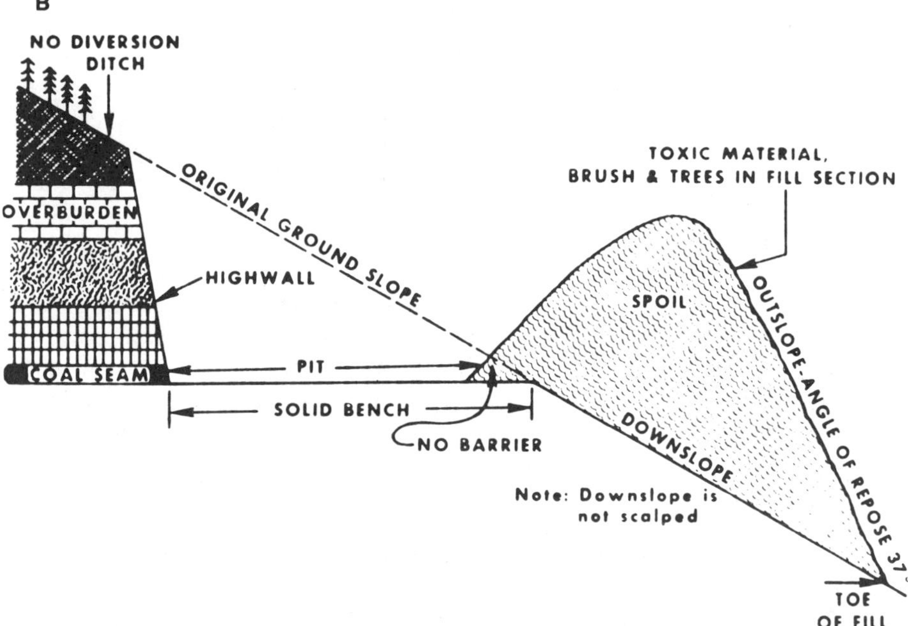

Figure 1. A, Three-dimensional view of contour mining (Grim and Hill, 1974). B, Profile view of contour mining (Grim and Hill, 1974).

or landforms. Each of the strip mining techniques has certain conditions under which it is the most efficient and favored technique, and topography or landform strongly influences the choice.

Contour strip mining is commonly utilized when coal is found in hilly or mountainous terrain. The technique is depicted in Figure 1 and described by Paone *et al.* (1978, p. 13):

> The contour stripping process involves removing the overburden over the coal seam, starting at the outcrop and proceeding around the hillside. After the coal is removed, successive cuts are made until the overburden becomes too great to permit the coal to be recovered economically.

At this point, auger mining may be used to extract additional resource from beneath the hilltop.

In the past, overburden was simply cast downslope. However, this resulted in serious environmental damage off-site and significant safety hazard. Recently, a procedure known as the "block-cut" method has been developed by which overburden is back-hauled along the cut to refill areas from which coal has already been removed. This minimizes the amount of unstable debris resting on the hillside downslope and facilitates reclamation.

Data compiled by Doyle (1976) from several reports reveal the severity of environmental problems created by conventional contour mining, prior to the adoption of the block-cut method and reclamation requirements. About 20,000 miles (32,180 km) of highwalls were produced by coal mining in Appalachia alone, and in some cases, the walls completely isolated entire mountaintops. It has been estimated that of the 25,000 miles (40,225 km) of contour bench in Appalachia, approximately 1700 miles (2,735 km) are the site of massive landslides. Slides coming off mining operations have uprooted trees, covered highways, destroyed farmland, filled reservoirs and streams, covered fish-spawning beds, caused flooding of adjacent lands, and destroyed farm buildings and homes. To this list can be added the detrimental environmental consequences of high erosion rates from unconsolidated materials resting on steep slopes without vegetation cover in a humid climatic region and toxic mine drainage. Hence, contour strip mines disturb an area of the earth's surface much greater than the area covered by the seam of coal extracted and produce environmental problems not experienced by area mining. Little wonder there evolved a public outcry to control this industry. It must be reiterated, however, that most of these impacts have been substantially alleviated by current mining and reclamation practices.

The mountain top removal technique, shown in Fig. 2, is described by Ramani and Grim (1978, pp. 259–260):

> The mountain-top removal method is an adaptation of the area mining method to contour mining. In this method, the entire mountain tops are removed down to the coal seam in a series of parallel cuts. Excess overburden that cannot be retained on the mined area is transported to head-of-hollow fills, stored on ridges, or placed in natural depressions. In either case, the

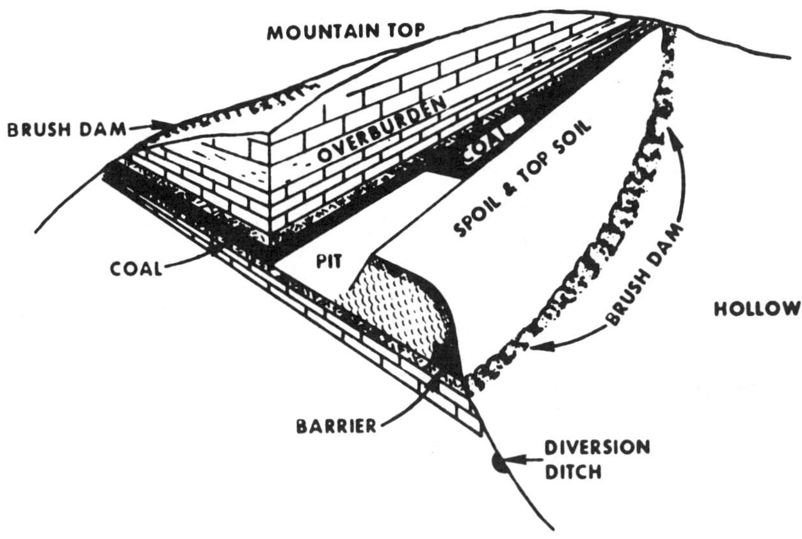

Figure 2. Mountain top removal method of surface mining (Grim and Hill, 1974).

> method produces plateaus of level, rolling land that may have great value in mountainous terrains.

Note the diversion, barrier, and brush dam, shown in Fig. 2, which are intended to minimize off-site impacts due to runoff and sediment discharge.

The area stripping technique is usually practiced in relatively flat terrain. This method is illustrated in Figs. 3 and 4 and described in a National Research Council (1981, pp. 133–134) report:

> After removal of topsoil and subsoil, where required, an initial cut is made down to the coal, and the spoil is cast into a stockpile to one side away from the planned mine. This cut—referred to as the box cut—is extended across the property to the limits of the planned mine and coal is removed as in a contour operation. Additional cuts are then made parallel to the first, with the spoil from each succeeding cut being placed in the previous mined-out pit. Each cut may be several hundreds or thousands of meters in length, and the final cut may be more than a kilometer from the initial cut.

Without reclamation, a surface mined by areal stripping would resemble an enormous washboard.

The information assembled by Doyle (1976) indicates that, while the rate of erosion on the spoils banks produced by this technique is comparable to that of contour mining, a large percentage of the sediment is retained in the depressions on the site. Thus, streams and adjoining lands are not affected as severely as in contour stripping areas. Nevertheless, large tracts are disturbed annually by this

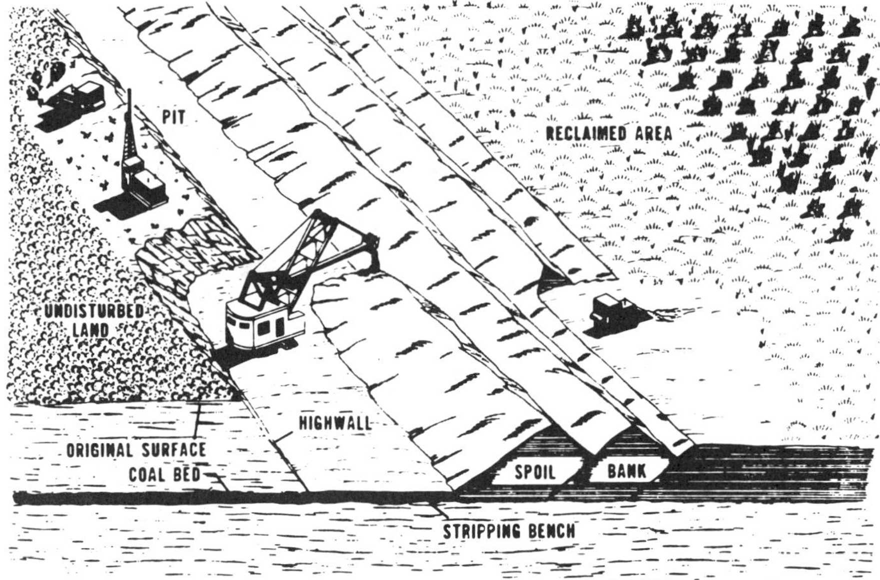

Figure 3. Area stripping method of surface mining (Grim and Hill, 1974).

Figure 4. Drag line method of removing overburden above coal seam.

technique, and the resulting surface would be of little economic or aesthetic value without effective reclamation.

To conclude, premining landscapes and landforms influence the type of strip mining and equipment utilized and, thereby, the economics of operation. Likewise, the type of mining and equipment employed influence the character of the postmining landscapes and landforms and, as a result, the design of the reclamation plan necessary to produce landscape stability. Of course, topography is not the only factor involved in decision making. Overburden properties are also highly important, especially thickness and physical and chemical characteristics. The National Academy of Sciences (1974) observed that the selection of a mining technique is usually made by engineers whose primary objectives are the most efficient removal of the overburden and transport of coal. Today, with the implementation of controlling legislation, reclamation has become a cost of operation and thus commands consideration in the selection of mining technique and equipment.

Assessment of Disturbance

A critical assessment of land disturbance due to surface mining rests upon three factors: (1) areal extent, (2) intensity or rate, and (3) duration of disturbance or impact.

Areal Extent

Areal extent may be examined in absolute and relative terms. Data compiled by Paone *et al.* (1974) are commonly used to evaluate past disturbances. They found that from 1930 to 1971, the surface mining of fossil fuels disturbed a total of 520,444 ha (1,286,000 acres), including both the area of excavation and areas used for disposal of overburden and other mine waste; a total of 405,509 ha (1,002,000 acres) or 78% was reclaimed during that same period. Thus, a total of 114,935 ha (284,000 acres) remained untreated. Averaged over a 41-yr record, this would amount to about 2803 ha (6927 acres) per year. Of course, the quality of reclamation was variable in the absence of federal legislation. More recent figures provided by the National Research Council (1981) indicate that about 40,470 ha (100,000 acres) of land are surface mined each year, and most of this land is eventually returned to some level of crop production through reclamation. Since the enactment of the Surface Mining Control and Reclamation Act of 1977 (U.S. Congress, 1977), all lands disturbed by the surface mining of coal must be satisfactorily reclaimed; and performance bonds are posted to ensure this. Johnson and Paone (1982) provided data for land disturbance from 1930 to 1980.

Unfortunately, it is not possible from this information to separate lands disturbed by surface mining from underground mining in the production of fossil fuels. It is observed in this report, however, that the quantity of land used for overburden coal waste as a distinct category is negligible during the past 10 yr (1971–1980) because of state and federal laws requiring reclamation concurrent with mining.

Areal extent may also be expressed in relative terms. The data above could be compared with the total surface area of the United States totaling 919.2 million ha (2271.3 million acres). Hence, the amount of land disturbed by surface mining of fossil fuels over a 41-yr period represents only 0.057% of the total surface area of the nation. The average annual disturbance for those years would have been 0.0003%.

A relative comparison might also be constructed by listing the surface areas devoted to various land uses, as provided in Table 2. The 2.3 million ha used by mining industries includes surface mining of all materials as well as underground mining. Paone *et al.* (1974) found that about 40% of the land disturbed by mining was accounted for by the coal industry. Even if we assume that this percentage has grown to 45%, then coal production would be responsible for the use of about 1.04 million ha. This figure may then be reasonably compared to other land uses. Thus, the amount of land engaged in coal mining is about equal to that of airports and railroads, about one-eighth that of highways, and about one-tenth that of forest service lands. Accepting the aforementioned disturbance rate of 40,470 ha/yr by surface mining, it would take about 745 yr before the surface mined area would equal the area contained within national parks. And most of the surface mined lands would have been reclaimed. The rapidly expand-

Table 2 Land Use by Various Activities[a]

Activity	Million acres	Million hectares
Agriculture	1589.0	643.1
Wildlife refuge	88.7	35.9
National parks	77.0	31.2
Urban areas	68.7	27.8
Forest service	25.1	10.2
Highways	21.5	8.7
All mining	5.7	2.3
Airports	4.0	1.6
Railroads	3.0	1.2
Other	388.6	157.3
Total	2271.3	919.2

[a]From Johnson and Paone (1982).

Table 3 Projected Coal Production in the Conterminous United States (in millions of tons)[a]

	1977	1985	2000
Surface mines			
Appalachia	184	130–155	130–175
Midwest	91	75–95	95–135
West	141	415–495	700–1005
Total	417	620–745	925–1315
Underground mines			
Appalachia	205	225–260	380–505
Midwest	54	60–80	120–180
West	13	50–60	80–110
Total	272	355–400	580–795
All mines			
Appalachia	390	355–415	510–680
Midwest	145	135–175	215–315
West	154	460–510	780–1115
Total	689	955–1145	1505–2110

[a]Source: U.S. Office of Technology Assessment (1979).

ing coal industry envisioned in the late 1970s has not materialized, largely due to the availability of oil and, in some years, hydroelectric power.

Projected coal use for the future has also been a source of concern. The United States possesses an estimated 31% of the world's recoverable coal reserves (Mountain West Research *et al.*, 1979) and a move toward energy self-sufficiency has generated numerous scenarios with regard to the future of the coal industry. The "typical" projection included in a National Research Council Report (1981) is provided in Table 3. Also contained in that report is the statement that the total landarea that will eventually be surfaced mined for coal is unlikely to exceed 4 million ha (10 million acres). This would still amount to only 0.44% of the surface area of the United States.

However, it is rather unrealistic to compare the areal extent of disturbance to the entire area of the United States. Coal reserves are generally concentrated in a few regions. The U.S. Department of Energy (1981) reported that the states with the largest coal reserves—Montana, Illinois, Wyoming, West Virginia, Kentucky, and Pennsylvania—contain approximately 76% of the national total. Montana and Wyoming alone account for about 40% of the total. Only 29% of the reserves in these six states is likely to be surface mined.

Due to variation in seam thickness and mining methods, the amount of land disturbance does not necessarily correspond directly to the amount of reserves. Table 4 shows the estimated acreage overlying strippable coal reserves. Here, the

Table 4 Estimated Average Overlying Strippable Coal Reserves by State[a]

State	Acres[b]	State	Acres
Alabama	365,200	Montana	1,001,000
Alaska	41,500	New Mexico	125,700
Arizona	18,400	North Dakota	362,300
Arkansas	37,500	Ohio	852,800
Colorado	175,500	Oklahoma	118,100
Georgia	100	Pennsylvania	238,100
Illinois	2,417,800	South Dakota	48,700
Indiana	246,500	Tennessee	59,400
Iowa	86,200	Texas	259,700
Kansas	303,000	Utah	10,600
Kentucky	1,294,400	Virginia	141,000
Maryland	25,700	Washington	12,600
Michigan	400	West Virginia	673,100
Missouri	998,900	Wyoming	200,100
		Total	10,113,200

[a]Source: Data from U.S. Bureau of Mines (1971, 1977).

[b]Based on computations using average seam thickness and strippable reserve data for each state. Data may not add to total because of rounding.

states of Illinois, Kentucky, Missouri, Montana, Ohio, and West Virginia account for 71.6% of potential land disturbance.

From the foregoing, it could be concluded that land disturbance resulting from the surface mining of coal is really not a significant environmental problem. However, there is a decided geographical dimension to the potential impact with a few states likely to be relatively "hard-hit" in the future.

Intensity of Disturbance

The second component to an assessment of disturbance, intensity, is examined through a comparison of geomorphic process rates before and after surface mining. Ideally, field measurements obtained during each time period would constitute the quantitative basis for such an evaluation; however, these data are rarely available. Alternative approaches involve comparisons between (1) rates on disturbed lands and nearby undisturbed lands; (2) rates generated by predictive models under a set of assumed conditions for the disturbed and undisturbed lands; or (3) observational evidence of rate differentials, such as the magnitude and frequency of rilling, gullying, or sedimentation. From the geomorphic perspective, mass movement and soil erosion are the processes of principal concern and, as such, are the focus of our assessment. In both cases, there are remarkably few actual data available.

It is probably accurate to suggest that mass movements resulting from surface-mining activities are a problem more typical of the eastern United States. The rugged terrain of Appalachia and the conventional contour mining technique extensively utilized in the past, which cast unconsolidated materials onto steep hillsides or "outslopes," provided the ingredients for frequent failure. Periodic blasting may well have served as the trigger. Conversely, the relatively flat terrain and areal stripping technique of the western United States generally produced rather stable hillslopes, except for spoils banks resting at the angle of repose. Here, however, occasional slumping is a nuisance and perhaps a safety hazard in some cases, but poses little environmental threat off-site.

For both regions, highwall failure is a significant safety hazard, but again would likely be considered a major environmental threat only in exceptional cases. In fact, highwall failure would help "soften the starkness of the highwall," as described by Plass (1971).

There are numerous descriptions of mass movements associated with past surface mining operations and their consequences in the eastern United States. Curtis (1971, p. 436) commented:

> Slides are important sources of sediment. In water-shed A, which is still being mined, a number of spoil slides occurred that extended all the way to the valley bottom. These slides blocked the channel and caused sediment deposition. A check dam was constructed above the weir to trap this sediment. On June 1, 1969, this dam was full, with an estimated 400 cubic feet of sediment.

Curtis and Superfesky (1977) remarked that the slump or instability of the slope that destroyed many of their erosion measurement stations at a Caryville, Kentucky, site indicates that not only surface erosion but also mass stability must be considered when returning spoils to approximate original contour.

Unfortunately, there is seemingly no systematic study of the changes in the magnitude and frequency of mass movement attributable to this industry. It has been suggested that mass movement processes would be fairly active in this climatic region anyway (Peltier, 1950). Nevertheless, recalling the statistic that of 25,000 miles of contour bench in Appalachia, approximately 1700 miles are the site of massive landslides, it appears improbable that nearly 7% of the hillslope surface of this region would be subjected to large-scale mass movements under natural conditions.

Many writers have suggested that soil erosion is the primary environmental problem caused by surface mining and erosion control is the principal challenge (e.g., Hodder, 1975; Dollhopf *et al.*, 1977; Stiller *et al.*, 1980; Wells and Rose, 1981; Toy, 1984b). Extensive rilled and gullied hillslopes or entrenched chan-

Figure 5. Accelerated erosion in the form of rilling on reclaimed hillslope in Wyoming.

nels, together with sites of rapid sedimentation, are usually taken as evidence of accelerated erosion. Although such conditions may occur naturally (Schumm and Hadley, 1957; Patton and Schumm, 1975), they are very often caused by human disturbance of the geomorphic system. On surface mined lands, raw spoils banks or hillslopes produced by early reclamation efforts frequently reveal physical evidence of high erosion rates, as shown in Fig. 5.

The U.S. Environmental Protection Agency (1973) has compiled the data contained in Table 5 which indicate that active surface mines may generate

Table 5 Representative Rates of Erosion from Various Land Uses[a]

	Erosion rates		
Land use	(metric tons/sq km/yr)	(tons/sq mile/yr)	Relative to forest (equal to 1)
Forest	8.5	24	1
Grassland	85.0	240	10
Abandoned surface mines	850.0	2,400	100
Cropland	1,700.0	4,800	200
Harvested forest	4,250.0	12,000	500
Active surface mines	17,000.0	48,000	2,000
Construction	17,000.0	48,000	2,000

[a]Source: U.S. Environmental Protective Agency (1973).

sediment at a rate 2000 times that of undisturbed forest lands. While such extremes are possible, the findings of specific research studies, contained in Table 6, suggest dramatic but considerably more modest changes in erosion rates. Sometimes, erosion from undisturbed areas may be greater than from reclaimed lands [Hadley *et al.*, 1981 (Montana); and Mitchell *et al.*, 1983 (Illinois and Indiana)]. It should be noted that the research reported in Table 6 was undertaken for a variety of reasons; methodologies and environmental settings varied widely.

Collectively, there is abundant evidence to document substantial increases in the amount of work performed by geomorphic processes as a result of radical changes to the environment caused by surface mining. To these impacts could be added those to biotic communities, but this goes beyond the scope of this analysis. In a final note, there are several reports providing erosion rates on mined lands alone, but not including rates on undisturbed areas for comparison (e.g., Curtis and Superfesky, 1977; Haigh, 1979a,b, 1980; Haigh and Wallace, 1982; Hartley and Schuman, 1983; Drake, 1980; Mandel *et al.*, 1982).

Duration of Disturbance

The third component to this assessment of disturbance, duration, refers to the length of time that process rates remain accelerated and impacted areas continue to be affected by the consequences of the surface mining operation. The adage "time heals all wounds" does not necessarily apply to surface mined lands. In a discussion of strip mining of western coal, Atwood (1975) asserted that surface mining without reclamation can remove the land forever from productive use; such land can best be classified as a national sacrifice area. Conversely, with proper reclamation planning and practice, lands may be taken out of original production for a relatively short time, perhaps a decade or less. If the postmining land use does not require reestablishment of vegetation, such as residential or industrial development, the length of disturbance may extend to only 4 or 5 yr. Examination of duration, therefore, must consider whether or not the land has been reclaimed.

Mined lands disturbed prior to the Surface Mining Control and Reclamation Act of 1977 or before applicable state legislation were sometimes left untreated; these areas are referred to as orphaned, abandoned, or derelict mine lands. McKenzie (1980) observed that some of these lands were reclaimed by nature to become forest, recreational, and wildlife resources. Although they were unnatural landscapes, they were neither a hazard to the public nor a source of pollution. Curtis (1971) and Curtis and Superfesky (1977) found that maximum sediment yield occurs during active mining operations and drops off within 1 to 2 yr after completion of the mining in some watersheds.

Table 6 Some Changes in Erosion and Sedimentation Rates Due to Surface Mining

Location of study	Surface mined land	Natural land	Reference
Eastern United States			
Kentucky	47 metric tons/ha from mined watershed	0.7 metric tons/ha from comparable unmined watershed	Collier *et al.* (1970)
Kentucky	Maximum sediment concentrations of 46,400, 26,900, and 9,600 ppm for three watersheds during active mining	Sediment concentration of 150 ppm from adjacent unmined watershed	Curtis (1971)
	During 5-month period (5/68–10/68) 2,905, 22,403, 4,943 cu ft/sq mile of sediment trapped in and removed from three mined watersheds	During same period no sediment removed from weirs in nearby unmined watersheds	Curtis (1971)
Alabama	Estimated (USLE) soil loss 54.7–331 tons/acre from selected basin during active mining	Estimated soil loss 0.04 tons/acre from same basin prior to mining	Shown *et al.* (1982)
Illinois (Mine S)	Soil loss 0.803 kg/m^2 for reclaimed site with vegetation removed and disked	Soil loss 1.054 kg/m^2 for cultivated unmined site with vegetation removed and disked	Mitchell *et al.* (1983)
Illinois (Mine M)	Soil loss 0.170 kg/m^2 for reclaimed site with vegetation removed and disked	Soil loss 0.777 kg/m^2 for cultivated unmined site with vegetation removed and disked	Mitchell *et al.* (1983)
Indiana (Mine A)	Soil loss 0.728 kg/m^2 for reclaimed site, A-horizon over B-horizon over graded overburden, with vegetation removed and disked	Soil loss 1.978 kg/m^2 for cultivated unmined site with vegetation removed and disked	Mitchell *et al.* (1983)
Western United States			
Wyoming	Sediment yield of 529 (dry) and 1200 (wet) tons/sq mile from rehabilitated site	Sediment yield of 249 (dry) and 283 (wet) tons/sq mile from nearby natural site	Lusby and Toy (1976)
Wyoming	Sediment yield of 1729 (dry) and 2437 (wet) tons/sq mile from rehabilitated site	Sediment yield of 6.6 (dry) and 39.7 (wet) tons/sq mile from nearby natural site	Lusby and Toy (1976)

Table 6 *(Continued)*

Location of study	Surface mined land	Natural land	Reference
North Dakota	Soil loss from bare spoils averaged 15,000 kg/ha and 21,000 kg/ha for several cultivated and non-cultivated spoil sites	Sediment production of 200 kg/ha from unmined rangeland site	Gilley *et al.* (1977)
Wyoming	Sediment yield for mined basin 185 $m^3/km^2/yr$	Sediment yield for unmined basin 16 $m^3/km^2/yr$	Ringen *et al.* (1979)
Wyoming	Sediment yield from reclaimed sites averaged about 430 g for 2 yr	Sediment yield from native sites averaged about 80 g for same period	Gifford (1983)
Wyoming	Estimated (USLE) soil loss 0.18 tons/acre/yr from mined area	Estimated soil loss of 0.03 tons/acre/yr from nearly identical unmined area	Frickel *et al.* (1981)
Montana	Estimated (USLE) soil loss 1.15 tons/acre/yr from reclaimed area	Estimated soil loss of 1.59 tons/acre/yr from area before mining	Hadley *et al.* (1981)
North Dakota	Soil loss from various reclaimed sites: 16 lb/acre for ungrazed, 57 lb/acre for lightly grazed, 138 lb/acre for moderately grazed, and 940 lb/acre for heavily grazed	Soil loss from ungrazed native lands 7 lb/acre	Hofmann *et al.* (1983)

However, McKenzie (1980) also remarked that not all the lands have the capacity to recover rapidly from the severe disturbance of mining; decades after mining, these areas, with little recognized amenity value, are still barren and undergoing rapid erosion. Effective natural reclamation is certainly the exception rather than the rule. Commercial airline flights over virtually any of the nation's mining districts will reveal abandoned mines of various ages to the observant. The National Academy of Sciences (1974, p. 60) describes turquoise mine workings in the Cerrillos Hills near Santa Fe, New Mexico, which have been exploited periodically since the fourteenth century:

> These deposits were last worked about seventy years ago. Previously, they had been mined during Spanish colonial times in the 16th century, and long before then by Indians. No restoration was attempted. The mines all look much the same: the Anglos' diggings, the conquistadores', and the Indians'. No plants grow in abundance. The signs of disturbance are plainly visible, even to the untrained eye.

In that report, it was also observed that in Montana and the Dakotas the ungraded surface mined land of the 1920s to 1950s resembles the kettled topography of the terminal moraine left by the retreat of the continental ice sheet 10,000 yr ago, where no stream drainage networks have formed. It is suggested that in these environments vegetation slowly establishes itself, with adequate cover developing in 40 to 60 yr and that it is highly unlikely that such time requirements for natural revegetation will be acceptable to most of society.

There are numerous interesting references to soils development on unreclaimed mined lands. The National Research Council (1981) submits that if relatively long periods are acceptable (50 yr or more), natural soil-forming processes may be relied upon to recover the full productive potential of mined lands. This must assume, however, a transport-limited erosion system, wherein soils are not stripped away by wind or water nearly as fast as they develop. In this report, it is also suggested that under normal conditions, the buildup of organic matter in disturbed soils to predisturbance levels is a process that takes 100 yr or more.

Smith *et al.* (1971, p. 35) compared natural soils and old (70- to 130-yr-old) iron ore spoils in West Virginia drawing the following conclusions:

> Natural soils had distinctly lower bulk densities; higher porosity at all depths; stronger aggregation or soil structure development; higher nitrogen and organic matter contents; generally higher water intake (both wet and dry); silt loam, loam or light clay loam surface soil textures versus shaly clay loams for the spoils; less mica and more vermiculite in strongly acid soil than in comparable spoil; similar field moisture tensions in the top two feet of depth; plant rooting zones of 26 to 36 inches in natural soils versus rooting depths of 72 inches or more in spoils; higher cation exchange capacities and higher basic nutrients (Ca, Mg, K) in the top one or two inches of natural soils; but below two inches, higher exchange capacities and higher basic nutrients in the spoil; significantly more phosphorous in plants grown on spoils than on natural soils; deeper and more abundant total rooting of forest trees on spoils than on natural soils . . .

In some ways, the spoils were actually superior to natural soils for plant growth; but, of interest to us, the lower bulk densities, higher porosities, higher infiltration capacities, higher organic matter contents, and better soil structures indicate that natural soils would likely produce less runoff and possess lower inherent erodibility than spoils, other factors being equal. Better root development on the spoils would counteract the effects of greater runoff and erodibility to some extent.

Schafer *et al.* (1979) examined soil genesis on spoils of various ages at a site in Montana surface mined for coal since 1924. They concluded as follows:

1. Processes that have formed natural soils in the Colstrip area are also occurring in mine-soil. Only a small part of the change that created natural soils has occurred in 50-yr-old mine soils.
2. Some soil properties change rapidly in mine soil and may reach equilibrium in 200 to 400 yr. These include organic matter content, pH, and soil

structure. Other properties such as $CaCO_3$ distribution may require as much as 10,000 yr to resemble native soil. Because of their unique origin, mine soil will probably always remain different than undisturbed soils in some ways.

3. Four to six yr were required for root systems to develop on new mine soils that are similar in biomass and distribution to root systems of native plant communities.
4. Three to four yr were required for microbiological activity to reach the levels common in natural soils.

Numerous other references concerning soil development on mine spoils are included in the National Research Council (1981) report.

Additional research has centered upon the influence of abandoned mines on streams and drainage systems. For example, in Virginia, Matter *et al.* (1978) found that sulfate, total hardness, and silt indices were elevated in streams draining abandoned mines. As a result, total abundance and taxonomic richness of fish and benthic invertebrate populations were reduced in streams draining orphaned mine areas, even 10–20 yr after abandonment. Interestingly, from our viewpoint, these reductions bore some relationship to the degree of sedimentation, but not to pH.

Mandel *et al.* (1982) examined gully development at two coal refuse sites abandoned during the late 1940s in Kansas. They concluded that such refuse piles are still a major sediment source. For 10 locations at these two sites, sediment yield averaged 12.6 lb/sq ft (61.5 kg/m^2) of gully surface. Figure 6 shows the gully development in this area.

From the foregoing, it is apparent that abandoned surface mined lands are capable of significant environmental impact long after the termination of mining operations and that natural regenerative processes operate slowly in most cases. The inclusion of Title IV, Abandoned Mine Reclamation, in the Surface Mining Control and Reclamation Act of 1977, with provisions to reduce the environmental and human hazards of these sites through the creation of funds and fees attached to current coal production, attests to society's general unwillingness to await eventual natural repair to the disturbed land.

Although there has been no systematic study of the issue, it seems reasonable to suggest that the rate of natural recovery of abandoned surface mined lands is largely a function of climate and the physical and chemical properties of the overburden. Under ideal conditions of positive water budgets supplied by relatively nonerosive rainfall regimes, coupled with materials that are high in water-holding capability, low in erodibility, high in required plant nutrients, and low in toxic elements and compounds, autoreclamation may proceed at a maximum rate. Soil and vegetation development may satisfy the demands of society. However, the landscape configuration will remain dissimilar to the surrounding

Figure 6. Accelerated erosion in the form of gully development on coal refuse pile in Kansas.

undisturbed land for hundreds and perhaps thousands of years. The reduction of erosion rates as vegetation cover develops may, in fact, prolong the time necessary for landscape adjustment. The question becomes, Does this unreclaimed topography represent a new condition of equilibrium for disturbed surface materials, or does it represent disequilibrium merely waiting for a triggering force to bring about a major modification of landforms and, in the process, cause substantial off-site environmental damage?

The duration of impact for reclaimed mine land is quite different. Existing federal and state legislation, with extensive guidelines and regulations, ensure effective reclamation. Performance bonds are posted by mine operators and may be released by the regulatory authority only after the stability of reclaimed land has been demonstrated over a period of years. Reclamation programs must progress concurrently with coal production. Timeliness is further encouraged by the desire of the operator to secure bond release so that encumbered securities may be applied to newly disturbed lands or used for other purposes.

Reclamation methodology has evolved rapidly in the past decade. Each year professional journals and symposia disseminate the results of copious reclamation research. These fora, together with the trade magazines, provide photographic and statistical evidence of reclamation successes. The controlling legislation met considerable resistance from the surface mining coal industry, but the progress in the state of the reclamation art might rank among the more notable technological achievements of recent years. A refreshing new attitude has emerged that is expressed by the contention of some that restrictive legislation precludes the opportunity to actually improve upon premining land characteristics through the reclamation process. Noteworthy is a volume resulting from a 1982 conference in North Dakota entitled "Can Mined Land Be Made Better Than Before Mining" (Land Reclamation Research Center *et al.*, 1982).

Further discussion of successful reclamation practices will be reserved for a later section of this chapter. Suffice it to say at this point that it is reasonable to expect successful revegetation and erosion control of reclaimed lands within a few years after coal extraction.

Summary

In conclusion, this assessment of land disturbance due to surface mining of coal has shown that: (1) contrary to popular perception, areal extent is not really the principal component of environmental impact; (2) intensity, however, is very significant because geomorphic processes may function at rates several times those of adjacent undisturbed lands; and (3) duration is likewise an important component because recovery times for various watershed variables may be measured in tens, hundreds, or even thousands of years without effective reclamation. The environmental problem of land disturbance due to surface mining, then, is largely a geomorphic problem. Historically, many people have been rightfully concerned about the impacts of surface mining of coal, but for the wrong reason. The data compiled by Paone *et al.* (1974) simply do not support the devastation of vast areas by fossil fuel extraction.

The Nature of Disturbance

An understanding of the means by which surface mining alters the physical environment is a prerequisite to an appreciation of the remaining major topics to be discussed in this chapter. We shall now describe the stages of the mining process which cause modification of vegetation, soil, topography, and surface and groundwater regimes, along with the geomorphic consequences.

The coal extraction procedure by surface mining techniques consists of several steps. First of all, trees, shrubbery, and other obstructions are cleared and grubbed from the mine site. Next, topsoil is removed by motor scrapers and carried to stockpile locations. Any unconsolidated materials are likewise removed and deposited in spoils piles. Exposed consolidated geologic strata overlying the coal are drilled and blasted. These are also moved to spoils piles by dragline or shovel and truck. Excavation has now extended to the depth of the coal resource which is drilled and blasted into blocks and fragments of manageable size. These are loaded by shovels or "front loaders" onto trucks for transport to rail or conveyor head. There are numerous variations of this process, but the above is fairly common.

Surface Disturbance

Vegetation stripping, by itself, would be enough to substantially disrupt the natural geomorphic system in operation at a particular site. As described in an earlier chapter, concerning geomorphic processes, vegetation cover is the first line of defense, resisting the erosive forces of rainsplash and overland flow. The importance of vegetation cover on soil surfaces can be empirically examined through the cover (C) factor incorporated into the universal soil loss equation (USLE) (Wischmeier and Smith, 1965). Soil loss from base surfaces is computed as the product of rainfall energy, inherent soil erodibility, topography, and any erosion control practice factors. The cover factor adds another multiplier into the equation which is usually a small decimal fraction. Gray and Leiser (1982, p. 20) remarked

> Factor C values range as low as 0.003 for well established plant cover. This corresponds to almost a thousand-fold reduction in erosion losses over the a continuous-fallow or bare-ground case. Few other variables or factors are amenable to management with such dramatic results (i.e., reduction in erosion losses) as this one.

Surface infiltration capacities are also usually directly proportional to vegetation cover, so as vegetation cover decreases, infiltration capacity decreases. This would result in greater volume, depth, and velocity of runoff for a given rainfall event, which in turn would result in greater shear and tractive forces impinging upon the soil surface. However, in this case, the relationship is not especially relevant because the topsoil itself is removed in the next step.

Prior to the enactment of federal or state legislation and regulation, topsoil was not generally removed and stockpiled in a separate mining operation. Ellison (1976) indicated that topsoil removal could account for up to 5% of the total cost of mining. Most early miners elected to realize this savings by handling the topsoil right along with other spoils material. This resulted in a mixing of materials, some of which were satisfactory for plant growth and some of which possessed nutrient deficiencies or toxicities. Today, Section 515 (b)(5) of the Surface Mining Control and Reclamation Act of 1977 specifies removal of topsoil from the land in a separate layer, along with guidelines for subsequent handling. Occasionally, the topsoil may be removed in two "lifts," taking the A- and B-horizons separately. However, Schuman and Power (1981) assert that unless subsoil characteristics are problematic, perennial crops do not benefit enough from soil profile segregation to warrant the higher handling cost. Figure 7 shows a "topsoil island" which remains after stripping for inspection by a regulatory authority.

The effects of topsoil removal and handling have been the focus of considerable research. Most have dealt with comparisons of plant growth on redistributed topsoil and graded spoils material (e.g., Power *et al.*, 1976; Schuman *et al.*, 1980, 1985) or the effect of stockpiling on biological activity within the soil (e.g., Miller and Cameron, 1976; Singleton and Williams, 1979).

A few studies have examined the consequences of topsoil handling that are of geomorphic interest. Miller and Cameron (1976) found that stockpiled topsoil, at a site in North Dakota, had increased bulk density, decreased water-holding capacity, and smaller amounts of organic matter than undisturbed topsoil. Vogel (1981) suggested that replaced topsoil may erode more easily than some spoils materials. It should be noted that changes in bulk density and related properties

Figure 7. Residual topsoil island for inspection following the stripping of topsoil.

are frequently a function of handling practices and particularly the amount of equipment traffic to which the topsoil is subjected.

Paone *et al.* (1978, pp. 20–21) described some of the differences between topsoil and spoils materials thusly:

> One of the characteristics distinguishing soils and spoils is the location of horizons that retard the infiltration and percolation of water, and that direct the development of root growth. Where topsoils are intact, surface infiltration rates are usually higher than the percolation rates of deeper horizons, and the root distribution of plants is mostly in or near the surface. The opposite is generally true of lands disturbed by mining. The contrast has recognized significance in reference to infiltration of moisture for plant use but otherwise has received little attention for applications that may be beneficial in reclamation.

Thus, topsoil removal that mixes the eluviated and illuviated horizons or breaks the various types of "pan" horizons that may develop in soils is likely to increase the infiltration capacity of the soil. This should reduce the volume, depth, and velocity of overland flow from a given rainfall event, other factors being equal. Stearns *et al.* (1984) examined infiltration capacities on several plots at the Rosebud Mine near Colstrip, Montana. The average infiltration rate, using a ring infiltrometer, was 10.2 cm/hr for undisturbed sites. Using a rainulator on two reclaimed sites showed an infiltration rate of 3.5 cm/hr, while the same technique applied to two undisturbed sites produced a rate of 3.0 cm/hr. From the uniform application of water by rainulator at sites in North Dakota, Gilley *et al.* (1977) found that runoff averaged 60 and 70% of rainfall applied on bare topsoil and spoils, respectively, in contrast to 12% on an undisturbed rangeland site. Thus, 40 and 30% of the rainfall entered the soil on the topsoil and spoils site, respectively, while 88% infiltrated the soil on the rangeland site. High sodic spoils such as those used in this study tend to disperse and crust resulting in low infiltration capacities. High sodic soils, of course, have the same tendency. Hofmann *et al.* (1983) examined runoff from plots using rainfall simulation at another North Dakota mine. They obtained 0.11 in. from an ungrazed, reclaimed plot and 0.04 in. from an ungrazed, native plot when water was applied at a rate of 1.8 in./hr for 1 hr. Hence, it can be again inferred that infiltration capacity was higher on the native land.

Lusby and Toy (1976) also used rainfall simulation to examine both runoff and sediment yield from an undisturbed site and a reclaimed site at each of two surface mines in Wyoming. At one location, runoff was less from the reclaimed site (0.60 in.) than from the undisturbed site (0.78 in.) when antecedent soil moisture was low. However, under wet soil conditions, runoff was greater from the reclaimed site (1.23 in.) than from the undisturbed site (0.85 in.). At the second location, runoff was always considerably greater from the reclaimed site. With low antecedent soil moisture, the undisturbed site produced 0.03 in. of runoff, while the reclaimed site produced 0.64 in. With wet soil conditions, the undisturbed site produced 0.13 in., while the reclaimed site produced 0.82 in. In all cases, the reclaimed sites generated greater sediment yields than the undisturbed site.

Because rainfall applications were uniform throughout these experiments of Lusby and Toy, it may again be inferred that infiltration capacities were generally greater on the undisturbed sites as compared to the reclaimed sites. These reclaimed sites were usually not well topsoiled, and the clay content of the surface material was always higher at the reclaimed sites. Nevertheless, differences in infiltration capacities of the surface materials probably only partially explain these differences in runoff. For example, the average slope was greater for the reclaimed sites as well.

There has been considerable concern regarding changes in soil erodibility as a consequence of removal, stockpiling, and replacement. Lang *et al.* (1983) concluded that (1) stockpiled topsoils apparently maintain some of the properties that influence erodibility (aggregate stability, dispersion, and crusting) found in soils before they were disturbed, and (2) stockpiled topsoils that were cultivated prior to stockpiling are potentially more erodible after respreading than topsoils that were in native vegetation, other factors being equal. These experiments were also conducted at surface mines in North Dakota.

The *K* factor of the USLE (Wischmeier and Smith, 1965) expresses the inherent erodibility of a soil. Hence, changes in *K* factor values or comparisons between these values on disturbed and undisturbed topsoil may be used as a basis for determining the effects of disturbance on erodibility. Table 7 contains a compilation of *K* factor data collected under various conditions using rainulators. Gilley *et al.* (1977) found that topsoil was more erodible than spoils, although they unfortunately did not provide a *K* factor value for the topsoil of an undisturbed location. Mitchell *et al.* (1983, p. 1421) found that mean soil erodibilities for the respective runs were generally greater on the unmined sites than those calculated for the reclaimed sites. They concluded

> One of the concerns that prompted this study was that reclaimed mined sites were perceived to be more erodible than adjacent unmined land. It is obvious from this study that newly reclaimed topsoils are not inherently more erodible than unmined soils; in fact, they are apparently less erodible. Thus, any increased soil loss and, hence, sediment yield occurring on newly reclaimed mine land is apparently due to lack of cover and long slope lengths.

On the other hand, Stein *et al.* (1983) found that erodibility of reclaimed A-horizon plots was higher than unmined plots at two out of three sites and that reclaimed B-horizon plots generally had erodibilities lower than reclaimed A-horizon plots but higher than unmined topsoil plots. In unmined soils at this locale, B-horizons generally had higher erodibilities than A-horizons. This suggests that it is the erodibility of the reclaimed A-horizon that is most affected by mining.

Table 7 Soil Erodibility for Various Mine Materials

Location of study	Surface material	*K* factor	Reference
North Dakota			Gilley *et al.*
	Spoil, sandy clay loam, 4.6% slope	0.10	(1977)
	Spoil, sandy clay loam, 17% slope	0.02	
	Spoil, clay loam, 10% slope	0.08	
	Spoil, silty clay loam, 12.9% slope	0.04	
	Topsoil, sandy loam, 10.4% slope, 25 cm	0.25	
	Topsoil, sandy loam 9.9% slope, 61 cm	0.29	
Illinois			Mitchell *et al.*
(Mine S)	Reclaimed including topsoil-ing, silt loam	0.37	(1983)
	Unmined, silt loam	0.48	
Illinois			Mitchell *et al.*
(Mine M)	Reclaimed including topsoil, silty clay loam, and silt loam mix	0.21	(1983)
	Unmined, silty clay loam	0.31	
Indiana			Mitchell *et al.*
(Mine A)	Reclaimed including topsoil, silt loam	0.65 (A-horizon on overburden) 0.45 (A-horizon on B-horizon on overburden)	(1983)
	Unmined, silt loam	0.72	
Indiana			Stein *et al.*
(Ayrshire Mine)[a]	Reclaimed including topsoil A over B horizon, 6% slope	0.051	(1983)
	Reclaimed including topsoil, A-horizon, 6% slope	0.069	
	Reclaimed including topsoil, A-horizon, 12% slope	0.067	
	Reclaimed including topsoil, A-horizon, 18% slope	0.074	
	Reclaimed including topsoil, B-horizon, 6% slope	0.076	
	Reclaimed including topsoil, B-horizon, 12% slope	0.082	
	Overburden, 6% slope	0.031	
	Unmined, silt loam, 6% slope	0.091	

Table 7 *(Continued)*

Location of study	Surface material	*K* factor	Reference
Indiana (Solar Sources Mine)[a]	Reclaimed including topsoil, A over B horizon, 6% slope	0.107	Stein *et al.* (1983)
	Reclaimed including topsoil, A-horizon, 6% slope	0.126	
	Reclaimed including topsoil, A-horizon, 12% slope	0.131	
	Reclaimed including topsoil, A-horizon, 18% slope	0.125	
	Reclaimed including topsoil, B-horizon, 6% slope	0.085	
	Reclaimed including topsoil, B-horizon, 12% slope	0.119	
	Overburden, 6% slope	0.110	
	Unmined, silt loam, 6% slope	0.059	
Indiana (Chinook Mine)[a]	Reclaimed including topsoil, A over B horizon, 6% slope	0.111	Stein *et al.* (1983)
	Reclaimed including topsoil, B-horizon, 6% slope	0.061	
	Overburden, 6% slope	0.026	
	Unmined, silt loam, 6% slope	0.058	

[a]Textural classification not reported on disturbed and reclaimed sites.

Collectively, these results indicate that the change in erodibility is site specific and that broad generalizations concerning the impact of removal on topsoil erodibility may be unwarranted. Although subject to field verification, it would appear that particular handling techniques, including compaction by equipment traffic and surface manipulation after respreading (to be discussed later), and the length of time stockpiled may influence changes in inherent erodibility. During prolonged stockpiling, there is sometimes a decrease in organic matter content due to decay and this is one soil constituent related to erodibility.

The next step in the mining process is drilling and blasting of consolidated geologic materials and their removal to spoils piles. Thereafter, the coal resource is extracted and transported to loading facilities for market. The geomorphic consequence is that hillslopes, channels, and drainage systems, probably heretofore in equilibrium with their environmental settings, are replaced by spoils piles, contour benches, highwalls, and minepits. The ramifications of such drastic landscape modification are numerous and may be discussed in terms of materials, processes, and forms.

The original consolidated geologic strata possessed material strength much greater than the fragmented spoils. As a result, it was capable of better resisting gravity and other forces which cause mass movement. In addition to reduced material strength, high porosity and permeability of spoils material allows absorption of rainfall and runoff which increases the weight and pore pressures of this earthen mass. Subsurface flow along the interface between the buried hillside surface and the bottom of the spoils material may serve to lubricate the slide plane as well. Periodic explosions associated with mining operation can provide the trigger to set unstable masses in motion. With areal stripping, this colluvium is likely contained on-site; with contour stripping, the landslide and colluvium may initiate a chain reaction of off-site environmental impacts.

These unconsolidated materials also offer considerably less resistance to erosion processes driven by wind or water. While erosion may have been weathering limited before disturbance, it is more likely to be transport limited afterward. The data provided above suggest that whether the spoils are more or less erodible than the original topsoil depends on site-specific characteristics; in general, it probably tends to be less erodible. The "stoniness" and armoring of the spoils surface might explain this decreased erodibility, although DePloey (1981) discovered that stone covers can sometimes increase sheet erosion by increasing the turbulence of overland flow under certain conditions.

Alteration of the landscape also has a profound effect on hydrologic processes. Changes in the infiltration capacity of topsoil have been acknowledged. Although there was variation in the results, in the majority of cases infiltration rates were greater for undisturbed soils than for reclaimed soil. Smith *et al.* (1971) and Gilley *et al.* (1976) also found infiltration to be less into spoils material than into undisturbed soil. This is confirmed by research conducted by Arnold and Dollhopf (1977) in Montana, together with several other studies in various locations by various investigators. Interestingly, Schafer *et al.* (1979) found no difference between infiltration rates on spoils and undisturbed soil.

Handling may be chiefly responsible for lower infiltration rates in spoils as compared to natural soils. Limstrom (1960) found infiltration rates up to seven times greater on ungraded spoils as compared to graded spoils in Ohio. Grandt and Lang (1958) found infiltration rates on ungraded spoils from 2 to 40 times greater than on graded spoils in Illinois. Compaction due to heavy equipment traffic can substantially reduce infiltration rates.

Usually, barren spoils piles are capable of generating runoff faster than the undisturbed surrounding areas. Erosion rates are a function of erodibility as well as the forces produced by this runoff, but there is ample evidence that spoils piles are major sources of sediment from most surface mining operations. This runoff and sediment is capable of considerable off-site damage; hence, the rationale for diversions, dams, and barriers along the perimeters of mines, as shown in Figure 2. Similar water control structures are utilized with contour and areal stripping.

Drainage System Disturbance

The U.S. Environmental Protection Agency (1976) recognized that surface mining disrupts the natural drainage system and that if this disruption prevents runoff from leaving the disturbed site, the likelihood of sediment being carried into adjoining waterways is greatly reduced. However, should surface drainage from the disturbed area have uninterrupted access to the adjoining drainage system, then serious downstream sedimentation problems are a distinct possibility. In this report, written prior to the enactment of the Surface Mining Control and Reclamation Act of 1977, it was the opinion of the EPA that on-site containment was more likely with areal stripping, while off-site discharge of runoff and sediment was more likely with contour stripping. They commented:

> Several factors contribute to the magnitude of the problem. The most significant one is that contour mines have a narrow, linear geometry and, therefore, more spoil area drains directly into the offsite drainage system. Also, the bench area being actively mined, unlike the pit area of an area strip mine, often drains directly into the offsite drainage system. Another factor, one of extreme significance, is that the receiving waterway is generally closer to the source of sediment and separated from the source by relatively steep terrain. Additionally, the contour strip mine site receives more potentially erosive runoff from undisturbed areas at higher elevations due to the shallow soils and the linear exposure of the mined area to drainage areas above it. (p. 10)

Concerning mountain top removal mining, the EPA remarked that, considering the areal nature of these operations and the overall reduction in relief that is achieved, the potential for off-site sediment damage is likely to be less than for a contour strip mine disturbing an equal area of land. This, of course, would be true when surface drainage is controlled internally.

With the passage of strict federal and state legislation, the above observations now relate to the potential for off-site damage due to accidental release of water and sediment or failure of water control structures. The legislation and regulation in force requires management of water and sediment discharges from the mine site.

All types of surface mining modify the drainage pattern of the impacted area. With the area types, including areal stripping and mountain top removal, the premining drainage pattern is completely obliterated. Until reclaimed, these lands possess a form of "trellis–centripetal" drainage pattern. The linear arrangement of spoils piles produces the trellis features, and the lack of outlets between the rows of spoils piles results in a centripetal effect. The ponding of drainage internally, along with subsequent evaporation, results in on-site consumptive use of water. While this poses no problem in humid regions, it has been a minor concern in arid or semiarid western regions (National Academy of Sciences, 1974). These reservoirs may augment groundwater recharge, however (Ringler, 1983). An evaluation of an impoundment at the Big Sky Mine in Montana produced the rather surprising finding that, contrary to having any adverse effect on surrounding water resources, it appears to have improved them (Goering and Dollhopf, 1982).

Keefer and Hadley (1976) encountered an interesting drainage problem near Gillette, Wyoming. Donkey Creek is the only stream that flows through the area, and its eastward course is nearly perpendicular to the strippable tract of the Wyodak–Anderson coal. Because this seam is 75–100 ft thick over much of this area and overburden ranges from 0–200 feet in thickness, the ground surface will be lowered appreciably by coal extraction. How will the passage of Donkey Creek be maintained? Perhaps, it will be perched above the surrounding landscape as an aqueduct.

With contour strip mining there is produced a discontinuity between drainage patterns above and below the bench and headwall. Diversion ditches and engineered channels connect the two segments of drainage patterns while maintaining water and sediment discharges within specified limits.

Channel Disturbance

Historically, surface mining has engendered changes in the hydrologic regime of both disturbed and undisturbed downstream areas. The National Research Council (1981, p. 113) stated:

> An alteration of infiltration/runoff relationships in a watershed by mining can have two very different effects. Reduced infiltration during periods of high precipitation increases the incidence, duration, and intensity of floods, and less water enters the groundwater reservoirs so that during periods of low precipitation there is not enough ground water to maintain the base flow of streams. Increased infiltration, conversely, reduces flooding and increases base flow.

The Council observes that a common result of postsurface mining practices in Appalachia has been an increase in flooding intensity, probably due to a combination of reduction in vegetation cover and a reduction in channel capacity due to sedimentation. Collier *et al.* (1970) found greater peak flows in mined watersheds when compared to undisturbed watersheds in Kentucky, and Minear and Tschantz (1974) found the same in Tennessee. Similarly, Curtis (1972, 1979) recorded higher peak flows in mined watersheds as compared to undisturbed watersheds in Kentucky. However, following reclamation, peak flows were actually less than in undisturbed watersheds, presumably because of greater water-holding capabilities of spoils. Different results came from an examination of drainages in flat to gently rolling terrain of a glacial till plain in Indiana (Harza

Engineering, 1975). Here, flood peaks were reduced, due to the creation of depressions between ungraded spoils piles that did not allow runoff to leave the site. Increases in base flow indicate increases in infiltration at this locale. Lastly, Touysinhthiphonexay and Gardner (1984) found that the morphologic response of streams to disturbance by strip mining in Pennsylvania takes the form of enlarged channel dimensions and increases in the occurrence and size of moving blocks in the channel.

The National Research Council (1981) submitted that, in western states where ephemeral streams recharge groundwater reservoirs seasonally, mined-induced changes often decrease base flows in nearby streams because more water is lost by surface flows in shorter periods of time, leaving less to recharge the groundwater system. Streams that discharge small amounts throughout the year may dry up more frequently and flood more dramatically after establishment of a postmining contour and soil mosaic. However, there does not appear to be specific documentation for these assertions at present.

Additionally, surface mining also affects the sediment load of streams. Some evidence of these changes was provided in Table 6. The National Research Council (1981, p. 115) remarked in this regard that:

> Increased sediment discharge caused by mining may fill stream channels and reduce the volume of flow that streams can carry. The result is to increase the frequency of flooding. Again, this effect has been observed most often in Appalachia (Boccardy and Spaulding, 1968).

Although channel sedimentation may occur periodically due to the deposition of colluvium into the stream by landslides, water and sediment control structures have surely reduced the magnitude of this problem due to runoff from mined sites. Instead, another problem may be created by mandated release of water with very low sediment loads.

The consequences of changes in discharge and sediment characteristics on channel morphology are summarized by Schumm (1977) and were discussed in an earlier chapter. The relationships below have usually been developed on the basis of alluvial channels. The plus or minus exponent indicates how various channel parameters would likely respond to an increase or decrease in discharge or sediment load. If both a plus and minus appear, then the nature of change for this parameter is uncertain. Schumm (1977) proposed, based upon his own research and numerous references, the following relationships:

$$Q^{+}Q_{s}^{+} \simeq b^{+}, d^{\pm}, \lambda^{+}, S^{\pm}, P^{-}, F^{+} \quad (1)$$

$$Q^{-}Q_{s}^{-} \simeq b^{-}, d^{\pm}, \lambda^{-}, S^{\pm}, P^{+}, F^{-} \quad (2)$$

$$Q^{+}Q_{s}^{-} \simeq b^{\pm}, d^{+}, \lambda^{\pm}, S^{-}, P^{+}, F^{-} \quad (3)$$

$$Q^{-}Q_{s}^{+} \simeq b^{\pm}, d^{-}, \lambda^{\pm}, S^{+}, P^{-}, F^{+} \quad (4)$$

where Q is the water discharge, Q_s the bedload, b the width, d the depth, λ the meander wavelength, S the slope of energy line, P the sinuosity, and F the width/depth ratio. For example, if both discharge and bedload increase as a result of uncontrolled water and sediment discharge, one would expect channel width to increase, depth to increase or decrease depending upon proportional changes in water and sediment, meander wavelength to increase, slope or channel gradient to increase or decrease depending upon the proportional changes in water and sediment, sinuosity to decrease, and the width/depth ratio of the channel to increase. These would be the sorts of changes that one would expect in the channels draining the watersheds examined by Curtis (1971, 1972, 1979) during active mining and before effective reclamation. Following reclamation, if delayed sufficiently to allow the full adjustments anticipated by Eq. (1), there could be another series of alterations, as proposed in Eq. (2). The results of the various studies presented above caution us that the exact character of change will be site specific.

Alluvial channels are relatively free to adjust their form to accommodate the water discharge and sediment loads that they are obliged to transport because hydraulic forces are sufficient to move the unconsolidated materials composing their beds and banks. When in equilibrium, we might regard them as a transport-limited system because the graded condition determines that the energy and forces are balanced with the imposed resistances.

Rock outcrops in the beds or banks of the channel will effectively restrict the options for adjustment. It may simply not be possible for a channel to increase width, depth, or meander wavelength in response to increasing discharge. Morphologic change will have to be directed toward whichever avenues remain available in a particular setting. We might think of this as a weathering-limited situation because hydraulic forces are of insufficient magnitude to move the consolidated rock, except over extended time periods as a result of corrasion and corrosion. Hence, the nature of materials to be encountered needs to be considered when anticipating morphologic changes in stream channels.

Disturbance of Groundwater Regime

The excavation of overburden and coal resources also disturbs the groundwater regime of an area. Both the chemical (quality) and physical (quantity) characteristics of these flows are often modified. Groundwater quality is altered when it comes into contact with the minerals contained within the spoils. In the past, pyrite has caused serious environmental problems because the reaction between it, oxygen, and water produces a highly acidic drainage. Caruccio and Geidel (1978) provided a useful summary of the chemical process and its occurrence.

The physical properties of groundwater flow will be altered whenever the depth to coal exceeds the depth to ground water. The mining operation may effectively dewater an aquifer for some distance on either side of the excavation. Gardner and Woolhiser (1978) depict the consequences in Fig. 8. The mine acts as a large diameter well, and the water level in the aquifer is drawn down on all sides. The base flow of stream channels nearby could be affected by the general lowering of the water table, together with any shallow wells in the area. The response of the groundwater regime to reclamation is described by Gardner and Woolhiser (1978, p. 181):

> After the pit is backfilled, the piezometric surface will eventually approach the original piezometric surface. However, the transmissibility characteristics of the area may change and the piezometric surface may be much nearer the ground surface. If the hydraulic conductivity of the spoil is less than the effective conductivity of the coal seam, wells will require a greater draw-down to produce the same amount of water (McWhorter and Rowe, 1976). If the spoil is more permeable, the aquifer will be improved insofar as water quantity is concerned.

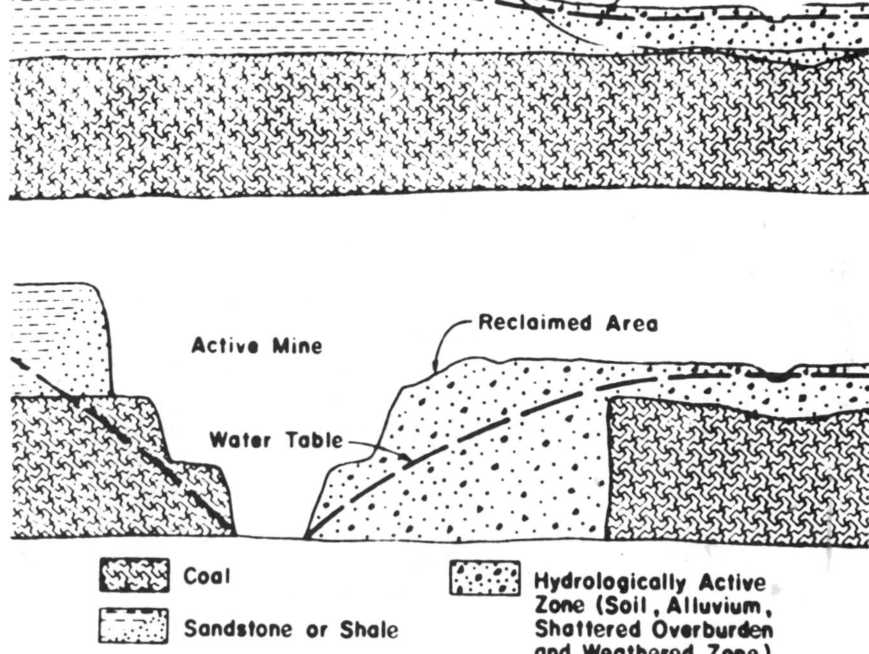

Figure 8. Effect of surface mining on ground water [Reproduced from "Reclamation of Drastically Disturbed Lands," 1978, Chapter 10, Gardner and Woolhiser, Hydrologic and climatic factors, pp. 173–191, by permission of the American Society of Agronomy, Inc., Crop Science Society of America, Inc., and Soil Science Society of America, Inc.]

Keefer and Hadley (1976) presented another example illustrating potential regional impacts.

Ringler (1983) suggested that the rate of spoils resaturation and thickness of the resaturated layer depends on physical characteristics of the spoils, method of replacement, and availability of ground water for recharge, based on his experience at the Rosebud and West Decker Mines in Montana. He also stated that three major sources of ground water for spoils recharge are adjacent unmined coals, underlying aquifers undisturbed by mining, and surface infiltration. Goering and Dollhopf (1982) found that surface impoundments may also provide recharge. Banaszak (1980) concluded that area coal mining should not cause irreparable harm to coal aquifers in the eastern United States, southwestern Indiana specifically, and probably will enhance recharge of this type of aquifer because the replaced spoil could act as an infiltration gallery. However, Helgesen and Razem (1980) found that 2 yr of postmining data reflected a slow rate of resaturation of overburden spoils for an eastern Ohio location. The notion of a spoils surface constituting an effective zone of recharge is challenged somewhat by information discussed earlier, wherein spoils had a lower infiltration rate than undisturbed soils in many cases. The top few meters of spoils can determine the infiltration capacity for the entire column and, hence, the rate of resaturation. If these have been compacted by grading, then overall resaturation could be slow. Proper handling is obviously critical if spoils are expected to effectively recharge aquifers and raise the piezometric surface.

Shallow groundwater resources in alluvial valley floors have been a controversial subject since the enactment of the Surface Mining Control and Reclamation Act of 1977. Fundamentally, alluvial valley floors are geomorphic entities, landforms created by the geomorphic processes of fluvial erosion and deposition over extended periods of time. Additionally, they are hydrologic entities. Perennial, intermittent, or ephemeral streams flow across their surface, while ground water flows below the surface through unconsolidated valley fill, or alluvium. For many years in the West, they have been economic entities because flood irrigation or subirrigation of vegetation on their surface provided forage for the livestock of the ranching industry. Since 1977, they have been legal entities as well because provisions of the Surface Mining Control and Reclamation Act sought to exclude areas of alluvial valley floors from disruption by surface mining despite the presence of valuable coal beneath them. While alluvial valley floors had often been an interface between surface water and groundwater hydrologic systems, they became an interface between government and industry. Beach (1980) remarked that the intent of the statutory language was clear; Congress decided that lands containing alluvial valley floors were of greater value to the agricultural industry of the West than were the coal reserves underlying these lands to the nation's energy demand. We cannot dwell upon this complex and sometimes emotional issue, but brief discussion of some geomorphic perspectives is appropriate. Thorough treatment of the subject is available in a U.S. Office of Surface Mining Reclamation and Enforcement (1983) publication entitled "Alluvial Valley Floor Identification and Study Guidelines."

Figure 9 depicts an alluvial valley floor. The geomorphic history may include numerous stages, but must include the formation of a bedrock valley, usually by fluvial erosion, followed by the filling of this valley with unconsolidated alluvium. Stratigraphic examinations of these valleys commonly show several cycles of cutting and filling.

Hadley and King (1980) described the geomorphic and hydrologic characteristics of alluvial valley floors in the western United States. First, they noted that alluvial valley floors and the ephemeral stream channels that frequently occupy them in arid and semiarid environments are unstable and are susceptible to erosion if the longitudinal profile is disturbed. Even if undisturbed, stratigraphic evidence suggests that gullying would eventually occur due to natural processes, although this might not happen for many years. Begin and Schumm (1979) proposed a methodology for assessing instability in alluvial valley floors based

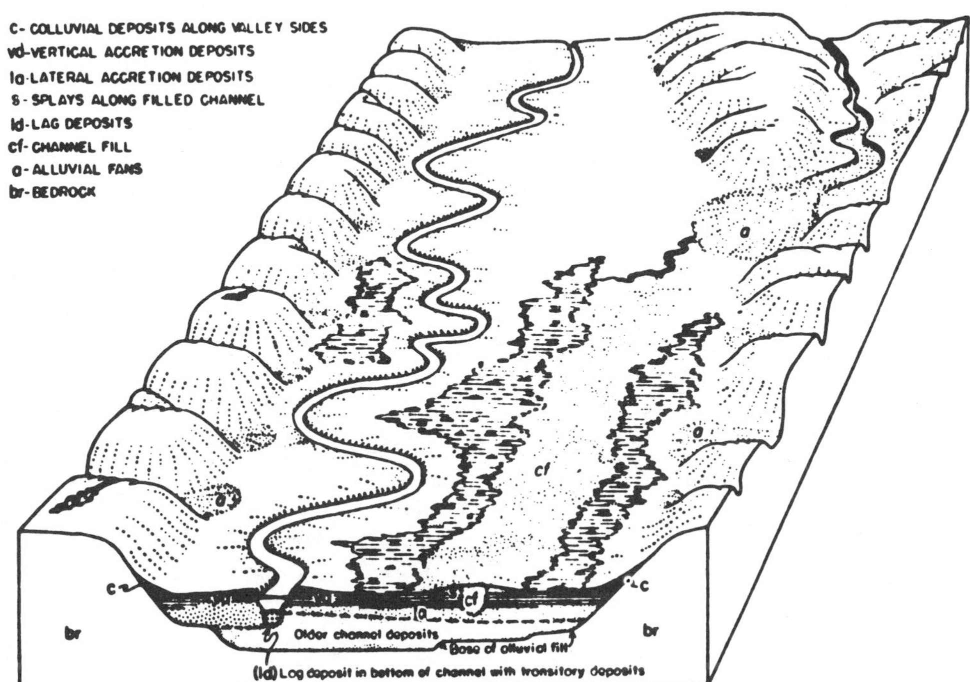

Figure 9. Three-dimensional view of an alluvial valley (Happ *et al.*, 1940).

upon relationships between drainage area, discharge, and flow depth, but their procedure requires additional refinement.

The entrenchment of channels into alluvial valley floors can cause a local lowering of the water table and draining of soil moisture, thus depriving subirrigated riparian vegetation of an adequate water supply. Loss of vegetation cover results in a decrease in infiltration capacity but an increase in vulnerability of the surface to erosion. Consequently, the proportion of rainfall from given events that enters the channel will increase and the sediment load would also be greater. For a perennial stream, the alterations of channel morphology indicated in Eq. (1) would apply; for ephemeral streams, the sediment load will probably be deposited in-channel downstream and, perhaps, contribute to the development of instability once again. Additionally, the entrenchment of the channel reduces the likelihood of natural flooding of the valley floor with the natural irrigation of vegetation.

Hadley and King (1980) stated that the potential geomorphic impacts of surface mining on alluvial valley floors with regard to fluvial processes are twofold: (1) disruption of the longitudinal profile of the channel followed by valley trenching upstream and high sediment yields downstream, and (2) replacement of the excavated alluvium with rock and unconsolidated material that does not have similar physical characteristics and may be susceptible to erosion following reclamation. Also, the postmining terrain may be subject to local subsidence as a result of compaction. The disruption of the longitudinal profile of the channel produces an unstable steepened reach or an incipient headcut. Replacement or mixing of alluvium with finer textured overburden material often increases the erodibility of the surface due to reduced infiltration capacities. Heavy equipment traffic during reclamation can exacerbate this problem unless countermeasures are taken.

The change in position of the water table, shown in Fig. 8, demonstrates that actual surface mining operations need not pass directly through the alluvial valley floor to have an impact on subirrigated vegetation. The potential for this form of near-site impact should be recognized prior to the initiation of mining.

Finally, leaching of soluble elements and compounds from the replaced spoil may have an adverse effect on groundwater quality. Although water may be present in sufficient quantity to allow vegetal growth, the quality may be problematic due to the presence of toxic substances or very low pH.

It is not inevitable that all surface mining will permanently destroy the function of all alluvial valley floors. Procedures prior to, during, and following the mining operation can substantially reduce potential problems and alleviate unavoidable consequences; these may be expensive, however.

From all of the foregoing, it is apparent that surface mining causes severe disturbance to the environment and attendant geomorphic systems. The ramifications of these disturbances are not always the same. General tendencies exist, but there are often exceptions. Infiltration capacities for spoils or disturbed topsoil are not always lower than for undisturbed sites. Soil erodibility is not always greater for disturbed topsoil than for undisturbed sites. Peak flood flows are not always increased by surface mining operations. In some cases, spoils may serve as an infiltration gallery for ground water allowing fairly rapid recharge; in other circumstances, it seems that recharge is a prolonged process. It is clear, then, that impact analysis is a site-specific undertaking. There appears to be conflicting data available for nearly every generalization one might propose or every preconception that one might hold. We will comment further on the problems created by the range of environmental response to disturbance in a subsequent section addressing the subject of cumulative impact.

Spatial Aspects of Disturbance

The discussion thus far has alluded to two basic categories of environmental impact, on-site and off-site. While admitting that there is no simple procedure for projecting their magnitude, the National Academy of Sciences (1974) suggested that, as a rule of thumb, the land area affected by mining might be twice the average disturbed in mining itself. Elsewhere, Box (1978, p. 7) remarked:

> In most cases the on-site disturbances are receiving attention, but, unfortunately, many times the amount disturbed for the actual mining operation is usually disturbed in roads, transmission lines, dams, etc., necessary for water for the project. These associated projects generally create many more off-site effects than those on the site itself.

He then continued, describing the indirect consequences of the "boom town" effect wherein large numbers of people move into an area to provide for the labor demands of the industry. As a result, there are many land use changes because these new residents require homes, parks, sewer systems, schools, etc. In his remarks concerning indirect off-site land disturbance due to surface mine development, Box (1978, p. 7) commented:

> For instance, the amount of off-road vehicle use has been shown to increase around such mining towns as Rock Springs, Wyoming; Vernal, Utah; or other western towns associated with energy development. In these arid environments, the drastic disturbance caused by four-wheel drive vehicles climbing the surrounding mountains may cause a greater contribution to sediment load in the surrounding rivers than the mining operation where the drivers of the vehicles work. The loss through vandalism of archaelogical and historic artifacts cannot be estimated. However, these effects were not anticipated in the reclamation reports submitted by mining companies prior to the development of their mines.

While these sorts of land disturbance are very real, they are clearly not limited to surface mining but probably accompany any economic development and concomitant population growth in virtually any locale. Box (1978) seemed to suggest that mining companies have some special obligation to function "in loco parentis" and assume responsibility for any actions of their employees which are anywhere detrimental to the environment. Until all industries are made liable for the actions of their employees, this may extend the list of off-site impacts beyond reason.

Nevertheless, it is possible now to identify several categories of land disturbance associated with surface mining: (1) primary on-site disturbance, directly associated with the coal extraction process; (2) secondary on-site disturbance, due to support operations, including haul roads, railheads, maintenance facilities, office space, etc.; (3) primary off-site disturbance, resulting from activities directly related to the mining industry, such as highways and roads, power lines and transmission corridors, railroad rights-of-way, etc.; (4) secondary off-site disturbances, related to the demands of mine employees, including houses, schools, parks, etc.; and, perhaps, (5) tertiary off-site disturbances, caused by the activities, recreational and otherwise, of these employees. In each case, land disturbance translates into alteration of geomorphic systems, commonly through changes in hydrologic systems. The National Academy of Sciences (1974) stated that several on-site impacts must be considered in the total framework of rehabilitation of surface-mined areas and of primary consideration are the effects of soil erosion, channel erosion, and the disruption of surface drainage and groundwater aquifers. These have been described in some detail earlier. In the same report, the National Academy of Sciences (1974, p. 45) identified several off-site disturbances:

> Some of the conspicuous hydrologic impacts of surface mining that occur away from the site of mining operation are: (1) changes in volume of surface flow, both increases and decreases; (2) loss of groundwater; (3) deterioration of water quality; and (4) channel changes caused by an increase in sediment load; (5) destruction of aquatic habitats; and (6) increase in endemic diseases among users of water that has been contaminated by mining.

The U.S. Environmental Protection Agency (1976) commented that sediment is widely regarded as the greatest source of water pollution in the United States and that surface mining operations, like all other large-scale, earth-moving operations, have the potential to generate large volumes of sediment. They have compiled the following list of detrimental impacts resulting from sedimentation:

1. Occupies water storage in reservoirs
2. Fills lakes and ponds
3. Clogs stream channels
4. Settles on productive land
5. Destroys aquatic habitat
6. Creates turbidity that detracts from recreational use of water and reduces photosynthetic activity
7. Degrades water for consumptive use
8. Increases water treatment costs

9. Damages water distribution systems
10. Acts as a carrier of other pollutants (plant nutrients, insecticides, herbicides, heavy metals)
11. Acts as a carrier of bacteria and viruses

Recall that the U.S. Environmental Protection Agency (1976) also suggested that the potential for off-site sediment damage was greatest with contour strip mining and least with areal strip mining, while mountain top removal falls somewhere in between.

These lists of impacts also partially explain the breadth and basis of environmental concerns that precipitated the Surface Mining Control and Reclamation Act of 1977. The phrase "both on- and off-site" can be found repeatedly throughout this legislation. The intent of the law clearly extends well beyond the permit area.

Cumulative Impacts

In much of the above, the consequence of land disturbance due to surface mining was examined at a specific site and at a specific point in time. This picture does not necessarily reveal the total amount of environmental damage caused by surface mining. However, such an evaluation was mandated by the Surface Mining Control and Reclamation Act of 1977 wherein:

> no permit or revision application shall be approved unless . . . the assessment of the probable cumulative impact of all anticipated mining in the area on the hydrologic balance specified in Section 507(b) has been made by the regulatory authority and the proposed operation thereof has been designed to prevent material damage to hydrologic balance outside permit area [510, b, 3].

Lumb (1982, p. 4) provided clarification regarding the intent of this legislative directive:

> Cumulative in the context of impact assessments has been interpreted as cumulative in time and cumulative in space. As used herein, cumulative in space refers to the cumulation of flows and dissolved or suspended matter from all mine-permit sites and land uses to common downstream channels. Cumulative in time refers, for example, to the gradual change in stream biota from a more acidic stream or the gradual loss of reservoir capacity from sediment deposition. Some impacts take time to produce material damage.
>
> Generally, cumulative will refer to the spatial cumulation of impacts of several permit areas being mined. Yet some of those cumulative impacts also may cumulate in time.

Such an evaluation centers upon the same three fundamental components described earlier, that is, (1) areal extent, (2) intensity, and (3) duration, as they apply to hydrologic systems specifically.

Although few would contest the desirability of cumulative impact assessment, implementation of this legal requirement is very difficult. Lumb (1982, p. 46) carefully examined the situation and concluded:

> A review of the Federal legislation on surface mining, a review of the available hydrologic techniques, and a case study using the proposed techniques has led to the following conclusions:
>
> (1) Standard hydrologic methods for storm hydrograph analysis do not meet the needs of a regulatory authority for making cumulative impact assessment.
> (2) A simplified yet conservative approach can be used as a minor impact analysis to screen many cases where the impacts are judged to be of minor consequence.
> (3) Simulation with a rainfall-runoff, water-quality, and routing model for a period of a year or more can be used for a major impact analysis, meets the needs of a regulatory authority, is feasible, but may be too complex or costly.

It appears, then, that with current technology we can fairly easily determine the existence of negligible or minor impact which would not preclude the issuance of a mining permit. However, there is great difficulty associated with an assessment of a potentially major impact which would serve as a basis for permit denial. A point not made by Lumb (1982), but very important, is that the methodology employed in an assessment of impact which leads to the rejection of a permit application must be defensible in a court of law because it is likely to be challenged.

It is not possible to fully discuss the combination of techniques tested by Lumb (1982). As stated above, these were standard procedures for hydrologic simulation and hydrographic analysis developed by the U.S. Geological Survey and other agencies. Usually, however, these procedures were intended for use on perennial flow in undisturbed streams. Now, they must be adapted for use under the drastically disturbed conditions caused on-site by surface mining and for use in various climatic regions with various flow regimes.

Some fundamental problems exist in adapting these methodologies. First, as observed by Gardner and Woolhiser (1978), one of the difficult problems in hydrologic research in disturbed areas is that the system is in a transient state during the research period; this invalidates many traditional research procedures which assume the system is in a quasi-equilibrium condition. Thus, the concept of "cumulative in time" is problematic.

Second, much of the discussion in this chapter has shown that the impact of surface mining is frequently site specific, depending not only on environmental characteristics but also mining methods. Hence, it follows that "cumulative in space" must largely consist of an aggregate of site-specific impacts linked through a drainage network or a cascading hydrologic system. Thus, it is not enough to understand the attributes and behavior of the linkages alone, but comprehension of each site-specific impact is also necessary.

Lastly, the mine operators and their reclamation staffs have the capability to manipulate water and sediment releases from their sites. As a result, it becomes necessary to go beyond physical principles of hydrologic modeling and assign a value to the effectiveness of these operators.

From the nature of the disturbance, the climatic conditions encountered, and the requirements of the quantitative approaches, several important questions emerge:

1. Can the model accommodate significant changes in stream base flow due to disruption and resaturation of a shallow aquifer?
2. Is the model applicable to intermittent and ephemeral streams of arid and semiarid regions?
3. Can the model differentiate between and assess the impact on a stream of several mines in the same area, each of which is in a different stage of development?
4. Can time decay and distance decay functions be developed to accurately estimate impact changes for each of these several mines which could occupy somewhat different environmental settings or positions in the drainage basin?

For each question, an affirmative answer is possible because the technology exists to cope with such complex problems. Commonly, however, the data upon which to build and verify the models are the limiting factor. The U.S. Geological Survey, through the Coal Hydrology Program, is attempting to fill this gap to facilitate assessment of probable cumulative impact as described by Kilpatrick and Rollo (1980). In all, there will be more than 60 reports describing hydrologic conditions in parts of the nation's coal provinces. Kilpatrick and Rollo (1980) remarked that it remains to be seen if the data input requirements for such models are reasonable in light of the thousands of mines that must be permitted and the resulting cost of such efforts. They further observed that the present state of the art does not permit precise evaluations of cumulative impacts. Lumb's (1982) report does not refute this comment. Clearly, the regulatory authority has been presented a formidable task in attempting to assess probable cumulative impact, as required by law.

Baseline Data Requirements

In order to obtain a surface mining permit, baseline data are required for several variables in numerous categories. This information is used by the regulatory authority to evaluate the feasibility of the mining operation, the probable environmental impact, and the adequacy of the reclamation plan. As a result, the mining permit application tends to be a rather voluminous document. Although specific data requirements can vary from state to state, a general overview of the

Table 8 Baseline Data Requirements of Geomorphic Significance[a]

1. Type and method of surface mining
2. Amount of land to be affected (including map)
3. Hydrologic data
   a. Name of watershed
   b. Location of surface stream or tributary into which surface and pit drainage will be discharged
   c. Hydrologic regime of area
      (1) Quantity and quality of water in surface and ground water systems
      (2) Dissolved and suspended solids under seasonal flow conditions
4. Climatologic factors (when requested)
   a. Average seasonal precipitation
   b. Seasonal temperature ranges
   c. Average direction and velocity of prevailing winds
5. Geologic information
   a. Cross-section maps or plans of land to be affected and showing:
      (1) Pertinent elevation
      (2) Location of test borings or core sampling
      (3) Nature and depth of the various strata of overburden
      (4) Location of sub-surface water and its quality
      (5) Nature and thickness of any coal or rider seam above the coal seam to be mined
      (6) Nature of stratum immediately beneath the coal seam to be mined
      (7) All mineral crop lines and the strike and dip of the coal to be mined
      (8) The location of aquifers
      (9) The estimated elevation of the water table
      (10) Constructed or natural drainage ways and the location of any discharges to any surface body of water on the area of land to be affected or adjacent thereto
   b. Profiles at appropriate cross sections of the anticipated final surface configuration that will be achieved pursuant to the operator's proposed reclamation plan
   c. Statement of the result of test borings or core samples including
      (1) Log of the drill holes
      (2) Thickness of the coal seam
      (3) An analysis of the chemical properties of such coal
      (4) The sulfur content of any coal seam
      (5) Chemical analysis of potentially acid or toxic forming sections of the overburden
      (6) Chemical analysis of the stratum lying immediately beneath the coal to be mined
   d. If disturbed land may be prime farm land, a soil survey shall be made or obtained
6. Reclamation plan
   a. Identification of the lands subjected to surface coal mining operations
   b. Condition of the land to be covered by the permit prior to any mining including:
      (1) The land uses existing at the time of permit application or prior to any previous mining
      (2) Capability of the land prior to any mining to support a variety of uses giving consideration to:
         (a) Soil and foundation characteristics
         (b) Topography
         (c) Vegetation cover
      (3) Productivity of the land prior to mining including appropriate classification as prime farm lands, as well as the average yield of food, fiber, forage, or wood products from such land under high levels of management

Table 8 *(Continued)*

c. Proposed land use following reclamation including a discussion of the utility and capacity of the reclaimed land to support a variety of alternative uses
d. The engineering techniques proposed to be used in mining and reclamation and a description of major equipment
e. A plan for the control of surface water drainage and of water accumulation
f. A plan, where appropriate, for backfilling, soil stabilization, and compacting, grading, and appropriate revegetation
g. A plan for soil reconstruction, replacement, and stabilization
h. A detailed description of the measures to be taken during the mining and reclamation process to assure the protection of
(1) The quality of surface and ground water systems, both on- and off-site, from adverse effects of the mining and reclamation process
(2) The quantity of surface and ground water systems, both on- and off-site, from adverse effects of the mining and reclamation process or to provide alternative sources of water where such protection of quantity cannot be assured

7. Other data (required by implication under "Environmental Protection Performance Standards")
a. Topography of land to be affected prior to mining (approximate original contour)
b. Depth of topsoil (removal, segregation, and replacement of topsoil)
c. Topsoil physical and chemical analysis (suitability of topsoil or other available materials)
d. Recharge characteristics of replaced overburden (restoration of recharge capacity of the mined area to approximate premining conditions)
e. Channel geometry of streams (avoid channel deepening or enlargement in operations requiring the discharge of water from mines)
f. Existence and characteristics of alluvial valley floors (preserve throughout the mining and reclamation process the essential hydrologic functions of alluvial valley floors in arid and semiarid areas of the country)
g. Vegetation diversity as well as cover prior to mining (establish on the regraded areas, and all other lands affected, a diverse, effective, and permanent vegetation cover of the same seasonal variety native to the area)
h. Mechanical properties of spoils (protect off-site areas from slides or damages occurring during the surface coal mining and reclamation operations)

[a]Compiled from Surface Mining Control and Reclamation Act of 1977, Sections 507, 508, and 515.

requirements can be gleaned from the contents of the Surface Mining Control and Reclamation Act of 1977. Sometimes, data for a particular variable are specifically required; other times, they are required by inference to accommodate the intent or mandate of the law. It would be impractical and unnecessary to reproduce all pertinent sections of this legislation. However, Table 8 provides a list of common baseline data requirements of geomorphic interest compiled from that source. This list is intended to provide only a general overview. More specific information is best obtained from the Office of Surface Mining Reclamation and Enforcement, although it can be obtained from various entries in the Federal Register from 1979 to the present.

Table 9 Basic Data Collection for Surface Mined Lands

Environmental characteristics	References
Climatology	*Development of Climate Profiles for Reclamation,* McKee *et al.*, Colorado State University, 1981
Hydrology	*Methodology for Hydrologic Evaluation of a Potential Surface Mine: Loblolly Branch Basin, Tuscaloosa County, Alabama,* Shown *et al.*, U.S. Geological Survey, 1982 (available for 3 other locations)
	Alluvial Valley Floor Identification and Study Guidelines, U.S. Office of Surface Mining Reclamation and Enforcement, 1983
Vegetation	*Methods for Vegetation Sampling and Analysis on Revegetated Mined Lands,* Chambers and Brown, U.S.D.A., Forest Service, 1983
	Measuring Species Diversity on Revegetated Surface Mines, Chambers, J. C., U.S.D.A., Forest Service, 1983
	Models to Estimate Revegetation Potentials of Lands Surface Mined for Coal in the West, Packer *et al.*, U.S.D.A., Forest Service, 1982
	A Guide for Revegetating Coal Minespoils in the Eastern United States, Vogel, W. G., U.S.D.A., Forest Service, 1981
	Characteristics of Plants Used in Western Reclamation, Long, S. G., Environmental Research and Technology, 1978
Soil erosion	*Preliminary Guidance for Estimating Erosion on Areas Disturbed by Surface Mining Activities in the Interior Western United States,* U.S. Soil Conservation Service and Environmental Protection Agency, 1977
Wildlife	*A Guide for Vegetating Surface-Mined Lands for Wildlife in Eastern Kentucky and West Virginia,* Rafaill, B. L., and Vogel, W. G., Fish and Wildlife Service, 1978
	Wildlife: Users Guide for Mining and Reclamation, U.S. Forest Service, 1982
	An Environmental Guide to Western Surface Mining, Part Two, Moore, R., and Mills, T., Fish and Wildlife Service, 1977
Combinations of characteristics	
Surface water, ground water, soil, and overburden	*Procedures Recommended for Overburden and Hydrologic Studies of Surface Mines,* Barrett, J. *et al.*, U.S. Forest Service, 1980
Climate, physiography, geology, coal resources, overburden, soils, hydrology, and vegetation	*Resource and Potential Reclamation Evaluation,* EMIRA Report 12, Hanging Woman Creek Study Area, U.S. Bureau of Land Management and U.S. Geologic Survey, 1977 (available for 11 other locations)
Geology, soils, land use, surface drainage, water use, climate, coal reserves, hydrologic network, surface water quantity and quality, groundwater, and data sources	*Hydrology of Area 54, Northern Great Plains, and Rocky Mountain Coal Provinces, Colorado, and Wyoming,* Kuhn, G., Daddow, P. B., Craig, G. S. *et al.*, U.S. Geological Survey, 1983 (available for over 60 other locations)

Collectively, this information defines the environmental setting in which surface mining is to be undertaken and, as a consequence, also defines the character of geomorphic and hydrologic systems in operation. Mining permit applications could become a valuable source of geomorphic and hydrologic data for earth scientists in the future.

Most are familiar with the handbooks prepared by professional societies and government agencies for collecting and organizing geomorphic, hydrologic, and related data. It is sometimes useful to modify the recommended procedures to accommodate the circumstances encountered at a surface mine or the specific demands of a regulatory authority. Guidelines for data acquisition and manipulation at mine sites are usually available but often difficult to locate for those not intimately involved with the industry. Thus, the references contained in Table 9 may be useful. This does not constitute an exhaustive list; the references given, along with guidelines and suggestions provided by regulatory authorities, give additional direction.

Land Disturbance by Surface Mining: A Geomorphic Perspective

Under natural environmental conditions, it is expected that usually there exists a balance, variously described as an equilibrium, quasi-equilibrium, or steady state, between landforms and geomorphic processes. Geomorphic work generally proceeds at modest rates as determined by natural environmental conditions because there also exists a balance between the forces applied by geomorphic processes and the resistances provided by surface materials. Although this work results in the entrainment and transport of materials, the landforms themselves will change very slowly, if at all. Spatially and temporally, the system is adjusted to accommodate the products of geomorphic work with only infrequent substantial realignment of component parts. The landforms and landscape that they comprise are stable.

Surface mining alters virtually all of the components of the environment that determine the characteristics of geomorphic systems, except climate. Vegetation cover is removed, topsoil is disturbed and probably mixed with subsoil to some extent, surface flows are disrupted, geologic structures are fragmented, litholgies become mixed, and subsurface flows are interrupted. Consequently, a significant disequilibrium between altered landforms and geomorphic processes is created. Immediately following disturbance, these geomorphic processes set about to reestablish the balance. Their work is made easy because the effect of drastic disturbance is to increase force while decreasing resistance. Hence, geomorphic thresholds frequently are exceeded and often by large magnitudes. Through time, these processes will inevitably succeed in establishing the balance with form, although process rates will vary in accordance with numerous positive and negative feedback mechanisms operating within the system.

Geomorphologists are also keenly aware of the complex interactions among systems components: (1) processes are related to other processes, (2) processes are related to landforms, and (3) landforms are related to other landforms. These interrelations also extend beyond permit boundaries. As a result, changes in processes or landforms in one area are likely to produce additional changes in the processes and landforms of that area, but also are likely to produce subsequent changes in the processes and landforms in adjacent areas, despite the fact that the latter may be far removed from the initial disturbance. In other words, the transmission of on-site impacts to off-site locations is to be expected as a normal characteristic of geomorphic systems.

Taken together, the geomorphic mechanisms and supporting data indicate the necessity of proactive measures to mitigate the potential impacts of land disturbance by the surface mining of coal. Although several techniques are employed during the mining process to reduce the severity and areal extent of impact, effective reclamation provides the best long-term solution to this environmental problem.

Reclamation and the Geomorphic Process–Response System

A review of the literature reveals a range of goals for surface mine reclamation, from rather broad general principles to rather specific requirements. The National Research Council (1981) suggested that an appropriate goal for reclamation is to ensure that society does not lose important land use opportunities that were available prior to soil disturbance or that can be generated in the reclamation process. Hodder (1978) remarked that reclamation or rehabilitation of surface mined lands is the process of returning disturbed lands to a topography, productivity, and use capability that conform to a predesignated purpose; reclamation must be compatible with or supplementary to use and aesthetics of surrounding land and must minimize environmental degradation. Similarly, Narten *et al.* (1983) contended that planned reclamation is the process of returning mined lands to their premining productive capability or to a greater productive capability; it involves reshaping the disturbed lands into stable landforms and establishing a self-sustaining vegetative cover. An array of earth scientists have submitted that erosion control is either a major goal or *the* major goal of reclamation (Hodder, 1975; Dollhopf *et al.*, 1977; Stiller *et al.*, 1980; Wells and Rose, 1981; Hodder, 1983; Colbert, 1983; Toy, 1984b). The control of mass movement would be an equally important goal, especially in the eastern United States because of the steep terrain in which surface mining takes place and the mining methods utilized. Most agree that complete restoration of disturbed lands is not possible.

Narten *et al.* (1983) observed that the primary goal of early reclamation was to stabilize spoils in order to reduce erosion. This resulted in the planting of mostly nonnative species that were tough and adaptable but not necessarily useful for grazing nor conducive to the establishment of a diverse native plant community. Subsequently, the grass species being planted were upgraded to include more productive strains and more palatable species and included some native species. The broadened objective then became both erosion control and establishment of grazing areas for livestock on rangelands. The above obviously applies most directly to arid and semiarid regions.

It is safe to conclude that as erosion control became recognized as an attainable objective, the general goal of reclamation was expanded to include returning the land to productive use rather than merely preventing it from causing additional environmental degradation. Nevertheless, control of all geomorphic processes has always been and will continue to be a fundamental necessity because productivity cannot long endure on any unstable surface.

The regard for postmining land use is endorsed by many. Hansen (1976, p. 1) asserted:

> The term 'mined land reclamation' is also equated with revegetation which is directed at the question: 'how high can we get grass to grow on the spoils piles?' To achieve this primary and supreme objective, any and all means are to be employed and no expense spared. This 'Green Grass Syndrome' is probably the principal obstacle to effective and imaginative programs to restore land to an economically productive or socially desirable ultimate use.

Times have changed and as we learn to do better, we attempt to do more.

The Legal Requirements of Reclamation

Contrary to popular perception, the surface mining of coal is among the most highly controlled economic activities in the United States. While there has been frequent reference to the Surface Mining Control and Reclamation Act of 1977 throughout this chapter, it constitutes only the latest in a long succession of legislation. The first surface mining legislation was passed by West Virginia in 1939 and recognized the disruptive environmental effects of the industry. Bowling (1978) reported 38 states with environmental controls on surface mining. The contents of these state laws varied considerably as shown by the compilation of inclusions assembled by Imhoff *et al.* (1976). The National Research Council (1981) observed that individual state laws were generally directed at the critical problems specific to that state. For example, where moist climates and overburden characteristics produced acid runoff, the focus of legislation was on water quality control, while in arid regions, the problems of erosion and revegetation received greatest attention.

This variability of state law resulted in various degrees of environmental protection from state to state. Further, less stringent reclamation requirements could lead to lower overall costs of production for operators within a particular state and thereby afford them a competitive advantage in the marketplace. This, in turn, might well work to the detriment of the general development of the industry, as well as the environment.

The Surface Mining Control and Reclamation Act of 1977 (PL 95–87) established minimum national standards for the surface mining of coal and required reclamation. Among the purposes listed in the Act are the following of interest to us:

1. Establish a nationwide program to protect society and the environment from the adverse effects of surface coal mining operations.
2. Assure that surface mining operations are not conducted where reclamation as required by this Act is not feasible.
3. Assure that surface coal mining operations are so conducted as to protect the environment.
4. Assure that adequate procedures are undertaken to reclaim surface areas as contemporaneously as possible with the surface coal mining operations.
5. Assure that the coal supply essential to the nation's energy requirements, and to its economic and social well being, is provided and strike a balance between protection of the environment and agricultural productivity and the nation's need for coal as an essential source of energy.
6. Promote the reclamation of mined areas left without adequate reclamation prior to the enactment of this Act and which continue, in their unreclaimed condition, to substantially degrade the quality of the environment, prevent or damage the beneficial use of land or water resources, or endanger the health and safety of the public.
7. Assure that appropriate procedures are provided for the public participation in the development, revision, and enforcement of regulations, standards, reclamation plans, or programs established by the Secretary or any state under this Act.

Table 10, taken from the National Research Council (1981) report, provides a comparison between the content of the Surface Mining Control and Reclamation Act of 1977 and the provisions of applicable laws in 26 states prior to the 1977 enactment. The solid circles indicate that there was no specific state requirement in a particular category before federal legislation was passed; hence, even a cursory examination of this table reveals the rigor of the 1977 Act.

There is insufficient space to fully describe the provisions of this Act and subsequent permanent regulations that, in one way or another, influence the geomorphic system functioning at a potential or active mine site. The entries

Table 10 Comparison of Key Federal Provisions[a] and Pre-PL 95-87 State Surface-Mining Regulations[b]

LEGEND

- ☐ (empty box) State Requirement Essentially Same as Federal
- ○ State Requirement Less Stringent
- ● No Specific State Requirement
- NA Not Applicable

STATE	Segregated Topsoil Handling	Road Standards	Effluent Limitations	Runoff and Stream Diversion Criteria	Sedimentation Pond Criteria	Stream Buffer Zone Limit	Blasting Standards and Requirements	Head-of-Hollow, Valley Fill Standards	Coal Waste Dam and Embankment Criteria	Elimination of Highwalls	Return to Approximate Original Contour	Covering of Coal and Toxic and Acid Forming Material	Revegetation Success Standards	Subsidence Controls for Underground Mining	Maximum Recovery for Auger Operations	Alluvial Valley Floor Mining Standards	Prime Farmland Mining Standards	Woody Material Disposal Restrictions	Downslope Spoil Disposal Restrictions	Application of Surface Standards to Underground Mining
ALABAMA	●	●	○	●	●	●	○	NA	○	●	●	○		●	●	●	●	●	●	●
ALASKA[c]																				
ARIZONA[c]																				
ARKANSAS	●	●	○	●	●	●	●	NA	○	●	○	○	●	●	●	NA	●	●	●	●
COLORADO	○	●	○	●	●	●	●	NA	○	●	○	○	○	●	●	●	●	●	●	●
GEORGIA	○	●	○	●	●	●	●	NA	○	●	●	●	●	●	●	NA	●	●	●	●
ILLINOIS		○		○	○	●	●	NA	○	●			●	●	●	NA	○	●	●	●
INDIANA	●	●	○	●	●	●	●	NA	○	●	○	○	●	●	●	NA	●	●	●	●
IOWA	○	●	○	○	●	●	●	NA	○	●	○	●	○	●	●	NA	●	●	●	●
KANSAS	●	●	○	●	●	○	●	NA	○	●	○	●	●	●	●	NA	●	●	●	●
KENTUCKY	●	●	○	○	○		●	○	○	○	○	●		●	●	NA	●	●	●	●
MARYLAND	●	●	○	●	●	●	●	NA	○	○		●	○		●	NA	●	●	●	●
MISSOURI	●	●	○	●	●	●	●	NA	○	○	○	○	○	●	●	NA	●	○	●	●
MONTANA	○	●	○	●	●	●	●	NA	○	○		○	○	●	●	●	●	●	●	●
NEW MEXICO	●	○	○	●	●	●	●	NA	○	●	○	●	○	●	●	●	●	●	●	●
NORTH DAKOTA	○	●	○	●	●	●	●	NA	○	○		●	○	●	●	●	●	●	●	●
OHIO	○	○		●	●	○	○	NA	○	○		○	○	●	●	NA	●	○	●	●
OKLAHOMA	●	●	○	●	●	●	●	NA	○	○	○	○	●	●	●	●	●	●	●	●
PENNSYLVANIA			○	○	○		○	NA	○			○	○		●	NA	●	●	●	●
TENNESSEE	○		○	○	○		○	○	○	○	○			●	●	NA	●	○	○	●
TEXAS	○	●	○	●	●	●	●	NA	○	○	○	○	○	●	●	●	●	●	●	●
UTAH	●	●		●	●	●	●	NA	○	●	●	●		●	●	●	●	○	●	●
VIRGINIA	●		○	○	○	●	○	○	○	●	●		●	●	●	NA	●	●	●	●
WASHINGTON	●	●	○	○	●	●	●	NA	○	○	○	○	●	●	●	NA	●	●	●	●
WEST VIRGINIA	○		○	○	○		○		○	○				●	●	NA	●	●	○	
WYOMING	○	○	○	○	●	●	●	NA	○	●	○	●	○	●	●	●	●	●	●	

[a]Based on Office of Surface Mining's proposed permanent program regulations.

[b]Source: Adapted from Skelly and Loy (1979) by the National Research Council (1981).

[c]Mining in Alaska and Arizona is on federal, state, or Native American land. Before PL 95-87, reclamation requirements were determined on a case-by-case basis; federal coal-leasing regulations applied, and local governments also exercised some control.

contained in Table 8 and the list of key federal provisions affecting mining performance contained in Table 10, however, provide a useful summary. Some specific requirements will be discussed in subsequent sections.

These are not the only controls that have been placed upon the surface mining industry in general. The National Research Council (1981) also found that since 1970 at least 21 federal laws affecting the coal industry have been enacted and 5 pre–1970 federal laws are still in effect or have been replaced or amended by one of these laws. However, only about a half dozen of these laws consistently apply to all mines. Table 11 contains the list of other federal legislation affecting the coal industry. The frequent amendment of some laws causes compliance by the mine operator to be tantamount to shooting at a moving target. Finally, the National Research Council (1981) found that counties and municipalities may

Table 11 Federal Legislation Affecting the Coal Industry[a]

Antiquities Act of 1906
Archaeological and Historic Preservation Act of 1974
Bald Eagle Protection Act of 1904, amended 1969, 1979
Clean Air Act of 1955, amended 1970, 1973, 1974, 1977
Clean Water Act of 1972, amended 1973, 1974, 1975, 1976, 1977, 1978; replaced the Federal Water Pollution Control Act of 1948, amended 1952, 1956, 1959, 1960
Department of Energy Organization Act of 1977
Endangered Species Act of 1973, amended 1976, 1977, 1978
Energy Supply and Environmental Coordination Act of 1974, amended 1975, 1978, 1979
Federal Coal Leasing Amendments Act of 1976
Federal Land Policy and Management Act of 1976
Federal Mine Safety and Health Act of 1977, replaced the Federal Coal Mine Health and Safety Act of 1969
Fish and Wildlife Improvement Act of 1960, amended 1978
Migratory Bird Treaty Act of 1918, amended 1936, 1939, 1960, 1969, 1974
Mineral Leasing Act of 1920
Mining and Minerals Policy Act of 1970
Multiple-Use Sustained Yield Act of 1960
National Environmental Policy Act of 1969
National Forest Management Act of 1976
National Historic Preservation Act of 1966, amended 1979
Powerplant Industrial Fuel Use Act of 1978
Resource Conservation and Recovery Act of 1976, amended 1978; replaced the Resource Recovery Act of 1970, amended 1973, 1975; replaced the Solid Waste Disposal Act of 1965
Rivers and Harbors Act of 1899
Safe Drinking Water Act of 1974, amended 1977, 1979
Soil and Water Resources Conservation Act of 1977
Surface Mining Control and Reclamation Act of 1977
Toxic Substance Control Act of 1976

[a]Source: National Research Council (1981).

also influence and regulate mining and reclamation through zoning laws and that local taxes may be imposed on a mining company in addition to the fees and royalties assessed by the state and federal government.

In evaluating the history of legislation governing surface mining, Borchert (1980) concluded that the older laws regulating the industry addressed the problems but did not provide coordination of the separate authorities at any given time and place. The Surface Mining Control and Reclamation Act of 1977 concentrates and channels the authority but lacks flexibility to deal with widely different environmental settings. Some geographic aspects of surface mine reclamation have been described elsewhere by Toy (1984a).

Reclamation Practices

In essence, the various practices used in surface mine reclamation take the disturbed surface, with its highwalls, benches, pits, and spoils banks, and recreate entire drainage basins or parts of drainage basins with stable and useful hillslopes and channels. The reclamation process actually begins well in advance of any land disturbance because a detailed reclamation plan is an integral part of the mining permit application submitted to the regulatory authority.

Today, the contents of the reclamation plan are controlled by laws, guidelines, and regulations. Full consideration must be given to the site-specific environmental conditions at the potential mines, the particular demands and preferences of the regulatory authority, and the postmining land use. Additionally, performance bonds force the mine operator to assume responsibility for the quality of their reclamation work for a period of years, which varies depending upon geographic location.

Reclamation procedures will usually vary slightly from mine to mine within a given region and are likely to differ even more between regions. It is necessary, therefore, to describe practices in somewhat generic terms. Brown and Hallman (1984) provided more detailed information including discussion of the equipment used at various stages of the reclamation program. A typical sequence begins with refilling the pit or rebuilding the hillside above the bench with spoils materials. During this process, it is usually possible to dispose of overburden materials containing toxic or acid-producing substances so that they are isolated from both the surface and subsurface hydrologic systems and below the root zones of the plant communities to be reestablished.

The shaping of spoils replaced into the contour bench or pit simulates the action of geomorphic processes operating over the centuries to fashion a landscape composed of hillslopes and channels. Natural geomorphic systems function in such a way that the various components of the landscape are integrated into a drainage basin competent to efficiently remove the water and sediment produced; integration of components is often lacking, however, in reconstructed landscapes.

The Surface Mining Control and Reclamation Act of 1977 requires that the land be returned to approximate original contour. This has become a highly controversial issue. There is abundant literature questioning the necessity, desirability, and feasibility of this requirement. Robinson *et al.* (1980) discussed the issue and proposed a methodology for determining approximate original contour for area mining of a mountain top.

Generally, it can be agreed that "approximate original contour," as required by law, means that the shape of the land should be about the same after mining as it was before, although not necessarily at the same elevation. In the western United States the relatively low stripping ratios often preclude the opportunity to raise the landscape to its former elevation through the reclamation process, as was described by Keefer and Hadley (1976) in the case of Donkey Creek in Wyoming. In contrast, the relatively high stripping ratios in the eastern United States frequently result in excess spoils which must be placed in head-of-hollow fills or other depressions.

Based upon personal observation at several mines in the western United States, it appears that "approximate original contour" may be a readily achievable goal in many cases. In fact, it seems that motor scraper operators are better at this than grading hillslopes to a specific gradient such as 20 to 25%. One could speculate that these operators intentionally or unintentionally interpret this legal mandate to mean "make it look like the surrounding undisturbed area so that it blends in." Robinson *et al.* (1980) have suggested that this was the intent of the law from the beginning. Clearly, however, there are cases where configurations other than "approximate original contour" better serve the intended postmining land use.

The National Academy of Sciences (1974) remarks that, in most cases, the geologic materials should be reshaped to achieve: (1) the best ecologic conditions for the land use objectives, (2) the proper hydrologic requirements, and (3) the most pleasing aesthetic experience for the viewer. They recognize, however, that these criteria may or may not be compatible. Narten *et al.* (1983) commented that the reshaped form of the mined land surface strongly influences the potential for reclamation success.

Hillslopes

As presented earlier, surface form of hillslopes includes the one-dimensional features of profile and plan, as well as the three-dimensional configuration. The first profile attribute which one commonly considers is gradient. The National Academy of Sciences (1974, p. 53) commented:

> Overburden removal techniques need to take account of the fact that maximum vegetative stability cannot be obtained on slopes steeper than 33 percent (3:1). Optimum vegetative

stability requires slopes of less than 25 percent (4:1) (U.S. Environmental Protection Agency, 1972), and the use of agricultural machinery may require that slopes be no greater than 20 percent (5:1).

Most people realize that steep hillslopes must be avoided in order to minimize erosion (Stiller *et al.*, 1980). Similarly, it is commonly recognized that the propensity for mass movement increases with hillslope angle, other things being equal. Many understand that infiltration rates often decrease with increasing hillslope angle, and this would result in reduced soil moisture.

The Surface Mining Control and Reclamation Act of 1977 does not specify exact hillslope gradients in recognition of variation among different environmental settings. However, the permanent regulatory program accompanying the Act states that disturbed areas shall be backfilled and graded to achieve the approximate original contour (with few exceptions) and preparation of final graded surfaces shall be conducted in a manner that minimizes erosion and provides a surface for replacement of topsoil that will minimize slippage. One inference of this is that mine operators must measure hillslope angles prior to disturbance. Construction of a topographic map from aerial photography is the common practice herein. Some regulatory authorities have apparently adopted a 25% (4:1) gradient as the maximum generally allowable. However, there does not seem to be any established procedure for measuring hillslope angle. Does the above standard refer to the angle of the straight segment of a convex–straight–concave hillslope, or does it apply to an average gradient taken from the crest to the base? These values can be considerably different, depending upon the percentage of the hillslope contained within the convex, straight, and concave elements and segments. If a specific angle is to be designated as a maximum limit, it probably should apply to the straight segment because this is where erosion tends to be greatest (Hadley and Toy, 1977).

Despite the several benefits associated with grading hillslopes to the lowest possible angles, mining companies will resist such a practice because of the expense. The National Research Council (1981) has compiled reclamation cost data showing that backfilling and grading accounts for the greatest proportion of total reclamation costs, up to 90% in some instances. As hillslope angle comes down, the amount of material moved usually goes up and, along with it, reclamation costs.

Hillslope length is another profile attribute that has been shown to influence erosion rates. Zingg (1940) found the exponent relating hillslope length to soil loss was larger than that relating hillslope angle to soil loss. Nevertheless, Stiller *et al.* (1980) observed that during the grading phase of reclamation it is common to grade the topography to long straight hillslopes. It would seem shortsighted of regulatory authorities to control hillslope angle but not hillslope length. In fact, control of hillslope angle, together with the costs of material handling, encourages the construction of excessively long hillslopes.

Generally, there is an inverse relationship between hillslope length and hillslope gradient. However, in areas where excess spoils is not a problem, it could be possible to create a greater number of hillslopes with both reduced gradient and length, although this would certainly increase reclamation costs. In areas of deficient spoils, this could assist in returning the surface to its premining elevation, thus avoiding the problems described by Keefer and Hadley (1976).

Meyer and Romkens (1976, p. 2.71) remarked that during massive land reshaping, the shape of a hillslope's final surface may greatly affect subsequent sediment yields:

> A convex slope is more erodible than a uniform slope, because it is steepest near the toe where runoff is greatest. A uniform slope will yield more sediment than a concave one, because the concave slope is steepest where the flow is least and because some of the sediment eroded from upper portions of the concave slope may be deposited as it flattens near the toe. If a concave shape is not acceptable, because of safety problems on cut or fill sideslopes along a roadway or because of disharmony on building sites, shaping to a complex slope may be a satisfactory solution. A complex slope that is convex along its upper portion and concave along its lower portion will generally yield less sediment than a uniform slope. A flat section at the toe of a slope will also reduce the sediment yield. Proper slope shaping during massive topographic modification can significantly reduce downslope sediment problems.

Many geomorphologists regard the complex (convex–straight–concave) profile form as the "equilibrium" form under many environmental conditions. If this is correct, then hillslopes constructed of a single element (convex or concave) or segment (straight or uniform) should develop toward this configuration through time. This may result in accelerated erosion rates for a while, although some of the sediment will be deposited near the base of the slope in most cases. Meyer and Kramer (1968), however, feel that the convex, straight, and complex profiles tend to evolve toward the concave form. There are geomorphologists who could support this proposition, observing that the greatest proportion of the hillslope profile is often found in the concave element near the base. Thus, it appears that complex or concave profiles would be appropriate grading objectives, while convex or straight (uniform) profiles are generally undesirable.

Other geomorphic research should be considered in topographic design and reconstruction. For example, the results of hillslope profile examination in various climatic regions but with similar geologic structure and lithology (Toy, 1977) suggest that hillslopes in arid regions tend to be shorter, steeper, and have a smaller radius of curvature of the convex element than those in humid areas. Because of the destruction of geologic structure and the mixing of lithologies in the mining process, these findings may apply to reconstructed landforms.

Hadley and Toy (1977) found that erosion rates were greatest in the straight segment of complex hillslopes developed on easily erodible materials and least in the upper convex element; the lower concave element experienced either erosion or deposition, depending on proximity to a channel capable of transporting the sediment carried downslope. Again in this study there was no effective geologic control on hillslope form or erosion, so the results may be applicable to mined lands. These results, then, suggest that erosion must be controlled in the straight segments of reclaimed hillslopes. This, in turn, suggests that the depth of topsoil applied to the regraded spoils should be greatest in the straight segment if a plant growth medium is to be maintained; perhaps, mulching and seeding rates should also be increased therein to stabilize this segment as quickly as possible.

The relationship between hillslope form and aspect has been investigated by Hadley (1962), among others. Frequently, south-facing hillslopes are found to be longer but less steeply inclined than their north-facing counterparts. If such a configuration represents "equilibrium," then it appears desirable for the grading process to replicate the shape.

The plan of a hillslope refers to its linearity or curvilinearity along the horizontal dimension or width. The plan may be straight, as with a valley side hillslope, or curved, as with spur end or valley head hillslopes. When viewed from a two-dimensional perspective, the plan is composed of an infinite series of adjacent profiles. Hence, it could be argued (as in an earlier chapter) that plan form by itself does not directly affect stability and specifically erosion.

However, when plan and profile are combined to produce three-dimensional hillslope configurations, erosion rates are significantly influenced. The relationship between shape and erosion is simplified somewhat on newly reclaimed surfaces where soil and vegetation characteristics are relatively uniform. Here, the hydrologic properties associated with these forms can dominate the geomorphic system. As a result, it may be inferred that the plan and profile concavity of valley head hillslopes will tend to concentrate flow producing greater erosion, while the convex profile and plan of spur end hillslopes will tend to disperse flow producing less erosion. One might infer, then, a hierarchy wherein valley heads produce the greatest erosion, straight or planar valley sides occupy an intermediary position, and spur ends produce the least erosion, based on three-dimensional form features.

Of course, the entire reconstructed landscape cannot be composed of valley sides, valley heads, or spur ends. Each is necessary in various proportions. However, the above does suggest that erosion control techniques should be intensified at the base of valley head hillslopes where flow is likely to be concentrated. The base of a valley head hillslope is a transition zone between the overland-flow processes on the hillslope and the channel flow processes in swales or valleys.

There is probably little likelihood that all of the suggestions offered above will ever be incorporated into the actual design and implementation of topographic reconstruction. The technical and logistical problems associated with such an attempt would be substantial. But it cannot be overemphasized that the regraded surface is the foundation upon which subsequent reclamation and eventual land use rests. It is reasonable to suggest that the extent to which the new landscape includes the attributes of natural landscapes will determine the extent to which a stable, "equilibrium" landscape has been re-created. In other words, a fundamentally unstable form is unlikely to be rectified by clever techniques for very long, assuming that a maintenance-free geomorphic system is the goal.

After the shaping of the topography, several reclamation practices are employed to improve the hydrologic characteristics of the surface and the potential for revegetation success. Usually before topsoiling, but sometimes afterward, one or more types of surface manipulations are applied.

The several types of surface manipulations range from terracing and dozer basins, which markedly alter the reclaimed surface, to common agricultural techniques of plowing and disking. The U.S. Environmental Protection Agency (1976) and Hodder (1977) provided concise descriptions of these methods. Dollhopf *et al.* (1977) presented some research results for various surface treatments at surface mines in Montana and North Dakota.

Hodder (1978) listed the following benefits of these practices: (1) improved erosion control, (2) increased infiltration, (3) conservation of soil moisture, and (4) protection of seedlings from the harsh climates in semiarid and arid regions. The geomorphic effect is to reduce the force impinging on the surface by reducing runoff through improved infiltration capacity and to increase resistance by increasing microtopographic relief, or roughness of the surface. The depressions produced also serve as small sediment traps capturing entrained materials for a few years.

Some workers favor using surface manipulations following topsoiling. Among the arguments offered to support this sequence are (1) the manipulations can reduce any surface compaction resulting from heavy equipment traffic in the spreading of topsoil, and (2) the techniques that penetrate more deeply tend to bind the topsoil and spoils by mixing the materials across the interface. The latter may encourage deeper root development as well.

Topsoiling is the next stage in the reclamation process. Soil materials, which preferably have only recently been disturbed but may have been stockpiled, are spread across the graded and manipulated reclaimed surface. Although the permanent regulatory program of the Surface Mining Control and Reclamation Act of 1977 indicates that a relatively uniform thickness should be applied across the surface, some regulatory authorities specify variable thicknesses depending upon topographic position. The benefits of topsoiling are well documented (Schuman *et al.*, 1976, 1980; Packer and Aldon, 1978; Power, 1978; Huntington *et al.*, 1980; Schuman and Power, 1980, 1981; Power *et al.*, 1981; Merrill *et al.*, 1983; Narten *et al.*, 1983; Barth, 1984). Packer and Aldon (1978) listed the following benefits: (1) provides fertility not usually encountered in raw spoils, (2) furnishes a source for renewed microbiological activity to improve soil-building processes, and (3) has better infiltration and soil stability characteristics. Schuman and Power (1981) remarked that the use of topsoil for reclamation provides a good rooting medium, improves infiltration, reduces runoff, encourages faster reestablishment of nutrient cycles, and can also increase the species diversity of vegetational cover. Narten *et al.* (1983) commented that topsoil seems to be more important to long-term reclamation success than various cultivational techniques that are useful for improving the physical characteristics and water-holding capacity of the growth medium.

The principal direct geomorphic effect of topsoiling is the reduction of erosive force as a result of increased infiltration capacity and complementary reduction in runoff. To the extent that topsoil facilitates revegetation, it will eventually have the indirect consequence of increasing surface resistance. Given the common variability of erosion rates at various parts of the landscape (Hadley and Toy, 1977; Meyer and Romkens, 1976), it does not seem appropriate that topsoil should be applied uniformly across the reclaimed hillslopes, if the objective is to maintain a satisfactory growth medium for a period of years. Because erosion rates tend to be greatest in the straight segment and least in the convex element and because deposition frequently takes place in the basal concave element, it would seem to follow that the depth of applied topsoil could be relatively less in the convex and concave elements and greatest in the straight segment. Some of the soil placed in the convex element and straight segment will naturally find its way to the concave element anyway.

After the available topsoil has been distributed, seedbed preparation begins. Standard agricultural practices are employed to loosen the soil which has likely experienced some compaction during the spreading operation. Thereafter, soil amendments are added to enhance the prospect of successful revegetation and soil stabilization. There are two general categories of amendments: (1) those intended to improve the chemical properties of the soil and (2) those intended to provide a measure of surface protection until a vegetation cover has been established. The first group includes additives to alter the alkalinity, acidity, sodicity, or salinity of the soil and to ensure adequate amounts of necessary plant nutrients. A useful synopsis of several spoil amendment studies is provided by Doyle (1976). For the eastern United States, valuable discussions of soil amendments can be found in U.S. Environmental Protection Agency (1976), Mays and Bengtson (1978), Plass (1978), Capp (1978), Halderson and Zenz (1978), Vogel (1981), and numerous papers contained within the annual symposia on surface mining hydrology, sedimentology, and reclamation sponsored by the University of Kentucky and others. For the western United States, information can be found in Packer and Aldon (1978), Bauer *et al.* (1978), Kay (1978), Sandoval and Gould (1978), DePuit and Coenenberg (1979), and numerous papers contained within the symposia "Reclamation of Mined Lands in the Southwest" (Aldon and Oaks, 1982) and "Soils and Overburden in Reclamation of Arid/Semiarid Mined Lands" (DePuit and Schuman, 1983).

Soil stabilizers and mulches are of particular interest to us because they are specifically intended to reduce soil erosion. Stabilizers are chemical substances that bind together soil particles (tackifiers). These materials are usually quite expensive, so their use is generally limited to problem areas. Israelsen *et al.* (1980) presented test results for several soil stabilizers and mulch tackifiers.

A number of natural and man-made materials have been used as mulches. These materials vary considerably in cost with the most expensive mats, nets, and mulches reserved for highly erosive conditions in selective areas. However, the less expensive types, such as straw, have been used extensively on reclaimed lands to provide temporary protection. Hodder (1977) lists the benefits of mulching: (1) encourages infiltration by impeding runoff; (2) lowering soil temperature; (3) reducing evaporation; and (4) minimizing raindrop splash, surface puddling, and sealing. Noteworthy, also, is the support practice of crimping mulches into the surface or, occasionally, topping with a tackifier sprayed onto the mulch to reduce loss from the surface due to high wind velocities.

From the geomorphic perspective, soil amendments that protect the surface or alter the physical properties of the soil have the collective effect of reducing erosive force and increasing surface resistance. In addition, mulches will often trap some of the sediment that has been entrained.

Revegetation is the most important process in the stabilization of the reshaped topography. The permanent regulation program of the Surface Mining Control and Reclamation Act of 1977 requires the establishment on all affected land of a diverse, effective, and permanent vegetation cover of the same seasonal variety native to the area of disturbed land or species that support the approved postmining land use. Further, these regulations specify that the vegetative cover shall be capable of protecting the soil surface from erosion.

The National Research Council (1981) remarked that the revegetation of reclaimed soils can serve a variety of purposes, and the selection of species and timing of planting depend upon the purpose under consideration. The short-term goal of rapid establishment of vegetation cover is to prevent erosion and, in some cases, to serve as a live mulch. This may demand different plant species than the long-term goal of establishing a level of productivity similar to that of premining conditions. There are abundant references to assist in all aspects of revegetation, plant selection, fertilizer requirements, time of planting, method of planting, etc. Plass (1974) and Doyle (1976) are useful general sources. For the eastern United States, the U.S. Environmental Protection Agency (1976), Bennett *et al.* (1978), Rafaill and Vogel (1978), and Vogel (1981) provide valuable guidance. For the western United States, pertinent information can be found in Thornburg and Fuchs (1978), Packer and Aldon (1978), Long (1978), several papers contained in the symposia "The Reclamation of Disturbed Arid Lands" (Wright, 1978) and "Reclamation of Mined Lands in the Southwest" (Aldon and Oaks, 1982), and lastly, Packer *et al.* (1982).

From the geomorphic perspective, it is difficult to overestimate the importance of revegetation. The relationship between vegetation cover and erosion has been described repeatedly. In essence, vegetation serves to reduce the force impinging upon the soil surface and concurrently increase the resistance. Raindrop impact is intercepted, infiltration capacities are improved, frictional resistance is provided by vegetal surfaces, and soil particles are bound together by root networks. Materials that have been detached and entrained elsewhere are often trapped by vegetation cover.

Research by DePloey and associates, however, is of interest (DePloey, 1981). They have found that there are circumstances where vegetation covers contribute to relatively high erosion rates. Of these conditions, one is noteworthy with regard to reclaimed lands. Laboratory experiments indicate that low vegetation covers of steppe grasses on an otherwise bare soil surface cause increased turbulence of overland flow and, hence, increased erosion. As the cover increases, the protective attributes become dominant. It is commonly assumed that erosion rates follow a simple decay function as vegetation cover increases (Smith and Woolhiser, 1978); however, if these experimental results are indeed applicable to field situations, erosion may increase somewhat as vegetation emerges and then decrease in some regular fashion.

Land management is the final component of the reclamation program. Lang (1975) stated that management after revegetation is equal in importance to all prior reclamation procedures. Narten *et al.* (1983) asserted that the greatest biological detriment to reclamation is overgrazing by domestic livestock. Thames and Verma (1975) recognized this problem at Black Mesa, Arizona, and commented that the regraded mine spoils was not vegetated despite several repeated seedings because of unrestricted grazing. They contend that revegetation will continue to be unsuccessful as long as grazing remains uncontrolled. The National Academy of Sciences (1974) indicated that management during seedling establishment includes the protection of new seedlings from grazing by appropriate fencing or chemical repellents for at least the first two growing seasons and frequently for as long as three or four growing seasons, the criterion being that seedlings should not be grazed until they are firmly rooted.

While fencing will control domestic livestock, wildlife poses a more difficult problem. Narten *et al.* (1983) remarked that in some areas, heavy degradation of reclaimed lands by seed- and seedling-eating rodents, rabbits, and birds has retarded reclamation efforts until artificial habitats and perches for predatory species were added. Clearly, both domestic animals and wildlife regard reclaimed lands as "improved pasture." Also, based upon personal impression, the number of game animals seems especially large on mine property during hunting season, perhaps because firearm restrictions create a *de facto* refuge for them. G. Atwood (personal communication, 1985) notes that some mining companies in Utah now try to revegetate with plant species that are less delectable than the original species in order to avoid the "ice cream store" phenomena.

The extent of postreclamation management is often limited by the liability period established in the Surface Mining Control and Reclamation Act of 1977. Herein, the duration of extended liability for reclaimed lands begins when ground cover matches an approved standard following the last year of augmented seeding, fertilizing, irrigation, or other works which ensure success, and continues for 5 yr in areas receiving more than 26 in. of average annual precipitation or for 10 yr in areas receiving 26 in. or less of average annual precipitation.

Redente *et al.* (1983) found that some management practices are not allowed without reinitiating the liability period, and this would, of course, delay eventual bond release. They recommended that there needs to be a clear separation between the management practices that restart the timeclock of the liability period and those that do not. Further, it is suggested that the criteria for this determination should be flexible enough to allow regulatory authorities and operators to make this judgment on a site-specific basis. In the absence of definitive guidelines and the prospect of considerable time investment and delays in securing a ruling from the regulatory authority, mine operators may elect to merely comply with the minimum requirement of the laws and regulations rather than manage the reclaimed lands to the best of their ability.

In some cases, special reclamation practices are utilized to improve the probability of success. Irrigation is one such option. Ries and Day (1978) commented that irrigation as a tool in the reclamation of surface disturbed lands can be divided into two primary uses: (1) supplemental water to the new plant community during its establishment period, and (2) excess water to leach undesirable water soluble constituents to greater depth in the soil–spoil complex. Packer and Aldon (1978) and DeRemer and Bach (1977) provided detailed discussion of various irrigation methods, together with their advantages and disadvantages. Generally, sprinkler irrigation has been favored.

For much of the mined land west of the 100th meridian, water is the principal limiting factor for revegetation and, hence, successful reclamation. The National Academy of Sciences (1974) concluded that the best techniques in revegetating disturbed landscapes must include, among other things, the sparing addition of irrigation water, if needed, during the period of seedling establishment because the erratic climate conditions in the West still remain the predominant factor determining success or failure on arid rangeland ecosystems. Likewise, Aldon and Springfield (1977) concluded that supplemental irrigation will be necessary for stand establishment when annual precipitation is less than 20 cm (7.9 in.) based upon their experience in the Southwest.

Narten *et al.* (1983) observed that irrigation allows planting over a greater time period and assures the highest percentage of seed germination and initial plant survival. In the northern Great Plains, Ries (1980) concluded that: (1) the use of irrigation to supplement natural precipitation can help ensure rapid and successful perennial plant establishment, especially when natural precipitation is low; (2) the use of irrigation to supplement natural precipitation will permit planting later into the growing season, which can help control weedy species; and (3) the application of different levels of water after seeding can provide for some control of the final species composition in newly established stands. Rosso (1980) found that both treated and untreated acid mine drainage used as irrigation water increased the harvest yields of fescue/alfalfa and sweet clover/bermuda grass at an experiment site in Kentucky.

There are data demonstrating that irrigation is essential to revegetation in arid regions and that the practice benefits plant establishment in semiarid and even humid regions. However, there remains some concern regarding the long-term consequences of this reclamation practice. Aldon and Springfield (1975) identified a fundamental question: What happens when you turn off the water? Some suggest that "die back" largely negates the initial benefits of irrigation, while others believe that vegetation will stabilize at a level supportable by the natural soil moisture regime. Narten *et al.* (1983) observed that on some irrigated mine spoils in New Mexico, vegetation thinned out significantly after irrigation was stopped. However, Young (1983) found that none of the data collected on study plots in Montana indicated plant dependence on irrigation for future growth. The answer to the question posed probably depends upon the plant species utilized, the irrigation procedures employed, including the volume and frequency of application, and the severity of subsequent climatic conditions.

The source of irrigation water is another concern. In arid regions where irrigation may be critical to reclamation success, there is an obvious water shortage and that which is available is likely allocated through historic water rights. In New Mexico, power plant cooling water has been used with apparent success. In North Dakota, Ries (1980) found that even poor quality water was useful in plant establishment. Because irrigation usually takes place only during the first year or two of revegetation, the buildup of toxic materials in the soil will usually not become a problem even if water of poor quality is applied.

Irrigation may also be used to reclaim surface materials containing excessive amounts of soluble salts and exchangeable sodium. Ries and Day (1978) observed that leaching with irrigation water can move soluble soil constituents, if hydraulic conductivity and drainage of the materials permit sufficient water intake and permeability to accomplish the translocation along with deep percolation. As a result of the current legislation requiring topsoil salvage and replacement, leaching is generally not necessary for newly disturbed lands. However, this practice may be important in the reclamation of raw spoils occurring on abandoned mine lands.

Channels

Channels are the other landform created by the reclamation process. As previously discussed, there appear to be two basic philosophies for the reclamation of channels, the geomorphic and the engineering approaches. With the geomorphic approach, the reclamation process attempts to replicate channel characteristics, such as channel gradient, sinuosity, and channel geometry. With the engineering approach, the reclamation process attempts to construct a conveyance channel of sufficient capacity to transport the water and sediment delivered to it from the surrounding hillslopes and, at the same time, neither aggrade nor degrade its bed, nor enlarge its width. Substantial problems are encountered in attempting to utilize either alternative.

There are surprisingly few references addressing the geomorphic approach to channel reconstruction on lands surface mined for coal. Stiller *et al.* (1980), however, provide some valuable commentary. They asserted that the chief consideration in the design of postmining landscapes should be given to relocating the stream courses originating on or crossing the reclaimed surface. Stream gradients and longitudinal profiles should be graded to form a smooth, concave-upward configuration that is integrated with both upstream and downstream drainage. Similarly, tributaries to a reclaimed stream channel should grade into the master stream so that no knickpoints are created in either channel at the confluence. They also suggest that, while regulations may set the design storm for reclaimed channels, the actual dimensions should be based upon the geometry of natural streams; in this way, the reconstructed streams approximate natural stream geometry, which is a function of natural flow regime and the nature of bed and bank materials. Earlier, we discussed the potential problems which may arise from this approach; however, given the current state of the art, this could be the best available option. Nevertheless, Touysinhthiphonexay and Gardner (1984) submit that streams disturbed by mining will not return to their former equilibrium conditions, but will rapidly adjust their morphology in order to achieve a new equilibrium that is compatible with the conditions imposed by mining and reclamation. Stiller *et al.* (1980) concluded that if channels are well designed and quickly revegetated with riparian species, structural types of sediment control should not be necessary.

Channel design using an engineering approach can be no more precise than the estimates of water and sediment discharges that the channel must transport. Shown *et al.* (1982) provided a valuable discussion of various streamflow estimation methods for both premining and postmining applications. Generally, in this report and elsewhere (Shown *et al.*, 1981) they tended to favor the SCS approach (U.S. Soil Conservation Service, 1972). In reviewing the two aforementioned reports, it is interesting to note that in some situations, estimates suggest that postmining peak discharges and runoff volumes may actually be less than those of premining periods (Shown *et al.*, 1981). Here, the replication of premining channel properties would likely result in overdesign. In other situations, the estimates suggest that postmining peak discharges and runoff volumes are likely to be several times greater than during premining periods (Shown *et *al.*, 1982). Here, the replication of premining channel properties would likely result in underdesign. Of course, the assumptions made concerning the postmining topography, depth, and infiltration capacities of topsoil and subsoil, vegetation cover, and land use will determine the estimates of discharge. The eventual extent to which these assumptions prove valid following reclamation will determine the accuracy of the predictions upon which channel design is based. And, finally, recall the comment by Gardner and Woolhiser (1978) that one of the difficult problems in hydraulic research in disturbed areas is that the system is in a transient state during the research period and that this invalidates many traditional research procedures which assume the system is in a quasi-equilibrium condition. The same problem equally pertains to the application of most research to these conditions.

Channel geometry is also determined by the characteristics of the sediment transported by the streamflow. Khanbilvardi *et al.* (1983) presented a modeling technique for predicting erosion and deposition on strip mined and reclaimed land. Shown *et al.* (1982) also offered a useful discussion of various procedures for estimating the erosion and sediment yield rates for surface mined lands. There does not appear to be an efficient method, however, for determining the particle size distribution of those sediments nor the proportion likely to be transported as suspended load and bedload.

Unwittingly and wittingly, those engaged in channel reclamation find themselves working with safety nets in the form of streambed armoring and settling ponds. In the first case, the heterogeneous size distribution of mine spoils can result in the development of streambed armoring should the channel entrench into it. Admittedly, this has not been well documented in reclamation literature, but there seems little reason why it would not occur. The effect of such armoring is described by Begin *et al.* (1980, 1981) who found that it could have a dominant role in preventing channel degradation.

The Surface Mining Control and Reclamation Act of 1977 specified that all surface drainage from disturbed areas, including disturbed areas that have been graded, seeded, or planted, shall be passed through a sedimentation pond or a series of sedimentation ponds before leaving the permit area. The conditions under which a regulatory authority could grant exemption from this regulation were also clearly listed and likely to result in only rare exception. Additionally, design and construction requirements were rather detailed and produced considerable debate concerning necessity, adequacy, universal applicability, feasibility, and cost. Del Rio (1980), Chakrabarti *et al.* (1980), Krishnamurthi *et al.* (1980) have provided useful summaries of the controversies.

Although sedimentation ponds can certainly offer a measure of environmental protection for undisturbed lands adjacent to surface mining, the sediment effluent standards and eventual disposition of the pond are also issues of geomorphic concern. The permanent regulation program for the Surface Mining Control and

Reclamation Act of 1977 originally established total suspended solids (TSS) effluent limitations of 70.0 mg/liter for the maximum allowable from sedimentation ponds and 35.0 mg/liter for the average of daily values for 30 consecutive discharge days. In Colorado, Montana, North Dakota, South Dakota, Utah, and Wyoming, total suspended solids limitations were to be determined on a case-by-case basis, but they must not be greater than 45 mg/liter (maximum allowable) and 30 mg/liter (average of daily value for 30 consecutive discharge days) based upon representative sampling. Both the limitations themselves and this entire approach to regulation will probably amaze many geomorphologists. First, it is not uncommon for natural streams and rivers, draining virtually undisturbed watersheds, to exceed the aforementioned limitations, especially in the western United States (Wells and Rose, 1981). Second, the limitations are actually more stringent in states experiencing semiarid climatic conditions, and these, according to the frequently cited research by Langbein and Schumm (1958), will probably experience some of the highest natural erosion and sediment yield rates of any region. Third, the discharge of water from sedimentation ponds with very small sediment loads may well result in downstream gullying and channel entrenchment. This commonly occurs below dams (Leopold and Maddock, 1953). Stiller *et al.* (1980) discussed this potential problem on mined lands. Fourth, the development and application of uniform standards throughout areas of diverse geology, climate, topography, soils, and vegetation appear ill-advised at best, from the geomorphic viewpoint. These limitations presented above have been challenged repeatedly and have been recently replaced with other sediment discharge performance standards. Nevertheless, the foregoing illustrates some of the pitfalls encountered in attempting to devise simplistic and uniform legislative requirements for implementation in complex environmental systems.

The eventual reclamation of these sedimentation ponds poses additional potential problems. The regulations specify that sedimentation ponds shall not be removed until the disturbed area has been restored, the vegetation requirements have been met, and the drainage entering the pond has met the applicable water quality requirements for the receiving stream. When the sedimentation pond is removed, the affected land shall be regraded and revegetated unless the pond has been approved by the regulatory authority for retention as being compatible with the approved postmining land use. By virtue of their function, sedimentation ponds are the foci of channels draining the reclaimed lands. Consequently, it is likely that a segment of the postmining drainage network will cross the reclaimed sedimentation pond. Any abrupt change in the longitudinal profile of this channel will become the site of channel readjustment. Of particular concern is the creation of a knickpoint in the profile that might migrate upstream through the reclaimed lands, resulting in the entrenchment of large parts of the drainage network and the production of large amounts of sediment to be carried down to the undisturbed channels off-site. A very recent amendment of the regulations has relaxed the absolute necessity for sediment ponds. It is probably fair to suggest that the work of past and present geomorphologists has assisted in the modification of the regulations pertaining both to sediment ponds and effluent standards.

Limited discussion with reclamation personnel concerning the design of reclaimed channels has not been very enlightening. It appears that in actual practice some combination of the geomorphic and engineering approaches, together with a large measure of past experience, provides the basis of design.

Stiller *et al.* (1980) stated that postmining stream gradients are often chosen arbitrarily and commonly contain knickpoints and reaches that are steeper than their premining gradients. Further, oversteepened reaches of alluvial channels are susceptible to channel incision and gully development producing increased sedimentation downstream.

Drainage Basins

The grading and topsoiling of reclaimed landscapes produce hillslopes and channels which together necessarily constitute drainage basins. Although in natural undisturbed systems these hillslopes and channels are integrated for the efficient transportation of water and sediment, this mutual adjustment is often lacking in reclaimed landscapes, as pointed out by Wells and Rose (1981). Hopefully, sedimentation ponds will contain the products of the evolution of reclaimed drainage basins as they move toward the regeneration of an equilibrium condition.

As discussed earlier, geomorphologists have discovered numerous significant statistical relationships among a considerable number of basin parameters. Collectively, all of these relationships document the existence of an adjustment, balance, or equilibrium among drainage basin parts in natural, undisturbed landscapes.

Theoretically, successful reclamation should re-create these relationships in order to reestablish the equilibrium. Of course, this would be practically impossible and enormously expensive. Realistically, then, we must settle for a good approximation and allow nature to "fine tune" the new system for us during the legally mandated liability period.

Once again, there is a paucity of literature concerning the geomorphic approach to drainage basin reconstruction. Stiller *et al.* (1980, pp. 275–276) discussed drainage densities on reclaimed lands and remark:

> Given sufficient time and the fact that reclaimed surfaces have lower infiltration rates, such a surface will not only develop the original drainage density, but will in all probability develop a drainage density greater than that of the pre-mining surface. The redevelopment of this drainage network would take the form of accelerated erosion; and given the unconsolidated nature of the reclaimed surface material, gullying is likely to occur.

Later, these writers presented their suggestion for designing the drainage density of a reclaimed surface and the rationale.

> Until the state-of-the-art has advanced, one possible solution is reclaiming the surface to a post-mining drainage density at least equal to the pre-mining drainage density.
>
> It may appear contradictory to recommend returning a post-mining drainage density to the pre-mining equivalent when the post-mining density is likely to exceed the pre-mining density. However, this recommendation is based on the assumption that concurrent efforts will be made to reduce runoff from the reclaimed surface by promoting infiltration with techniques currently in use, for example, mulching or deep ripping. Furthermore, over time, revegetation and soil development will restore organic matter to the reclaimed soil and naturally promote reduced bulk density. A gradual return of the surface should result in something approaching the pre-mining hydrologic properties.

Wells and Rose (1981) offered a somewhat different approach that might be employed in designing the drainage network of a reclaimed surface, based upon work by Melton (1958). The latter suggested that watersheds with a relative network density, defined as $F/D^2 = 0.694$, are in an equilibrium condition where F is the number of stream segments per unit area (stream frequency) and D is the drainage density. This proposal may have utility but would be somewhat complicated to implement in the field. Further, the concept probably requires additional verification and refinement before widespread adoption, although Melton (1958) utilized a fairly large sample of basins in developing the aforementioned relationship. While these have been criticisms of this relationship, as mentioned in the Chapter 11 of this volume, the general approach may be of value.

A major problem for the design of reclaimed mined lands concerns the boundary between the disturbed and undisturbed surfaces. The National Research Council (1981) asserted that reclaimed landforms should be designed to mesh hydrologically with local natural landforms. However, as noted by Keefer and Hadley (1976), the removal of a coal seam may result in a general lowering of the postmining surface level where area stripping is employed. Similarly, contour stripping and mountain top removal are generally going to cause some changes in the landscape, despite the reclamation goal of "approximate original contour." Additionally, the infiltration capacities, erodibilities, and transmissibilities of surface and subsurface materials, together with vegetation composition and cover, are likely to differ somewhat on either side of the boundary. To the extent that landforms and surface properties vary, one could expect concomitant variation in drainage networks and channel characteristics such as gradient and geometry.

These boundaries deserve particular attention because any disequilibrium existing here can easily be transmitted off-site in the downstream direction, off-site in the upstream direction, or even back onto the reclaimed surface. The boundaries, then, may be the locus of incipient instability of reclaimed lands.

Some have suggested (Wells and Rose, 1981) that hillslope gradients on reclaimed land should be 1% or less. However, if the surrounding terrain is gently undulating, it may be just a matter of time before gullying attacks this flat surface along the perimeter in an attempt to integrate the area into the drainage network. Such is the nature of drainage basin evolution.

The alluvial valley floors found in the western United States offer a special challenge to reclamation. The Surface Mining Control and Reclamation Act of 1977 contains several provisions to protect these geomorphic and hydrologic features. Generally, these requirements serve to preclude mining through any significant alluvial valley floor. Hadley and King (1980) and, especially, the U.S. Office of Surface Mining Reclamation and Enforcement (1983) provide complete discussion of this controversial issue.

A couple of items, however, deserve mention. These comments are taken from the report of the U.S. Office of Surface Mining Reclamation and Enforcement (1983). First, an applicant for a mine permit must affirmatively demonstrate that the proposed operation will not interrupt, discontinue, or preclude farming on alluvial valley floors (with two exceptions) and the proposed operation will not materially damage the water supply of those alluvial valley floors not excepted. Second, the hydrologic function of all alluvial valley floors outside the mine area must be preserved and all alluvial valley floors disturbed by mining must be reclaimed. The reclamation of alluvial valley floors should be approached with the intent of reestablishing the essential hydrologic functions of the original alluvial valley floor. Commonly, reestablishing these functions can be accomplished most completely by restoring all components of the valley floors to their preexisting condition. In reality, it is not always possible to do this, and alternative design of some components is necessary owing to the drastic disturbance of the original alluvial valley materials. The test of reclamation success is whether the functions of the valley are reestablished.

It is also observed that no designated alluvial valley floor has yet been mined and successfully reclaimed; however, several plans for reclamation have been proposed. Insufficient time has elapsed for successful reclamation to have been demonstrated. With time, some types of alluvial valley floors will be successfully mined and restored to their premining essential hydrologic functions.

In addition, successful reclamation must re-create an aquifer near the surface. This poses considerable difficulty many times. It was observed in the Office of Surface Mining report that the science of aquifer reconstruction is in its infancy.

One final but highly significant point concerning drainage basin reclamation must be made: a reclaimed surface ultimately can be no more stable than the adjacent, undisturbed landscape. Wells and Rose (1981) correctly observed that geomorphic processes occurring outside of reclaimed areas can adversely affect those areas causing instability despite proper engineering and revegetation. For example, gully incision and headcut migration from undisturbed to reclaimed lands sometimes poses a substantial threat to reclaimed lands in semiarid regions. Wells and Rose (1981) submitted that when the valley fill component of the landscape is unstable, all other landscape components are potentially unstable. We shall expand upon the concept of landscape stability in the chapter concerning uranium mining.

Reclamation Successes

While it is fashionable to dwell upon the environmental destruction caused by surface mining, it is fair to state that, following enactment of the Surface Mining Control and Reclamation Act of 1977, environmental damage will be temporary and abandonment of mined land is highly unlikely. In the event of abandonment, the bond forfeited by the permitee should ensure reclamation. Whereas the industry had clearly earned societal criticism for past exploitive practices, most companies are now entitled to a measure of praise for the recent changes in attitude as well as operational methods. Through personal observation, it is evident that many companies are willing to go well beyond the letter of the law. It is not uncommon for operators to retreat lands reclaimed prior to the federal legislation, although they are usually not required to do so.

Successful reclamation, resulting in stable productive postmining landscapes is, in the vast majority of cases, more than an expectation; it is reality. Given the establishment of the permanent regulatory program in March 1979 (with subsequent modifications) and the required liability period for reclamation, there are not as yet a large number of cases where bonds have been released for particular parcels of reclaimed lands. As time passes, the accumulation of these data will attest to the accomplishments of the industry in reclaiming the areas that they

Figure 10. Successful reclamation at a coal mine in the western United States.

disturb. Figure 10 shows a successfully reclaimed landscape in the western United States.

Reclamation Failures

Careful examination at most mines will reveal some areas of reclamation failure. Often these are prelaw attempts, but sometimes even contemporary efforts do not achieve the desired goals. The absence of vegetation cover, rilled and gullied hillslopes, and entrenched channels are commonly regarded as evidence of reclamation failure. A variety of factors may preclude the establishment of an effective and diverse vegetation cover, including deficiencies in the selection or application of reclamation practices, the development of soil toxicities, and extreme climatic conditions.

Rilling and gullying imply that the forces of flowing water are rather easily overcoming the resistances of the surface or subsurface materials. The cause may be traced to one or more of the following: (1) excessive drainage area, (2) lack of vegetation cover, (3) high clay content of the reclaimed soil, or (4) hillslopes that are too steep or too long. A recent study by Soulliere and Toy (1986) addressing the problem of rill development on prelaw hillslopes at a surface mine in Wyoming found that the width, depth, length, and frequency of rills could be estimated with reasonable accuracy on the basis of three to five characteristics of the environmental setting, including hillslope age, hillslope length, soil bulk density, soil shear strength, Bouyoucos clay ratio, soil compaction, and soil reaction (pH). It was surprising that hillslope gradient and vegetation cover were not used in the predictive equations; however, previous reclamation practices caused these features to be similar for most hillslopes so that they tended to behave statistically more like constants than variables.

Again, several factors may account for entrenched channels dissecting reclaimed lands. The engineered channel gradient may have been too steep or even too gentle. Channel geometries may provide insufficient conveyance for the water and sediment discharges to be transported. Similarly, inadequate drainage density may result in very large drainage areas generating and delivering runoff to a paucity of incompetent channels. Off-site gullying and knickpoint migration may traverse the reclaimed surface.

Regardless of cause, areas of reclamation failure represent sites of disequilibria between the landforms produced by reclamation and geomorphic processes. Rilling, gullying, and channel entrenchment are surficial expressions of the adjustment toward the establishment of a balance between form and process, force and resistance.

In order to comply with federal and state legislation and regulation, it is usually necessary to retreat any reclaimed land that shows signs of failure. Remedial reclamation efforts range from mere replication of previously utilized practices and intensification of standard procedures, to complete restructuring of the reclaimed landscape. The tacit assumption in the first case is that the initial practices employed were essentially adequate but unusual circumstances, such as atypical climate conditions or operator inconsistency, resulted in the failure. The intensification of procedures implies recognition that initial practices were somehow insufficient. The improved methodology frequently allows the development of vegetation cover that can effectively resist the erosive forces impinging on the surface.

Sometimes, however, fundamental instability exists within the reclaimed landscape and permanent solutions to this problem require complete restructuring of the area. Hillslopes must be recut to reduce gradients and lengths. Channels must be redesigned to appropriate gradients and channel geometries. Drainage basin area may have to be reduced or drainage density increased. Where fundamental instability exists, "more of the same" is unlikely to engender success and, hence, becomes a poor investment. The addition of special engineering structures to artificially support hillslopes and channels may give the appearance of success and create a false sense of accomplishment. Such devices are inevitably temporary and even then require periodic maintenance. For these reasons, regulatory authorities often prohibit this approach to the stabilization of reclaimed lands.

Complete restructuring of previously reclaimed lands is very expensive and usually occurs only after repeated remedial actions have been attempted without success. Because material handling generally comprises most of the total reclamation cost, any substantial regrading will nearly double the final cost of reclamation per unit area. Nevertheless, in the long run, restructuring may prove less expensive than repetitive retreatment, which will usually restart the timeclock for the period of extended liability.

Summary

Surface mining completely changes the environmental system that functioned at a site under natural conditions. Vegetation cover is removed; soil profiles are disturbed; and landforms, including hillslopes, channels, and entire drainage basins, are eradicated. Collectively, the geomorphic and hydrologic characteristics of the land are drastically modified.

Under current legislative regulations, the disturbance to the natural system, albeit drastic, must be temporary. Reclamation is required to proceed along with the mining operation. Periodic inspections by regulatory authorities who are empowered to levy fines or even close down the mine, together with the posted performance bonds, should ensure environmental protection.

It is apparent that surface mine reclamation is a combination of science and art; with the complexity and variability of natural systems, it could hardly be otherwise. As time progresses and research results accumulate, the field is moving toward the scientific end of the spectrum. And while it is impossible to guarantee success, it is entirely possible to avoid serious mistakes. Should they occur, stringent legislation and vigilant regulatory authorities dictate rectification.

Although it is fashionable among geomorphologists and others to discuss the sensitivity of environmental systems, the common success of reclamation attests to a certain element of resiliency in these systems as well. Unfortunately, we are no better able to evaluate resiliency than sensitivity at the present time.

References

Aldon, E. F., and Oaks, W. R. (eds.), 1982, Reclamation of mined lands in the Southwest—a symposium: Albuquerque, New Mexico, Soil Conservation Society of America, 218 pp.

Aldon, E. F., and Springfield, H. W., 1975, Problems and techniques in revegetating coal mine spoils in New Mexico: *in* Wali, M. K. (ed.), Practices and Problems of Land Reclamation in Western North America: Grand Forks, University of North Dakota Press, pp. 122–132.

Aldon, E. F., and Springfield, H. W., 1977, Reclaiming coal mine spoils in the Four Corners: *in* Thames, J. L. (ed.), Reclamation and Use of Disturbed Lands in the Southwest: Tucson, University of Arizona Press, pp. 229–237.

Arnold, F. B., and Dollhopf, D. J., 1977, Soil water and solute movement in Montana strip mine spoils: Vol. 1, Research Report 106, Bozeman, Montana, Montana Agricultural Experiment Station, 129 pp.

Atwood, G., 1975, The strip-mining of Western coal: Scientific American, v. 233, no. 6, pp. 23–29.

Banaszak, K. J., 1980, Coals as aquifers in the Eastern United States: *in* Graves, D. H. (ed.), Proceedings of 1980 Symposium on Surface Mining Hydrology, Sedimentology, and Reclamation: Lexington, Kentucky, Office of Engineering Services, University of Kentucky, pp. 235–241.

Barrett, J., Deutsch, P. C., Ethridge, F. G., Franklin, W. T., Heil, R. D., McWhorter, D. B., and Youngberg, A. D., 1980, Procedures recommended for over burden and hydrologic studies of surface mines: U.S.D.A., Forest Service, Intermountain Forest and Range Experiment Station, Ogden, Utah, General Technical Report INT–71, 106 pp.

Barth, R. C., 1984, Soil-depth requirements to reestablish perennial grasses on surface-mined areas in the Northern Great Plains: Mineral and Energy Resources, v. 27, no. 1, pp. 1–20.

Bauer, A., Berg, W. A., and Gould, W. L., 1978, Correction of nutrient deficiencies and toxicities in strip-mined lands in semiarid and arid regions: *in* Schaller, F. W., and Sutton, P. (eds.), Reclamation of Drastically Disturbed Lands: Madison, Wisconsin, American Society of Agronomy, Crop Science Society of America, Soil Science Society of America, pp. 451–466.

Beach, G. G., 1980, Do I have an alluvial valley floor?: *in* Graves, D. H. (ed.), Proceedings of 1980 Symposium on Surface Mining Hydrology, Sedimentology, and Reclamation: Lexington, Kentucky, Office of Engineering Services, University of Kentucky, pp. 97–101.

Begin, Z. B., and Schumm, S. A., 1979, Instability of alluvial valley floors: a method for its assessment: Transactions, American Society of Agricultural Engineers, v. 22, pp. 347–350.

Begin, Z. B., Meyer, D. F., and Schumm, S. A., 1980, Knickpoint migration due to base-level lowering: Journal of the Waterway Port Coastal and Ocean Division, American Society of Civil Engineers, v. 106, no. WW3, pp. 369–388.

Begin, Z. B., Meyer, D. F., and Schumm, S. A., 1981, Development of longitudinal profiles of alluvial channels in response to base-level lowering: Earth Surface Processes and Landforms, v. 6, pp. 49–68.

Bennett, O. L., Mathias, E. L., Armiger, W. H., and Jones, J. N., 1978, Plant materials and their requirements for growth in humid regions: *in* Schaller, F. W., and Sutton, P. (eds.), Reclamation of Drastically Disturbed Lands: Madison, Wisconsin, American Society of Agronomy, Crop Science Society of America, Soil Science Society of America, pp. 285–306.

Boccardy, J. A., and Spaulding, W. M., 1968, Effects of surface mining on fish and wildlife in Appalachia: Bureau of Sport Fisheries and Wildlife, Resource Publication 65, 20 pp.

Borchert, J. P., 1980, Geography and environmental controls in the United States: The case of surface mining: *in* I.G.U. Symposium of Commission on Environmental Problems, Study and Control of Anthropogenic Transformation of Natural Systems [1979]: Moscow, no publisher identified, pp. 188–197.

Bowling, K. C., 1978, History of legislation for different states: *in* Schaller, F. W., and Sutton, P. (eds.), Reclamation of Drastically Disturbed Lands: Madison, Wisconsin, American Society of Agronomy, Crop Science Society of America, Soil Science Society of America, pp. 95–116.

Brown, D., and Hallman, R. G., 1984, Reclaiming disturbed lands: U. S. Forest Service, 1454.1-Technical Services, Range, 91 pp.

Box, T. W., 1978, The significance and responsibility of rehabilitating drastically disturbed land: *in* Schaller, F. W., and Sutton, P. (eds.), Reclamation of Drastically Disturbed Lands: Madison, Wisconsin, American Society of Agronomy, Crop Science Society of America, Soil Science Society of America, pp. 1–10.

Capp, J. P., 1978, Power plant fly ash utilization for land reclamation in the Eastern United States: *in* Schaller, F. W., and Sutton, P. (eds.), Reclamation of Drastically Disturbed Lands: Madison, Wisconsin, American Society of Agronomy, Crop Science Society of America, Soil Science Society of America, pp. 339–353.

Caruccio, F. T., and Geidel, G., 1978, Geochemical factors affecting coal mine drainage quality: *in* Schaller, F. W., and Sutton, P. (eds.), Reclamation of Drastically Disturbed Lands: Madison, Wisconsin, American Society of Agronomy, Crop Science Society of America, Soil Science Society of America, pp. 129–148.

Chakrabarti, S., Shaw, E. E., and Almes, R. G., 1980, Probabilistic analysis of sedimentation pond requirements, *in* Graves, D. H. (ed.), Proceedings of 1980 Symposium on Surface Mining Hydrology, Sedimentology, and Reclamation: Lexington, Kentucky, Office of Engineering Services, University of Kentucky, pp. 261–268.

Chambers, J. C., 1983, Measuring species diversity on revegetated surface mines: an evaluation of techniques: U.S.D.A., Forest Service, Intermountain Forest and Range Experiment Station, Ogden, Utah, Research Paper INT–322, 15 pp.

Chambers, J. C., and Brown, R. W., 1983, Methods of vegetation sampling and analysis on revegetated mined lands: U.S.D.A., Forest Service, Intermountain Forest and Range Experiment Station, Ogden, Utah, General Technical Report INT–151, 57 pp.

Colbert, T. A., 1983, Non-quantitative and non-vegetative reclamation success criteria and the requirements of the U. S. Surface Mining Control and Reclamation Act: *in* Redente, E. F., *et al.* (eds.), Symposium on Western Coal Mining Regulatory Issues: Land Use, Revegetation, and Management, Colorado State University, Range Science Dept., Science Series No. 35, pp. 56–59.

Collier, C. R., Pickering, R. J., and Musser, J. J., 1970, Influences of strip-mining on the hydrologic environment of parts of Beaver Creek Basin, Kentucky, 1955–1966: U. S. Geological Survey, Professional Paper 427-C, 80 pp.

Curtis, W. R., 1971, Surface-mining, erosion, and sedimentation: Transactions of American Society of Agricultural Engineers, v. 14, no. 3, pp. 434–436.

Curtis, W. R., 1972, Strip-mining increases flood potential of mountain watersheds: *in* Proceedings of National Symposium on Watersheds in Transition, American Water Resources Association and Colorado State University, pp. 357–360.

Curtis, W. R., 1979, Surface mining and hydrologic balance: Mining Congress Journal, v. 65, no. 7, pp. 35–40.

Curtis, W. R., 1981, Reclamation research needs in relation to Public Law 95–87: *in* Proceedings of the 1980 Convention of Society of American Foresters, Society of American Foresters, pp. 70–76.

Curtis, W. R., and Superfesky, M. J., 1977, Erosion of surface-mine spoils: *in* Proceedings of the 32nd Annual Meeting of Soil Science Society of America, New Directions in Century Three, Strategies for Land and Water Use, pp. 154–158.

del Rio, J. R., 1980, Alternatives for proposed regulations on effluent limitations and sedimentation pond standards, *in* Graves, D. H. (ed.), Proceedings of 1980 Symposium on Surface Mining Hydrology, Sedimentology, and Reclamation: Lexington, Kentucky, Office of Engineering Services, University of Kentucky, pp. 215–217.

DePloey, J., 1981, The ambivalent effects of some factors of erosion: Institute of Geology, University of Leuven, v. 21, pp. 171–181.

DePuit, E. J., and Coenenberg, J. G., 1979, Responses of revegetated coal strip mine spoils to variable fertilization rates, longevity of fertilization program and season of seeding, Reclamation Research Unit, Montana Agricultural Experiment Station, Montana State University, Research Report 150, 81 pp.

DePuit, E. J., and Schuman, G. E. (eds.), 1983, Soils and overburden in reclamation of arid/semiarid mined lands: Proceedings of the Symposium, Society for Range Management, Wyoming Mining Association, University of Wyoming (Agricultural Experiment Station), Agricultural Research Service (High Plains Grasslands Research Station), 75 pp.

DeRemer, D., and Bach, D., 1977, Irrigation on disturbed lands: *in* Thames, J. L. (ed.), Reclamation and Use of Disturbed Lands in the Southwest: Tucson, University of Arizona Press, pp. 224–228.

Dollhopf, D. J., Jensen, I. B., and Hodder, R. L., 1977, Effects of surface configuration in water pollution control on semi-arid mined lands: Montana Agricultural Experiment Station, Research Report 114, 179 pp.

Doyle, W. S., 1976, Strip mining of coal, environmental solutions: Park Ridge, New Jersey, Noyes Data Corporation, 352 pp.

Drake, L. D., 1980, Erosion control with prairie grasses in Iowa strip-mine reclamation: *in* Kueera, C. L. (ed.), Proceedings of the 7th North American Prairie Conference, Missouri State University, pp. 189–196.

Ellison, R. D., 1976, Effects of mining methodology on reclamation planning: *in* Vories, K. C. (ed.), Reclamation of Western Surface Mined Lands, Workshop Proceedings, Ecology Consultants, Inc., pp. 38–52.

Frickel, D. G., Shown, L. M., Hadley, R. F., and Miller, R. F., 1981, Methodology for hydrological evaluation of a potential surface mine, the Red Rim site, Carbon and Sweetwater Counties, Wyoming: U. S. Geological Survey, Water Resources Investigations, Open-File Report 81–75, 59 pp.

Gardner, H. R., and Woolhiser, D. A., 1978, Hydrologic and climatic factors: *in* Schaller, F. W. and Sutton, P. (eds.), Reclamation of Drastically Disturbed Lands: Madison, Wisconsin, American Society of Agronomy, Crop Science Society of America, Soil Science Society of America, pp. 173–191.

Gifford, G. F., 1983, Rainfall simulation studies—1981: written communication.

Gilley, J. E., Gee, G. W., Bauer, A., Wiliis, W. O., and Young, R. A., 1976, Water infiltration at surface-mined sites in western North Dakota: North Dakota Farm Research, v. 34, no. 2, pp. 32–34.

Gilley, J. E., Gee, G. W., Bauer, A., Willis, W. O., and Young, R. A., 1977, Runoff and erosion characteristics of surface-mined sites in western North Dakota: Transactions of American Society of Agricultural Engineers, v. 20, pp. 697–700, 704.

Goering, J. D., and Dollhopf, D. J., 1982, Evaluation of a mine pit impoundment in the semiarid Northern Great Plains, Big Sky Mine, Montana: Montana Agricultural Experiment Station, Bulletin 748, 62 pp.

Grandt, A. F., and Lang, A. L., 1958, Reclaiming Illinois strip coal land with legumes and grasses: Illinois Agricultural Experiment Station, Bulletin 628, 64 pp.

Gray, D. H., and Leiser, A. T., 1982, Biotechnical slope protection and erosion control: New York, Van Nostrand Reinhold Company, 271 pp.

Grim, E. C., and Hill, R. D., 1974, Environmental protection in surface mining of coal: U. S. Environmental Protection Agency, Office of Research and Development, EPA–670/2–74–093, 277 pp.

Hadley, R. F., 1962, Some effects of microclimate on slope morphology and drainage basin development: U. S. Geological Survey, Geological Survey Research 1961, pp. B–32 to B–33.

Hadley, R. F., and King, N. J., 1980, Geomorphic and hydrologic problems associated with surface mining on alluvial valley floors, Western United United States, Transaction Society of Mining Engineers (AIME), v. 268, pp. 1818–1823.

Hadley, R. F., and Toy, T. J., 1977, Relation of surficial erosion on hillslopes to profile geometry: U. S. Geological Survey, Journal of Research, v. 5, no. 4, pp. 487–490.

Hadley, R. F., Frickel, D. G., Shown, L. M., and Miller, R. F., 1981, Methodology for hydrologic evaluation of a potential surface mine, the East Trail Creek Basin, Big Horn County, Montana: U. S. Geological Survey, Water Resources Investigations, Open-File Report 81–58, 73 pp.

Haigh, M. J., 1979a, Ground retreat and slope evolution on plateau-type colliery spoil mounds at Blaenavon, Gwent: Transactions, Institute of British Geographers, v. 4, no. 3, pp. 321–328.

Haigh, M. J., 1979b, Ground retreat and slope evolution on regraded surface-mine dumps, Waunafon, Gwent: Earth Surface Processes, v. 4, pp. 183–189.

Haigh, M. J., 1980, Slope retreat and gullying on revegetated surface-mine dumps, Waun Hoseyn, Gwent: Earth Surface Processes, v. 5, pp. 77–79.

Haigh, M. J., and Wallace, W. L., 1982, Erosion of strip-mine dumps in LaSalle County, Illinois, preliminary results: Earth Surface Processes and Landforms, v. 7, pp. 79–84.

Halderson, J. L., and Zenz, 1978, Use of municipal sewage sludge in reclamation of soils: *in* Schaller, F. W., and Sutton, P. (eds.), Reclamation of Drastically Disturbed Lands: Madison, Wisconsin, American Society of Agronomy, Crop Science Society of America, Soil Science Society of America, pp. 355–377.

Hansen, R. P., 1976, Statutory and regulatory aspects of mined land reclamation: *in* Vories, K. C. (ed.), Reclamation of Western Surface Mined Lands, Workshop Proceedings, Ecology Consultants, Inc., pp. 1–7.

Happ, S. C., Rittenhouse, G., and Dobson, G. C., 1940, Some principles of accelerated stream and valley sedimentation: U. S. Department of Agriculture, Technical Bulletin 695, pp. 22–31.

Hartley, D. M., and Schuman, G. E., 1983, Soil erosion potential of reclaimed mined lands: Presented at 1983 summer meeting of American Society of Agricultural Engineers, paper no. 83–2146, 20 pp.

Harza Engineering Company, 1975, Hydrologic effects of strip mining, Busseron Watershed, Indiana: Prepared for U. S. Soil Conservation Service, 29 pp.

Helgesen, J. O., and Razem, A. C., 1980, Preliminary observation of surface-mine impacts on ground water in two small watersheds in eastern Ohio: *in* Graves, D. H. (ed.), Proceedings of

1980 Symposium on Surface Mining Hydrology, Sedimentology, and Reclamation, Lexington, KY, Office of Engineering Services, University of Kentucky, pp. 351–360.

Hodder, R. L., 1975, Montana reclamation problems and remedial techniques: *in* Wali, M. K. (ed.), Practices and Problems of Land Reclamation in Western North America: Grand Forks, University of North Dakota Press, pp. 90–106.

Hodder, R. L., 1977, Dry land techniques in the semiarid West: *in* Thames, J. L. (ed.), Reclamation and Use of Disturbed Lands in the Southwest: Tucson, University of Arizona Press, pp. 217–223.

Hodder, R. L., 1978, Potentials and predictions concerning reclamation of semiarid mined lands: *in* Wright, R. A. (ed.), The Reclamation of Disturbed Arid Lands, Albuquerque, University of New Mexico Press, pp. 149–154.

Hodder, R. L., 1983, Alternatives to established systems for meeting post-mining land use goals: *in* Redente, E. F., *et al.* (eds.), Symposium on Western Coal Mining Regulatory Issues: Land Use, Revegetation, and Management, Colorado State University, Range Science Department, Science Series no. 35, pp. 23–25.

Hofmann, L., Ries, R. E., and Gilley, J. E., 1983, Runoff and erosion from grassland reestablished on mined-land soils: *in* Reclamation Research Summaries, Agricultural Research Service and North Dakota State University Land Reclamation Research Center, March 1983, Northern Great Plains Research Center, pp. 8–12.

Huntington, T. G., Barnhisel, R. I., and Powell, J. L., 1980, The role of soil thickness, subsoiling and lime incorporation methods on the reclamation of acid surface mine spoils: *in* Graves, D. H. (ed.), Proceedings of 1980 Symposium on Surface Mining Hydrology, Sedimentology, and Reclamation, Lexington, Kentucky, Office of Engineering Services, University of Kentucky, pp. 9–13.

Imhoff, E. A., Fritz, T. O., and LaFevers, J. R., 1976, A guide to state programs for the reclamation of surface-mined areas: U. S. Geological Survey Circular 731, 33 pp.

Israelsen, C. E., Clyde, C. G., Fletcher, J. E., Israelsen, E. K., Haws, F. W., Packer, P. E., and Farmer, E. E., 1980, Erosion control during highway construction: Research Report, National Cooperative Highway Research Program Report 220, Transportation Research Board, National Research Center, 30 pp.

Johnson, W., and Paone, J., 1982, Land utilization and reclamation in the mining industry, 1930–1980: U. S. Bureau of Mines, Information Circular 8862, 22 pp.

Kay, B. L., 1978, Mulch and chemical stabilizers for land reclamation in dry regions: *in* Schaller, F. W., and Sutton, P. (eds.) Reclamation of Drastically Disturbed Lands: Madison, Wisconsin, American Society of Agronomy, Crop Science Society of America, Soil Science Society of America, pp. 467–483.

Keefer, W. R., and Hadley, R. F., 1976, Land and natural resource information and some potential environmental effects of surface mining of coal in the Gillette area, Wyoming: U. S. Geological Survey Circular 743, 27 pp.

Khanbilvardi, R. M., Rozowski, A. S., and Miller, A. C., 1983, Predicting erosion and deposition on a strip-mined and reclaimed area: Water Resources Bulletin, v. 19, no. 4, pp. 585–593.

Kilpatrick, F. A., and Rollo, J. R., 1980, Coal hydrology program of the U. S. Geological Survey: *in* Graves, D. H. (ed.), Proceedings of 1980 Symposium on Surface Mining Hydrology, Sedimentology, and Reclamation: Lexington, Kentucky, Office of Engineering Services, University of Kentucky, pp. 113–117.

Krishnamurthi, N., Humphreys, D. L., and Balzer, J. L., 1980, Is sediment control worth a dam?: *in* Graves, D. H. (ed.), Proceedings of 1980 Symposium on Surface Mining Hydrology, Sedimentology, and Reclamation: Lexington, Kentucky, Office of Engineering Services, University of Kentucky, pp. 383–385.

Kuhn, G., Daddow, P. B., Craig, G. S., 1983, Hydrology of Area 54, Northern Great Plains, and

Rocky Mountain Coal Provinces, Colorado, and Wyoming: U. S. Geological Survey, Water Resources Investigations, Open-File Report 83–146, 80 pp.

Land Reclamation Research Center (North Dakota State University), North Dakota Mining and Mineral Resources Research Institute (University of North Dakota), and North Dakota Energy Development Impact Office, 1982, Can mined land be made better than before mining?: The North Dakota Energy Development Impact Office, Bismarck, North Dakota, 99 pp.

Lang, K. J., Schroeder, S. A., Prunty, L. D., and Disrud, L. D., 1983, Interrill erosion as an index of mined land soil erodibility: Presented at 1983 summer meeting of American Society of Agricultural Engineers, paper no. 83–2145, 18 pp.

Lang, R. L., 1975, Strip mine rehabilitation problems and research in Wyoming: *in* Wali, M. K. (ed.), Practices and problems of land reclamation in Western North America: Grand Forks, University of North Dakota Press, pp. 182–189.

Langbein, W. B., and Schumm, S. A., 1958, Yield of sediment in relation to mean annual precipitation: American Geophysical Union Trans., v. 30, no. 6, pp. 1076–1084.

Leopold, L. B., and Maddock, T., 1953, The hydraulic geometry of stream channels and some physiographic implications: U. S. Geological Survey, Professional Paper 252, 57 pp.

Limstrom, G. A., 1960, Forestation of strip-mined land in the Central States. U.S.D.A. Agricultural Handbook 166, 74 pp.

Long, S. G., 1978, Characteristics of plants used in western reclamation: Ft. Collins, Colorado, Environmental Research and Technology, Inc., 146 pp.

Lumb, A. M., 1982, Procedures for assessment of cumulative impacts of surface mining on the hydrologic balance: U. S. Geological Survey, Open-File Report 82–334, 50 pp.

Lusby, G. C., and Toy, T. J., 1976, An evaluation of surface-mined spoils area restoration in Wyoming using rainfall simulation: Earth Surface Processes, v. 1, pp. 375–386.

Mandel, R. D., Sorenson, C. J., and Jackson, D., 1982, A study of erosion- sedimentation processes on abandoned coal refuse piles in Southeastern Kansas: *in* Graves, D. H. (ed.), Proceedings of 1982 Symposium on Surface Mining, Hydrology, Sedimentology and Reclamation, Lexington, Kentucky, Office of Engineering Services, University of Kentucky, pp. 663–669.

Maneval, D. R., 1979, Coal mining vs. environment, a reconciliation in Pennsylvania: *in* Rowe, J. E. (ed.), Coal Surface Mining, Impacts of Reclamation, Boulder, CO, Westview Press, pp. 107–150.

Matter, W. J., Ney, J. J., and Maughan, O. E., 1978, Sustained impact of abandoned surface mines on fish and benthic invertebrate populations in headwater streams of Southwestern Virginia: *in* Samuel, D. E. *et al.* (eds.), Proceedings of a Symposium, Surface Mining and Fish/Wildlife Needs in the Eastern United States, West Virginia University, Fish and Wildlife Service FWS/OBS–78/81, pp. 203–216.

Mays, D. A., and Bengtson, G. W., 1978, Lime and fertilizer use in land reclamation in humid regions: *in* Schaller, F. W., and Sutton, P. (eds.), Reclamation of Drastically Disturbed Lands, Madison, Wisconsin, American Society of Agronomy, Crop Science Society of America, Soil Science Society of America, pp. 307–328.

McKee, T. B., Doeskin, N. J., Smith, F. M., and Kleist, J. D., 1981, Development of climate profiles for reclamation: Dept. of Atmospheric Science, Colorado State University, Climatology Report 81–2, 58 pp.

McKenzie, G. D., 1980, Resource potential of abandoned surface mines: *in* Graves, D. H. (ed.), Proceedings of 1980 Symposium on Surface Mine Hydrology, Sedimentology, and Reclamation, Lexington, Kentucky, Office of Engineering Services, University of Kentucky, pp. 311–314.

McWhorter, D. B., and Rowe, J. W., 1976, Inorganic water quality in a surface mined watershed: Symposium on Methodologies for Environmental Assessment in Energy Development Regions, American Geophysical Union.

Melton, M. A., 1958, Geometric properties of mature drainage systems and their presentation in an E phase space: Journal of Geology, v. 66, no. 1, pp. 35–56.

Merrill, S. D., Doering, E. J., and Sandoval, F. M., 1983, Reclamation of sodic minespoils with topsoiling and gypsum: presented at 1983 summer meeting of American Society of Agricultural Engineers, paper no. 83–2141, 17 pp.

Meyer, L. D., and Kramer, L. A., 1968, Relation between land-slope shape and soil erosion: Transactions of American Society of Agricultural Engineers, v. 11, p. 1–14.

Meyer, L. D., and Romkens, M. J. M., 1976, Erosion and sediment control on reshaped land: Proceedings of the Third Interagency Sedimentation Conference, PB–245–100, Water Resources Council, Washington, D. C., p. 2–65 to 2–76.

Miller, R. M., and Cameron, R. E., 1976, Some effects on soil microbiota of topsoil storage during surface mining: Fourth Symposium on Surface Mining and Reclamation, National Coal Association, pp. 131–135.

Minear, R. A., and Tschantz, B. A., 1974, Contour coal mining overburden as solid waste and its impact on environmental quality: Appalachian Resources Project Report No. 30, University of Tennessee, 32 pp.

Mitchell, J. K., Moldenhauer, W. C., and Gustavson, D. D., 1983, Erodibility of selected reclaimed surface mined soils: Transactions of American Society of Agricultural Engineers, v. 26, no. 5, pp. 1413–1417, 1421.

Moore, R., and Mills, T., 1977, An Environmental Guide to Western Surface Mining, Parts I and II: U. S. Fish and Wildlife Service, Biological Services Program, FWS/OBS–77/20, Part I 70 pp., Part II, 482 pp.

Mountain West Research, Missouri River Basin Commission, and Resource and Land Investigations Program (RALI), 1979, Fact book for western coal/energy develop ment, section II, Western Coal Planning Assistance Project: publisher not identified, 516 pp.

Narten, P. F., Litner, S. F., Allingham, J. W., Foster, L., Larsen, D. M., and McWreath, H. C., 1983, Reclamation of mined lands in the western coal region: U.S. Geological Survey, Circular 872, 56 pp.

National Academy of Sciences, 1974, Rehabilitation potential of western coal lands: Cambridge, Ballinger Publishing Co., 198 pp.

National Research Council, 1981, Surface mining, soil, coal, and society: Washington, D. C., National Academy Press, 233 pp.

Packer, P. E., and Aldon, E. F., 1978, Revegetation techniques for dry regions: *in* Schaller, F. W., and Sutton, P. (eds.), Reclamation of Drastically Disturbed Lands: Madison, Wisconsin, American Society of Agronomy, Crop Science Society of America, Soil Science Society of America, pp. 425–450.

Packer, P. E., Jensen, C. E., Noble, E. L., and Marshall, J. A., 1982, Models to estimate revegetation potentials of lands surface mined for coal in the west: U. S. D. A., Forest Service, Intermountain Forest and Range Experiment Station, Ogden, Utah, General Technical Report INT–123, 25 pp.

Paone, J., Morning, J. L., and Giorgetti, L., 1974, Land utilization and reclamation in the mining industry, 1930–71, U. S. Bureau of Mines, Information Circular 8642, 61 pp.

Paone, J., Struthers, P., and Johnson, W., 1978, Extent of disturbed lands and major reclamation problems in the United States: *in* Schaller, F. W., and Sutton, P. (eds.), Reclamation of Drastically Disturbed Lands: Madison, Wisconsin, American Society of Agronomy, Crop Science Society of America, Soil Science Society of America, pp. 11–22.

Patton, P. C., and Schumm, S. A., 1975, Gully erosion, Northern Colorado, a threshold phenomena: Geology, v. 3, pp. 88–90.

Peltier, L. C., 1950, The geographical cycle in periglacial regions as it is related to climatic geomorphology: Association of American Geographers, Annals, v. 40, pp. 214–236.

Plass, W. T., 1971, Highwalls—an environmental nightmare: Proceedings of Revegetation and Economic Use of Surface-Mined Lands and Mine Refuse Symposium, pp. 9–13.

Plass, W. T., 1974, Revegetating surface-mined land: Mining Congress Journal, pp. 53–59.

Plass, W. T., 1978, Use of mulches and soil stabilizers for land reclamation in the Eastern United States: *in* Schaller, F. W., and Sutton, P. (eds.), Reclamation of Drastically Disturbed Lands, Madison, WI, American Society of Agronomy, Crop Science Society of America, Soil Science Society of America, pp. 329–337.

Power, J. F., 1978, Reclamation research on strip-mined lands in dry regions: *in* Schaller, F. W., and Sutton, P. (eds.), Reclamation of Drastically Disturbed Lands, Madison, Wisconsin, American Society of Agronomy, Crop Science Society of America, Soil Science Society of America, pp. 521–535.

Power, J. F., Ries, R. E., and Sandoval, F. M., 1976, Use of soil materials on spoils—effects of thickness and quality: North Dakota Farm Research, v. 34, no. 1, pp. 23–24.

Power, J. F., Sandoval, F. M., Ries, R. E., and Merrill, S. D., 1981, Effects of topsoil and subsoil thickness on soil water content and crop production on a disturbed soil: Journal of Soil Science Society of America, v. 45, no. 1, pp. 124–129.

Rafaill, B. L., and Vogel, W. G., 1978, A guide for vegetating surface-mined lands for wildlife in eastern Kentucky and West Virginia: U. S. Dept. of Interior, Fish and Wildlife Service, FWS/OBS–78/84, 89 pp.

Ramani, R. V., and Grim, E. C., 1978, Surface mining—a review of practices and progress in land disturbance control: *in* Schaller, F. W., and Sutton, P. (eds.), Reclamation of Drastically Disturbed Lands, Madison, Wisconsin, American Society of Agronomy, Crop Science Society of America, Soil Science Society of America, pp. 241–270.

Redente, E. F., Sowards, W. E., Steward, D. G., and Ruiter, T. L., 1983, Symposium on Western Coal Mining Regulatory Issues, Land Use, Revegetation, and Management: Colorado State University, Range Science Dept., Science Series no. 35, 110 pp.

Ries, R. E., 1980, Supplemental water for the establishment of perennial vegetation on strip-mined lands: North Dakota Agricultural Experiment Station, Farm Research, v. 6, pp. 21–23.

Ries, R. E., and Day, A. D., 1978, Use of irrigation in reclamation in dry regions: *in* Schaller, F. W., and Sutton, P. (eds.), Reclamation of Drastically Disturbed Lands, Madison, Wisconsin, American Society of Agronomy, Crop Science Society of America, Soil Science Society of America, pp. 505–520.

Ringen, B. H., Shown, L. M., Hadley, R. F., and Hinkley, T. K., 1979, Effect on sediment yield and water quality of a nonrehabilitated surface mine in north-central Wyoming: U. S. Geological Survey, Water Resources Investigations, 79–47, 23 pp.

Ringler, R. W., 1983, Spoils aquifer resaturation following coal strip mining: Dept. of State Lands, Helena, Montana, written communications.

Robinson, M. K., Borda, A. B., Holbrook, J. A., Dougherty, M. T., 1980, Area mining of a mountaintop, a methodology for determining approximate original contour: *in* Graves, D. H. (ed.), Proceedings of 1980 Symposium on Surface Mine Hydrology, Sedimentology, and Reclamation, Lexington, Kentucky, Office of Engineering Services, University of Kentucky, pp. 395–403.

Rosso, W. A., 1980, Revegetation augmentation by reuse of treated acid mine drainage: *in* Graves, D. H. (ed.), Proceedings of 1980 Symposium on Surface Mine Hydrology, Sedimentology, and Reclamation, Lexington, Kentucky, Office of Engineering Services, University of Kentucky, pp. 1–8.

Sandoval, F. M., and Gould, W. L., 1978, Improvement of saline- and sodium-affected disturbed lands: *in* Schaller, F. W., and Sutton, P. (eds.), Reclamation of Drastically Disturbed Lands, Madison, Wisconsin, American Society of Agronomy, Crop Science Society of America, Soil Science Society of America, pp. 485–504.

Schafer, W. M., Nielsen, G. A., Dollhopf, D. J., and Temple, K., 1979, Soil genesis, hydrological properties, root characteristics and microbial activity of 1- to 50-year-old stripmine spoils: Montana Agricultural Experiment Station, Montana State University, U. S. Environmental Protection Agency, EPA–600/7–79–100, 212 pp.

Schuman, G. E., and Power, J. F., 1980, Plant growth as affected by topsoil depth and quality on mined lands: Soil Conservation Society of America, Adequate Reclamation on Mined Lands?—A Symposium, Ankeny, Iowa, Soil Conservation Society of America, pp. G–1 to G–9.

Schuman, G. E., and Power, J. F., 1981, Topsoil management on mined lands: Journal of Soil and Water Conservation, v. 36, no. 2, pp. 77–78.

Schuman, G. E., Berg, W. A., and Power, J. F., 1976, Management of mine wastes in the Western United States: Land Application of Waste Materials, Ankeny, Iowa, Soil Conservation Society of America, pp. 180–194.

Schuman, G. E., Rauzi, F., and Taylor, E. M., 1980, The effect of topsoil depth on forage production, water infiltration and water storage: Agronomy Abstracts, 1980 annual meeting of American Society of Agronomy, p. 36.

Schuman, G. E., Taylor, E. M., Rauzi, F., and Pinchak, B. A., 1985, Revegetation of mined land: influence of topsoil depth and mulching method: Journal of Soil and Water Conservation, v. 40, no. 2, pp. 249–252.

Schumm, S. A., 1977, The fluvial system: New York, John Wiley and Sons, Inc., 338 pp.

Schumm, S. A., and Hadley, R. F., 1957, Arroyos and the semiarid cycle of erosion: American Journal of Science, v. 255, pp. 161–174.

Shown, L. M., Frickel, D. G., Hadley, R. F., and Miller, R. F., 1981, Methodology for hydrologic evaluation of a potential surface mine, the Tsosie Swale Basin, San Juan County, New Mexico: U. S. Geological Survey, Water Resources Investigations, Open File Report 81–74, 57 pp.

Shown, L. M., Frickel, D. G., Miller, R. F., and Branson, F. A., 1982, Methodology for hydrologic evaluation of a potential surface mine, Loblolly Branch Basin, Tuscaloosa County, Alabama: U. S. Geological Survey, Water Resources Investigations, Open File Report 82–50, 93 pp.

Singleton, P. C., and Williams, S. E., 1979, Effects of long-term storage on the fertility and biological activity of topsoil: Institute of Energy and Environment, Laramie, Wyoming, University of Wyoming. 36 pp.

Skelly and Loy, Inc., 1979, Analysis of the impact of Public Law 95-87 on mining performance: Final report. U.S. Department of Energy, Contract No. ET-77-CO1-8914, Harrisburg, Pennsylvania, Skelley and Loy.

Smith, R. E., and Woolhiser, D. A., 1978, Some applications of hydrologic simulation models for design of surface mine topography: *in* Wright, R. A. (ed.), The Reclamation of Disturbed Arid Lands: Albuquerque, University of New Mexico Press, pp. 189–196.

Smith, R. M., Tyron, E. H., and Tyner, E. H., 1971, Soil development on mine spoils: Agricultural Experiment Station, West Virginia University, Bulletin 604T, 47 pp.

Soulliere, E. J., and Toy, T. J., 1986, Rilling of hillslopes reclaimed before 1977 Surface Mining Law, Dave Johnston Mine, Wyoming: Earth Surface Processes and Landforms, v. 11, pp. 293–305.

Stearns, M. W., Norbeck, P. M., and Mellbom, P. M., 1984, Assessment of infiltration rates for disturbed and undisturbed sites at the Rosebud Mine, Colstrip, Montana: Western Energy Co., Colstrip, Montana, written communication, 18 pp.

Stein, O. R., Roth, C. B., Moldenhauer, W. C., and Hahn, D. T., 1983, Erodibility of selected Indiana reclaimed strip mined soils: *in* Graves, D. H. (ed.), Proceedings of the 1983 Symposium on Surface Mining Hydrology, Sedimentology and Reclamation, Lexington, Kentucky, Office of Engineering Services, University of Kentucky, pp. 101–106.

Stiller, D. M., Zimpfer, G. L., and Bishop, M., 1980, Application of geomorphic principles to surface mine reclamation in the semi-arid West: Journal of Soil and Water Conservation, v. 35, no. 6, pp. 274–277.

Thames, J. L., and Verma, T. R., 1975, Coal mine reclamation on the Black Mesa and the Four Corners areas of Northeastern Arizona: *in* Wali, M. K. (ed.), Practices and Problems of Land Reclamation in Western North America, Grand Forks, North Dakota, University of North Dakota Press, pp. 48–64.

Thornburg, A. A., and Fuchs, S. H., 1978, Plant materials and requirements for growth in dry regions: *in* Schaller, F. W., and Sutton, P. (eds.), Reclamation of Drastically Disturbed Lands, Madison, Wisconsin, American Society of Agronomy, Crop Science Society of America, Soil Science Society of America, pp. 411–423.

Touysinhthiphonexay, K. C. N., and Gardner, T. W., 1984, Threshold response of small streams to surface coal mining bituminous coal fields, central Pennsylvania: Earth Surface Processes and Landforms, V. 9, pp. 43–58.

Toy, T. J., 1977, Hillslope form and climate: Bulletin of the Geological Society of America, v. 88, p. 16–22.

Toy, T. J., 1984a, Geographic aspects of surface-mine reclamation in the United States of America: *in* Lydolf, P. E. (ed.), Review of geographical research in America on the interrelations among society, economy, and the environment, Institute of Geography, Academy of Sciences of the U. S. S. R., pp. 60–78.

Toy, T. J., 1984b, Geomorphology of surface-mined lands in the Western United States: *in* Costa, J. E., and Fleisher, P. J. (eds.), Developments and Applications of Geomorphology, New York, Springer-Verlag, pp. 133–170.

U.S. Bureau of Land Management and U.S. Geological Survey, 1977, Resource and potential reclamation evaluation, Hanging Woman Creek study area: EMRIA Report No. 12, 309 pp.

U.S. Bureau of Mines, 1971, Strippable reserves of butuminous coal and lignite in the United States: Information Circular 8531, 148 pp.

U.S. Bureau of Mines, 1977, The demonstrated reserve base of coals in the United States on January 1, 1976: Mineral Industry Survey, 121 pp.

U.S. Congress, 1977, Surface mining control and reclamation act of 1977: public law 95–87.

U.S. Department of Energy, 1981, Demonstrated reserve base of coal in the United States on January 1, 1979: Energy Information Administration, Office of Coal and Electric Power Statistics, 121 pp.

U.S. Environmental Protection Agency, 1972, Guidelines for erosion and sediment control planning and implementation: Environmental Protection Technology Series, EPA R2–72–015.

U.S. Environmental Protection Agency, 1973, Methods for identifying and evaluating the nature and extent of nonpoint sources of pollutants: EPA–4030/9–73–014, 261 pp.

U.S. Environmental Protection Agency, 1976, Erosion and sediment control, surface mining in the Eastern U. S.: EPA–625/3–76–006, 91 pp.

U.S. Forest Service, 1982, Wildlife, users guide for mining and reclamation: U.S.D.A., Forest Service, Intermountain Forest and Range Experiment Station, Ogden, Utah, General Technical Report INT–126, 77 pp.

U.S. Office of Surface Mining Reclamation and Enforcement, 1983, Alluvial valley floor identification and study guidelines, 293 pp.

U.S. Office of Technology Assessment, 1979, The direct use of coal—prospects and problems of production and combustion: Congress of the United States, 411 pp.

U.S. Soil Conservation Service, 1972, Hydrology, National Engineering Handbook, sec. 4.

U.S. Soil Conservation Service and U.S. Environmental Protection Agency, 1977, Preliminary guidance for estimating erosion on areas disturbed by surface mining activity in the interior Western United States, Interim Final Report, EPA–908/4–77–005, 57 pp.

Vogel, W. G., 1981, A guide for revegetating coal minesoils in the Eastern United States: U.S.D.A.,

Forest Service, Northeastern Forest Experiment Station, General Technical Report, NE–68, 190 pp.

Wells, S. G., and Rose, D. E., 1981, Applications of geomorphology to surface coal-mining reclamation, Northwestern New Mexico: New Mexico Geological Survey, Special Publications No. 10, pp. 69–83.

Wischmeier, W. H., and Smith, D. P., 1965, Predicting rainfall-erosion losses from cropland east of the Rocky Mountains: U.S.D.A., Handbook 282, 44 pp.

Wright, R. A. (ed.), 1978, The reclamation of disturbed arid lands: Albuquerque, New Mexico, University of New Mexico Press, 196 pp.

Young, S. A., 1983, Native vegetation responses to temporary supplemental irrigation: presented at 1983 summer meeting of American Society of Agricultural Engineers, paper no. 83–2139, 16 pp.

Zingg, A. W., 1940, Degree and length of land slope as it affects soil loss in runoff: Agricultural Engineering, v. 21, pp. 59–64.

9

Lands Disturbed by Surface Mining of Uranium

Introduction

Uranium oxide was discovered nearly 200 years ago, in 1789, by the German chemist Martin Klaproth. However, it was not until 1896 that the French physicist, Henri Becquerel, revealed the natural radioactivity of the metal. With the first splitting of the uranium atom in 1938, this element became more than merely a source of radium.

The nuclear age began on December 2, 1942, in a squash court at the University of Chicago when Enrico Fermi and his associates demonstrated the first nuclear reactor (Stoker *et al.*, 1975). Exactly 15 years later, December 2, 1957, the nuclear energy industry was born with the operation of the first commercial reactor at Shippingport, Pennsylvania. In the past 40 years the uranium industry has experienced numerous transformations, with uranium changing almost overnight from a commodity of only minor commercial interest to one vital for nuclear weapons and, now, to its important peaceful use as a fuel for generation of electrical energy. With each change there has been a surge of interest in ore exploration and development, and in new expanded production facilities [Office of Nuclear Material Safety and Safeguards (ONMSS), 1980].

In a message to Congress on April 18, 1973, President Richard Nixon declared that nuclear energy was the major alternative to fossil fuels for the remainder of this century. By the end of that year, 42 nuclear power generation plants were licensed, another 56 were in various stages of construction, and 56 others were under review for construction permits [U.S. Atomic Energy Commission (AEC), 1974]. Throughout that decade nearly all projections forecast progressive and rapid growth of the nuclear power industry. None could deny or ignore the enormous supply of energy contained in uranium. A pound of enriched fuel holds nearly 3 million times the energy in a pound of coal (National Geographic

Society, 1981). However, these optimistic prognostications have not been realized due to several economic and political developments in recent years. Environmental issues surrounding the mining, enrichment, transportation, use, and disposal of nuclear fuels have substantially contributed to the current stagnation of this industry.

The military demand for uranium during the 1940s was met from known sources of high-grade ore in the Belgian Congo and Canada, supplemented by production from the treatment of old tailings dumps and a few small mines in the Colorado Plateau.

Strong emphasis was placed on the discovery and development of new worldwide uranium resources with enactment of the Atomic Energy Act of 1946 (U.S. Congress, 1946). Concurrently, research efforts were expanded in the search for improved processing technologies capable of utilizing low-grade ores to a greater extent. As a result, several new uranium sources were located and processing costs steadily declined. The AEC encouraged prospecting and mining of uranium through guaranteed fixed prices for ore, bonuses, haulage allowances, establishment of ore-buying stations, and other forms of assistance. These incentives led directly to an increase in the known mineable reserves of ore in the western United States. By 1958, it was estimated that ore production would shortly exceed government requirements, and the AEC withdrew its offer to purchase uranium from any ore reserves developed in the future. This, of course, caused the shutdown of several mills when their contracts expired and also resulted in a concomitant reduction in production rates. Eventually, commercial demand for uranium provided a sufficient market to sustain and permit the renewed growth of uranium production.

The Mining of Uranium

Open pit and underground mining are the two most common methods of extracting uranium ore from the earth, and each has contributed approximately an equal share to total ore production. Solution or *in situ* mining has accounted for only a very small proportion of production, although it has been suggested that this may increase somewhat in the future. The choice of method basically depends upon unit costs of production, and these are influenced by the size, shape, grade, depth, and thickness of the ore body.

Open pit mining is usually preferred where deposits of ore are relatively close to the surface (90–150 m depth) and covered with poorly consolidated overburden. Compared with the mining of other minerals, the ratio of overburden to ore in uranium mining is unusually large, ranging from 8:1 to 35:1. Some of the characteristics of openpit mining and the equipment utilized are presented in the chapter concerning the surface mining of coal. In 1976, openpit mining accounted for about 51% of uranium ore produced in the United States but only about 40% of the total uranium concentrate production because of the relatively lower grade of the ore obtained (ONMSS, 1980).

Ore deposits placed deeply within the earth's crust necessitate underground mining. Again, a variety of techniques may be employed depending upon the shape, size, depth, and grade of the ore. Basic options include room-and-pillar, longwall retreat, and panel methods. Groundwater intrusion is a common problem associated with underground mining, and this requires pumping, which adds to the total cost of these methods. Additionally, underground uranium mines must be well ventilated to prevent the build up of radon–222 gas to concentrations hazardous to the health of miners. In 1976, underground mining accounted for about 49% of the uranium ore produced in the United States and about 57% of the concentrate (ONMSS, 1980).

Solution or *in situ* mining is employed to produce uranium from low-grade ores when conventional mining methods would be uneconomical. The process consists of injecting leaching solutions into an underground ore body to dissolve the uranium minerals and then recovering the enriched solution by pumping it to the surface for processing.

Although there are some advantages to this method, including miner safety and elimination of the need to handle large volumes of waste materials, there are serious environmental concerns, chiefly focused upon potential groundwater pollution. These issues will not be addressed herein because solution mining is unlikely to become a major mining method for the uranium industry. In 1976, solution mining accounted for about 2% of the total uranium concentrate production in the United States. Further discussion of this procedure can be found elsewhere [U.S. Nuclear Regulatory Commission (NRC), 1978].

Processing of Uranium Ore

The mined ores are transported to a mill where the uranium content is separated from the waste rock. The recovered uranium values are then concentrated into an intermediate, semirefined product known as "yellowcake" [U_3O_8, $(NH_4)_2U_2O_7$, or $Na_2U_2O_7$]. The milling process involves the following basic steps: (1) ore handling and preparation, (2) mill concentration, and (3) product recovery.

Ore handling and preparation consists of ore blending to ensure a mill feed of uniform physical and chemical properties, crushing of ore to a size between −1.9 and −3.8 cm, temporary fine ore storage, ore grinding to effect a size reduction of the ore to approximately 28 mesh (0.589 mm or 0.023 in.) for the acid leach concentration process or 200 mesh (0.074 mm or 0.0029 in.) for the alkaline leach process, and possibly drying or roasting to improve handling or solubility properties.

The uranium content of processed ores is recovered using hydrometallurgical leaching techniques together with concentration and purification processes such as ion exchange, solvent extraction, or the Elvex process (ONMSS, 1980). The choice of acid or alkaline leaching techniques depends largely on the lime content of the ore. When the amount of lime is 12% or less, the acid process is preferred, while lime contents exceeding this amount usually dictate the alkaline process because of the economic factor associated with need for quantities of acid for neutralization. A study in 1976 found that about 82% of the total concentration capacity of this industry employed the acid leach process, whereas only 18% used the alkaline leach process (Midwest Research Institute, 1976).

Occasionally, a "heap-leaching" recovery process is utilized for the treatment of low-grade ore dumps or when the ore body is small and situated far from the milling site. In this case, economic considerations preclude the construction of another mill or transportation of the ore itself to existing mills. The ONMSS (1980) offers detailed discussion on this topic as well.

By-Products of Uranium Ore Processing

The processing of uranium ore in concentrating mills produces solid, liquid, and gaseous effluents that may be released to the environment, depending upon the process controls and waste management practices instituted by the mill operator. Tailings account for nearly all of the wastes from a uranium mill. These are a mixture of solids and solutions varying in chemical and physical compositions depending on the nature of the ore and the concentrating process employed. The fractions comprising the tailings are (1) the sands, consisting of solids greater than 200 mesh; (2) the slimes, consisting of solids less than 200 mesh; and (3) the liquids, which are solutions of chemicals, dissolved ore solids, and water. The weight of dry tailings approximately equals the weight of ore as mill feed because of the small percentage of uranium recovered and the addition of various substances in the recovery process. A representative average ore grade for nine mines in New Mexico, Wyoming, Utah, and Colorado is 0.16%. About 10% of the uranium and virtually all of the other members of the uranium series in the original ore are discharged with the tailings.

For the acid leach process, the dry tailings generally consist of 20 to 37% slimes (29% as the average for the nine aforementioned mills), with the remainder within the sand fraction. The chemical composition of the sand component varies but consists largely of silica with minor amounts of complex silicates of Al, Fe, Mg, Ca, Na, K, Mn, Ni, Mo, Zn, U, and V. Also present are small amounts of sulfates, phosphates, and chlorides.

The slime component of tailings also consists of various complex silicates of the elements listed above but are regarded as "claylike" due to their minute particle sizes. The slimes will contain higher concentrations of the radioactive elements as well (Misaqi, 1976; Moffett and Tellier, 1978).

The tailings are transported to impoundments in slurry form with about equal parts of solid and solution, by weight. From 0 to 70% of the liquid may be recycled. These solutions contain various concentrations of SO_4^{2-}, NaCl, NH_4^+, and PO_4^{3-}, together with minor amounts of flocculants, kerosene, and other organic materials used in processing. Ions of Cu, Mo, U, V, Zn, Mg, Ca, Be, Al, Ni, Sb, and Fe are usually present in the solution also. The pH of the liquid ranges from 1.2 to 2.0, and total dissolved solids may approach 1%.

Generally, the physical properties of alkaline leach tailings are similar to those of acid leach mills with about 35% of the total solids within the slimes fraction. These again consist generally of silica and various complex silicates of Na, Ca, Mn, Mg, Al, and Fe. In minor amounts are sulfates, chlorides, and carbonates.

The sand component consists of approximately 99% silica and various minor amounts of U, Mo, and Se. For the alkaline-leach process, radiological analyses are not readily available for the individual sand and slime separates.

Again these tailings are deposited in an impoundment at about the same pulp density and are 50% liquid, by weight. These liquids can be nearly 100% recycled, with only losses due to evaporation and seepage during the decant process. The solution contains various concentrations of CO_3^{2-}, HCO_3^-, Se, SO_4^{2-}, Cl^-, Mo, V_2O_5, and U_3O_8. The pH of the liquid ranges from 10 to 10.5.

In addition to mill tailings, both radioactive and nonradioactive materials are released from processing areas of the concentrator facilities in amounts varying with the extent of controls utilized. The sources of radioactive emissions include the ore hauling and storage pad, ore crushing and grinding, and yellowcake drying and packaging.

The nonradioactive emissions include the products of fuel combustion, fumes released from the leach tank vent system, and vaporized organic solvents. If natural gas is used as a fuel, then carbon dioxide, nitrogen oxide and water vapor are commonly emitted, whereas if fuel oil is used, then release of SO_2 and NO_2 will often take place. The emissions from the leaching system, usually in very low concentrations, are again sulfur dioxide together with sulfuric acid fumes. The most common organic solvent released in gaseous phase is kerosene (92%).

General Environmental Impacts

Mining and concentration of uranium produces all of the short-term environmental impacts associated with any surface mining operations and, in addition, several long-term impacts as a result of the deposition at or near the earth's surface of toxic substances and radioactive materials in the mill tailings impoundments. Toxic elements will remain indefinitely at these locations and the half-lives of some isotopes are measured in hundreds or thousands of years. General categories of impact include: (1) air quality, (2) land use, (3) surface and ground water, (4) soil and terrestrial biota, and (5) socioeconomic conditions.

Although these environmental issues are significant, by far the greatest concern focuses upon the possible radiological impacts associated with mining and milling of uranium ore and especially the disposal of tailings that possess some residual radioactivity. The U.S. Environmental Protection Agency (EPA, 1981) states that tailings are hazardous primarily because (1) breathing radon and its decay products exposes the lungs to alpha particles, (2) the body may be exposed to gamma rays, and (3) radioactive materials and nonradioactive toxic elements from tailings may be swallowed with food and water. The alpha particles from inhaled radon decay products can cause lung cancer. Also, gamma rays can cause cancers, teratogenic effects, and genetic damage. Further, the EPA observes that the radiation hazard from tailings lasts for many thousands of years.

In fairness, it must be noted that many of those involved in the industry vigorously challenge these assertions. They submit that the probabilities for the occurrence of some of these hazards are very small or that actual effects of exposure are unproven. For example, the National Research Council (1979, p. 40) stated:

> Of unknown consequences is the threat of radioactive contamination by radon emission and wind-blown dust from uranium mill tailings, ore haulage, ore piles, and waste dumps. Whether radon-bearing ventilation air discharged from underground uranium mines and wind-blown radon from surface uranium mines poses a threat to public health is not now known.

Misaqi (1976) found that the total amount of radon–222 released into the atmosphere by the United States uranium mines and mills is only an infinitely small fraction of the total amount of radon released by the top layers of soil with a high concentration of uranium, such as found in Grants, New Mexico. Even when compared to areas where mineable deposits of uranium are unknown, the amount of radon released by all U.S. mining and milling operations is still relatively insignificant. Of course, this author concedes that higher local concentrations of airborne radioactivity may be observed around stockpiled ore and tailings areas under conditions of limited air movement. The ONMSS (1980) observed that in a case involving the operation of twelve 1800 tons/day mills in a 50-mile region (the assumed worst possible concentration of milling in year 2000), lifetime risks to the average individual in the region from milling would be about 0.015% of natural cancer risks. However, occupational health effects due to radiation exposures (from 1979–2000) at the mills may result in a total of 32 premature deaths. This is equivalent to about a 20% increase in risk of cancer among occupationally exposed mill workers over a career of 47 years.

Those contending that uranium mining and milling poses little health threat can also question the mathematical procedures used to generate risk estimates. Commonly, the models utilized necessitate numerous assumptions, some of which may be suspect. Second, computations of regional risk to populations usually require multiplication of very small risk estimates by very large population figures, and many mathematicians admit that probabilities so obtained are inherently unstable. Finally, the basic research providing the data for risk estimation may be called to question. Sometimes data are collected in situations very dissimilar to uranium mining and milling operations but transposed for the purpose of analyses. For example, the ONMSS (1980) noted a study that bases a procedure for estimating lung cancer risk in human populations due to radiation exposure in part on a study of lung tumor mortality for beagles exposed to alpha irradiation.

Nevertheless, the EPA has concluded, following thorough examination, that existing evidence is sufficient to warrant public health standards for permissible exposure. Further, the EPA (1981) reaffirms their belief that a linear, non-threshold dose–effect relationship is a reasonable basis for deriving estimates of radiation risk to the general public and for establishing regulations. This model assumes that any radiation dose presents some risk to humans and that the risk of low doses is directly proportional to the risk demonstrated at higher doses. Thus said, however, they recognize that the data preclude neither a threshold for some types of radiation below which there is no damage to people, nor the possibility that low doses may do more damage to people than the linear model implies. Indeed, the presence or absence of dosage thresholds seems to lie at the core of much of the controversy concerning the impacts of radiation.

The EPA (1981) observed that the radiation hazard from mill tailings lasts for many thousands of years, and this persistence played a major role in determining their tailings disposal standards. The ONMSS (1980) remarked that radioactivity in tailings, unlike high level nuclear waste, poses a chronic as opposed to an acute hazard. Long and sustained exposure to radioactivity in the tailings pile would be required to produce detectable adverse effects. The longevity of the radiation hazard is, of course, a consequence of half-lives of isotopes in the uranium–238 decay series. Uranium-238 supports 13 lineal radioactive decay products (excepting two minor branches in the decay scheme) traditionally referred to as daughters. The chain terminates in stable lead–206.

With respect to overall health impacts, the critical mill-released radionuclides and their primary sources are, in descending order of importance, radon-222 from the tailings pile, radium-226 and lead-210 from the tailings pile, and U-238 and U-234 from yellowcake operations (ONMSS, 1980). These yellowcake emissions essentially terminate when the mill shuts down. Radon gas has a half-life of only 3.8 days; however, its production is supported by radium-226 that has a half-life of 1620 yr and thorium-230 with a half-life of about 80,000 yr. Of the radionuclides found in tailings, radium has the most potential for groundwater contamination (Buelt *et al.*, 1981).

The principal pathways by which radioactivity from the above sources reaches human beings and irradiates their tissue include (1) direct, external exposure from radionuclides in the air or on the ground; (2) inhalation of radioactivity into the lungs, possibly followed by redistribution to other organs of the body; and (3) ingestion of radioactivity in foodstuffs. Figure 1 illustrates the flow of radioactivity to humans.

Table 1 contains a summary of unavoidable impacts of the conventional uranium milling industry compiled by the ONMSS (1980). Although the number of operating mills in the year 2000 now appears rather optimistic (or pessimistic, depending on one's viewpoint), these estimates reveal the major areas of concern. It should also be understood that the actual environmental impact of particular uranium mining and milling operations is highly site specific.

Purpose of Chapter

In many ways uranium mining is similar to any surface mining enterprise. However, the persistence of radiation hazard associated with ore processing and waste disposal creates additional serious problems. As a result, those engaged in

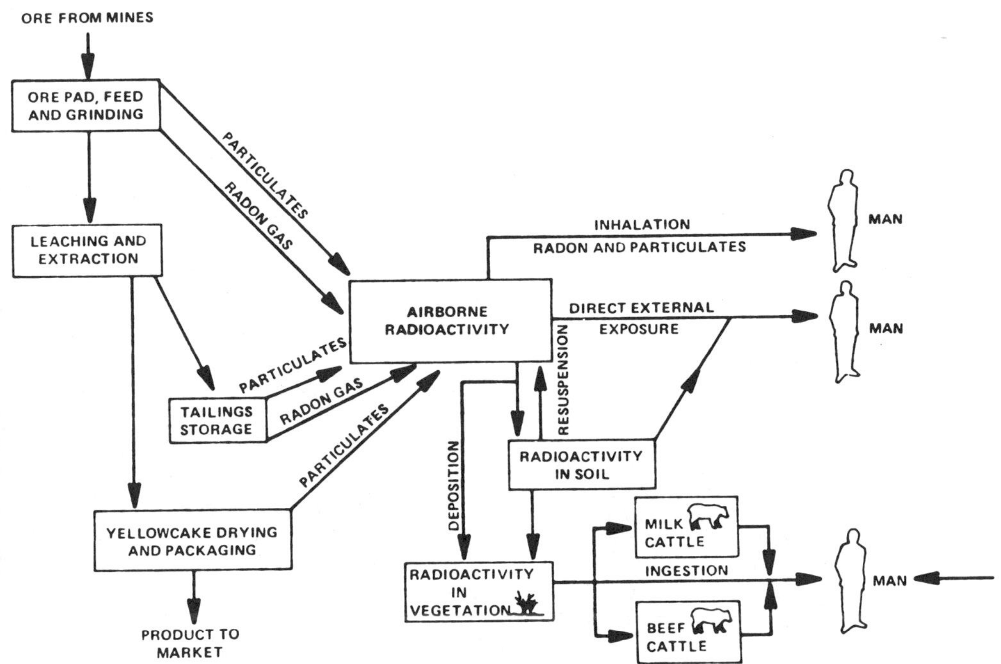

Figure 1. Pathways of radioactive effluents to a human population (ONMSS, 1980).

Table 1 Summary of Integrated Unavoidable Impacts of Conventional Uranium Milling Industry Through the Year 2000[a,b]

	Estimated value	Estimated range
Production (metric tons $U_3O_8 \times 10^3$)	440	410–490
Number of operating model mills in year 2000	55	47–69
Number of model mill years of operation	840	780–940
Natural resources use		
Land temporarily disturbed by milling (ha $\times 10^3$)	23	20–29
Tailings disposal land permanently committed to restricted use (ha $\times 10^3$)	6.1	3.4–7.6
Land temporarily disturbed—mining (ha $\times 10^3$)	6.6	5.6–8.2
Water lost to evaporation ($m^3 \times 10^8$)[c]	5.5	3.1–6.9
Effluents		
Tailings solids generated (metric tons $\times 10^8$)	4.7	4.0–5.9
Dusts released (metric tons $\times 10^3$)[d]	330	180–410
Fumes (metric tons)	22	20–25
Gases (SO_x, NO_x) (metric tons $\times 10^3$)	8.3	7.7–9.3
Radon—mills (1979 to 2000) (Ci $\times 10^6$)[e]	5.3	3.0–6.6
Radon—mines (1979 to 2000) (Ci $\times 10^6$)	6.2	5.3–7.7
Persistent radon releases from tailings (KCi/yr)	3.9	1.6–4.9
Radiological impacts		
Milling		
Health effects—1979 to 3000 (premature deaths)	92	38–120
Life-shortening—1979 to 3000 (years lost)	1750	720–2200
Persistent health effects—beyond 3000 (premature deaths/yr)	0.043	0.018–0.054
Milling Occupational		
Health effects—1979 to 2000 (premature deaths)	32	30–36
Life-shortening—1979 to 2000 (years lost)	610	570–680
Mining		
Health effects—1979 to 2000 (premature deaths)	56	48–70
Life-shortening—1979 to 2000 (years lost)	1060	910–3100

[a]Source: Office of Nuclear Material Safety and Safeguards (1980).

[b]Estimated values and ranges of impacts presented here are based primarily on information provided in Chapters 3, 5, 6, 9, and 12 and Appendix S. Ranges are based on the types and magnitudes of uncertainties as described and assessed in Appendix S. Uncertainties in health effect conversion factors are not included; as described in Appendix G-7, this would extend the ranges for health effect estimates by about a factor of two in either direction. The average life shortening per health effect is taken to be 19 yr (see Appendix G-7).

[c]An additional 20–50% of this amount would be lost due to mining operations. This is counterbalanced by the fact that some evaporation of soil moisture would occur anyway; this evaporative loss would roughly equal precipitation intercepted by tailings areas, which amounts to 42% of the evaporation losses shown.

[d]Only dust losses during mill operation are shown. Tailings dust losses during assumed 5-yr prereclamation drying periods would add 50,000 metric tons during 1979–2000.

[e]The estimate shown is conservative in that the effects of radon attenuation by interim stabilization or reclamation are not included; this overestimates total radon releases by approximately 10% for the 1979–2000 period.

the development and implementation of protective controls are forced to think in terms of extended time frames. Rafferty (1982) commented that the period of time for which tailings should be isolated from the environment should effectively be indefinite. The NRC (1980) cited an inescapable fact that tailings will remain hazardous for hundreds of thousands of years. In an attempt to confront reality, the EPA (1983) submitted that the design life of uranium mill tailings impoundments is to be 1000 yr where practicable, but at least 200 yr.

As a result, both regulatory and industrial personnel became aware of the need for geomorphic inputs to the decision-making process. It was recognized that the geomorphic history of uranium mining and milling sites could have a profound influence on the long-term stability of these locations, in addition to the contemporary geomorphic processes. Hence, it is probably permissible to suggest that the uranium industry has systematically drawn upon the science of geomorphology to a greater extent than any other human activity causing disturbance of the natural system.

The purpose of this chapter is to demonstrate the intimate association between uranium mining, ore processing, reclamation, and geomorphic principles. This discussion will be structured around the following topics: (1) an assessment of land disturbance resulting from this industry, (2) an examination of the specific nature of the disturbance to the environmental system, (3) discussion of the legal constraints placed upon mining and milling of uranium resources and some geomorphic ramifications, and (4) consideration of various common reclamation practices employed by this industry and their geomorphic basis.

Assessment of Disturbance

An assessment of the disturbance to the environmental system caused by the development of uranium resources centers on three factors as described in Chapter 1: (1) areal extent, (2) intensity or rate, and (3) duration of disturbance or impact. The treatment of areal extent must consider the amount of land disturbed in both an absolute and relative perspective, the geographic distribution of the industry, and the prospect for future expansion of this primary economic activity.

Areal Extent

Data compiled by Johnson and Paone (1982) show that about 6880 ha (17,000 acres) of land were utilized by the uranium industry from 1930 to 1980 and about 125 ha (310 acres) in that final year. During that 1930 to 1980 period, only 526 ha (1300 acres) were reclaimed and only about 28 ha (70 acres) in 1980. Thus, about 7.6% of the land used by this industry was reclaimed over a 51-yr period. The remaining 92.4% or 6354 ha (15,700 acres), exists in a drastically disturbed state. Additionally, it should also be recognized that much of the early reclamation would not meet the requirements of current standards. The data above include the area of surface mine excavation, area used for disposal of surface mine wastes, surface area subsided or disturbed as a result of underground workings, surface area used for disposal of underground wastes, and surface area used for disposal of mill or processing wastes. Excluded are areas used for haul roads, freshwater reservoirs, railroads and public highways to the edge of the mine properties, and streams affected by acid drainage and sedimentation; these would markedly increase the total area of disturbance.

The areal extent of disturbance includes mining and milling sites still active as well as those now inactive. Until recently, with the enactment of federal legislation, these inactive sites were a particular environmental concern. Most were potential hazards and many were daily discharging air, water, and ground pollutants to nearby areas. Additionally, there was often no one legally responsible for even temporary remedial action, let alone full reclamation, although state or federal agencies might intervene to the extent permitted by available resources.

The ONMSS (1980) provides a summary of the events and actions preceding passage of PL 95–604, the Uranium Mill Tailings Radiation Control Act of 1978 (UMTRCA), which authorized the Department of Energy (DOE), along with the

Table 2 Inactive Processing Sites, Priorities for Remedial Action, and Quantity of Tailings[a]

Location and processing site	Remedial priority	Amount of tailings (10^6 metric tons)[b]	Last year of operation
Arizona			
Monument City[c]	Low	1.0	1968
Tuba City[c]	Medium	0.7	1966
Colorado			
Durango	High	1.8	1963
Grand Junction	High	2.5[b]	1970
Gunnison	High	0.6	1962
Maybell	Low	2.4[b]	1964
Naturita	Medium	0.6	1963
Rife (New Rifle)	High	2.4[b]	1972
Rifle (Old Rifle)	High	0.3	1958
Slick Rock (NC)	Low	0.2	1957
Slick Rock (UC)	Low	0.2	1961

Table 2 *(Continued)*

Location and processing site	Remedial priority	Amount of tailings (10^6 metric tons)[b]	Last year of operation
Idaho			
Lowman	Low	0.08	1960
New Mexico			
Ambrosia Lake	Medium	2.3	1963
Shiprock[c]	High	1.9	1968
North Dakota[d]			
Belfield	Low	0.6	1968
Bowman	Low	0.5	1967
Oregon			
Lakeview	Medium	0.1	1960
Pennsylvania			
Canonsburg[e]	High	0.2	1966
Texas			
Falls City	Medium	2.2[b]	1973
Utah			
Green River	Low	0.1	1961
Mexican Hat[c]	Medium	2.0	1965
Salt Lake City	High	2.0	1968
Wyoming			
Baggs[d]	Low	0.01	—
Converse County	Low	0.17	1965
Riverton[f]	High	0.8	1963

[a]Source: Office of Nuclear Material Safety and Safeguards (1980).

[b]At three sites, tailings from commercial sales were comingled with those from government defense operations. At Grand Junction, CO, the ratio is 80% government, 20% commercial; at New Rifle, CO, 99% government, 1% commercial; and at Falls City, TX, 34% government, 66% commercial. All other piles contain tailings from government operations only. (Based on personal communication from J. S. Themelis, Director, Engineering and Safety Division, Grand Junction Office, DOE, March 14, 1980.)

[c]Processing site on tribal lands owned by the Navajo Nation.

[d]The sites in North Dakota and the Baggs site in Wyoming were identified and designated subsequent to the enactment of the Act. All other sites were identified prior to the Act and designated by the Act.

[e]The Canonsburg, PA, site was started in 1911 and was used as a custom mill to extract radium and uranium from ores. (Based on "First Annual Status Report on Inactive Mill Tailings Sites; Remedial Action Program, December 1979," U.S. DOE, 1979.)

[f]Processing site located on private property within the boundaries of the Wind River Indian Reservation.

affected states, Indian tribes, and persons who owned or controlled inactive uranium tailings mills, to establish assessment and remedial action programs at inactive uranium mill tailings sites.

As a result of the enactment of PL 95–604, the 22 previously identified hazardous sites were designated by Congress for inclusion in the Remedial Action Program and three more inactive sites were later added by the Department of Energy. This agency is currently preparing remedial action plans for the first several high-priority sites. Table 2 contains some basic information for all 25 sites, and Figure 2 shows the geographic distribution of the 23 western sites, excluding those at Canonsburg, Pennsylvania, and Falls City, Texas.

The areal extent of land disturbance may also be expressed in relative terms. Comparing the 6880 ha (17,000 acres) of land disturbed by the uranium industry over a 51-yr period to the total surface area of 919.2 million ha (2.27×10^9

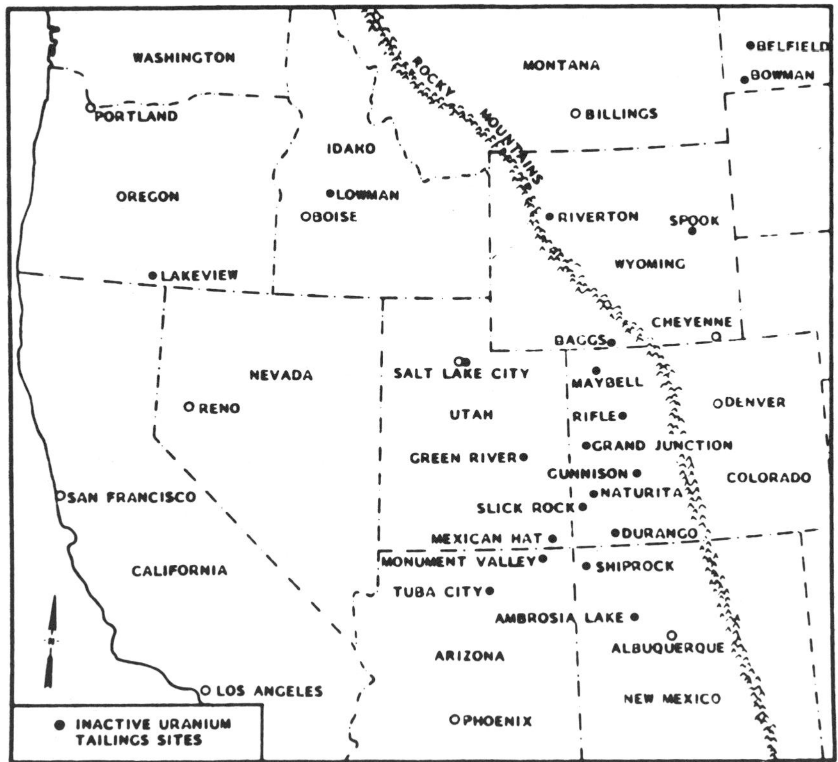

Figure 2. Location of inactive uranium mill tailings piles in the western United States (Beedlow, 1984b).

Table 3 Share of Potential Resources of Uranium in Individual States[a]

State	Share of probable resources[b] (%)
New Mexico	30
Wyoming	15
Colorado	11
Utah	14
Texas	10
California	2
Arizona	4
South Dakota	1
Nevada	2
Washington	2

[a]Source: Office of Nuclear Material Safety and Safeguards (1980) taken from Hetland (1978).

[b]Conventional sources only.

acres) for the entire United States, we find that the extent of disturbance constitutes only about 0.0007% of the national land base. The average annual disturbance for those years would have been approximately 0.000015%.

Recall also that the average amount of land disturbed due to the surface mining of coal, based on the period 1930 to 1970 (Paone *et al.*, 1974), was 0.057% of the total surface area of the nation or about 0.0003% per year. Thus, the amount of area disturbed by uranium recovery constitutes only a small percentage of that disturbed by surface mining of coal nationwide.

A relative comparison might also be constructed using Table 3 from the chapter concerning the surface mining of coal, which shows the area devoted to various categories of land use in the United States as of 1980. Here, all mining accounted for about 2.3 million ha (5.7 million acres). The 6880 ha (17,000 acres) used by the uranium industry comprises only 0.3% of this total. According to that table, land disturbed by uranium mining and milling would amount to about 0.6% of that committed to railroads and 0.02% of that enclosed within national parks.

The cumulative environmental impacts resulting from the operation of the uranium milling industry through the year 2000 as projected by the ONMSS (1980) were presented in Table 1. These estimates suggest that about 6600 ha (16,309 acres) will be temporarily used by mining, 23,000 ha (56,833 acres) temporarily disturbed by milling, and 6100 ha (15,073 acres) of land permanently committed to restricted use due to tailings disposal thereupon. Recent developments throughout the industry in the United States may cause these to become overestimations; however, they constitute the best information readily available.

Although future rates of land disturbance and areal extent escape prediction at present, the geographic location of future disturbance is determined by the existence of known uranium resources and thus can be anticipated. Further, as noted for other types of land disturbance, it is somewhat inappropriate to compare the areal extent to the entire surface area of the United States because uranium resources are also concentrated in a few regions.

Most of the nation's known uranium resources are located in the West, as depicted in Figure 3, and all of the 21 conventional uranium mills now operating or currently planned for operation are west of the Mississippi River (ONMSS, 1980). Future uranium production is likely to be centered in NURE regions A and B, and New Mexico and Wyoming in particular. Table 3 shows the share of potential uranium resources for individual states. These data reveal that the environmental impact of the uranium industry is very likely to be concentrated at a relatively few locations.

Another aspect of the areal extent of disturbance is worthy of mention. Because of the residual radioactivity and toxicity of mill tailings, there has been concern that the failure of active or reclaimed disposal sites by flooding could result in the widespread dissemination of these contaminants in the downstream direction. It is virtually impossible to predict this possibility and its potential severity.

Collectively, the foregoing suggests that the areal extent of land disturbed by the uranium industry is relatively insignificant. Even when geographic concentration of mining and milling is considered, the amount of land impacted is very small. Consequently, we must conclude that areal extent is not a basis for alarm.

Intensity of Disturbance

Intensity of the disturbance is the second component of this assessment. Within the geomorphic context, this is determined through a comparison of process rates before and after impact. As suggested earlier, field measurements collected for both time periods would constitute the ideal base for such an evaluation. In the absence of such data, recourse may be given to a comparison between the following: (1) rates on disturbed lands and adjacent undisturbed lands; (2) rates generated by accepted predictive models for the disturbed and undisturbed lands; or (3) observational evidence of rate differentials, such as magnitude and frequency of rilling, gullying, or sedimentation.

Uranium production results in two categories of land disturbance, that associated with the actual mining of ore and that caused from the tailings deposits

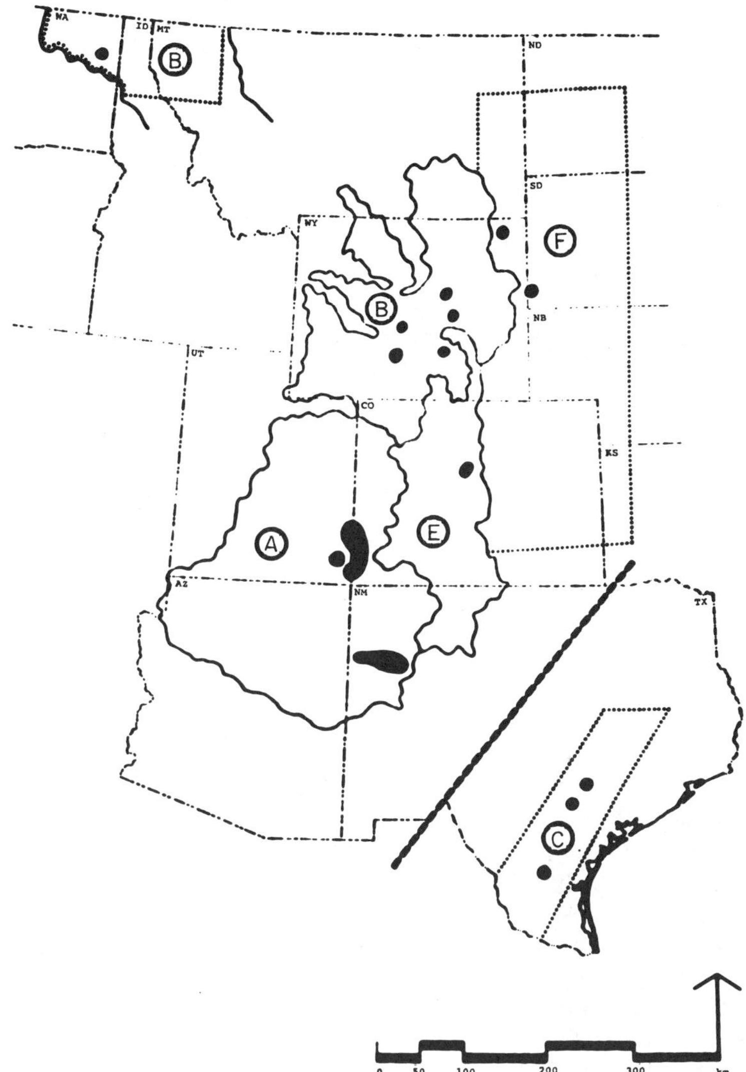

Figure 3. Uranium reserves and resources in the western United States. Circled letters are NURE region designations (ONMSS, 1980).

produced by the uranium concentrating procedures. In the foregoing discussion of areal extent of disturbance, these could be considered together. However, because the surface materials and landforms produced by these two categories of disturbance differ considerably and because available literature tends to focus on one particular type or the other, it is appropriate to consider each separately.

From a geomorphic perspective, accelerated soil erosion, including sheet, rill, and gully erosion, and disruption of drainage networks are probably the most serious consequences of disturbance. While there were relatively little actual field data for geomorphic processes on lands disturbed by the surface mining of coal, there is virtually none for the uranium industry. Most discussions describe what "may" or "could" happen based upon geomorphic principles (Nelson and Shepherd, 1978; Nelson *et al.*, 1983). While the observations contained therein are doubtlessly valid, they do not provide a firm foundation for an assessment of intensity through a comparison of real process rates.

First, let us address the disturbance associated with the surface mining of uranium ore. The open pit method is usually favored by those engaged in surface extraction. Anaconda's mine on the Laguna Indian Reservation in New Mexico is reportedly the largest open pit uranium mine in the world. The pit itself encompasses 486 ha (1201 acres) and over 181,560,000 metric tons (200,079,120 tons) of overburden or mine wastes have accumulated in the form of 28 dump sites covering over 445 ha (1100 acres) (Reynolds *et al.*, 1978). It was projected that the total disturbed surface area would be approximately 1052 ha (2599 acres), including roads, storage areas, and miscellaneous disturbances, at cessation of mining operations. As this has been one of the most intensively studied uranium mines in the United States, it can offer valuable information.

The intensity of the disturbance caused by mining itself will likely be roughly proportional to the size of the operation. Thus, the Anaconda mine probably represents an extreme case. Nevertheless, the general environmental and geomorphic consequences are similar for most mines. The excavation of the pit disrupts at least lower order drainage networks. Larger perennial streams will be diverted around the site, although in most cases it would be merely coincidental if the channel geometry, gradient, and sinuosity of these diversions matched those of the premining channel. Of course, the pit will possess a centripetal drainage pattern and so the total drainage area of downstream tributaries will be diminished by the size of the depression.

The overburden and mine wastes (spoils) are generally dumped on the surface in close proximity to the pit in order to minimize the cost of transportation; this again commonly disturbs at least lower order drainage systems. Fuller (1981) observed that the burial of meanders of the Rio Moquino just north of its confluence with the Rio Paguate by mine dumps at the Anaconda mine has introduced instability into this drainage and nearby arroyos. The Rio Moquino flows nearly year round with waters derived from a drainage basin of about 18,130 ha (70 sq miles). It is also suggested by Fuller that the increase in channel gradient from 0.007 to 0.01, which has resulted from the shortening of this reach of stream due to burial, has tended to accelerate headcutting upstream. Field studies of those arroyos that pose the most immediate threat to reclamation areas revealed that piping around headcuts is one of the major processes responsible for headward erosion, especially in areas where efforts have been made to minimize headcutting by dumping material into the headcut area.

Smith (undated) examined erosion on the outslopes of spoils dumps at Anaconda mine. The material comprising the dumps consists of stripped overburden (Mancos Shale and Dakota Sandstone), barren Jackpile Sandstone, and ore associated with the Jackpile Sandstone. These piles are constructed by dumping overburden to produce mesalike landforms with outslopes usually approximating the angle of repose for the materials. Outslope heights range from 7.6 m (25 ft) to 70 m (230 ft). Hillslope gradients range from 25 to 100%. Hillslope lengths range from 10.7 m (35 ft) to 129 m (423 ft).

Visual inspection by Smith (undated) revealed that these hillslopes are steep, heights are large, and erosion is occurring at a rapid rate. All waste dump hillslopes are traversed by numerous rills (less than 15 cm in depth) and many have been cut by gullies (greater than 15 cm in depth). Deep gullies have developed on the outslopes of the largest dumps, some up to 2.4 m (8 ft) wide and 1.8 m (6 ft) deep.

In an attempt to quantify erosion rates using the best procedures and data available, Smith (undated) employed the Universal Soil Loss Equation (USLE) to estimate soil loss by sheet and small rill erosion, together with field measurements of the frequency, width, and depth of larger rills and gullies to approximate soil loss through those channel processes. The USLE computations for 19 locations on 15 waste dumps are shown in Table 4. Values range from a low of 50.45 metric tons/ha/yr (22.50 tons/acre/yr) to a high of 441.45 metric tons/ha/yr (196.89 tons/acre/yr).

Soil loss through rilling and gullying was determined for 13 transects of 61 m (200 ft) length on seven waste dump outslopes. The morphometric dimensions and the U.S. Soil Conservation equation (1981) were used to calculate the erosion estimates contained in Table 5. Here, values range from a low of 8.97 metric tons/ha/yr (4.00 tons/acre/yr) to a high of 1257.76 metric tons/ha/yr (560.96 tons/acre/yr). Smith also notes that one site appears to experience somewhat less erosion than might be expected on the basis of physical characteristics, and he attributes this to the possible effects of surface roughness caused by boulder-sized material.

Although the above analyses cannot be considered an ideal basis for an assessment of the intensity of disturbance caused by uranium mining, it should be sufficient, together with our intuition, to allow the conclusion that the change in the rate of geomorphic processes is probably quite similar to those measured at areas disturbed by the surface mining of coal. Assuming the absence of reclamation in both cases and some comparability in the mining techniques, there do not seem to be compelling reasons why it should be otherwise. Hence, the data contained in Table 6 of Chapter 8 (on coal) may well be transferable.

Tailings disposal sites are composed of materials and landforms that differ substantially from those of spoils dumps. Tailings are the products of extensive physical and chemical processing of uranium ore. These sands and slimes are slurried to an impoundment where they settle from suspension.

Geomorphic processes of concern on the tailings impoundment surface itself include water and wind erosion, and the possibility of subsidence due to differential bearing strengths of tailings materials within the impoundment. The downstream dam and peripheral berms, where constructed, are also subject to wind and water erosion as well as mass movements, such as sliding, slumping, and creep.

Apparently, there has been no field measurement of geomorphic processes at uranium tailings disposal sites, which is rather surprising in light of their potential environmental impact. It becomes possible only to cite expert observation in this regard. Nelson *et al.* (1983) remarked that because of the fine-grained,

Table 4 Estimated Soil Loss (USLE) from Waste Dump Outslopes[a]

Waste dump	% Slope	Slope length (m)	Soil loss (metric tons/ha/yr)
South	100	38.71	308.95
South	100	60.35	409.84
SP-1	78	15.54	138.78
R	100	10.67	138.78
T	77	49.99	283.84
N	89	36.58	271.06
N	77	23.16	176.67
N	50	27.13	107.17
N2	60	17.68	107.17
FD-3	90	59.44	371.95
U	75	30.48	201.78
V	80	105.16	441.45
V	71	78.64	346.84
SP-2	89	18.29	176.67
Y	72	59.74	277.56
Y2	75	75.90	334.28
I[b]	75	60.96	302.67
(Three segments)	80	19.51	163.89
	67	10.97	94.61
FD-2	65	128.93	378.45
S	25	49.99	50.45
$\bar{X}$	76.16	48.26	247.83

[a]Adapted from Smith (undated) for Anaconda Mine.
[b]Treated as one slope in computation of means.

Table 5 Mean Rill and Gully Erosion Sampled Waste Dumps[a]

Waste dump	Slope age	% Slope	Slope length (m)	Erosion (metric tons/ha/yr)
Y	Mid-1950s	72	59.74	1257.76
Y2	Early 1960s	75	75.90	385.62
V	~1977	71	105.16	363.20
FD-3	~1979	90	59.44	35.87
J	~1977	73	39.62	60.53
T	~1978	77	49.99	53.81
S	~1979	25	49.99	8.97
$\bar{X}$		69	62.83	309.39

[a]Adapted from Smith (undated) for Anaconda Mine.

noncohesive nature of tailings material in general—and uranium tailings in particular—these materials have a high potential for sheet erosion by water. Figure 4 shows the deep dissection of gold mill tailings in Colorado. Later, these authors submitted that uranium tailings are also highly susceptible to wind erosion. Likewise, Nielson and Peterson (1978) contended that the major environmental problem that develops as a result of tailings ponds is dust.

Figure 4. Erosion of gold mill tailings with small channel to confine sediment in Colorado.

Nelson and Shepherd (1978) provide a comprehensive, albeit semiquantitative, evaluation of possible modes of tailings disposal site failure and the possible consequences. Concerning the cap or upper surface of the impoundment, they concluded that the likelihood of differential settlement is great, of wind erosion is high, of sheet erosion is almost a certainty, and of gully formation within the cap itself is low due to slight gradients. Concerning the berms or embankments, they concluded that the likelihood of wind erosion is high and increases with time, of sheet erosion is again almost a certainty, of gully formation is higher than for the surface or cap due to steeper slopes and over long periods of time will be relatively high, of slope failure is low to moderate under conditions of minimal maintenance, and of soil creep is minimal if residual strength of the embankment soil is used in the design.

In 1974, the AEC, EPA, and state representatives visited 22 inactive uranium mill tailings sites to determine the condition of each, any need for corrective action, ownership, proximity to populated areas, and prospects for future population increases near the site. All of the sites at the time of the study showed signs of deterioration, such as erosion due to wind or water and loss of cover.

Taken together, there is ample evidence that both mining and milling of uranium ore cause significant increases in the amount of work performed by geomorphic processes. Geomorphic principles suggest that off-site areas will also be impacted in the absence of affirmative remedial action.

Duration of Disturbance

The third component of this assessment of disturbance, duration, refers to the length of time that process rates remain accelerated as a consequence of uranium mining and milling operations. Because uranium production is a relatively new industry, there are few old mines at which to observe the effects of disturbance through time. The Canonsburg, Pennsylvania, site was started in 1911 as a custom mill to extract radium and uranium from ores, and operations continued until about 1966. This facility is included in the list of inactive sites with a high priority for timely remedial action.

If abandoned, the open pit will probably remain a topographic depression from decades to millennia depending upon the rates of erosion and deposition in a particular locale and the volume of the excavation. In geomorphic terms, it becomes a localized baselevel for lower order streams and, perhaps infrequently, for higher order intermittent channels. There are no data concerning the natural rates of filling for these abandoned open pit mines, although mass budget estimates could be computed.

Information concerning the duration of uranium spoils instability under natural conditions is likewise sparse. At the Anaconda mine in New Mexico, Kelley

(1978) and Reynolds *et al.* (1978) found that some spoils surfaces about 25 yr old had been revegetated with several volunteer species of grasses and forbs, while other sites had remained virtually barren, depending largely on the characteristics of the geologic material dumped at the surface. Higher erosion rates would be predicted for these areas lacking any protective vegetation cover.

Smith (undated) found that rill and gully development on spoils outslopes increases, without exception, as hillslope age increases at the Anaconda mine. In this arid environment with infrequent rainfall and runoff, it could well be hundreds or thousands of years before the large spoils dumps are fully dissected and equilibrium hillslopes and channels are re-created by natural processes, even under prevailing conditions of accelerated erosion rates.

Generally, it is reasonable to expect that the duration of environmental disturbance caused by the surface extraction of uranium ore would be about the same as for other surface mining industries. Of course, any attempted reclamation is likely to reduce the duration of disturbance, even if not entirely successful.

In a subsequent section we shall describe various options for uranium tailings disposal. From the geomorphic perspective, the above-grade alternative is usually the most problematic and so this mode will provide the focus for a brief consideration of duration of disturbance. As noted by Toy (1984), the position of the disposal facility above-grade imparts additional potential energy to the materials contained therein. Hence, their vulnerability to geomorphic processes can exceed that of the adjacent landscape on which they are situated.

The question of duration of disturbance for above-grade tailings disposal centers on the rate that the deposit is degraded by geomorphic processes. If abandoned in a raw condition, the material will revegetate very slowly, if at all, due to the physical and chemical properties of the tailings, and erosion rates will be very high in comparison to nearby undisturbed lands. The contents of the tailings impoundment are likely to be dissected and disseminated faster than spoils material due to this absence of protective vegetation cover and relatively high erodibility of the fine-grained tailings.

In conclusion, this assessment of land disturbance resulting from the uranium industry has revealed that (1) areal extent is a relatively minor part of the problems produced; (2) intensity of disturbance is very important because geomorphic processes commonly operate at rates many times those of nearby, undisturbed lands; and (3) duration is a very significant component of the environmental impact because the time required for natural geomorphic processes to reestablish a semblance of equilibrium within these disturbed landscapes may extend to tens, hundreds, or thousands of years, much like that of unreclaimed areas surface mined for coal. Further, the sediment produced and dispersed by accelerated geomorphic processes may contain toxic elements that remain ecologically dangerous for indefinite periods and/or radioactive substances with half-lives up to several thousands of years.

The Nature of Disturbance

The total environmental impact associated with the uranium industry may be divided into two basic parts, that attributable to the mining of ore and that caused by the subsequent milling of this resource. An understanding of these impacts and the mechanisms that created them provides the foundation for an appreciation of the current legal framework within which the industry must function and the concerns and characteristics of reclamation programs.

Open Pit Mining

As noted previously, the open pit method is usually preferred by those engaged in the surface mining of uranium ore. In most ways, this technique is quite similar to areal stripping in the sequence of steps employed at a specific site. The surface is cleared of man made or natural obstructions. Next, vegetation cover is removed. Topsoil is then stripped and stockpiled. The overburden above the uranium ore is blasted and piled to one side. Finally, thc rcsourcc is cxtracted and transported to the mill.

As a consequence of the similarity in method between the areal stripping of coal and open pit mining of uranium, the nature of the disturbance is also comparable. Thus, the discussion included in Chapter 8 concerning the surface mining of coal is generally applicable.

However, there are a few differences between areal stripping and open pit mining that are noteworthy. First, an areal stripping operation generally moves at a fairly rapid pace along the direction of the subsurface resource. This allows concurrent reclamation as the pit can be filled behind the extraction activities. Spoils can be graded and revegetated in a timely fashion. Under ideal conditions, the number of hectares disturbed annually should equal the number reclaimed.

Open pit mines are largely stationary, although they usually expand over the years. Large volumes of ore are obtained while disturbing a relatively small surface area. Depending upon changes in production costs and market conditions, mines may occasionally shut down and eventually reopen.

At a particular location the open pit mine may constitute an environmental disturbance for several years, even decades. Although highly site specific, many mine operators suggest a mine life of about 30 yr and this figure seems to renew itself annually; 30 yr ago the mine life was about the same. This may suggest that overall production costs and market conditions have retained approximately the same relationship.

As a result of extended mine life at one location, there is substantially less opportunity for concurrent reclamation. Spoils piles, although perhaps graded somewhat, remain positioned near the pit. Topsoil is stockpiled for several years, and the deleterious effects of this on quality of the growth medium has been described (Singleton and Williams, 1979). Water diversions around the site are maintained for long periods. The hydrologic properties of the site remain altered for many years and may cause consequent adjustment of drainage patterns and channel properties upstream and downstream unless carefully controlled on-site.

Historically, about one-half of the uranium production has resulted from underground mining methods. Although these also cause disturbance to the environmental system at the surface, it may be permissible to treat these as a subset of the entire suite of impacts produced by surface mines except where shallow underground mining results in collapse of the rooms with concomitant surface subsidence. Space does not permit discussion of this mining method and its environmental consequences further herein. Those interested are directed to the ONMSS (1980) for additional information.

Tailings Disposal

Tailings are usually transported by water in a pipeline from the mill to a retention pond. A peripheral discharge results in the deposition of tailings along the sides of the ponds with the water often recycled. On very large ponds the tailings may be discharged from a single point and allowed to run as a meandering stream across the pond surface. The solid components of the slurry are stratified by deposition with the coarse sandy material settling out near the point of discharge and the finer slime material moving to lower regions of the pond and settling out in deeper, quiescent water. Because of this stratification of materials with differing physical properties, instability can be a problem in the areas where the slimes have accumulated. These areas have high water retention capability and may remain unstable for years after a pond ceases to be active, according to Nielson and Peterson (1978). Many cases of equipment loss into these unstable areas during reclamation have been reported.

Uranium mill tailings constitute an industrial waste that must be removed from the mill site in order to allow continuous ore procesing and resource recovery. However, the method of disposal is a critical concern. The Congress of the United States, in Public Law 95–604 [Section 2(a)], The Uranium Mill Tailings Radiation Control Act of 1978 (UMTRCA), has found that:

> Uranium mill tailings located at active or inactive mill operations may pose a potential and significant radiation health hazard to the public, and that the protection of the public health, safety, and welfare and the regulation of interstate commerce requires that every reasonable effort be made to provide for the stabilization, disposal, and control in a safe and environmentally sound manner of such tailings in order to prevent or minimize radon diffusion into the environment and to prevent or minimize other environmental hazards from such tailings.

Herein, then, lie the goals and rationale for any tailings disposal program.

The nature of the environmental disturbance caused by uranium mill tailings disposal is largely a function of the disposal method and site characteristics. Tailings impoundments can be either above or below grade. There are several possible designs within each category. ONMSS (1980) remarked that surface emplacement of tailings is a convenient mode of disposal and has been the conventional practice to date. Such impoundments can be constructed as four-sided structures in relatively flat terrain, or they can be formed by constructing a dam or embankment in an existing natural drainage area. In the latter case, diversion ditches are constructed to divert runoff around the impoundment. Heights of tailings embankments vary from 10 to 30 m (30 to 100 ft) above the surrounding surface. Figure 5 shows a rectangular ring dike impoundment. Walters (1983) noted that ring dike designs allow for any shape of impoundment and can be adapted to blend with the surrounding topographic features. They also allow flexibility in site selection and can generally be located close to the mining operation. Figure 6 shows a large above-grade tailings impoundment in Wyoming. Ring dikes can be located away from floodplains to reduce their susceptibility to erosion by major floods, although Walters (1982) remarked that the desire to position these impoundments on nearly flat surfaces to permit a greater average depth for the same maximum dike height has led to frequent usage of floodplains for this purpose. Ecker (1984) provided detailed documentation for a few such inactive sites.

Figure 7 shows a cross-section through a typical valley dam impoundment. Here, a dam is placed across a valley so that the natural topography provides the remaining sides of the impoundment. The shape and location of impoundment depends upon suitable sites on the mine property. This option is economical and presents less risk of failure than do multiple dike designs, according to Walters (1983). Another advantage is that the shape of the valley restricts seepage and surface flow to a known path, greatly simplifying monitoring and control measures. There are also numerous disadvantages, largely resulting from the block-

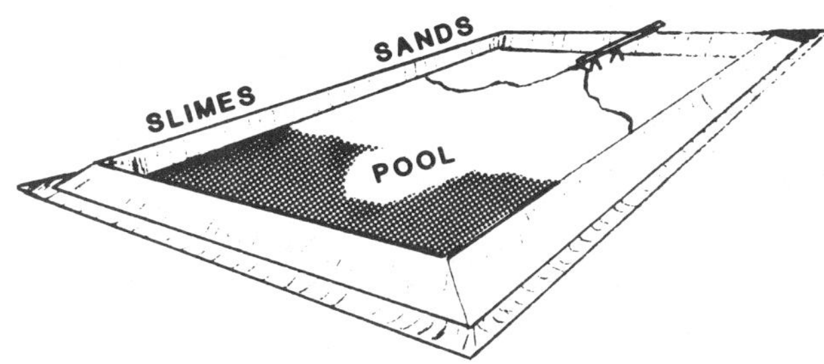

Figure 5. Diagrammatic representation of a ring dike impoundment.

Figure 6. Above-grade tailings pond in the western United States.

age of natural drainage systems. These will be discussed in a subsequent section concerning geomorphic implications of tailings disposal.

The below-grade disposal alternatives include shallow and deep burial of tailings. Shallow or in-pit disposal utilizes natural depressions, abandoned surface mines, or specially excavated pits as impoundments. The ONMSS (1980) presented descriptions of numerous possibilities. Figure 8 shows two examples of shallow disposal including an available open pit mine and a specially excavated disposal pit.

These designs eliminate the need for man made retention structures and do not produce a prominent topographic high. Walters (1983) identified the most critical disadvantage as possible proximity to the water table that may fluctuate in response to seasonal climatic changes.

Deep burial of tailings (deeper than 100 m, or 330 ft) may be an option where

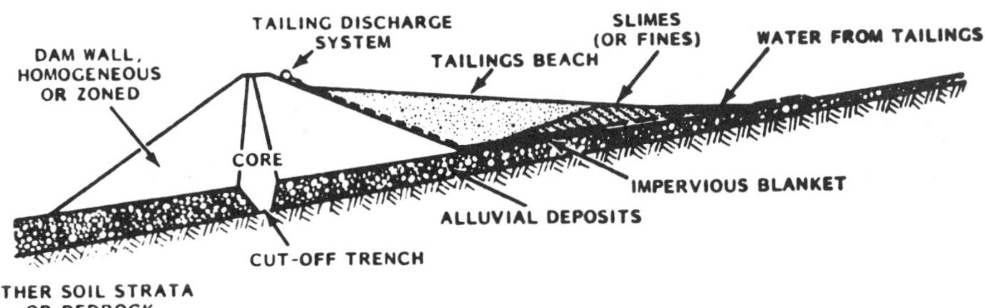

Figure 7. Typical cross section of a valley dam impoundment (Walters, 1982).

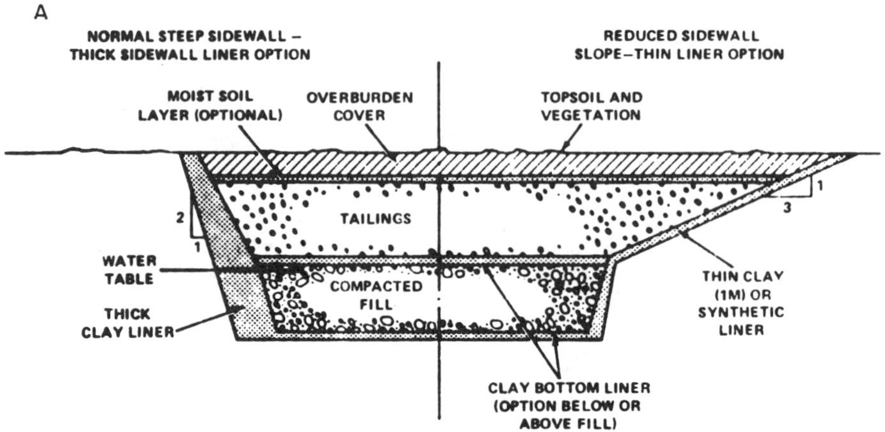

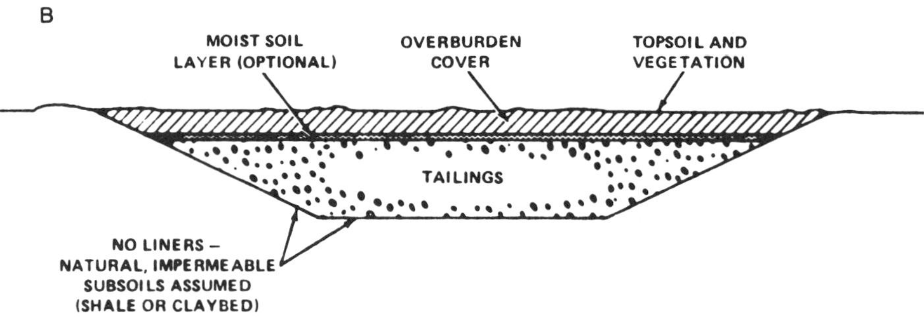

Figure 8. A, Disposal of tailings slurry in available open-pit (ONMSS, 1980). B, Disposal of tailings slurry in specially excavated below-grade pit (ONMSS, 1980).

abandoned deep mines are present near the mill site. The potential advantage of this method is that the depth of cover would eliminate all radon emissions and provide an enormous physical barrier to subaerial geomorphic processes so that the need for long-term institutional controls or monitoring could be greatly reduced or eliminated (ONMSS, 1980). In his analysis, Walters (1983) stated that the deep mine tailings disposal option was considered but rejected because of insurmountable problems caused by long transportation distances from existing mills, groundwater contamination potential, and institutional barriers. Only above-grade and shallow below-grade options have been used in the United States.

All tailings disposal methods are subject to various geomorphic processes, as we shall see. Most agree that the shallow below-grade alternatives constitute the "prime options" (Nelson and Shepherd, 1978; ONMSS, 1980; NRC, 1980; Rafferty, 1982; Toy, 1984).

In addition to the method of disposal, the selection of disposal site also influences the nature of the environmental disturbance. For example, a ring dike impoundment located atop a hillslope may add sufficiently to the total weight load on the underlying surface and adjacent hillslopes to trigger some type of mass movement. The same ring dike impoundment placed in a floodplain at the base of a hill can constrict floodwater flow, thereby increasing floodstage and scouring nearby.

The positioning of tailings deposits in upland areas is only likely to bury lower order channels that transport relatively small discharges through ephemeral or intermittent flows. Valley dam impoundments are likely to obstruct higher order streams, transporting greater discharges through intermittent or perennial flows. Further, as illustrated by Horton (1945) and discussed in a previous chapter, a large number of lower order streams will converge toward the valley dam impoundment.

From the foregoing, it is evident that the exact character of environmental impact associated with uranium tailings disposal is highly site specific, depending in large part on the method of disposal and the location within a particular landscape. Fortunately, it is often possible to predict the consequences of the disturbance once these characteristics are known.

In the absence of efforts to stabilize and reclaim inactive impoundments, contamination of surrounding areas can be expected due to wind and water transport of tailings material. Once entrained, the radioactive and toxic constituents of the impoundment can be scattered far and wide. Some would argue that this diffusion and dispersion renders these materials harmless; others contend that significant hazards are created as a result of accumulations through time and the minute amounts of some toxic elements necessary to have deleterious effects on flora or fauna (Horak and Olson, 1980; Nagy *et al.*, 1978; Rapaport, 1963).

Land Disturbance by Uranium Industry and Geomorphic Principles

While other disciplines, such as civil or agricultural engineering, can provide guidance for the control of geomorphic processes over relatively short periods, it was the geomorphologist who possessed an understanding of processes over both the short and long term. Because of the long-term hazards associated with the mill tailings, it is essential to consider the possible frequency and magnitude of all geomorphic processes that may operate through an extended time frame. Hence, the pertinent literature and legislation are replete with references to geomorphic processes, geomorphic factors, and geomorphic assessments of sites.

As described above, there are some differences in the types of disruptions to the natural systems resulting from the mining and milling operations. However, there are also several common effects, and we shall begin with a discussion of these. Both the mining and milling operations remove or bury existing surface covers and soil profiles at a site. New topographies and drainage systems are generated by excavation or deposition. As a consequence, the balance between force and resistance is altered. The equilibrium or steady state is destroyed locally and this may, in turn, change the character of the geomorphic system off-site as well, due to the transport of water and sediment. Both mining and milling will leave exposed, unconsolidated materials on sloping surfaces. The forces, driving geomorphic processes, that impinge on this surface, will at least remain the same and more often increase because of steeper hillslope gradients. Concomitantly, resistance is inevitably reduced and so geomorphic work is performed at accelerated rates. Through time, process rates should gradually return to those typical or "normal" for a given environmental setting as a semblance of a quasi-equilibrium is reestablished.

The mining process itself causes environmental disturbance through excavation and deposition. Except for the extended length of time that open pit uranium mines remain at a particular location, their impact on environmental systems and the geomorphic consequences can be expected to be similar to those generated by the areal stripping of coal. Thus, we can again turn our attention to uranium mill tailings.

With below-grade disposal, the deposited tailings possess the same potential energy as the earth material forming its confines. Because relatively flat terrain is usually favored in locating these disposal sites, it seems permissible to conclude that the forces impinging on the tailings surface will be about the same as those affecting the surrounding natural landscape. Barren surfaces of unconsolidated tailings will offer considerably less resistance to these forces than the vegetated soils of nearby areas. During this period of exposure, geomorphic work should progress at accelerated rates on these disposal sites unless action is taken to control them. Erosion by water and wind are likely to be the dominant geomorphic processes, assuming that a site has been selected that is physically stable, bearing no evidence of mass movement. Additionally, the high pH of tailings produced by alkaline leaching of ore and the low pH of tailings produced by acid leaching will accelerate the weathering of earthen material with which any seepage comes into contact.

Below-grade disposal sites may act as temporary baselevels for a few upstream channels. However, deposition of sediment herein is likely to be of minor concern and perhaps even as advantageous in contributing to the cover above the tailings. Generally, it would be rather easy to divert the flows away from these basins. The possibility of intersecting the groundwater table or an aquifer is usually the principal environmental concern with below-grade tailings disposal. Within a given landscape, one might suspect that this prospect would be somewhat less likely at upland locations and somewhat more likely in lowlands or floodplains, other things being equal. In either case, placing the impoundment over a zone of recharge could seriously damage a groundwater resource.

With above-grade disposal, the tailings possess greater potential energy than adjacent surfaces. The gradients of embankments (berms) may also be steeper than those of nearby hillslopes, although they could be constructed at lesser angles. Reducing these gradients will increase the area of disturbance somewhat. Additionally, the embankments are constructed of unconsolidated material and will remain virtually devoid of vegetation, at least for a time, despite efforts to prevent lateral diffusion of toxic seepage through this porous medium with linings. Thus, one may conclude that the forces impinging on these embankments are likely to increase while the resistance is decreased, in comparison to surrounding natural surfaces. During the prereclamation period, geomorphic work will progress at accelerated rates. With proper engineering design, the probability of mass movement along the embankments should be minimized; however, it will remain greater than for a properly selected below-grade disposal site. This is especially true until the impoundment has been dewatered. As with below-grade disposal, erosion by water and erosion by wind are likely to be the dominant geomorphic processes affecting the site. And again, weathering of embankment materials will be accelerated by alkaline or acid seepage.

Both ring dike above-grade and shallow below-grade disposal sites tend to be situated on relatively flat surfaces. When these occur in proximity to a stream channel, the site is subject to disturbance by flooding or the lateral migration of the channel. To the extent that the disposal site modifies the hydrologic properties of the surface, reducing infiltration and increasing runoff, or alters the gradient and geometry of tributary channels, it may also increase the probabilities of disruption by these fluvial processes.

As suggested earlier, tailings disposal behind valley dams creates additional geomorphic concerns. Not only do the tailings deposited possess greater potential energy than downstream surfaces, they are also located in a relatively high-energy geomorphic setting. Both channel flow and overland flow from the valley sides are directed toward the impoundment. Diversion ditches may be used to convey upstream discharge past the impoundment, but these are certainly not maintenance free in either the short or long term. Failure of the diversion, perhaps during flood events that exceeding design discharge, endangers the adjacent tailings deposit.

The deposition of tailings in valley dam impoundments also produces a perturbation in the longitudinal profile of the channel. As a result, the geomorphologist would expect initial deposition in the channels upstream, if the baselevel has been elevated, and erosion or scouring of the channel downstream, if the channel gradient has been increased by the impoundment. With time, the knickpoint so created will probably migrate upstream, rejuvenating the drainage system and removing the materials deposited earlier.

Other potential problems also exist. While the valley's topographic configuration may protect the tailings somewhat from winds blowing orthogonal to the valley orientation, those blowing up or down valley will continue to be erosive. Further, it is possible that a "venture effect" within the valley may increase wind velocity and erosion.

Lastly, the valley dam impoundment may be constructed upon valley alluvium that can be wholly or partially saturated during a humid season. This wetting and drying can affect the bearing strength of the impoundment foundation. Engineering design may accommodate this potential cause of embankment failure as long as the predisposal baseline data reveal the existence and extent of the fluctuating near-surface watertable beneath the site.

Thus, geomorphic principles support the proposition that, other things being equal, shallow below-grade tailings disposal systems are likely to maintain their integrity longer because they will be subjected to fewer and lower energy (force) geomorphic processes. Conversely, the valley dam disposal systems seem likely to be inherently less stable because they should be subjected to additional, higher energy (force) geomorphic processes. It is the prospect of diversion ditch failure, periodic flooding, or scouring along embankment or tailings perimeters over extended time periods in this relatively high-energy fluvial setting that is cause for concern. On this basis, then, one must conclude that the shallow below-grade burial options are generally preferable as long as groundwater resources are not threatened. Of course, this is theoretical, and site-specific conditions must be fully considered in the selection of disposal system alternatives.

Reclamation and the Geomorphic Process–Response System

As viewed from an earth science perspective, general goals for reclamation have been nicely articulated by Harwood (1979, p.10):

> . . .the (reclamation) project should be aimed at producing environmentally sound and stable conditions that will ultimately reintegrate the disturbed area into the general ecosystem, while alleviating as many of the initially perceived problems as possible. Minimally, the project objectives need to bring the site into compliance with applicable Federal, State, and local environmental quality regulations and existing legal and safety standards.

The National Research Council (1979, pp. 129–130) provided more specific direction:

> For most sites, reclamation is understood as rehabilitating the disturbed area by grading in a manner compatible with the surrounding landscape, or with the postmining use, and establish-

ing a self-sustaining vegetation cover. Reclamation is intended to control erosion, to restore the hydrologic balance, and to limit the off-site impacts, during and after mining. The ultimate purpose is to restore the mining site to productive use, for example, for agriculture, forestry, industry, housing, recreation, or fish and wildlife habitat. Desirably, this use should be at least as beneficial as the capability of the site before mining although not necessarily for the same purpose.

As a result of the physical and, especially, chemical differences between mine spoils and uranium mills tailings, there are different primary objectives for the reclamation of lands disturbed by these two materials. The reclamation of the open pit can proceed in a fashion similar to that for any type of surface mining. Accordingly, the two citations above are directly applicable. As observed by Toy (1984), uranium surface mining necessitates the handling of larger volumes of material than coal surface mining because of low grade ores and deposit depth; while this adds to the cost of reclamation, it should not exacerbate the technical difficulties of reclamation. Hence, it is tailings disposal site reclamation that requires special attention.

The ONMSS (1980, p. 1) commented that:

> Requirements regarding decommissioning (of mill sites) and mill tailings disposal are stated primarily as performance criteria. These criteria are intended to assure that mill and mill tailings disposal sites are returned to conditions which are reasonably near those of surrounding environs and to assure that on-going active care and maintenance programs, to redress degradation of the tailings isolation by natural weathering and erosion forces, will not be necessary. While continued surveillance of mill tailings disposal sites is prudent in order to confirm that sites are not disrupted by unexpected natural erosion or human activity, decommissioning of remaining sections of the mill site should assure their unrestricted use.

Nelson *et al.* (1983, p. xv) remarked:

> Long-term reclamation plans for uranium mill tailings impoundments must include engineering designs to protect against disruption of the tailings and the potential release of radioactive materials. The goals for such engineering designs should be to provide overall site stability for long-term periods with no need for planned on-going maintenance and to provide a repository for the tailings that will not place a burden on future generations.

Similarly, Beedlow (1984a) submitted that the goal of tailings containment is to minimize the escape of hazardous materials through leaching, surface runoff, gaseous diffusion, or biotic transport for the design life of the system. And lastly, Rafferty (1982) stated that the persistence of radioactivity within the tailings requires that the period of time for which tailings should be isolated from the environment should effectively be indefinite and that engineering solutions for tailings disposal should ideally be permanent and should also ideally provide for a very high degree of containment.

From the foregoing, it is clear that there are some fundamental differences between the goals of surface mine reclamation and uranium mill tailings disposal site reclamation. In the former instance postreclamation land productivity is a concern and the decision herein controls various other aspects of reclamation. In the latter case, the long-term isolation of hazardous radioactive and toxic materials is the principal concern. As we shall see reclamation practices aimed at ensuring long-term isolation may actually preclude most opportunity for subsequent productive landuses.

The re-creation of stable landscapes remains an objective common to both surface mine and tailings impoundment reclamation. However, these differences in goals produce somewhat different geomorphic implications for reclamation. For surface mine sites, it appears that the establishment of an equilibrium condition would be satisfactory, which allows some erosion at a rate "normal" for the prevailing environment. For tailings impoundments, it seems that there should be virtually no erosion in order to maintain the thickness of the radon-absorbing surface cover for very long time periods. The ability to absorb is proportional to cover thickness.

Nelson and Shepherd (1978) concluded on the basis of an exhaustive, albeit semiquantitative, study that for short long-term periods (several hundred years) the engineering design of the site will govern its performance. For long long-term periods (about 100,000 yr), natural geomorphic processes would predominate. The medium long-term periods (several thousand years) represent a transition from engineering dominance to the geomorphic dominance.

Although this conclusion may be justified on the basis of probabilities associated with the magnitude and frequency of geomorphic processes, it must be understood that the concept of recurrence interval and resulting probability offers no indication as to when an extreme event may occur. For example, the "100-yr" flood does not recur every 100 yr but has a probability of 0.01 for occurrence in each year. Hence, one cannot exclude the possibility of geomorphic dominance in any year.

Rafferty (1982, pp. 14–15) also acknowledged the uncertainties resulting from the extended time framework:

> . . .the difficulty of being unable to predict either quantitatively or qualitatively the nature and extent of all manner of geomorphological and climatological processes in the long-term, i.e., beyond the first few thousand years, makes it difficult to be confident that the release of contaminants will not exceed design performance standards or authorised limites, or that the containment system will retain its integrity. Over these time scales extreme events, such as extremely large earthquakes and floods, whose magnitude could exceed the design bases for such events, could also be expected to occur. These would more probably result in partial or substantial breeching of the containment system and consequent dispersal of tailings material. The engineering implications for the integrity of the containment system and the radiological consequences of such failures cannot be satisfactorily assessed within current perspectives. However, because of the low specific activity of the tailings, the consequences of partial or substantial breeching of the containment system and consequent dispersal of tailings would not lead to catastrophic or necessarily even significant radiological impact, since long and sustained exposure to radioactivity from tailings would be required to produce detectable adverse results.

From these comments by various writers, it seems reasonable to conclude that reclamation is the actual goal for the surface mine site, while geomorphic isolation is the primary goal for the tailings disposal site.

The Legal Requirements of Reclamation

The Surface Mining Control and Reclamation Act of 1977 (SMCRA) (PL 95-87) (U.S. Congress, 1977) estabished minimum national standards for the surface mining of coal and reclamation of lands disturbed in the process. Unfortunately, there is no single, unified legal structure that controls the surface mining of uranium, disposal of mill tailings, and reclamation of lands so disturbed. However, this is not to say that the industry is underregulated in any way. Indeed, a number of statutes and government agencies have a hand in the process. Perusal of the proposed, adopted, suspended, reproposed, and readopted standards during the past few years reveals a complex and somewhat confusing array of regulations. One must monitor the legal framework on a nearly daily basis in order to remain up to date. In recent months, considerable effort has been expended by government agencies in an effort to clarify and simplify this situation.

Nevertheless, it is possible to examine the legal foundations and the intent of enacted legislation and subsequent regulations. Generally, these may be categorized as federal, state, and local governance taking the form of laws, regulations, and guidelines, the latter of which are often treated as law or regulation by the responsible controlling agency.

The federal legislation listed in Table 11 of Chapter 8 (on the surface mining of coal) can also be applied to the uranium industry. In addition, a few laws are especially important and largely constitute the legal foundation for government regulation of the industry. First, the Atomic Energy Act of 1946 (U.S. Congress, 1946) emphasized the discovery and development of uranium resources worldwide for the purpose of satisfying military demands. Herein, the AEC was created to facilitate attainment of this objective. At this time, however, reclamation of lands disturbed by the industry was not a real concern.

The Atomic Energy Act of 1954 (U.S. Congress, 1954) established a civilian uranium industry by allowing private citizens to produce, possess, and use uranium for nonmilitary purposes. This legislation also authorized the development of generally applicable standards for the protection of the population from radiation. The AEC remained as the regulatory authority but now became responsible for the private sector of the industry as well. Thus, the AEC had become charged to both promote and regulate the uranium industry. Some viewed these two missions as fundamentally in conflict.

The Energy Reorganization Act of 1974 (U.S. Congress, 1974) abolished the

AEC and created two new agencies, the Nuclear Regulatory Commission (NRC) and the Energy Research and Development Administration (ERDA). The NRC carries out the licensing and regulatory functions formerly vested in the AEC.

Throughout the late 1960s and 1970s, Congress passed a number of laws designed to protect the natural environment. These, too, eventually became part of the governing structure of the uranium industry. The National Environmental Policy Act of 1969 (NEPA) (U.S. Congress, 1969) was among the most important laws. Rouse (1978) commented that this enactment is one of the most significant pieces of federal legislation influencing the development of the uranium industry. In broad terms, this statute requires careful examination of possible environmental impacts caused by proposed activities on federal or Indian lands or activities on privately owned land by industries subject to federal regulation. This law also created EPA to implement the objectives of this legislation. Summaries of other laws contributing to the legal structure for the uranium industry are provided by Wagner (1980), the National Research Council (1979), and ONMSS (1980).

Despite the above legislation, there remained some ambiguity concerning the authority of the NRC to effectively control the disposal of mill tailings through the licensing mechanism, although the NEPA (1969) had substantially strengthened their position. Therefore, Congress passed the Uranium Mill Tailings Radiation Control Act (UMTRCA, 1978) (U.S. Congress, 1978). This created a new category of licensable material. The definition of the term *by-product material* was expanded to include tailings. As a result of this act, the NRC's authority clearly continues upon cessation of milling operations and removal of licensable source material (ore), extending at least until tailings are finally disposed of according to Commission regulation.

As a consequence of these and other laws, three agencies play a role in the regulation of the uranium industry at the federal level. The NRC is responsible for the licensing and regulation of the uranium industry and, in addition, assuring compliance with the NEPA (1969) through environmental review of each proposed major uranium mill licensing action, which usually culminates in the preparation of an environmental statement that is circulated for public review and comment.

The EPA has been granted authority to establish environmental standards of "general applicability" covering radiological and nonradiological hazards associated with the uranium industry. The DOE and the NRC are usually designated to enforce these standards.

The DOE has been assigned primary responsibility for remedial actions at abandoned inactive mill tailings sites and long-term control of decommissioned sites. The DOE and appropriate states are to purchase abandoned tailings sites and undertake specific treatments that will bring these sites into compliance with EPA standards. The cost sharing between the DOE and state assumes a 90% to 10% proportion. Once cleaned, the state may sell the land.

The DOE also assumes custody of inactive sites that are ultimately owned by the federal government. This office may be required to monitor, maintain, or provide emergency measures at these sites to protect public health and safety. Because of its duty to perform remedial actions for the stabilization of sites, the DOE assumes a leading role in the development of technology applicable to tailings disposal at all sites.

The body of regulations resulting from the aforementioned legislation pertains to the initial step of license application, through the production phase of mine and mill operation, through clean-up and reclamation of disturbed lands, to provisions for long-term surveillance of the site. It is not necessary to list these regulations in their entirety; those interested may refer to the Code of Federal Regulations (1984a, Title 10, Parts 40, 50, 51, 55, 61) and in the Federal Register (U.S. Nuclear Regulatory Commission, 1985; Vol. 50, No. 200, Oct. 16, 1985, pp. 41852–41865). It is useful to present those regulations which are of particular geomorphic significance in order to better understand the forthcoming discussion of reclamation practices. These are contained in Table 6 and show that geomorphic principles have been incorporated into their formulation.

In addition, the Code of Federal Regulations (1984a, Title 10, Part 51), specifies the requirements for environmental reports submitted to the NRC by applicants for permits and environmental impact statements prepared subsequently by the Commission. Both follow the structure prescribed by NEPA (1969). The data required and examined will be essentially the same as listed in Table 8 of Chapter 8 (on the surface mining of coal).

Individual state governments usually participate in the regulation of the uranium industry as well. They can become an "Agreement State" whereby the NRC may transfer to the state certain regulatory authority over specified nuclear materials, when (1) a state desires to assume this authority, (2) the Governor certifies that the state has an adequate regulatory program, and (3) the Commission finds that the state's program is compatible with that of the Commission's and is adequate to protect health and safety. The standards imposed by the state may be more stringent but not less so than those developed by the NRC and the EPA. As of 1980, nearly 60% of "probable" uranium resources were located in Agreement States.

Of course, states may also enact their own legislation applicable to this industry. The ONMSS (1980) has prepared a summary of reclamation regulations for ten states possessing uranium resources. Control is exercised through requirements for permits and performance bonds.

Finally, counties or municipalities may regulate mining and reclamation of mined lands through zoning or other land use laws, in accordance with powers granted them by state statutes. Several Supreme Court decisions have upheld the legal authority of local governments to regulate private mining near metropolitan areas. Local land use controls naturally tend to emphasize local priorities and may ignore national needs on occasion.

Table 6 Some Regulations Pertaining to the Uranium Industry of Geomorphic Significance[a]

1. Licenses or applicants may propose alternatives to specific requirements which take into account local or regional conditions including geology, topography, hydrology, and meteorology.
2. The general goal or broad objective in siting and design decisions is permanent isolation of tailings and associated contaminants by minimizing disturbance and dispersion by natural forces, and to do so without ongoing maintenance. The following site features which will contribute to such a goal or objective must be considered in selecting among alternative tailings disposal sites or judging the adequacy of existing tailings sites:
   a. Remoteness from populated areas;
   b. Hydrologic and other natural conditions as they contribute to continued immobilization and isolation of contaminants from groundwater sources; and
   c. Potential for minimizing erosion, disturbance, and dispersion by natural forces over the long term.
   d. While isolation of tailings will be a function of both site and engineering design, overriding consideration shall be given to siting features given the long-term nature of tailings hazard.
   e. Tailings shall be disposed of in a manner that no active maintenance is required to preserve conditions of the site.
3. The prime option for disposal of tailings is placement below grade either in mines or specially excavated pits.
   a. The evaluation of alternative disposal sites, provided in a license applicant's environmental reports, shall reflect serious consideration of this disposal mode.
   b. In some instances below-grade disposal may not be the most environmentally sound approach such as where:
      (1) Groundwater formations lie relatively close to the surface or are not very well isolated by overlying soils and rocks and
      (2) Geologic and topographic conditions might make below grade disposal impracticable (i.e., bedrock near the surface requiring blasting for excavation).
   c. Where the below grade option is not practicable, the size of retention structures and size and steepness of slopes of associated exposed embankments shall be minimized by excavation to the maximum extent reasonably achievable or appropriate given the geologic and hydrologic conditions at a site.
      In these cases, it must be demonstrated that an above grade disposal program will provide reasonably equivalent isolation of the tailings from natural erosional forces.
4. The following site and design criteria shall be adhered to whether tailings or wastes are disposed of above or below grade.
   a. Upstream rainfall catchment areas must be minimized to decrease erosion potential and the size of the floods that could erode or washout sections of the tailings disposal area.
   b. Topographic features should provide good wind protection.
   c. Embankment and cover slopes shall be relatively flat after final stabilization to minimize erosion potential and to provide conservative factors of safety assuring long-term stability. The broad objectives should be to contour final slopes to grades that are as close as possible to those which would be provided if tailings were disposed of below grade; this could, for example, lead to slopes about $10H{:}1V$ or less steep. In general, slopes should not be steeper than $5H{:}1V$. Where steeper slopes are proposed, reasons why a slope less steep than $5H{:}1V$ would be impacticable should be provided, and compensating factors and conditions that make such slopes acceptable should be identified.

Table 6 *(Continued)*

d. A full self-sustaining vegetative cover shall be established or rock cover employed to reduce wind annd water erosion to negligible levels.
   (1) Where a full vegetation cover is not likely to be self-sustaining due to climatic or other conditions, such as in semiarid and arid regions, rock cover must be employed on slopes of the impoundment system. The NRC will consider relaxing this requirement for extremely gentle slopes such as those which may exist on the top of the tailings pile.
   (2) The following factors shall be considered in establishing the final rock cover design to avoid displacement of rock particles by human and animal traffic or by natural processes, and to preclude undercutting and piping:
      (a) Shape, size, composition, and gradation of rock particles (excepting bedding material average particle size shall be at least cobble size or greater),
      (b) Rock cover thickness and zoning of particles by size, and
      (c) Steepness of underlying slopes.
   (3) Individual rock fragments shall be dense, sound, and resistant to abrasion and shall be free from cracks, seams, and other defects that would tend to unduly increase their destruction by water and frost action. Weak, friable, or laminated aggregate shall not be used.
   (4) Rock covering of slopes may not be required where top covers are very thick (on the order of 10 m or greater), impoundment slopes are very gentle (on the order of $10H{:}1V$ or less), bulk cover materials have inherently favorable erosion resistance characteristics, and there is negligible drainage catchment area upstream of the pile and good wind protection.

e. Furthermore, all impoundment surfaces shall be contoured to avoid areas of concentrated surface runoff or abrupt or sharp changes in slope gradient.
   (1) In addition to rock cover on slopes, areas toward which surface runoff might be directed shall be well protected with substantial rock cover (rip-rap).
   (2) In addition to providing for stability of the impoundment system itself, overall stability, erosion potential, and geomorphology of surrounding terrain shall be evaluated to ensure that there are not ongoing or potential processes, such as gully erosion, which would lead to impoundment instability.

f. The impoundment shall not be located near a capable fault that could cause a maximum credible earthquake larger than that which the impoundment could reasonably be expected to withstand.

g. The impoundment, where feasible, should be designed to incorporate features which will promote deposition; the object of such a design feature would be to enhance the thickness of cover over time.

5. In disposing of waste by-product material, the licensees shall place an earthen cover over tailings or wastes at the end of milling operations and shall close the waste disposal area in accordance with a design that provides reasonable assurance of control of radiological hazards to:
   a. Be effective for 1000 yr to the extent reasonably achievable and, in any case, for at least 200 yr, and
   b. Limit release of radon-222 from uranium by-product materials and radon-220 from thorium by-product material to the atmosphere so as to not exceed an average release rate of 20 $pCi/m^2/sec$, to the extent practicable throughout the effective design life.

(continued)

Table 6 *(Continued)*

c. In computing required tailings cover thicknesses, moisture in soils in excess of amounts found normally in similar soils in similar circumstances shall not be considered.
d. If nonsoil materials are proposed as cover materials, it must be demonstrated that such materials will not crack or degrade by differential settlement, weathering, or other mechanisms over long-term time intervals.

6. At least one full year prior to any major site construction, a preoperational monitoring program shall be conducted to provide complete baseline data on a milling site and its environs. Throughout the construction and operating phases of the mill, an operational monitoring program shall be conducted to measure or evaluate compliance with applicable standards and regulations, to evaluate performance of control systems and procedures, to evaluate environmental impacts of operation, and to detect potential long-term effects.
7. The final disposition of tailings or wastes at mill sites should be such that ongoing active maintenance is not necessary to preserve isolation.

[a]Compiled from Code of Federal Regulations (1984a, Title 10, Part 40); Federal Register, Vol. 49, No. 228, pp. 46418–46425; and Vol. 50, No. 200, pp. 41852–41865.

Reclamation Practices: Open Pit

In this section we shall examine the practices commonly utilized to reclaim lands disturbed by the surface mining and processing of uranium ore. As late as 1978, Kelley (1978) remarked that literature dealing specifically with open pit uranium mine reclamation was either not available or nonexistent. However, this does not constitute a truly serious problem. Yamamoto (1982, p. 7) stated:

> From a geographical, ecological, or geological standpoint, the revegetation of uncontaminated uranium mine spoils should not pose problems unusually different from coal mine spoils. . . . The same techniques for dealing with physical and chemical characteristics of coal spoils can be used for uranium spoils (USDA Forest Service 1979b, Schaller and Sutton 1978). However, if the spoils are contaminated with radionuclides and their progenies or with potentially toxic trace elements, in greater amounts than those found in adjoining lands, these materials must be handled and treated in the same manner as mill tailings and analyzed for these contaminants at appropriate laboratories. . . . However, unless the spoils are contaminated by radio-activity or chemically toxic elements, priority should be given to research on mill tailings waste.

Two studies concerning reclamation problems at Anaconda's open pit uranium mine in New Mexico are noteworthy, however. Reynolds *et al.* (1978) found that all major plant nutrients, except calcium and magnesium, were very low in the spoil material on all waste dumps. Spoils derived from different geologic strata differed considerably in their ability to support vegetal growth. Soil properties that correlated closely with different degrees of plant establishment on the various waste dumps were pH, electrical conductivity, and soil texture.

Smith (undated) examined the existing conditions of waste dumps at the same mine. He observed that reclamation success has been quite varied. Waste dump tops have been revegetated in many cases, presumably due to gentle hillslope gradient and good water retention; however, revegetation of waste dump side slopes has failed due to steep gradients and resultant vulnerability to erosional soil loss. According to Smith, the severity of erosion on the dump side slopes dictates that, unless significant hillslope modifications are made, revegetation–reclamation success is highly unlikely. He submits that the reduction of hillslope gradient is the modification most crucial to reclamation success. Although edaphic and geomorphic site characteristics clearly influence the facility of revegetation at this location, the aridity of climate is the most serious limitation.

Reclamation Practices: Tailings Disposal Site

The decommissioning of a mill site requires decontamination of salvageable equipment; burial of contaminated, nonsalvageable equipment; disposal of fuel and chemicals; dismantling, decontamination, and removal of buildings and structures; burial of foundations; excavation and burial of contaminated ore pads, sludge or collection ponds, and soil surfaces; and reclamation of all disturbed lands. Burial of contaminated materials in tailings disposal sites serves to localize the potential radiation hazard.

Reclamation of these tailings disposal sites has been the focus of extensive research in recent years. The primary goal of long-term isolation of radioactive and toxic materials was discussed earlier. The question could be asked, however: Is this a realistic possibility in light of the manifold uncertainties surrounding our ability to predict the future environmental conditions at a site and the response of the reclaimed disposal facility through extended periods of time? Schumm *et al.* (1982) contend that within a landscape there are locations that are favorable for long-term disposal of radioactive materials and others that are potentially very hazardous. Several writers (Nelson and Shepherd, 1978; ONMSS, 1980) have cited the work of Hunt and Mabey (1966) which suggests that alluvial surfaces in Death Valley may have been stable since late Pleistocene time or about 20,000 yr. Mabbutt (1977) has also identified desert landforms that are believed to have been stable for several thousand years. It should be noted, however, that Hunt and Mabey (1966) also identified areas where there had been discernible damage to roads, flood control ditches, trails, and archeological features during a 25- to 50-yr period. Nevertheless, the ONMSS (1980) concluded that disposal sites could be stabilized for long periods if the "armoring" and "pavement" features, characteristic of some Death Valley regions, were incorporated into tailings disposal programs.

Lindsey *et al.* (1983) have taken a most interesting approach to the foregoing question. They examined numerous archeological sites and concluded that there was compelling evidence that large manmade earthen structures can remain sound and intact for time periods comparable to those required for the stabilization of tailings piles if properly constructed. Of particular note was the use of clay caps and layers of rock cover on some ancient mounds. As we shall see, these have also been proposed as components of long-term stabilization programs for uranium tailings impoundments.

Before turning our attention to specific reclamation practices and their geomorphic implications, we should first briefly consider the "ALARA" principle that is often found in the literature of the uranium industry. The term is an acronym that means "as low as reasonably achievable." As stated in the Code of Federal Regulations [1984a, Title 10, Part 20.1(C)], persons engaged in activities under licenses issued by the NRC should make every reasonable effort to maintain radiation exposures, and releases of radioactive materials in effluents to unrestricted areas, as low as is reasonably achievable. This means as low as is reasonably achievable taking into account the state of technology and the economics of improvements in relation to benefits to the public health and safety and other societal and socioeconomic considerations and in relation to the utilization of atomic energy in the public interest.

Disposal Site Selection

Long-term isolation of uranium mill tailings actually begins with the identification of the best disposal site in an area. Nelson *et al.* (1983) submitted that site selection and the physical location of an impoundment relative to natural disruptive forces are perhaps the most important consideration in the construction of stable reclamation alternatives.

Nelson and Shepherd (1978) examined potential failure modes of tailings impoundments that could cause release of radioactive components of tailings over long time periods. It is not possible to fully discuss their findings; however, Table 7 summarizes some salient components of the work. The first part of this table concerns failure due to specific natural phenomena. Of course, the magnitude and frequency of phenomena are site specific; the estimated likelihood proposed by Nelson and Shepherd (1978) represents a generalization for the areas currently engaged in uranium production. It should also be noted that although the likelihood of occurrence may be high or great for a particular mode of failure, the magnitude of release may be small. Perusal of Table 7 reveals the relationship between impoundment failure and geomorphic processes.

Collectively, siting requirements tend to favor locations near topographic divides, which serve to minimize the upstream catchment area and hence flood magnitude, as long as these surfaces are structurally stable, that is, devoid of mass movement evidence. However, these locations may experience somewhat higher erosion rates. Steeper average hillslope gradients could result in greater sheet erosion. Steeper channel gradients could result in greater stream erosion, and the elevated position in the landscape could result in greater wind erosion. Additionally, the probability of significant deposition to cover the site would be relatively small.

Nevertheless, the overriding consideration is the reduction of the flood hazard. Nelson and Shepherd (1978), as well as others, contend that floods appear to be the greatest potential cause of severe impoundment failure. Therefore, floodplain locations along major, higher order streams appear to be the worst choice for disposal sites. Unfortunately, the inactive site assessment leading to UMTRCA (1978) found that just such locations had been utilized on several occasions in the past.

Tailings Impoundment Construction

The mill tailing impoundment may be constructed above or below grade. The complete tailings disposal facility consists of several components or elements, as indicated in Table 7: (1) a lining on the surface beneath the tailings; (2) in the case of above-grade impoundments, embankments on one or more sides; and (3) a cap overlying the waste material. The lining is intended to prevent seepage of liquid from the tailings into the underlying soil and rock that would contaminate these materials, possibly diffuse into adjacent areas, and perhaps find its way into groundwater resources. Liners may be composed of compacted soil, clay layers, or synthetic sheets.

Embankments are built according to standard engineering practice. In the past, they have been constructed of tailings themselves, but local earth materials are commonly used now. The embankment or dam may have a clay core to prevent the seepage of liquids from the tailings pond. Because the inclined surfaces of the embankment are subject to sheet, rill, and possibly even gully erosion, vegetation or rock material (riprap) is used to control degradation.

The adequacy of engineering structures is largely dependent upon: (1) an understanding of the forces and resistances that will be in operation upon completion, (2) the accuracy of available data for those variables or the estimates produced by mathematical models, (3) competency of structure design, and (4) the implementation of design during construction. Problems commonly arise as a result of deficiencies in the first two considerations. The magnitude and frequency of forces and resistances cannot be exactly predicted, especially over long time periods. Schumm and Chorley (1983) commented with regard to potential geomorphic hazards that, although a great deal of information is available and can be reviewed, it is apparent that the data base is inadequate and that site evaluation from a geomorphic perspective will be generally qualitative, or at best semiquantitative. Because of such uncertainties, prudence dictates the use of conservative assumptions in engineering design.

Although all agree that impoundment design must be site specific, there are

Table 7 Modes of Uranium Tailings Impoundment Failure[a]

Failure modes	Estimated likelihood	Site consideration	Design consideration
A. Due to natural phenomena			
1. Earthquake	Generally high over very long time-periods	Proximity to a major fault	Use of nonliquefiable materials, drainage of tailings, consideration of stresses caused by earthquake
2. Floods	Definite possibility that even maximum probable flood can be exceeded over very long time periods	Minimize catchment area above impoundment	Magnitude of design flood (maximum probable flood)
3. Windstorms	With tailings cap or cover, failure is unlikely	Selection of site with topographic protection from winds, if possible, to reduce erosion	Cap or cover over drained tailings
4. Tornadoes	Tornado activity relatively low in areas of current milling	Of little concern, even over very long time-periods	
5. Glaciation	For continental type, remote in western United States even over very long time periods	Avoidance of previously glaciated mountain valleys	None available at present time
6. Fire and pestilence	Generally high over very long time periods	None mentioned	None mentioned (perhaps, use of other than vegetative means of stabilization)
B. Failure of impoundment elements			
1. Caps			
a. Differential settlement	Great	Compressibility of subsoil and rock profiles	Stabilization of foundation, use of self-healing cap material
b. Gullying	On cap itself, generally low due to gentle slopes, small drainage area, protection by vegetation or other materials (progression of gullies from embankment considered later)	Primarily, annual amount, frequency, intensity, duration of precipitation; also drainage patterns, area, and location of impoundment in relation to these	Gentle slopes, grade to avoid concentration of runoff, vegetation, or rock cover to resist erosion, avoid highly erodible soils on surface, maintain high infiltration capacity, diversion of external runoff

c. Sheet erosion	Almost a certainty	Precipitation characteristics, surrounding topography, local vegetation which would become established on impoundment (factors of USLE beyond control)	Slope characteristics, management practices, cover material (factors of USLE that can be manipulated)
d. Wind erosion	High	Wind speed and direction, precipitation as it affects soil moisture and cover, soil characteristics, topography with respect to prevailing winds, unsheltered length of surface exposed to wind	Soil cap characteristics, surface roughness, surface slope, unsheltered slope length, vegetation cover
e. Flooding	Relatively high, depends on whether or not floodwater can come into contact with impoundment cap	Minimize upstream catchment area	Use of natural or manmade structures to divert water away from cap, erosion-resistant cap of vegetation or other materials
f. Chemical attack	High	Soil type used for cap, climate	Neutralization of tailings, use of thicker cap, import of resistant material
g. Shrinkage and cracking	Depends upon clay content of cap, moisture content, and aridity of region	Aridity of region, materials available for cap	Nonplastic soils mixed with clayey soils, compaction of soils
2. Liners			
a. Differential settlement	Could range from very low up to almost certainty	Same as for cap	Foundation preparation as for cap, increase thickness of liner, use of self-healing materials such as compacted bentonite
b. Subsidence of soil and rock	Very site specific, causes differential settlement	Subsurface soil and rock, level and fluctuation of groundwater, area and depth of impoundment, avoid areas prone to subsidence, i.e., subterranean voids, collapsing soil, etc.	Same as for differential settlement above

(continued)

Table 7 *(Continued)*

Failure modes	Estimated likelihood	Site consideration	Design consideration
c. Chemical attack	High	Same as for cap	Same as for cap
d. Physical penetration	Great	Foundation profile, thickness of prepared subgrade, material used in subgrade, presence of structures such as drains, spillways, presence of voids below liner, climate	Removal of rock outcrops and sharp loose rocks, filling of voids below liner, cushioning of liner with prepared subgrade
3. Embankment			
a. Differential settlement	High	Same as for cap and liner, also aridity of climate	Use of homogeneous embankment material, compaction at water contents at or above optimum for AASHO[b] compaction
b. Slope failure	Low to moderate if properly designed and drained	Avoidance of geologic faults, characteristics of foundation material, possibility of floodwater erosion of toe	Construction of gentle slopes, consideration of residual strength of embankment soils
c. Gullying	Relatively high over longer periods of time	Same as for cap; in addition, slope, soil, and runoff conditions that have potential for gullying below embankment that could advance headward	Gentle embankment slopes without sharp breaks between cap and embankment, grading to avoid concentration of runoff, diversion of off-site runoff, vegetation, or rock material to stabilize, avoid highly erodible soil
d. Sheet erosion	Same as for cap	Same as for cap	Same as for cap; also, special techniques such as riprap and gravel cover due to slopes steeper than for cap, concave slope shape preferred

e. Wind erosion	High	Direction and force of prevailing winds	Gentle slopes; windbreaks, gravel or cobble in soil matrix, vegetation cover
f. Flooding	High	Climate, upstream catchment area, proximity to floodplain	Use of flood diversion structures and riprap
g. Weathering and chemical attack	Great	Soil types for embankment construction, climate	Testing and selection of materials for embankment construction
4. Revegetation			
a. Fire	High	Low topographic location to reduce risk of lightning ignition, fuel quality of native vegetation, slope aspect as related to moisture regime	Selection of fire resistant low fuel vegetation types, use of firebreaks to isolate impoundment
b. Climate change	High over long time periods	Any site feature which minimizes drought effect, such as exposure, elevation or relief may be useful to reduce effect of drought on vegetation	Drought-resistant plant species chosen, grading to conserve moisture could be used to reduce effect of drought
5. Water diversion structures			
a. Slope failure	Small, if structures are properly drained	A concern only for sites requiring relatively large slopes above diversion ditches or extensive water diversion systems, also where materials conducive to long-term instability exist, such as overconsolidated clays and clay shales	Slopes around diversion structures designed according to residual shear strength of the soil, stresses to be considered are those existing after drainage and complete dissipation of excess porewater pressure
b. Obstruction	Great	Climate, topography, geology, hydrology interrelate to determine obstruction potential	Should be designed to be self-cleaning, water velocities to purge excess bedload, requires nonerosive bottom and side material

[a]Taken from Nelson and Shepherd (1978).
[b]American Association of State Highway Officials.

many challenges confronting design engineers, an overview of which is provided by Young *et al.* (1982) and summarized in Table 8. Perusal of available information permitted a broad assessment of potential facility failure due to various natural phenomena within several uranium producing regions. Even at this scale, information was sometimes not available.

Tailings Impoundment Reclamation

Once the impoundment is filled with tailings or production has ceased, reclamation can begin. The ONMSS (1980) observed that the major factors that must be considered in designing a reclamation plan for uranium mill tailings are: (1) desired land use, (2) climate of the site or region, (3) materials available as cover and growth substrate, (4) backfilling and grading procedures, (5) revegetation procedures, (6) maintenance procedures, and (7) total cost. Nielson and Peterson (1978) noted that site preparation for planting a tailings pond can pose problems not encountered in agriculture. Because tailings ponds are stratified with slimes and sand, physical instability can be a serious problem for equipment where slimes accumulate.

Thus, the stabilization and reclamation process must begin with the dewatering of the tailings. The common practice has been to rely upon evaporation by solar energy and wind for this purpose; most of the uranium-producing regions possess high evaporation rates. There are other dewatering techniques, such as *in-situ* dewatering with underdrain systems, thermal evaporation, and filtration systems, but these are usually considered to be too expensive to enjoy widespread use.

Table 8 Disruption Parameters Summarized by Region[a]

	Eastern Washington	Wyoming basins	Colorado Plateau	Texas Gulf Coast	Southern Rockies
Number of mills	2	11	13	2	3
Runoff potential	Low to high	Moderately low to high	Medium to high	Low to moderately low	High
Maximum probable precipitation (point value)	Information not available	48 cm in 6 hr	50 cm in 6 hr	Information not available	33 cm in 1 hr
Tornado potential	Low	Low	Low	Moderate	Low
Wind erosion potential	High	Medium	Low to high	Medium	Medium
Maximum earthquake acceleration	0.1 *g*	0.03 to 0.05 *g*	0.05 to 0.4 *g*	0.06 *g*	0.2 *g*

[a]Source: Young *et al.* (1982).

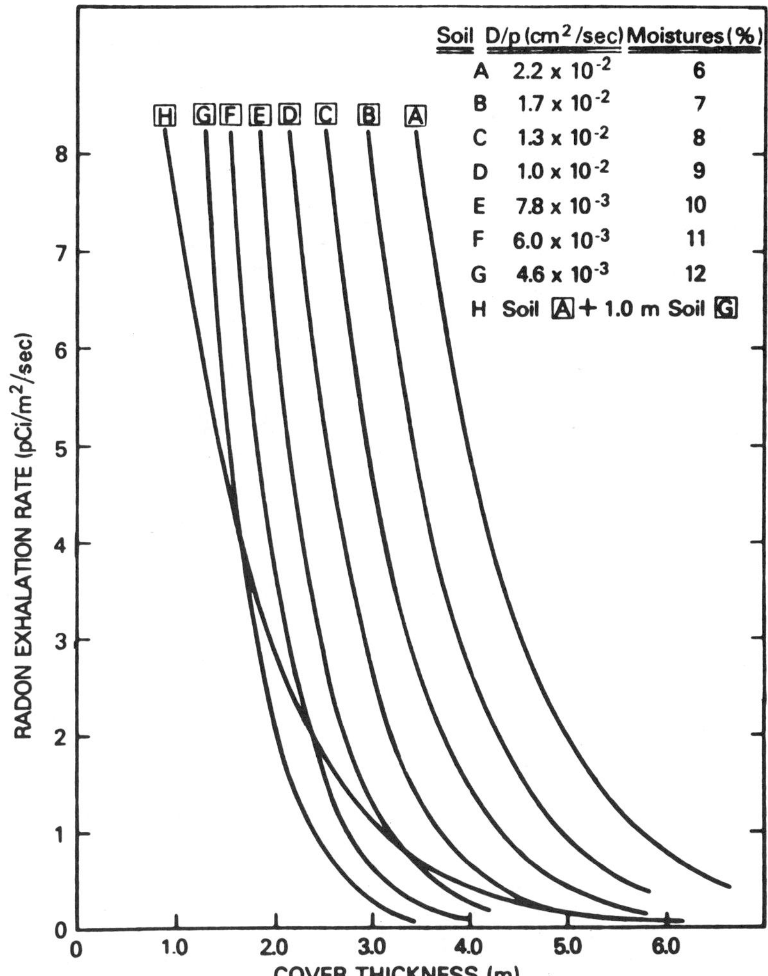

Figure 9. Effect of soil thickness and moisture on radon exhalation rate (ONMSS, 1980).

Following dewatering, the overburden radon suppression cover is placed atop the tailings. Although this material may improve the physical stability of the working surface, the primary objective is to limit radon emissions into the air to an annual average of 20 pCi/m²/sec for at least 200 yr and preferrably 1000 yr. This layer will mitigate gamma radiation flux as well. Figure 9 illustrates the effect of various thicknesses of cover material in reducing tailings radon exhalation. These curves also show the strong dependence of radon attenuation capability on residual soil moisture content.

The overburden surface can then be shaped and graded to the appropriate topographic configuration. Generally, the geomorphic principles guiding this process are the same as those discussed in the earlier chapter concerning hillslopes. If the covered area is relatively small, the topography may be broadly convex upward, to disperse overland flow, with concave or convex–concave sideslopes to encourage deposition around the perimeter. If the covered area is relatively large, then, as suggested by Nelson *et al.* (1983), shaping and contouring are recommended in order to reduce the size of drainage areas, thereby reducing the volume of flow in any one area and the potential for concentrated flow. Any channels produced by landscape construction on the cover surface, even those of low order, will be the loci of somewhat accelerated erosion, when compared to the adjacent hillslopes, unless infiltration capacity is extremely high or special erosion-control measures are taken.

Above the graded overburden radon barrier, layers of rock, clay, topsoil, or combinations thereof may be emplaced. A clay cap is generally employed to prevent infiltration of surface water into the tailings and, hence, the creation of a hydraulic head to drive seepage from the disposal site. As a result of its high moisture retention capacity, the clay layer will also assist in the reduction of radon gas diffusion. And, finally, it serves to resist biotic penetration into the tailings.

In most cases, the next step will be topsoiling. The surface layer of salvaged soil material will usually provide a better plant growth medium than tailings or spoils and will contribute somewhat to radon gas suppression and gamma ray blockage. The National Research Council (1979) commented that rebuilding the soil is thought to be the key to lasting reclamation success.

ONMSS (1980) observed that not all topsoil constitutes an optimal growth medium. Soils of the arid and semiarid regions that do not have high concentrations of salt or alkali may nevertheless be of low fertility because of low organic matter content (less than 1%), low nitrogen and plant-available phosphorus, and, as in certain areas of Wyoming, Texas, and South Dakota, relatively high concentrations of selenium. They also observe that soils of humid regions tend to be acid, and toxicities due to some trace metals may become important if previously buried material is brought to the root zone. Finally, they assert that chemical analysis of any material that is to be used as a medium for plant growth is imperative, not only to determine the types and quantities of fertilizers and amendments needed, but also to determine any concentrations of potentially toxic elements. Table 9 summarizes the characteristics of materials constituting a suitable growth medium.

The stripping, handling, long-term storage, and replacement of topsoil may have a deleterious effect on topsoil productivity. These issues were discussed with regard to coal land reclamation. The extended tenure of an uranium mine

Table 9 Selected Physical and Chemical Criteria for Earth Material[a] as a Medium for Plant Growth on Uranium Mill Tailings Disposal Sites[b]

Property[c]	Value or Concentration	
	Adequate	Poor
Texture	Wide range	Rocks, coarse sands, heavy clays
Bulk density (g/cm^3)	0.8–1.7	>1.8
Available water capacity (%)	5–30	<5
Salinity (mmhos/cm)	<4	>8[d]
Exchangeable sodium (%)	<15	>15[d]
Organic matter (%)	1–4	<1
Selenium (ppm)	0.1–2	>2[e]
Molybdenum (ppm)	<1	Variable[f]
Boron (μg/ml in sat. extr.)	<0.5	>1[d]
pH	4–9[g]	<4

[a]Earth material includes soil, overburden, and spoil.

[b]Source: ONMSS (1980).

[c]This table includes only those properties that are considered by the staff to be critical to revegetation of mill tailings and which are difficult to alter. Properties such as content of macro- and micronutrients are not included, since these can be added to the growth medium as required under the specific revegetation conditions. Elements such as vanadium and arsenic, whose toxicities are highly dependent on their chemical state in the particular growth medium and on the plant species growing in that medium, are not included. Specification of "adequate" or "poor" concentration ranges for such elements would be meaningless. For these elements, each site must be evaluated according to its particular conditions. In general, soils have been found to contain up to 38 ppm As and 20–500 ppm V.

[d]Certain plant species can adapt to more extreme conditions.

[e]Certain plant species can adapt to higher concentrations of soil selenium but then may become toxic to herbivores.

[f]Toxicity to foraging animals will depend on pH and drainage characteristics.

[g]Adequacy depends on the plant species.

and milling operation at a particular site exacerbates any deterioration caused by topsoil storage.

As described by Beedlow (1984a), the vegetation cover on a reclaimed tailings impoundment can serve several purposes, although erosion control is doubtlessly the principal function. Some estimate of the effectiveness of a presumed vegetation cover will commonly be incorporated into the determination of required overburden and topsoil depths to maintain the required level of radon suppression through extended periods of time.

The physical characteristics of vegetation development, both above and below the surface, influence the amount of resistance that the cover offers to erosive forces. The proportion of a surface covered by various plant species is the product of numerous climatic and edaphic variables operating in concert. Beedlow (1984a) remarked that precipitation is a primary controlling factor of plant production and, hence, cover. The relationship between annual precipitation and percentage of total vegetation cover for sites sampled in the vicinity of inactive western U.S. uranium mill tailings depositories supports this contention, as depicted in Fig. 10.

The revegetation technology described earlier with regard to lands disturbed by the surface mining of coal is applicable to tailings site reclamation as well. The surface manipulations, soil amendments, and seed selection processes are the same. Research and recommendations pertaining specifically to uranium lands are available through Murray and Moffett (1977), Kelley (1978), Reynolds *et al.* (1978), Nielson and Peterson (1978), and Beedlow (1984a). Several writers (Yamamoto, 1982; Nielson and Peterson, 1978; Beedlow *et al.*, 1982) have advocated irrigation of tailings sites in the Western United States to avoid dependency on sparse and erratic natural rainfall.

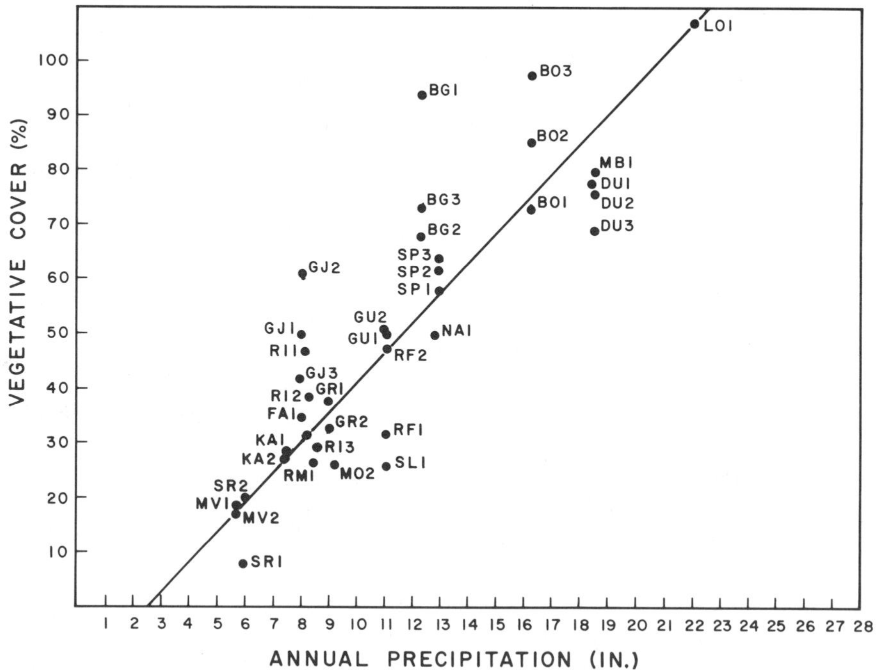

Figure 10. Relationship between vegetation cover and mean annual precipitation at sites sampled in the vicinity of the western inactive tailing piles (Beedlow, 1984a).

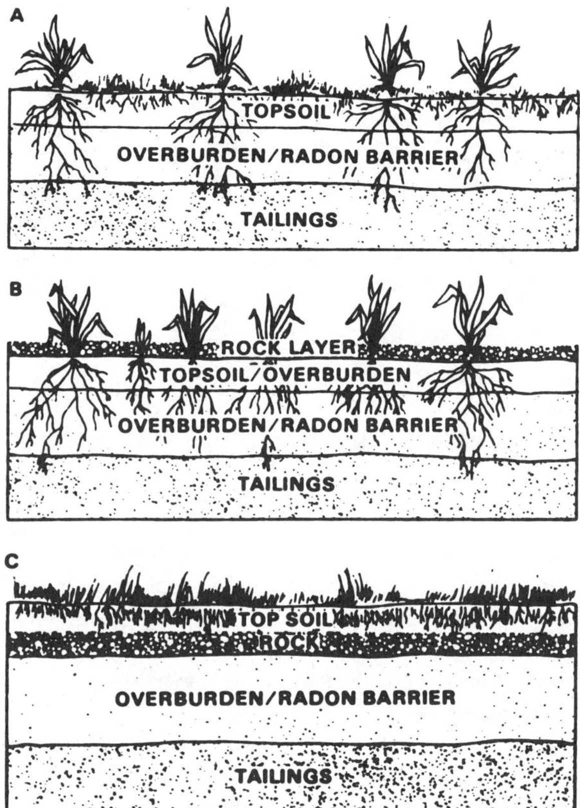

Figure 11. Some basic cover designs utilizing vegetation (Beedlow, 1984b).

ONMSS (1980) suggested that for impoundments in semiarid regions with generally low rainfall or those subjected to conditions that would not ensure a full and self-sustaining vegetation cover, recourse to a cover of soil protected by a layer of rock might be necessary. Additionally, the sideslopes of reclaimed impoundments may require extensive use of rock riprap where there is a possibility that they will be subjected to erosion due to gullying or flooding. In the latter case, tailings deposits located in the downstream parts of drainage basins, along higher order tributaries, would likely be susceptible to flooding over extended time periods. Figure 11 illustrates some basic tailings cover designs using vegetation and other materials.

Riprap is a substantial rock cover that is applied to control erosion of soil on slopes or other surfaces that are subject to potentially severe erosional forces.

Walters (1982) reported the following advantages of riprap used in stabilizing embankment surfaces:

1. Riprap is flexible and is not impaired or weakened by slight settlement movement of the embankment.
2. Local damage or loss is easily repaired by the addition of rock.
3. Construction is not complicated and no special equipment is necessary.
4. Appearance is natural.
5. If riprap is exposed to freshwater, vegetation will often grow through the rocks, strengthening the embankment material and restoring natural roughness.
6. Additional rock can be added at the toe to offset possible scour.
7. Wave run-up is reduced by as much as 70% with riprap as compared to smoother surfaces.
8. Riprap is salvageable and may be stockpiled and reused.
9. Riprap is less susceptible to ice damage than are other revetment protection methods.

He also stated that for sound riprap protection, a well-graded rock mixture should be placed in a well-knit, uniform layer without segregation of the rock. The structural integrity of riprap depends on the following design characteristics: (1) rock size, weight, and gradation; (2) rock shape and quality; (3) embankment slope; (4) riprap thickness; and (5) stability and effectiveness of the underlying filter layer of finer grained material.

ONMSS (1980) remarked that the sizing and design of rock cover to prevent wind and water erosion is difficult because it is not possible to accurately predict climatic, traffic, and other conditions that might influence erosion of the reclaimed tailings impoundment sideslopes and cover over the long term. Therefore, sizing and design of the rock cover for a particular tailings impoundment must provide a strong measure of conservatism, given these uncertainties.

Considerable research has been conducted on the topic of riprap design. Further elaboration on this subject is not warranted at this time; however, Walters (1982), Lindsey *et al.* (1982), Nelson *et al.* (1983), Ecker (1984), and Foley *et al.* (1985) provide detailed information.

Although it is convenient to discuss tailings containments as a series of discrete layers, it is important to recognize that they are interdependent in function. Beedlow (1984a,b) examined the dynamic character of containment systems through an analysis of soil moisture movement. Recall that water content of various layers commonly affects their behavior. To design effective liners, the amount and rate of water passing through the tailings must be known. Radon gas attenuation by the overburden barrier has been shown to be directly related to soil moisture content. Soil moisture storage also influences root growth; penetration into the tailings (sometimes known as biointrusion) provides a pathway for the escape of radon gas to the surface. The erodibility of topsoil covers is also related to soil moisture content. The mass stability of the reclaimed impoundment is often inversely related to the water content. And, finally, the probability of seepage from the containment and subsequent contamination of adjacent soil and ground water increases as the water content within the system increases to produce a hydraulic head. It is evident that containment design must address some "trade-offs" because there are good reasons to maximize and minimize the moisture content of reclaimed impoundments.

Beedlow (1984a,b) found that the moisture dynamics of the tailings pile can be altered significantly by surface treatments. Vegetation covers or vegetation and rock covers can produce a relatively stable moisture content. There was no significant increase in soil moisture under vegetation alone or vegetation and rock covers.

Rock covers that exclude plant growth, even for short time periods, can increase the moisture content of the tailings pile and cover. These rock covers increase infiltration, reduce evaporation, and preclude transpiration, to produce the reported effect on soil moisture. Simulation modeling suggested that drainage from tailings could occur within 3 yr if rock covers thick enough to prevent the growth of vegetation were applied.

Surface rock also affected vegetation structure by favoring the establishment of weeds, shrub, and forb plant species. It was noted that these plant types typically have deeper root systems than do grasses and, hence, may increase the possibility of root penetration into the tailings. Grass cover was highest on soil without any surface rock and these usually develop a shallower root network.

Therefore, even where the percentage of vegetation cover is likely to be low due to climatic factors and surface rock is employed to control erosion, it appears necessary, or at least prudent, to establish some vegetation on the surface in order to prevent the accumulation of water in the containment system. Thin rock covers (rock mulches) that permit plant growth, increase infiltration, and decrease evaporation make additional moisture available to plants and thereby support greater vegetation cover than would otherwise exist.

Following reclamation, the site must be monitored to ensure that health and safety standards have been met and are being maintained. It goes beyond the scope of this discourse to discuss the features of monitoring programs. Those readers so disposed may refer to Kuhaida *et al.* (1984), Kuhaida and Cotten (1985), and Kuhaida and Carnes (1983) for examples.

Geomorphology of Uranium Tailings Reclamation

The primary objective of uranium mill tailings disposal site reclamation is the creation of a stable landform that will retain its integrity over long time periods.

Although "stability" to engineers often refers strictly to the mass stability of a structure, natural or manmade, the term usually has a broader connotation for most geomorphologists. Toy (1982) suggested that stability is a state in which slight perturbations of the variables defining an environmental system do not lead to a progression toward a new steady or equilibrium state. Hence, it could be inferred that stable hillslopes are those that endure minor changes of the environmental setting without substantive alteration of morphometry. Stability may be examined in terms of both spatial and temporal scales.

Tricart and Cailleux (1972) commented that the concept of stability expresses a steady state: a particular hillslope is stable as far as a particular process is concerned. Thus, according to their approach, stability would be process specific. Wells and Rose (1981, p. 82) define stability thusly:

> Landscape stability is measured in terms of the rate of surficial processes (erosion rates), the potential for instability (threshold conditions), the complex response of the geomorphic system (interaction of landscape components to a disturbance) and the significant variables which affect stability at different time periods and different areas.

And finally, they asserted that for reclamation purposes landscape stability should be defined in terms of all regional landscape processes. Scale of study may reconcile the viewpoints of Tricart and Cailleux (1972) with those of Wells and Rose (1981); for individual landforms it might be possible to consider stability in terms of several or only one geomorphic process, whereas for entire landscapes, it might be more prudent to think in terms of all operational processes together.

Wells and Rose (1981) also identified some general attributes of stable landscapes: (1) high infiltration rates, (2) low sediment yields, (3) low relief, and (4) dense vegetation covers. They further observed that processes occurring outside of reclaimed landscapes, such as gullying, can cause unstable conditions in reclaimed areas as headcuts proceed upslope.

Carson and Kirkby (1972) distinguished between short-term and long-term stability. The former refers to instability caused by geomorphic events of large magnitude. For example, most stream undercutting of hillslopes occurs in times of large floods and is often accompanied by landslides. The interval between a triggering process and the response was short. In the latter case, instability may be the result of slow, methodical, and progressive changes in site characteristics that lead again to relatively rapid landform adjustments. For example, the weathering of geologic materials over thousands of years may eventually reduce the shear strength of hillslope materials to the point of failure regardless of the other geomorphic processes functioning in an area or on a particular surface. Therefore, it seems that short-term instability will commonly result when "extrinsic thresholds" have been exceeded and long-term instability will often occur due to "intrinsic thresholds" having been surpassed, to invoke the concepts developed by Schumm (1979).

Schumm and Chorley (1983) remarked that an important consideration during selection of or the evaluation of nuclear waste storage sites is the long-term geomorphic stability of the site, which depends on the risk associated with geomorphic hazards. A geomorphic hazard is any landform change, natural or otherwise, that may adversely affect site stability. Shepherd and Nelson (1978) suggested that while instability and change may be the overall pattern, specific locations within the broad landscape may be stable. However, reflecting upon the above observation of Wells and Rose (1981) concerning the migration of off-site instability, one must ask, Yes, but for how long?

As a consequence of the long-term stability objective, reclamation plans must include consideration of the containment failure potential due to both endogenic and exogenic processes. In the parlance of the geomorphologist, endogenic processes are those that originate within the earth itself, such as tectonism, diastrophism, or volcanism; the exogenic processes are those principally due to external forces, such as weathering, erosion, or mass movement. Therefore, reclaimed disposal sites could be examined and evaluated in terms of their endogenic stability or their exogenic stability.

Endogenic Stability

Active and reclaimed uranium tailings disposal sites may be subjected to forces produced by endogenic processes of various types. Diastrophism refers to the deformation of the earth's crust. The two major categories of diastrophism are (1) epeirogenesis, which applies to uplift or subsidence on a regional scale without internal disruption of original rock structures, and (2) orogenesis, pertaining to the creation of mountains with accompanying intense disruption of rocks due to folding and faulting of geologic structures in relatively localized areas. Tectonism is another term used in reference to movement and deformation of the earth's crust.

Schumm and Chorley (1983) remarked that everything we know about geomorphology and tectonics indicates that the earth's surface does not behave in a regular manner. Tectonic uplift is rarely uniform either over large areas, causing a problem as to the spatial scale over which uplift can be generalized, or over long time periods, presenting great difficulties in extrapolating rates of uplift through time. In particular, intense rates of uplift (i.e., 10 mm/yr or 0.39 in./yr) are very localized in space and time.

Nevertheless, precise releveling by the Coast and Geodetic Survey has discovered significant changes in the surface elevation of the United States in recent years. Figure 12 shows areas undergoing uplift and those subsiding. Schumm

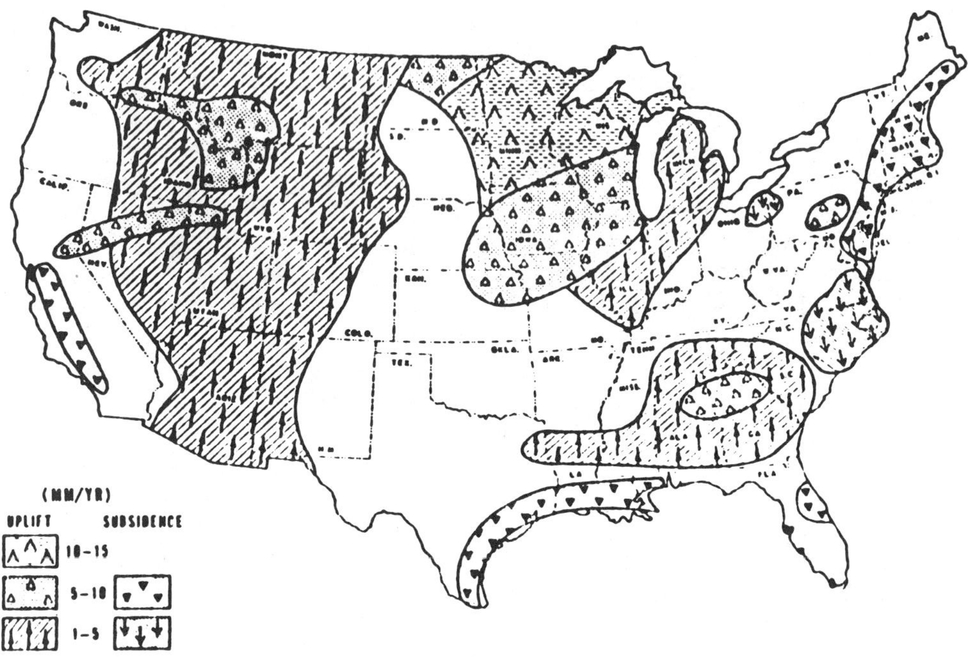

Figure 12. Vertical movement of the earth as estimated by the U.S. Department of Commerce in 1972 (Schumm and Chorley, 1983).

and Chorley (1983) concluded that uplift appears to be rapid and of short duration, allowing little erosional modification before its cessation. This view is broadly supported by a comparison between rates of orogeny (8.0 mm/yr or 0.31 in./yr) and maximum regional rates of denudation, showing that the former are about eight times the latter (Schumm, 1963).

Earthquakes produce rapid ground acceleration, generally over relatively small areas. However, major events may have regional impacts as well. The Alaska earthquake of 1964 generated surface elevation changes over an area of 170,000 to 200,000 km^2 (65,620 to 77,200 sq miles) in south-central Alaska. Uplifts of 10 to 15 m (32.8 to 49.2 ft) were recorded along with subsidence of 1 to 2.3 m (3.3 to 7.5 ft) in other places (Plafker, 1972).

Nelson and Shepherd (1978) described the mechanism by which earthquakes are created. Movement along one side of a fault results in elastic strain energy slowly accumulating in the crustal rock on either side of the fault. When the stress developed along the fault exceeds the strength of the rock, rupture occurs and spreads in all directions along the fault in a series of erratic movements. The elastic strain is liberated in the form of elastic stress waves that propagate outward from the fault and comprise the earthquake.

The prospect and intensity of earthquakes in a given area can be estimated on the basis of several factors, as discussed by Young *et al* (1982). Historic earthquake frequency data can be used as indicators of future activity. Because earthquakes typically occur along existing faults, another measure of potential disruption is the proximity of the tailings pile to existing faults, even if movement has not been recorded. When earthquakes occur along existing faults, the area of rupture is usually less than 50% of the fault length. Past studies have established relationships between fault rupture length and earthquake intensity (Bonilla and Buchanan, 1970). According to these data, faults would have to be at least 10 km (6.2 miles) in length to produce earthquakes capable of significantly disrupting a tailings pile. Nelson and Shepherd (1978) provided a useful summary of earthquake prediction procedures.

Young *et al.* (1982) also examined the potential for earthquake activity within five regions of past or present uranium production: (1) northeastern Washington State, (2) Wyoming basins, (3) Colorado Plateau, (4) Texas Gulf Coast, and (5) southern Rocky Mountains. They concluded that seismic risks occur in any of these areas; however, the extreme northwest part of the Colorado Plateau has the greatest seismic potential. At two sites in this area, estimated maximum ground acceleration caused by earthquakes is 0.2 *g* and 0.4 *g*, respectively. Figure 13 shows seismic risk throughout the United States. This map may be compared with Fig. 3. Of course, local factors may cause higher or lower ground accelerations on a local scale.

Any tectonic activity that significantly changes surface elevation at a disposal site can alter the operation of exogenic processes. Schumm and Chorley (1983) provided a concise summary of potential effects, as shown in Table 10. While discharge may remain unchanged, the sediment load of streams may increase if there is uplift in the upstream parts of the drainage basin because this will increase relief and likely stimulate erosion. Subsidence or downwarp in the upstream areas may decrease sediment yield by reducing relief and promoting deposition. Downstream changes in elevation result in baselevel changes. Lowering of baselevel in this part of the drainage basin could serve to rejuvenate the drainage network and cause substantial gullying from the point of baselevel lowering headward.

As described by Nelson and Shepherd (1978), the interaction of the earthquake with the tailings impoundment may be through accelerations, and hence stress, induced in the impoundment structure or actual displacement of some parts of the impoundment. They also contend that groundshaking is probably the largest earthquake hazard to impoundments because it occurs over wide areas and at great distances from the fault.

A major earthquake could lead to the release of tailings and radioactive materials by causing cracking or rupture of the cover or the embankment of an above-grade impoundment. Extensive failure of an embankment and liquefaction of the

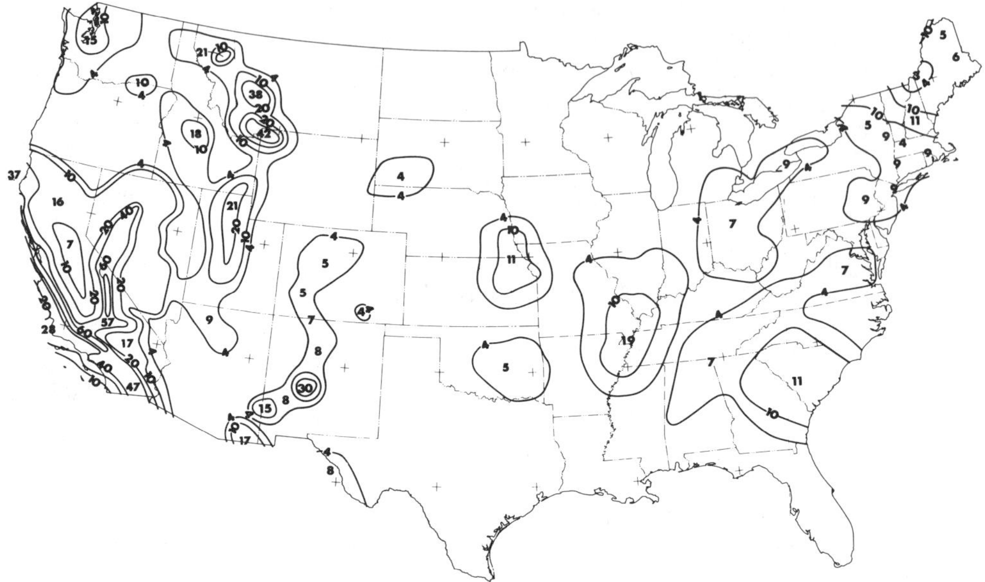

Figure 13. Horizontal acceleration of rock with 90% probability of not being exceeded in 50 yr, expressed as percentage of gravity. This may be interpreted as a map of relative hazard due to earthquakes (Algermissen and Perkins, 1976).

tailings could lead to dispersion of the tailings over large areas. A massive failure due to liquefaction, however, is only likely to occur while the tailings contain relatively large quantities of water. After the tailings dry and consolidate, liquefaction would not occur and failure of the impoundment would result in only localized exposure of tailings.

For tailings buried below-grade, there should be no adverse effect, that is, the

Table 10 Effects of Tectonics[a,b]

	Upstream		Downstream	
Variables	Uplift	Downwarp	Uplift	Downwarp
Discharge	0	0	0	0
Sediment load	+	−	0	0
Baselevel	0	0	+	−

[a]Source: Schumm and Chorley (1983).

[b]+, Increase; −, decrease; 0, no change.

most likely consequence would be further settling of the tailings and cover. Also, where retention structures are conservatively designed (such as to withstand a 1000-yr earthquake under saturated tailings conditions, as is commonly required by the NRC), the potential for loss of tailings from an impoundment during an earthquake over the long term, after the tailings dry and the impoundment is contoured to provide gradual slopes, is virtually eliminated (ONMSS, 1980).

As noted previously, the NCR (Federal Register, Oct. 16, 1985) specifies that an impoundment shall not be located near a capable fault that could cause a maximum credible earthquake larger than that which the impoundment could reasonably be expected to withstand. The "maximum credible earthquake" is defined to be that earthquake that would cause the maximum vibratory ground motion based upon an evaluation of earthquake potential considering the regional and local geology and the seismology and specific characteristics of local subsurface material.

Exogenic Stability

Although endogenic processes have the potential to adversely affect tailings impoundments, proper impoundment siting and engineering design can significantly reduce the hazard associated with these processes. The greatest danger occurs during active operation and shortly after reclamation, until the contained materials have dried; this will also be a period of intensive site monitoring so that any damages can be quickly recognized and rectified. Exogenic processes, on the other hand, will function at every site from the very first day of impoundment construction and thereafter, so long as there is gravity, wind, and water acting on the surface. Generally, the objective of exogenic processes is the rendering of the earth's surface to a common elevation. This is accomplished by erosion in some areas and deposition in others. While we cannot eliminate their presence at a site entirely, we can often control the rate at which they perform geomorphic work and plan for the consequences.

For our purposes, we will consider three categories of exogenic processes that can threaten the integrity of uranium containment tailings systems: (1) mass wasting (mass movement), (2) erosion, and (3) weathering. The mechanics of these processes and the environmental factors influencing their operation were discussed in the first chapters of this book. We shall thus limit the consideration herein to the context of tailings disposal sites.

Mass wasting is a generic term referring to the lowering of the earth's surface by the suite of gravity-driven, downslope mass movement processes. As noted in the earlier chapter concerning geomorphic processes, prediction of mass wasting or mass movements is difficult. Often, apparent causes are no more than triggers producing movement of hillslopes already on the threshold of stability. The problems, then, are the recognition of inherent hillslope instability and the prediction of effective triggering events or conditions.

Schumm and Chorley (1983) remarked that, although it can occur anywhere on the earth's surface, mass wasting in its most significant forms occurs in preferred locations. These locations consist of areas of high relief, shattered rocks, high precipitation, and tectonic activity where earthquakes can disrupt and trigger landslides. Under such conditions, mass movements can be the dominant geomorphic processes in the landscape. However, within a landscape, mass movement will occur at positions dependent upon the hillslope characteristics of aspect, angle, and shape, together with geologic features, such as zones of weak and/or closely jointed rock. Cooke and Doornkamp (1974) and Nelson and Martin (1981) have provided further information concerning identification of mass movement hazards in a particular area.

For an above-grade disposal site, one should consider the potential for (1) instability of the impoundment foundation, (2) instability of the constructed embankments, and (3) penetration of the cover due to rockfall from above. In the case of below-grade disposal sites, the first and third items are noteworthy. Foundation stability is examined prior to impoundment siting and with respect to the factors identified by Nelson and Martin (1981).

Nelson and Shepherd (1978) commented that embankment failure can occur as a result of several factors. During operation of the tailings impoundment, the primary cause of failure is the existence of a high phreatic line. After abandonment, however, the phreatic line becomes lower due to drainage. The exception would be if water collects over the impoundment and seep into the tailings. A general convex configuration of the reclaimed impoundment and, perhaps, the use of a clay layer in the cap, largely precludes this possibility. Another potential cause of failure is the buildup of pore water pressures due to precipitation entering cracks in the embankment surface. Such cracks might be caused by differential settlement, seismic activity, or the desiccation of some clay types contained within the topsoil cover of an embankment. A third cause of potential instability is creep failure occurring within the embankment hillslope. Creep failure results from continued deterioration of the soil strength as a result of creep strains that take place within the material.

In any case, embankment failure is the consequence of shear stress within the earth mass exceeding shear strength. Changes in water content, commonly associated with changes in climatic conditions, may accomplish both. Earthquakes or blasting operations that produce ground vibration may increase shear stress sufficiently to trigger movements. Weathering processes may eventually reduce shear strength sufficiently to permit displacement of material.

Nelson and Shepherd (1978) also discussed some methods of stability analysis for embankments. They noted various limitations for the available techniques. Further, these methods do not predict the extent of damage to an impoundment that results if instability takes place. It is their contention, however, that the prospect of failure is probably low if the impoundment is properly designed, drained, and maintained.

Finally, in some locations it is possible that significant rockfalls upon the reclaimed impoundment surface could breach the cover. If the depth of the radon suppression cover is substantial, only the impact of a very large rock mass would be detrimental. However, the creation of a depression or cracks in the surface could lead to the failures described above. The progressive burial of the impoundment under the cumulative effect of small rockfalls through time may be beneficial. Regardless, it should be quite easy to identify locations below cliffs or escarpments where these possibilities exist.

Erosion is the entrainment and transportation of earth materials by wind, water, or ice. It is reasonable to exclude glacial erosion from further consideration due to the geographical locations of past, present, and probable future uranium production. Additionally, wind erosion requires only modest discussion because, as Shepherd and Nelson (1978) assert, erosion control practices that are employed to control water erosion will also generally control wind erosion.

Nelson and Shepherd (1978) have shown that the most significant risks to stability would be posed by erosive forces of precipitation and flood events and/or wind. Smith (undated), in an examination of the Anaconda Uranium Mine in New Mexico, found that the waste dumps at this operation constituted a significant and definable hazard in the form of erosional instability and possible resultant exposure of radiological materials. All waste dumps had been cut by numerous rills and gullies. Recall from previous discussion of this mine that reclamation efforts had been sporadic and only marginally successful in most cases, although there is evidence that potential for adequate stabilization exists (Reynolds *et al.*, 1978). The ONMSS (1980) stated that erosion is expected to occur very slowly (on the order of a foot in tens of thousands of years), if it occurs at all, given the siting and design measures now being required to reduce erosional potential and avoid other long-term failure mechanism.

There are two broad categories of erosion by water: channel erosion and hillslope erosion. Gully erosion could be considered transitional between the two or exclusively within either category, depending upon the position of the conduit in the landscape. The flooding of an active or reclaimed tailings impoundment site, is a major concern in siting and stabilization of disposal facilities. Shepherd and Nelson (1978) suggested that the single most influential factor that could lead to dispersion of tailings is the effect of a major flood that could come in contact with an impoundment. The ONMSS (1980, p. 9.40) stated:

> If there is a flood of a magnitude larger than that which a surface impoundment could withstand, major portions of the impoundment that are contacted by the flood would be subject to severe erosion. Under the worst conditions, the entire impoundment could conceivably be washed away, dispersing the tailings over a wide area. This is probably the most severe failure mechanism in terms of catastrophic, irrevocable failure and wide dispersion of tailings.

Young *et al.* (1982) added that flooding appears to have the greatest disruptive potential of the phenomena considered in their report. Finally, Nelson *et al.*

(1983) contended that perhaps the most devastating potential failure mode for which uranium mill tailings sites must be designed is the occurrence of a major flood during which encroaching flood waters may come in contact with the embankment of the impoundment.

Nelson *et al.* (1983) provided comprehensive discussion of flooding in relation to the long-term stability of uranium tailings impoundments. They state that it is necessary to evaluate failure modes related to potential flooding of a site using two separate but interrelated approaches. The first is the possibility of flood intrusion if the design flood were to be exceeded and the river or stream course involved remains in its present location. The second is the geomorphic stability of the existing river course and the possibility that, over long-term design periods, the site may or may; not remain geomorphically stable in light of river course changes.

These investigators also observe that for reclaimed uranium tailings impoundments, public health and safety, not just property damage, are important concerns. Therefore, greater safety should be required than for situations where only property damage is involved. Thus, the probability of occurrence (of failure due to flooding) on which the designs are based should have a value lower than that used where only property damage is involved (i.e., 0.01), although this is not a regulatory requirement.

Their analysis revealed that even for 200-yr periods, the recurrence interval corresponding to a probability of failure due to flooding of 0.01 is about 20,000 yr. In other words, if the accepted level of risk (probability of occurrence or probability of failure) is 0.01 throughout a 200-yr period, then the design flood would have a recurrence interval of about 20,000 yr, the so-called 20,000-yr flood. To predict floods having recurrence intervals of thousands of years from limited data bases that extend over periods of only 50 to 100 yr is very unreliable and likely to produce considerable inaccuracy. A recent study (Kochel and Baker, 1982) has established a 10,000-yr paleoflood record for the lower Pecos and Devil rivers in southwestern Texas, using the evidence found in stratigraphic sequences of slack water deposits and radiocarbon dating. This method may be useful in determining the risk of catastrophic floods, if not precise magnitudes. Nelson *et al.* (1983) concluded that in view of the uncertainties associated with extrapolation of a limited data base and the accepted use of the "probable maximum flood" (PMF) concept, it is reasonable and prudent to use the PMF as the design flood where the stability time period of concern is 200, 500, or 1000 yr.

Unfortunately, this approach is not without difficulty as well. Schumm *et al.* (1982, p. 77) remarked:

> At present, the design of tailings site security from flooding is based on the concept of the probable maximum flood (PMF), which is derived from the probable maximum precipitation. The PMF was developed for situations where substantial risk of loss of life exists in the event

> of failure of critical structures such as dam spillways. The PMF is supposed to represent the worst possible flood conditions. No probability can be realistically attached to the PMF, and the implication is that complete protection is provided. The implication of no risk is false; in fact the risk is unknown. On numerous occasions PMF's have been exceeded and recurrence intervals for PMF's vary over at least four orders of magnitude. PMF's are calculated from data based on the historical period of rainfall and runoff records, and thus are iterative and change as the history of large rainstorms accrues.

Likewise, the ONMSS (1980, p. 9.40) stated:

> Prediction of the maximum possible flood is subjective because of uncertainties in the hydrological conditions that will contribute to the flood. Except where the upstream watershed is small, prediction of the maximum possible flood that could occur over a period of several hundred or a few thousand years is tenuous. For this reason, tailings disposal areas should be designed very conservatively to avoid flood damage. Therefore, even the determination of impoundment capacity required during the relatively short operational period is based on a Probable Maximum Flood (PMF) series which is defined in the NRC Regulatory Guide 3.11 as equivalent to 1.4 times the PMF followed by a 100-year storm, appropriately providing a strong measure of conservatism.

Nelson *et al.* (1983) also noted that in order to design an appropriate reclamation plan to withstand the PMF, judicious site selection is generally required. In many instances, only by such site selection can the forces imposed by the PMF be kept sufficiently small to allow practical designs to be implemented. The ONMSS (1980) commented accordingly that impoundments should be located at sites that have small tributary watersheds with a low potential for surface runoff. Schumm *et al.* (1982) commented that even if there is no geomorphic hazard, the location of a site in a valley renders it susceptible to extreme flood events. Normally, hydrologic records are too short to provide information on extreme events. Elsewhere, they assert that, as a general rule, floodplains and low terraces are not appropriate sites for the disposal of uranium tailings. Data obtained by geomorphic techniques have indicated that historic catastrophic floods, having recurrence intervals over 1000 yr, have inundated and eroded such surfaces.

The geomorphic basis for the disposal site locational preference toward upland areas becomes apparent. The size of the design flood (PMF) will be smaller and the accuracy of its prediction should be relatively better in these regions. It is also likely that error in estimation would have less catastrophic consequences.

Another potential stream-related problem that could jeopardize the long-term stability of the impoundment is a change in river course. Should an impoundment be located in a broad, rather flat landscape, a change in course could bring the river in closer proximity to the disposal site, thus subjecting it to direct erosion by common flows or increasing the potential for flood damage.

The term *avulsion* refers to an abrupt change in the course of a river. Schumm and Chorley (1983) observed that a new channel is formed below the point of avulsion. If the channel avulses into an existing smaller channel, a large increase in discharge and sediment load will result. If through avulsion the river takes a shorter course to the baselevel, it will steepen its gradient, and scour above the point of avulsion is likely unless a bedrock control prevents upstream degradation.

Nelson *et al.* (1983) commented that the rate at which meanders can traverse a floodplain is often relatively short compared to other geologic events and that river shifts may occur several times during the period of interest in long-term stability. Because of this, it is necessary to consider the possibility of river shift when developing reclamation plans.

Besides proper site location and conservancy in the estimation of the design flood, another measure of protection from stream erosion may be afforded the impoundment in the form of riprap placed along the embankment hillslopes. The benefits of riprap have been described earlier. Much of the research cited therein was undertaken to address stability problems associated with the inactive Uranium Mill Tailings Remedial Action Project (UMTRAP) sites, listed in Table 2, that were often poorly located in floodplains adjacent to higher order streams. Figure 14 shows the construction of riprap covers.

Walters (1982) identified three problem areas for the design of riprap covers which are of geomorphic interest: (1) the estimation of a flood magnitude, (2) local scour and forces on the structure during floods, and (3) rock durability over long time periods. The problem of flood magnitude estimation (considered

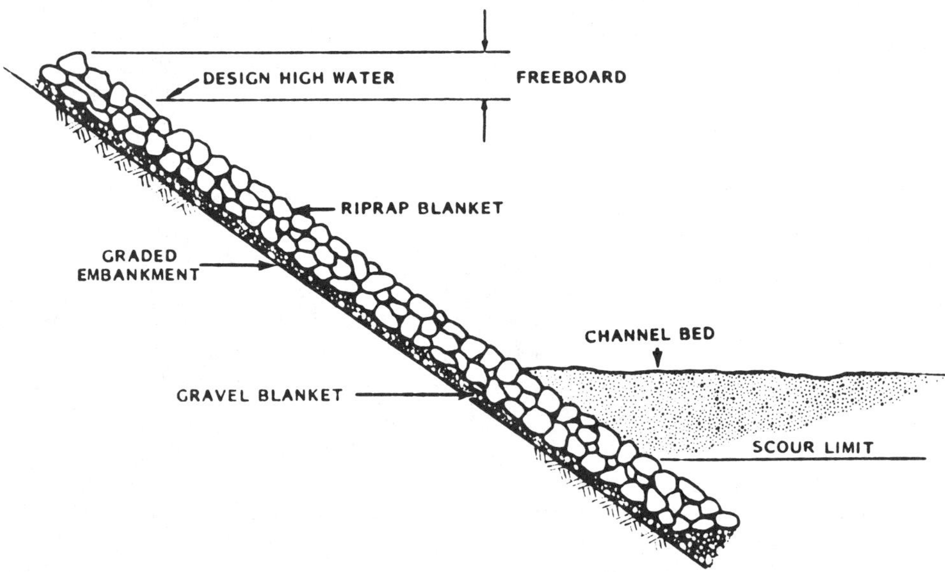

Figure 14. Cross section showing construction of riprap cover on streambank or tailings disposal site embankment (Walters, 1982).

above) is exacerbated by the prospect of climate change in the future; this will be discussed later. Local scour around the impoundment foundation, together with the associated hydrodynamic forces, could undermine the protective cap and cause failure of the entire impoundment structure. Rock durability is important because of the long-term protection requirement and the tendency of certain rock types to break down in a relatively short period of time as a result of weathering; rock weathering processes will also be discussed later.

Walters and Skaggs (1984) examined the effects of hydrologic variables on rock riprap design at two decommissioned tailings sites (Grand Junction and Slick Rock, Colorado). They observed that the probable maximum flood (PMF) can develop extremely high velocities and cause shear stresses that will require abnormally large-sized rocks to stabilize the armor cover. This study evaluated the sensitivity of three riprap design methods to hydrologic variables derived for a range of flood discharges. The three methods employed were the safety factor method, the Caltrans method, and the Corps of Engineers method. The variables considered are flood discharge, embankment sideslope, rock specific gravity, design safety factor, and hydraulic roughness.

The results indicate that the three design methods can yield significantly different rock sizes with the Corps of Engineers method being the most conservative. Other findings show that embankment sideslope angles flatter than about $4H{:}1V$ are probably not necessary because the effects on rock size is negligible. For values of specific gravity less than about 2.50, the rock size required is very sensitive, and therefore the use of low density rock may prove too costly. Further, using low values of channel roughness in the hydraulic calculations can significantly increase design rock size. Lastly, they concluded that safety factors greater than about 1.25 may not be warranted when the probable maximum flood is used as the design event.

Lindsey *et al.* (1982) found that the best available information indicates that certain rock types may be suitable for mill tailings armoring, including varieties of igneous rocks such as rhyolite and granite, metamorphic rocks such as quartzite and granitic gneiss, and perhaps a few sedimentary rocks such as limestone. The ability of these rocks to protect the mill tailings will depend on the weathering conditions to which they are exposed. In turn, these conditions depend on climatic and lithologic factors that usually require site-specific evaluations.

Of course, there are many other ways by which channel processes could affect the long-term stability of impoundments. As presented in Table 11, Schumm and Chorley (1983) identified 16 possible geomorphic hazards associated with rivers (Zone 2) themselves and an additional 4 related to the drainage networks of which they are a part. The zones indicated in this table refer to the portions of a drainage system described in the chapter concerning geomorphic concepts and processes. It is not feasible to fully discuss each of these hazards at this time;

Table 11 Geomorphic Hazards to Site Stability[a]

Zone 1. Landforms
- 1. Drainage networks
  - a. Erosion
    - (1) Rejuvenation: incision of all channels and headward lengthening into undissected areas of the drainage basin
    - (2) Extension: headward growth of the drainage network without rejuvenation
  - b. Deposition
    - (1) Valley filling: major deposition causing significant elevation of valley floor
  - c. Pattern change
    - (1) Capture: diversion of runoff and sediment from one channel to another with major channel adjustments (see 3a, b, d below)
- 2. Slopes
  - a. Erosion
    - (1) Denudation: normal slope erosion and retreat
    - (2) Dissection: incision of slope by rills, gullies, or by drainage network extension [1a(2) above]
    - (3) Mass failure: all forms of mass movement, slump, landslide, etc.

Zone 2. Landforms
- 3. Rivers
  - a. Erosion
    - (1) Degradation: general lowering of stream bed by incision
    - (2) Nickpoint formation and migration: development of break in longitudinal profile that migrates upstream
    - (3) Bank erosion: channel widening or local erosion as related to meander shift
  - b. Deposition
    - (1) Aggradation: general raising of stream bed by deposition
    - (2) Back and downfilling: deposition in channel that migrates upstream and downstream
    - (3) Berming: narrowing of channel by deposition of fine sediments on banks
  - c. Pattern change
    - (1) Meander growth and shift: increase of meander amplitude and downstream migration
    - (2) Island and bar formation and shift: bars form by local deposition and become islands when colonized by vegetation, islands as a result of in-channel erosion and deposition can shift position
    - (3) Cutoffs: neck and chute cutoffs of meanders locally steepen gradient
    - (4) Avulsion: major shift of channel position
  - d. Metamorphosis: a complete change of channel morphology as expressed by pattern change
    - (1) Straight to meandering: development of bends and probable increase of width, gradient decreases
    - (2) Straight to braided: development of bars and islands, increase of width and reduction of depth, may reflect aggradation
    - (3) Braided to meandering: narrowing, deepening, and development of bends, decrease of gradient
    - (4) Braided to straight: narrowing and deepening
    - (5) Meandering to straight: increase of gradient
    - (6) Meandering to braided: widening, shallowing, increase of gradient, may indicate aggradation

Table 11 *(Continued)*

Zone 3. Landforms
- 4. Piedmont and coastal plain (alluvial fan, delta, bajada, pediment)
  - a. Erosion
    - (1) Dissection: development of channels and incision of surface
  - b. Deposition
    - (1) Aggradation: general filling of channels and deposition of surface of fan or delta
    - (2) Progradation: lengthening of delta or fan with upstream deposition
  - c. Pattern change
    - (1) Development of drainage pattern: growth of drainage network on undissected surface
    - (2) Avulsion: channel shift and bifurcation on fan or delta

[a]Source: Schumm and Chorley (1983).

however, Schumm and Chorley (1983) have provided such a review. It is evident that some hazards can directly impact the impoundment while others alter the geomorphic and hydrologic characteristics of the surrounding landscape, and the consequences of these changes may eventually work their way to the impoundment.

Schumm and Chorley (1983) noted that the potential for major drainage network rejuvenation, extension, valley filling, and stream capture will normally depend on major climate and baselevel changes, or in the case of stream capture, on long-continued erosion. Where valleys contain abundant stored alluvium, where the bedrock is readily eroded, where the potential for a major baselevel change exists, and where relief is moderate to high, all of the drainage network hazards are possible. They conclude that disposal sites should be positioned in areas of low relief, well removed from stream channels and on resistant rock.

For our purposes, we have regarded gully erosion as transitional, between channel and hillslope erosion. According to Nelson *et al.* (1983), a gully may be defined as a relatively deep, recently created, eroding channel that forms on valley sides and on valley floors where no well-defined channel previously existed. Two major gully types have been recognized: (1) the valley side gully, which is an extension of the valley network and is incised into soil, colluvium, and weak bedrock; and (2) the valley floor gully, which may be discontinuous or continuous and is incised into alluvium.

Gully erosion may lead to tailings impoundment failure in two possible ways. First, gullies could form at a distance downstream from a tailings impoundment and eventually migrate upstream until they intrude upon the impoundment area. Second, gullies could form within the impoundment area itself and result in a similar failure mode. In either case, the gully can eventually cut through the embankment and cover material and disperse tailings downstream. Hence, it is necessary to consider both off-site and on-site gully potential in assessing the stability of inactive or active disposal sites, or evaluating the adequacy of reclamation plans.

Let us consider off-site gullying initially. The major site-specific parameters that influence gully development are topographical features such as hillslope angle and length, the existence of stable baselevels on or near the site, erodibility of the soil, and the flood flow velocity. Stable baselevels are levels below which no further erosion would be expected. Specific geomorphic and hydrologic conditions that increase the potential for gullying include steep hillslopes, narrow flow width, and large runoff volume as related to drainage basin area. Under such circumstances, the shear stresses of flowing water can exceed the resistance of most earth materials. More specific causes of gully development are listed in Table 12.

The threat of off-site gullying to reclaimed areas has been documented in several places. Wells and Rose (1981) discussed the problem for coal surface mines in northwestern New Mexico. Fuller (1981) noted that many arroyos adjacent to Anaconda's Uranium Mine in New Mexico are unstable and can be expected to advance their headcuts, probably during and immediately after periods of heavy rain. Here, piping around the headcuts was identified as a major process responsible for headward erosion.

Walters (1985) commented that there are no methods available to accurately predict rates of gully erosion or gully advance. However, there has been significant progress toward identifying specific areas where gully erosion potential exists. Two methods that may be useful are (1) empirical observation and (2) the tractive force method; both are described in some detail by Nelson *et al.* (1983). With the empirical method, critical locations in the drainage area of interest and similar neighboring areas are identified and mapped. Measurements of drainage

Table 12 Causes of Gully Development[a]

Causes
Decreased erosion resistance
Decreased vegetation cover owing to increased agricultural activity, urbanization, timbering, overgrazing, fires, and droughts
Surface disturbance causing decreased permeability and cohesion
Increased erosion forces
Constriction of flow by dikes, vegetation, etc.
Concentration of flow by roads, trails, and ditches
Steepening of gradient and energy slope by deposition of sediment (valley plug), channelization and meander cutoffs, baselevel lowering by meander cutoffs, shift of stream mouth, main channel incision, lowering of lake or reservoir level, lowering of discharge, and flood peak in main stream causing tributary adjustment
Increased discharge and flood peaks
Decrease of sediment load

[a]Source: Schumm *et al.* (1983).

areas above gullies and of the valley slopes at the point of gullying can be made and compared to similar measurements of ungullied areas. In this manner, the threshold relationship between drainage area and gradient of the valley can be established. For a given drainage area, it is possible to define a critical valley gradient, steeper than which the valley floor may be unstable.

With the tractive force method, the analysis of gullying potential consists of calculating the frictional forces at the soil–water interface and comparing the results to an allowable tractive force determined by an empirical evaluation of the geomorphology of an area. While the actual tractive force may be determined if the flow characteristics of the stream are known, the analysis of the allowable tractive force is not well established. Further description of this procedure is available in the aforementioned reference, and additional discussion regarding incised channels is provided by Schumm *et al.* (1983).

At the disposal site, gullying is most likely initiated at the points of abrupt change in slope or contour, such as the top edge of an embankment, and unless arrested it will present a much greater threat than sheet erosion for disruption and dispersion of tailings. Once initiated, gullying tends to become more severe with time because the concentration of flow resulting from the presence of the gully itself further accelerates the channeling process. Nelson and Shepherd (1978) suggested that the cause of gully formation on the embankment of a tailings impoundment will be a combination of hillslope gradient, concentration of runoff, and soil erodibility.

Walters and Skaggs (1985) suggested that rock riprap should be applied to the side slopes of earthen impoundment covers in order to ensure long-term protection from gully development. They further observed that procedures are not currently available for the design of rock riprap to prevent gully erosion, and therefore the rock protection of the side slopes will have to be based upon engineering judgment determined by particular site conditions.

Hillslope erosion on or near tailings impoundments can be of two types: rill and sheet erosion. Despite the propensity for unconsolidated material (such as that composing the embankment, cap, radon suppression barrier, and tailings themselves) to develop rill systems, there is surprisingly little research reported on this topic for uranium production facilities. Nelson *et al.* (1983) noted that even where relief is low, rills will form, incising immature and erodible lands to shallow depths. Figure 15 shows rill development on an above-grade tailings impoundment in Wyoming. Laboratory experimentation by Mosley (1972) has shown that the extent of rill development, expressed as drainage density, increases with increasing gradient for planar, convex, and concave surfaces.

One of the few quantitative examinations of rill development on waste materials at a uranium mine and mill was undertaken by Smith (undated). He found that all dumps were traversed by numerous rills, defined as channels less than 6 in. in depth. Hillslope gradient, length of slope, age, and surface roughness seem

Figure 15. Rilling on face of an above-ground tailings disposal impoundment in the western United States.

to be the factors controlling rill and gully development. Beyond this report, one may refer to the study by Soulliere and Toy (1986) concerning rill formation on reclaimed surfaces at an area surface mined for coal.

As observed by Walters (1983), a rill system does not generally last for very long but develops into a normal dendritic drainage network as a few dominant rills are enlarged to form gullies. The rilling of a hillslope is evidence that erosive forces are rather easily overcoming the resistances encountered. As such, rills identify areas of accelerated erosion and potential gully development. These local zones of instability require remedial treatment in order to secure the stability of the disposal site in general.

Young *et al.* (1982) remarked that, although erosion by flooding is a potentially more disruptive single event at a disposal site, sheet erosion can be expected to occur more often and eventually may be as damaging. Sheet erosion of the earthen cap is a gradual, nearly uniform erosion or wearing away of the impoundment surface. This process would be expected to occur on the crown surface as well as on the embankment side slopes. Nelson *et al.* (1983) suggested that the high potential for sheet erosion and the potential for transportation of eroded tailings away from the impoundment area are the principal concerns dictating the need for sound engineering design and proper construction of a stable cover over the tailings material to protect the impoundment against this possible failure mode.

Walters (1983, 1985) provided a detailed discussion of sheet erosion principles and methods of rate estimation for uranium mill tailings impoundments. He asserted that, although most of the information concerning overland erosion is derived from studies of agricultural soil losses, the same processes and computational methods generally apply to the uranium tailings erosion problem. In his assessment of various estimating models, he concluded that the Universal Soil Loss Equation (USLE) and Modified Universal Soil Loss Equation (MUSLE) models can be used to develop cover design schemes and identify the erosional characteristics of the site. A final cover design would then be developed using one of the physical process simulation models evaluated. Table 13 summarizes his findings. Walters and Skaggs (1985) concluded that it would be best to use the Chemical, Runoff, and Erosion from Agricultural Management Systems (CREAMS) model for an assessment of a final design.

The ONMSS (1980) stated that, despite its limitations, the USLE can provide some general guidance in the development of stable tailings disposal programs. It should be emphasized that the USLE provides average annual soil loss estimates based upon normal rainfall conditions. Beedlow (1984a,b) used this approach to calculate general soil loss for various uranium production regions and site characteristics. Of course, the particular attributes of a given site and its environs cause considerable variation from these average values; nevertheless, the soil loss rates obtained through this exercise offer some basis of comparison. Table 14 shows estimated soil loss for inactive tailings sites in three broad geoclimatic regions of the western United States. In very generalized terms, soil loss tends to decrease from the northern Great Plains to the west slope of the Colorado Plateau, although the range in each region is high.

Table 15 shows soil loss for various common vegetation types that may be expected to eventually inhabit a reclaimed site. Here, increasing soil loss rates were found from grassland to salt shrub types. The variability in erosion rates within vegetation types, however, is greater than between types.

Table 16 shows the effect of rock cover on soil loss calculations at several inactive sites. The dramatic reduction of erosion rates is apparent when 0.64–1.27 cm (0.25–0.50 in.) diameter material is applied to a surface. Taking an average value of 1.29 cm/1000 yr (0.51 in./1000 yr), a 3-m (9.84-ft) radon suppression barrier might be expected to last about 230,000 yr based upon this simplistic approach and in response solely to this one geomorphic process. It should be noted that these computations allow for a modest amount of rilling. However, the estimates are likely to be low if the surface is highly dissected.

Finally, it must be recognized that all forms of water erosion operate in concert to degrade the disposal site, whether abandoned or reclaimed and whether above grade or below grade. Accelerated rates of any erosional process may bring about impoundment failure and some degree of tailings dispersal. Hence, stabilization design must consider all processes simultaneously, with particular attention to those capable of exerting the greatest force on earth materials, namely, channel and gully processes.

Table 13 Summary of Selected Models for Evaluating Erosion of Uranium Tailings Impoundments[a]

Model[b]	Hydrologic component	Sediment component	Results	Application
USLE	Annual rainfall erosivity factor	Soil erodibility factor; sediment delivery ratio	Soil loss; sediment yield (with sediment delivery ratio)	Preliminary evaluation and screening
MUSLE	Peak discharge and runoff volume for single event; annual storm-weighted water yield	Uses USLE	Sediment yield	Preliminary evaluation and screening
CREAMS	SCS curve number; infiltration	Overland flow; channel flow; impoundment (pond) deposition	Sediment yield; particle size distribution	Final cover design
ARM	Runoff simulation; infiltration; interflow	Raindrop impact; overland flow; (clay/silt sizes only); groundwater flow	Sediment yield based on overland flow simulation	Final cover design
MULTSED	Infiltration; kinematic flow; channel routing	Raindrop impact; overland flow; channel flow	Sediment yield based on overland flow simulation; channel erosion, transport, and deposition; particle size distribution	Final cover design

[a]Source: Walters (1985).

[b]USLE, Universal soil loss equation; MUSLE, modified universal soil loss equation; CREAMS, chemical, runoff, and erosion from agricultural management systems; ARM, agricultural runoff management; MULTSED, multiple watershed water and sediment routing model.

Table 14 Soil Loss Calculated with the USLE for Inactive Tailings Sites on Various Hillslope Lengths and Gradients[a] in Three Geoclimatic Regions of the Western United States[b]

Region	Soil loss (cm/1000 yr)				
	LS-1	LS-2	LS-3	LS-4	LS-5
Colorado Plateau					
Median	1.52	2.62	7.32	17.93	0.69
Range	18.80	32.21	89.71	220.64	8.26
West Slope					
Median	1.02	1.73	4.80	11.79	0.43
Range	9.45	16.21	45.11	110.64	4.14
Northern Great Plains					
Median	0.76	1.32	3.63	8.94	0.33
Range	1.17	2.01	5.61	13.77	0.51

[a]LS-1, 3% slope for 100 ft; LS-2, 3% for 600 ft; LS-3, 10% for 100 ft; LS-4, 10% for 600 ft; LS-5, 2% for 1000 ft.

[b]Source: Beedlow (1984b).

Table 15 Soil Loss Calculated with the USLE for Inactive Testings Sites on Various Hillslope Lengths and Gradients[a] in the Western United States, Comparing Common Vegetation Types[b]

Vegetation type	Soil loss (cm/1000 yr)				
	LS-1	LS-2	LS-3	LS-4	LS-5
Grassland					
Median	0.84	1.42	3.99	9.45	0.36
Range	9.78	16.76	46.69	114.48	4.29
Shrub steppe					
Median	1.30	2.24	6.20	15.16	0.56
Range	3.30	5.64	15.72	38.56	1.45
Woodland					
Median	1.73	2.97	8.31	20.35	0.76
Range	2.59	4.42	12.34	30.28	1.14
Salt shrub					
Median	2.06	3.53	9.80	24.03	0.91
Range	18.42	31.60	87.96	215.70	8.10

[a]LS-1, 3% slope for 100 ft; LS-2, 3% for 600 ft; LS-3, 10% for 100 ft; LS-4, 10% for 600 ft; LS-5, 2% for 1000 ft.

[b]Source: Beedlow (1984b).

Table 16 Soil Loss as Reduced by Applying Rock to the Soil Surface[a]

Site	Soil loss (cm/1000 yr)	
	LS-3[b]	LS-3 + 539 metric tons/ha crushed rock
Shiprock, NM, Site 1	90.27	1.93
Rifle, CO, Site 1	47.24	1.68
Mexican Hat, UT, Site 2	18.42	0.43
Durango, CO, Site 3	14.45	1.92
Shiprock, NM, Site 1 (different from above)	14.33	0.49

[a]Source: Beedlow (1984b).
[b]LS-3, 10% hillslope gradient for 100 ft.

The possibility of wind erosion at tailings disposal sites cannot be ignored due to the nature of surface materials, the climatic conditions of uranium production regions, and the resultant vegetation cover. In their assessment, Nelson and Shepherd (1978) concluded that windstorms and tornadoes have the ability to cause minor erosion for short periods of time, but they do not appear to be of serious concern to the long-term stability of tailings impoundments. However, the ONMSS (1980) stated that the potential for wind erosion at a particular site may be high and if allowed to progress, wind erosion could remove cover material, expose tailings, and disperse them over a wide area. "Blowouts" or wind-caused gullies, as they described them, can occur where above-ground embankments are used and could lead to extensive erosion of tailings. Likewise, Young *et al.* (1982) found that wind erosion potential was medium to high in the various uranium producing regions, as reported in Table 8.

ONMSS (1980) states that the worst situation with respect to wind erosion involves (1) a loose, finely divided soil; (2) a smooth, bare soil surface; (3) topographic features, that do not provide wind breaks; and (4) long, steep, exposed embankment slopes. The same general conditions that contribute to poor vegetation growth also increase the potential for wind erosion, and thus rock hardening of exposed surfaces becomes essential in many cases. Nelson *et al.* (1983) remarked that the materials most susceptible to wind erosion are fine-grained noncohesive sands and silts; uranium tailings are therefore highly susceptible to wind erosion.

Beedlow (1984b) contended that the best model for predicting wind erosion is that developed by Chepil (Skidmore, 1974), which contains more wind erosion factors pertinent to tailings piles than any other model reviewed. Shepherd and Nelson (1978) noted that like gullying, there exists no model that can predict the potential for blow-outs. Nelson *et al.* (1983) suggested that wind erosion potential may be estimated in a manner similar to that for water. The critical shear stress concept is also applicable using air as the fluid medium. If the numerical analysis indicates that the critical D_{30} size for wind erosion is smaller than for water erosion, then a cover designed to resist the initiation of motion due to water flow will theoretically resist wind erosion. Later, they added that if one evaluates the potential soil loss from both water and wind, it will be observed that water sheet erosion typically produces a greater potential for soil loss than does wind for the same cover material and thickness. Consequently, if adequate protection against gully and sheet erosion on the side slopes and impoundment area is provided, losses by wind erosion should be minimal; nevertheless, computations should be performed to verify this conclusion, particularly in unprotected high wind areas.

The influence of weathering processes on the long-term stability of reclaimed disposal sites is also of geomorphic interest. Over extended time periods, weathering may reduce the shear strength of geologic materials and, hence, reduce the resistance opposing gravitational forces and those generated by other geomorphic processes. It is likely that basic weathering principles have applicability to the materials used in the construction of embankments and impoundment covers. Nelson and Shepherd (1978) are among the few who have discussed the effects of natural weathering processes on embankments, and then in rather general terms. They note that the likelihood of weathering is great and the propensity of materials to weather should be considered in construction design.

One might also consider the chemical alteration of materials within the reclaimed impoundment through time, and resultant change of physical characteristics, as a form of chemical weathering. Markos and Bush (1982) stated that studies indicate that tailings are in chemical disequilibrium and are reactive due to their high salt and moisture content. The chemical reactions redistribute the water and salts causing physical forces to operate within the tailings that are manifested by many observable features, such as the development of cracks and mounds on the tailings surface or pipes and tunnels within the waste material. The mechanisms by which these features are produced remain poorly understood, although there is evidence that mounding may be due to the generation of gases within the deposit. Cavities may be the consequence of salt and water redistribution processes or the flocculation and dispersion of colloidal size material.

Our interest lies in the creation of zones of physical weakness within the impoundment that may become the locus of surface water concentration and, in turn, sites of water infiltration or channelized runoff. An increase in the water content of the earthen material increases the possibility of mass instability, while channelized runoff leads to rilling and gullying of the impoundment surface. Subsurface flow through pipes results in surface subsidence.

Current reclamation procedures, particularly tailings drainage, may reduce the prospect of such scenarios. On the other hand, they may merely mask the internal workings of these processes within the reclaimed disposal site for some period of time.

Most research concerning weathering processes at uranium production sites has focused on the durability of rocks used as riprap along embankment hill-slopes. Lindsey *et al.* (1982) note that documentation of the uses and durability of riprap is limited for the most part to applications since the turn of the century. However, before riprap is to be used on tailings impoundments, rock durability must be evaluated for much longer time periods. They also observe that there is ample evidence of complete deterioration of many rock types in tens to hundreds of years and rocks that fail at this rate would not ensure the long-term stability of uranium tailings piles.

According to Lindsey *et al.* (1982), the durability of a rock is simply defined as its resistance to weathering processes. Because all rocks exposed at the earth's surface are experiencing attack by various weathering agents, the question pertinent to riprap durability is how long the rocks will remain in a suitably sound and stable state. They suggest that the answer to this question depends on three factors: (1) mineral composition and associated chemical properties, (2) physical rock properties, and (3) the climatic conditions to which the rocks are exposed.

The silicate content of igneous rocks is one indicator of their durability. Acidic igneous rocks (e.g., granite) contain high proportions of silica, whereas basic igneous rocks (e.g., basalt) contain less silica and more ferromagnesians. Silica is resistant to chemical weathering and abrasion, whereas ferromagnesians are softer and tend to oxidize readily under surface conditions. Although the minerals composing sedimentary rocks may be stable, durability of the rock itself is tied to the durability of the cementing material. For metamorphic rocks, durability is influenced by the same factors of mineral content, environment of mineral formation, and weathering conditions as for igneous rocks. In addition, mineral segregation and textural features produced during dynamic metamorphism can influence weathering rates by promoting differential weathering.

The physical rock properties that are relevant to durability include specific gravity, density, porosity, strength, and texture. Measurement of these characteristics helps to define a rock's susceptibility to weathering processes. High specific gravity and density tend to indicate resistance to physical weathering processes. Porosity serves as a measure of a rock's ability to sorb water; high sorption values are generally undesirable unless the water can drain freely from pores and capillaries. Grain texture is another measure of weatherability; coarse-grained rocks are usually less durable.

Climatic conditions determine the weathering environment to which the rock is exposed. Precipitation, temperature, wind, and airborne pollutants all contribute to defining this environment. Water is a significant factor in both chemical and physical weathering processes, and the rate of chemical reactions is a function of temperature.

Although the general principles mentioned above can be accepted in most cases, testing of specific rocks to determine their susceptibility to weathering usually has not produced conclusive results. Walters (1982), Lindsey *et al.* (1982), and Nelson *et al.* (1983) described the various tests available, together with their limitations. The latter group of authors concluded that current techniques for durability testing are inadequate because they do not consider the element of time. Voorhees *et al.* (1983) believe that an understanding of natural geomorphic patterns and processes may be more useful than past engineering experience as a guide to selection of weather-resistant materials.

Foley *et al.* (1985) completed a comprehensive examination of rock durability and concluded that conventional testing approaches are inadequate to address rock durability for periods longer than a century. However, they assert that good field technique followed by laboratory wet abrasion and wetting–drying tests could screen local rock types for those with the greatest potential durability. The expected decrease of rock mass with environmental stress (for example, flood impingement and diurnal wetting–drying cycles) can be estimated by this approach.

It would not be appropriate to further extend the discussion on the weathering of riprap. However, valuable consideration of weathering processes in general, including the compilation of contemporary literature, is available from the aforementioned references. It would be useful for geomorphologists interested in weathering processes to review these sources.

Conclusion

From the foregoing, it is clear that geomorphology and uranium mill tailings disposal are intimately related. Geomorphic processes operating both upon and near the impoundment, both during active operation and following reclamation, are capable of causing facility failure and the dispersion of toxic and radioactive materials into the environment. Warner (1982) remarked that the dump or dam represents a disturbance to the integrity of the natural site, where previously geomorphic processes may have been slowly modifying natural forms in some kind of dynamic equilibrium. In his estimation, the final manmade landform is a new, unconsolidated surface of unknown stability, particularly in terms of responses to contemporary geomorphic processes.

Schumm and Chorley (1983) stated that consideration of landscape change, as a result of climatic, tectonic, or baselevel influences, yields a list of 28 geomorphic hazards that may jeopardize the stability of a site (Table 11). For many places, only a few hazards will be of concern. For example, a site located on a bedrock surface that is not near a stream should be only affected by long-term landscape evolution, and only drainage network and hillslope hazards will be relevant. On the other hand, a site located on an alluvial terrace could be affected by drainage network, hillslope, and channel hazards.

Likewise, we must recognize that both endogenic and exogenic processes can work together to produce the eventual failure of the impoundment. Designs must consider each possibility concurrently and accommodate their potential over extended time periods, time periods rarely considered by engineers heretofore, but squarely within the traditions of geomorphology.

Not only must impoundment construction and reclamation programs accommodate the forces produced simultaneously by numerous geomorphic processes, but they must also acknowledge the possibility of changes in the magnitude and frequency of those processes over the long duration through which stability is necessary.

Changes in rates of tectonic activity, climatic conditions, or baselevel position can be expected to produce changes in the work performed by geomorphic processes. But while we know this to be true, the exact nature and extent of change that we should anticipate is considerably less certain. Nelson and Shepherd (1978, p. 4) offer general direction, however, through their adaptation of the principle of uniformitarianism to uranium mill tailings disposal:

> Viewed in another way, this concept (uniformitarianism) can be restated as follows: 'The past is a guide to the future.' Therefore, the history of the last 100,000 years should indicate what can be expected during the next 100,000 years. The predictions made this way may not be correct, but they do provide a basis from which plans for long-term storage of materials can be evaluated.

As discussed earlier, there are abundant data and physical evidence of tectonic activity and crustal movement within the United States and throughout most of the world. Schumm and Chorley (1983) discussed the effect of tectonic activity on drainage patterns and river systems in some detail. They commented that tectonic activity that either raises or lowers a portion of the landscape will affect a site differently depending on the location of the change in relief, as presented in Table 10.

Climate is a very important driving force and determinant of the rate and direction of geomorphic processes. Climatic changes that may significantly impact the stability of impoundments, through their effect on geomorphic processes, can be examined on two general time scales: short term and long term. Numerous investigators have noted the alteration of climate through time periods of various length and have commented on the implications for disposal site stability. The normal variability of climate conditions, as revealed in weather records, is usually taken into account through construction and reclamation plans. For some regions, reports are available that focus upon spatial and temporal variability of climate parameters, together with some implications for reclamation (Toy and Grim, 1980, 1984; Toy and Munson, 1978, 1979).

Cyclic changes, especially drought, can sometimes be anticipated as well, although their occurrence and duration cannot be precisely defined. Nelson and Shepherd (1978) observed that cyclic drought could be severe enough to destroy significant portions of the living vegetation; long-term drought will result in permanent changes in the vegetation. The possibility of increased erosion due to deterioration of vegetation cover, whether by fire, disease, or change in climate, is part of the rationale behind the use of stone or rock covers on reclaimed impoundment surfaces.

In a discussion of problems confronting flood estimation for riprap design based on contemporary information, Walters (1982) summarized significant climate changes that have occurred during the past 1000 yr. From about A.D. 1100 to 1400, there was a warm epoch. From about A.D. 1430 to 1850, a change took place, known as the Little Ice Age. This period was characterized in western Europe and western North America by short, wet summers and long, severe winters. During this period, many glaciers in Alaska, Scandinavia, and the Alps reached positions closest to their maximum advance since the last major glacial epoch, thousands of years before. The first half of the twentieth century was even warmer than the A.D. 1100 to 1400 epoch. A major point here is that a much wetter climate than that of today existed only about 150 yr ago and extended back in time for several hundred years.

Schumm and Chorley (1983) commented that a major concern for site stability is the effect of climate change on landforms. They suggested that there is a general consensus that the climate changes that are most probable during the next few thousand years are represented by the various climate episodes that have occurred during postglacial time. The postglacial climates mostly involved changes of temperature of a few degrees centigrade and changes of mean annual precipitation by as much as 20%. These temperature and precipitation changes were sufficiently large to affect the intensity of many geomorphic processes that determine the morphologic stability of landforms, especially river systems (Knox, 1982).

The geologic and climatic events that dominated the past 100,000 yr, from a geomorphic viewpoint, were the advance and retreat of the Wisconsin age continental ice sheet. The direct effects of glacial ice were great, but even more important were global changes of climate that brought about the ice ages. Significant changes of climate during the past 100,000 yr have drastically changed the hydrologic cycle and the erosional and depositional processes acting on landforms.

Schumm and Chorley (1983) suggested that for long time spans (greater than

10,000 yr) there is a very high probability for readvances of continental and alpine glaciers. Only during about 8% of the time in the last 70,000 yr has the earth experienced climates as warm as or warmer than the present (National Academy of Sciences, 1975). Nelson and Shepherd (1978), however, concluded that the likelihood of occurrence of continental glaciation in the western United States, even within a long long-term period, is remote. They also contend that a significant increase in valley (alpine) glaciation is considered to be remote within the short long-term period (few hundred years), possible within medium long-term periods (few thousand years), and likely within a long long-term period (100,000 yr).

Schumm and Chorley (1983) have carefully considered the consequences of climatic change on geomorphic processes. Based upon past research (Langbein *et al.*, 1949; Langbein and Schumm, 1958; Schumm, 1965), it was concluded that climate change can be expected to produce changes in runoff, sediment yield, and sediment concentration. The direction of sediment yield response to temperature and precipitation changes depends on conditions before the change, as noted by Langbein and Schumm (1958). For example, a change to warmer and drier climatic conditions will decrease annual sediment yield in the arid southwestern United States, but the same change probably will increase the magnitudes of mean annual sediment yields throughout much of the midcontinent grassland and grassland/forest border region. It is also accepted that glacial and periglacial activity increases sediment production as does an increase in the seasonality of precipitation. Schumm and Chorley (1983) summarized the effect of climate change on geomorphic characteristics of a region, as presented in Table 17.

Finally, it should be acknowledged that there is some disagreement as to the consequence of climatic change on the long-term stability of impoundments. Schumm and Chorley (1983) noted that climate change and even relatively minor

Table 17 Effects of Climate Change[a,b]

Variables	Arid to semiarid	Semiarid to subhumid	Subhumid to humid	Nonglacial to glacial	Uniform to seasonal
Discharge	+	+	+	+	+
Sediment load	+	−	−	+	+
Baselevel					
Lake	+	+	+	+	0
Ocean	0	0	0	−	0

[a]+, Increase; −, decrease; 0, no change.
[b]Source: Schumm and Chorley (1983).

climatic fluctuations have produced major changes of river morphology; we have seen in the foregoing that these could be transmitted, through the fluvial system, to the disposal site. On the other hand, Young *et al.* (1982) suggested that within a time span of 1000 yr, no major climate changes or crustal movements are expected that will greatly alter erosion rates. They contend that local anomalies, under current conditions, pose the greatest threat to stability.

Summary

The production of uranium involves two primary components, mining and milling of ore, and each results in substantial environmental impact at the site of operation. The consequences of open pit mining are generally similar to those associated with the surface mining of coal. However, the uranium mine usually remains functional at a specific location for several years, in contrast to the coal mine which follows along a particular seam. While the surface mining of coal allows contemporaneous reclamation, the open pit mining of uranium generally does not.

As a result, reclamation of the pit must await termination of mining. During this prolonged interval, both spoil and topsoil will be stockpiled. There is evidence that this will have an adverse effect on the potential of the topsoil to support revegetation, at least for a time.

The reclamation practices utilized by the coal industry are commonly appropriate for use by the uranium industry as well. The objectives of open pit reclamation are essentially the same as for the surface mining of coal. These procedures were discussed in Chapter 8.

From the geomorphic perspective, open pit mining results in an increase or, at best, constancy in the magnitude of the forces operating at a disturbed site, while at the same time there is a decrease in the resistance to those forces offered by surface or subsurface materials. The consequence is an increase in the amount of geomorphic work performed until a semblance of equilibrium or steady state has been achieved within the geomorphic system.

The milling of uranium ore produces tailings as a waste product that contain radioactive and toxic substances. These materials must be disposed near the processing sites. It is the reclamation of tailings disposal sites that distinguishes between coal and uranium production industries. Because these wastes retain their hazardous potential for thousands of years, the goal of reclamation is not merely the re-creation of a natural system in quasi-equilibrium, but the virtual isolation of the radioactive and toxic substances from the environment for a very long time period. In concept, this is the same apparent intent of the pyramids and other lesser burial mounds. The persistence of these monuments through the ages suggests that long-term isolation of tailings is a realistic possibility.

However, complications arise from the extended time frame for reclamation and stabilization. Both the active and decommissioned disposal site must be designed to contend with not only contemporary environmental conditions and resultant geomorphic processes, but also those of the distant future. At present, we can refer to the past in order to anticipate the future and employ conservative assumptions in our designs.

The situation presented in uranium mill tailings disposal provides a great challenge for geomorphologists and engineers alike. The geomorphologist is called upon to provide reasonable estimates of the geomorphic processes, their forces, and resistances or the resulting work that are likely to be experienced at a particular site through time. The engineer can then refine the values for probable forces and resistances and formulate plans to control the rate at which geomorphic work will be performed. It is the partnership that accomplishes the objective of long-term stability.

References

Algermissen, S. T., and Perkins, D. M., 1976, A probabilistic estimate of maximum acceleration in rock in the contiguous United States: U. S. Geological Survey, Open File Report 76–416, 45 pp.

Beedlow, P. A., 1984a, Designing vegetation covers for long-term stabilization of uranium mill tailings: NUREG/CR–3674 (PNL–4986), Nuclear Regulatory Commission, 76 pp.

Beedlow, P. A., 1984b, Revegetation and rock cover for stabilization of inactive uranium mill tailings disposal sites: DOE/UMT–0217 (PNL–5105), Final Report, Department of Energy, 74 pp.

Beedlow, P. A., McShane, M. C., and Cadwell, L. L., 1982, Revegetation/rock cover for stabilization of inactive uranium mill tailings disposal sites: UMT–0120 (PNL–4328), Status Report, Department of Energy, 30 pp.

Bonilla, M. G., and Buchanan, J. M., 1970, Interim report on worldwide historic surface faulting: U. S. Geological Survey, Open-File Report, 32 pp.

Buelt, J. L., Hale, V. Q., Barnes, J. M., and Silviera, D. J., 1981, An evaluation of liners for uranium mill tailings disposal sites—a status report: PNL–3679, Richland, Washington, Pacific Northwest Laboratories, 67 pp.

Carson, M. A., and Kirkby, M. J., 1972, Hillslope form and process: Cambridge, University Press, 475 pp.

Code of Federal Regulations, 1984a, Energy: Nuclear Regulatory Commission, Title 10, Parts 0–199.

Code of Federal Regulations, 1984b, Protection of Environment: Environmental Protection Agency, Title 40, Part 192.

Cooke, R. U., and Doornkamp, J. C., 1974, Geomorphology in environmental management: London, Oxford University Press, 413 pp.

Ecker, R. M., 1984, Effects of rock riprap design parameters on flood protection costs for uranium tailings impoundments: NUREG/CR–3751 (PNL–5068), Nuclear Regulatory Commission, 52 pp.

Foley, M. G., Kimball, C. S., Myers, D. A., and Doesburg, J. M., 1985, The selection and testing of rock for armoring uranium tailings impoundments: NUREG/CR–3747 (PNL–5064), Nuclear Regulatory Commission, 47 pp.

Fuller, H. K., 1981, Evaluation of arroyo-headcut erosion in the Jackpile—Paguate Mine area, Cibola County, New Mexico: U. S. Geological Survey, unpublished manuscript, 35 pp.

Harwood, G., 1979, A guide to reclaiming small tailings ponds and dumps: U.S.D.A., Forest Service, General Technical Report INT–57, 44 pp.

Hetland, D. L., 1978, Potential resources of uranium: Presented at Grand Junction Office Uranium Industry Seminar, Department of Energy.

Horak, G. C., and Olson, J. E. (eds.), 1980, Proceedings of the uranium mining and milling workshop: U. S. Fish and Wildlife Service, FWS/OBS–80/57, 107 pp.

Horton, R. E., 1945, Erosional development of streams and their drainage basins: Bulletin of Geological Society of America, v. 56, pp. 275–370.

Hunt, C. B., and Mabey, D. R., 1966, Stratigraphy and structure, Death Valley, California: U. S. Geological Survey, Professional Paper, 494-A, 162 pp.

Johnson, W., and Paone, J., 1982, Land utilization and reclamation in the mining industry, 1930–1980: U. S. Bureau of Mines, Information Circular 8862, 22 pp.

Kelley, N. E., 1978, Vegetation stabilization of uranium spoils areas, Grants, New Mexico: unpublished doctoral dissertation, University of New Mexico, 88 pp.

Knox, J. C., 1982, Responses of river systems to Holocene climates: *in* Porter, S. C., and Wright, H. E. (eds.), Late Quaternary of the United States, Minneapolis, University of Minnesota Press, pp. 26–41.

Kochel, R. C., and Baker, V. R., 1982, Paleoflood hydrology: Science, Vol. 215, No. 4531, pp. 353–361.

Kuhaida, A. J., and Carnes, D. P., 1983, Environmental monitoring of the DOE-owned surplus facility, Weldon Springs Site (WSS): Presented at Radiological Health Session of 111th Annual Meeting of American Public Health Association, Dallas, Texas.

Kuhaida, A. J., and Cotten, P. R., 1985, Environmental monitoring of the Middlesex Sites, Middlesex, New Jersey, under the Department of Energy formerly utilized sites remedial action program: Presented at Midyear Symposium of the Health Physics Society, Colorado Springs, Colorado.

Kuhaida, A. J., Liedle, J. M., Liedle, S. D., Cotten, P. R., 1984, The development of environmental monitoring programs at four formerly utilized sites remedial action program (FUSRAP) sites: Presented at 5th Department of Energy Environment Protection Information Meeting, Albuquerque, New Mexico.

Langbein, W. B., and Schumm, S. A., 1958, Yield of sediment in relation to mean annual precipitation: American Geophysical Union, Transactions, V. 39, pp. 1076–1084.

Langbein, W. B., *et al.*, 1949, Annual runoff in the United States: U. S. Geological Survey, Circular 52, 14 pp.

Lindsey, C. G., Long, L. W., and Begej, C. W., 1982, Long-term survivability of riprap for armoring uranium mill tailings and covers: a literature review: NUREG/CR–2642 (PNL–4225), Nuclear Regulatory Commission, 78 pp.

Lindsey, C. G., Mishima, J., King, S. E., and Walters, W. H., 1983, Survivability of ancient man-made earthen mounds: implications for uranium mill tailings impoundments: NUREG/CR–3061 (PNL–4541), Nuclear Regulatory Commission, 33 pp.

Mabbutt, J. A., 1977, Desert Landforms, an Introduction to Systematic Geology, Vol. II, Cambridge, Massachusetts, The MIT Press, 340 pp.

Markos, G., and Bush, K. J., 1982, Physico-chemical processes in uranium mill tailings and their relationship to containment: *in* Proceedings of Workshop on Uranium Ore Processing/Tailings Conditioning for Minimizing Long-Term Environmental Problems in Tailings Disposal, OECD, Nuclear Energy Agency, Paris, pp. 99–114.

Midwest Research Institute, 1976, A study of waste generation, treatment, and disposal in the metals mining industry: EPA No. 68–01–2665, NTIS PB–261–052, 407 pp.

Misaqi, F. L., 1976, U. S. uranium mines and mills emit negligible radiation: Mining Engineering, Vol. 28, No. 8, pp. 47.

Moffett, D., and Tellier, M., 1978, Radiological investigation of an abandoned uranium tailings area: Journal of Environmental Quality, Vol. 7, No. 3, pp. 310–314.

Mosley, M. P., 1972, An experimental study of rill erosion: unpublished M.S. thesis, Colorado State University. Fort Collins, Colorado, 118 pp.

Murray, D., and Moffett, D., 1977, Vegetating the uranium mine tailings at Elliot Lake, Ontario: Journal of Soil and Water Conservation, Vol. 32, No. 4, pp. 171–174.

Nagy, J. G., Haufler, J., and Meydani, M., 1978, A literature review of five trace elements in the environment and in animals: unpublished manuscript, Colorado State University, Fort Collins, Colorado.

National Academy of Sciences, 1975, Understanding climate change: A program for action, National Research Council, Committee for Global Atmospheric Research Program, Washington, D. C., 239 pp.

National Geographic Society, 1981, Energy: Facing up to the problem; getting down to solutions: A Special Report in the Public Interest, 115 pp.

National Research Council, 1979, Surface mining non-coal minerals: Committee on Surface Mining and Reclamation, National Academy of Sciences, Washington, D. C., 339 pp.

Nelson, J. D., and Martin, J. P., 1981, Slope stability as related to stream hazards to bridges: *in* Shen, H. W., Schumm, S. A., Nelson, J. D., Doehring, D. O., Skinner, M. M., and Smith, G. L., Methods for Assessment of Stream-Related Hazards to Highways and Bridges, Federal Highway Administration, FHWA/RD–80/160, pp. 87–143.

Nelson, J. D., and Shepherd, T. A., 1978, Evaluation of long-term stability of uranium mill tailings disposal alternatives, Final Report: Argonne National laboratories, Argonne, Illinois, 337 pp.

Nelson, J. D., Volpe, R. L., Wardell, R. E., Schumm, S. A., and Staub, W. P., 1983, Design considerations for long-term stabilization of uranium mill tailings impoundments: NUREG/CR –3397 (ORNL–5979), Nuclear Regulatory Commission, 163 pp.

Nielson, R. F., and Peterson, H. B., 1978, Vegetating mine tailings ponds: *in* Schaller, F. W., and Sutton, P. (eds.), Reclamation of Drastically Disturbed Land, American Society of Agronomy, Crop Science Society of America, Soil Science Society of America, Madison, Wisconsin, pp. 645–652.

Office of Nuclear Material Safety and Safeguards, 1980, Final generic environmental impact statement on uranium milling: NUREG–0706 (3 volumes), Nuclear Regulatory Commission.

Paone, J., Morning, J. L., and Giorgetti, L., 1974, Land utilization and reclamation in the mining industry, 1930–71, U. S. Bureau of Mines, Information Circular, 8642, 61 pp.

Plafker, G., 1972, Tectonics: In the great Alaska earthquake of 1964, seismology and geodesy: National Academy of Sciences, Washington, D. C., pp. 112–188.

Rafferty, P. J., 1982, Background to the nuclear energy agency program on the long-term aspects of the management and disposal of uranium mill tailings: *in* Proceedings of Workshop on Geomorphological Evaluation of Long-term Stability of Uranium Tailings Disposal Sites, OECD, Nuclear Energy Agency, Paris, pp. 11–18.

Rapaport, I., 1963, Uranium deposits of the Poison Canyon Ore Trend, Grants District: *in* Kelley, V. C. (ed.), Geology and Technology of the Grants Uranium Region, New Mexico Bureau of Mines and Mineral Resources, Memoir 15.

Reynolds, J. F., Cwik, M. J., and Kelley, N. E., 1978, Reclamation at Anaconda's open pit uranium mine, New Mexico: Reclamation Review, Vol. 1, pp. 9–17.

Rouse, J. W., 1978, Environmental considerations of uranium mining and milling: Mining Engineering, Vol. 30, pp. 1393–1436.

Schaller, F. W., and Sutton, P. (eds.), 1978, Reclamation of drastically disturbed lands: American Society of Agronomy, Crop Science Society of America, Soil Science Society of America, Madison, Wisconsin, 742 pp.

Schumm, S. A., 1963, Disparity between modern rates of degradation and orogeny: U. S. Geological Survey, Professional Paper 454-H, 13 pp.

Schumm, S. A., 1965, Quaternary paleohydrology: *in* Wright, H. E., and Frey, D. G., The Quaternary of the United States, Princeton, New Jersey, Princeton University Press, pp. 783–794.

Schumm, S. A., 1979, Geomorphic thresholds: the concept and its applications: Institute of British Geographers, Trans., Vol. 4, pp. 485–515.

Schumm, S. A., and Chorley, R. J., 1983, Geomorphic controls on the management of nuclear waste: NUREG/CR–3276, Nuclear Regulatory Commission, 137 pp.

Schumm, S. A., Costa, J. E., Toy, T. J., Knox, J. C., Warner, R. F., and Scott, J., 1982, Geomorphic assessment of uranium mill tailings disposal sites: *in* Proceedings of Workshop on Geomorphological Evaluation of Long-term Stability of Uranium Tailings Disposal Sites, OECD, Nuclear Energy Agency, Paris, pp. 69–79.

Schumm, S. A., Harvey, M. D., and Watson, C. C., 1983, Incised channels: morphology, dynamics and control: Littleton, Colorado, Water Resouces Publications, 200 pp.

Shepherd, T. A., and Nelson, J. D., 1978, Long-term stability of uranium mill tailings: *in* Symposium on Uranium Mill Tailings management, Fort Collins, Colorado, pp. 155–172.

Singleton, P. C., and Williams, S. E., 1979, Effects of long-term storage on the fertility and biological activity of topsoil: Institute of Energy and Environment, University of Wyoming, Laramie, Wyoming, 36 pp.

Skidmore, E. L., 1974, A wind erosion equation: development, application, and limitations: *in* Proceedings of the Atmosphere-Surface Exchange of Particulate and Gaseous Pollutants Symposium, National Technical Information Services, Springfield, Virginia.

Smith, G. W., undated, Existing conditions at waste dumps, Jackpile-Paguate Mine, Cibola County, New Mexico: unpublished manuscript, U. S. Geological Survey, 39 pp.

Soulliere, E. J., and Toy, T. J., 1986, Rilling of pre-law reclamation hillslopes at the Dave Johnston Coal Mine: Earth Surface Processes and Landforms, v. 11, pp. 293–305.

Stoker, H. S., Seager, S. L., and Capener, R. L., 1975, Energy from source to use: Glenview, Illinois, Scott, Foresman and Company, 337 pp.

Toy, T. J., 1982, Hillslope and scarp morphology: *in* Proceedings of Workshop on Geomorphological Evaluation of Long-Term Stability of Uranium Tailings Disposal Sites, OECD, Nuclear Energy Agency, Paris, pp. 19–27.

Toy, T. J., 1984, Geomorphology of surface-mined lands in the Western United States: *in* Costa, J. E., and Fleisher, P. J., Developments and Applications of Geomorphology, New York, Springer Verlag, pp. 133–170.

Toy, T. J., and Grim, D. S., 1980, Climatic appraisal maps of the rehabilitation potential of strippable coal lands in the Green River Basin, Colorado, Utah, and Wyoming, U. S. Geological Survey Miscellaneous Field Study Map MF–1212.

Toy, T. J., and Grim, D. S., 1984, Climatic appraisal maps of the rehabilitation potential of strippable coal lands in the Williston Basin, Montana, North Dakota, and South Dakota, U. S. Geological Survey, Miscellaneous Field Studies Map MF–1675.

Toy, T. J., and Munson, B. E., 1978, Climatic appraisal maps of the rehabilitation potential of strippable coal lands in the Powder River Basin, Wyoming and Montana: U. S. Geological Survey, Miscellaneous Field Study Map MF–932.

Toy, T. J., and Munson, B. E., 1979, Climatic appraisal maps of the rehabilitation potential of strippable coal lands in the Four Corners Area, Colorado, New Mexico, Arizona, and Utah, U. S. Geological Survey, Miscellaneous Field Study Map MF–1098.

Tricart, J., and Cailleux, A., 1972, Introduction to Climatic Geomorphology: New York, St. Martin's Press, 295 pp.

U. S. Atomic Energy Commission, 1974, The Nuclear Industry, 1973: U. S. Atomic Energy Commission, WASH 1174–73, Washington, D. C., 197 pp.

U. S. Congress, Atomic Energy Act of 1946, PL 585 (Aug. 1, 1946).

U. S. Congress, Atomic Energy Act of 1954, PL 703 (Aug. 30, 1954).

U. S. Congress, Energy Reorganization Act of 1974, PL 93–438 (Oct. 11, 1974).

U. S. Congress, National Environmental Policy Act of 1969, PL 91–190 (Jan. 1, 1970).

U. S. Congress, Surface Mining Control and Reclamation Act of 1977, PL 95–87 (Aug. 3, 1977).

U. S. Congress, Uranium Mill Tailings Radiation Control Act of 1978, PL 95–604 (Nov. 8, 1978).

U. S. Department of Agriculture, Forest Service, 1979, User guide to soils mining and reclamation in the West: Intermountain Forest and Range Experiment Station, Ogden, Utah, 80 pp.

U. S. Environmental Protection Agency, 1981, Proposed disposal standards for inactive uranium processing sites: Federal Register, Vol. 46, No. 6, Friday, January 9, 1981, pp. 2556–2563.

U. S. Environmental Protection Agency, 1983, Environmental standards for uranium and thorium mill tailings at licensed commercial processing sites, 40 CRF 1 and 2, Final Action. U. S. Environmental Protection Agency, Washington, D. C.

U. S. Nuclear Regulatory Commission, 1978, Groundwater elements on in situ mining of uranium: NUREG/CR–0311, 194 pp.

U. S. Nuclear Regulatory Commission, 1980, Uranium mill licensing requirements, final rules: Federal Register, Vol. 45, No. 194, Friday, October 3, 1980, pp. 65521–65538.

U. S. Nuclear Regulatory Commission, 1985, Uranium mill tailings regulations: conforming NRC regulations to EPA standards: Federal Register, Vol. 50, No. 200, Wednesday, October 16, 1985, pp. 41852–41865.

U. S. Soil Conservation Service, 1981, Guide for water erosion control: Conservation Agronomy Technical Note 28, 28 pp.

Voorhees, L. D., Sale, M. J., Webb, J. W., and Mulholland, P. J., 1983, Guidance for disposal of uranium mill tailings: long-term stabilization of earthen cover materials: NUREG/CR–3199, ORNL/TM–8685, Nuclear Regulatory Commission, 101 pp.

Wagner, P., 1980, EPA activities for the control of uranium mining and milling effluents: *in* Proceedings of the Uranium Mining and Milling Workshop, FWS/OBS-80/57, pp. 77–87.

Walters, W. H., 1982, Rock riprap design methods and their applicability to long-term protection of uranium mill tailings impoundments: NUREG/CR–2684 (PNL–4252), Nuclear Regulatory Commission, 62 pp.

Walters, W. H., 1983, Overland erosion of uranium mill tailings impoundments: physical processes and computational methods: NUREG/CR–3027 (PNL–4523), Nuclear Regulatory Commission, 53 pp.

Walters, W. H., 1985, Computational methods for estimating overland erosion of uranium tailings: Journal of Nuclear Safety, Vol. 26, No. 2, pp. 192–199.

Walters, W. H., and Skaggs, R. L., 1984, Effects of hydrologic variables on rock riprap design for uranium tailings impoundments: NUREG/CR–3752 (PNL–5069), Nuclear Regulatory Commission, 53 pp.

Walters, W. H., and Skaggs, R. L., 1985, The protection of uranium tailings impoundments against overland erosion: NUREG/CR–4323 (PNL–5520), Nuclear Regulatory Commission, 48 pp.

Warner, R. F., 1982, Summary of geomorphological considerations associated with the siting of uranium mill tailings disposal dams and their abandonment after decommissioning and rehabilitation: *in* Proceedings of Workshop on Geomorphological Evaluation of Long-Term Stability of Uranium Tailings Disposal Sites, OECD, Nuclear Energy Agency, Paris, pp. 64–67.

Wells, S. G., and Rose, D. E., 1981, Applications of geomorphology to surface coal-mining reclamation, northwestern New Mexico: New Mexico Geological Survey, Special Publications, No. 10, pp. 69–83.

Yamamoto, T., 1982, A review of uranium spoil and mill tailings revegetation in the Western United States: U.S.D.A. Forest Service, General Technical Report RM–92, 20 pp.

Young, J. K., Long, L. W., and Reis, J. W., 1982, Environmental factors affecting long-term stabilization of radon suppression covers for uranium mill tailings: NUREG/CR–2564 (PNL–4193), Nuclear Regulatory Commission, 67 pp.

10

Lands Disturbed by Construction Activities

Introduction

We will never know the time in earliest prehistory when humans ceased merely adapting to the character of the natural landscape and began modifying it to better suit their needs. One might speculate that the demand for shelter was a major motivating factor as groups moved into areas where topography did not provide adequate protection or as population densities grew in a given area to exceed the capacity of habitable sites. Nevertheless, archeological evidence clearly reveals that human alteration of the environment through construction activities extends thousands of years into the past.

Although many types of construction can be enumerated, the discussion herein shall focus upon only two that have greatly impacted the landscape through time, namely: (1) the building of edifices, both residential and commercial, and (2) road and highway construction. While the structures built by scattered groups probably resulted in minor environmental impact, disturbance of the natural system intensified with the development of urban areas. Jordan and Rowntree (1979) suggested that in the Near East, where the first cities appeared, a network of permanent agricultural villages emerged about 10,000 yr ago. These settlements were small in size, rarely having more than 200 occupants. But from such humble beginnings grew towns, cities, and megalopolises, all the while increasing in population, complexity of function, spatial expanse, and, surely, the severity of environmental impact. Undoubtedly, the capability of people to alter the landscape parallels technological advancement; however, the pyramids of Egypt and Latin America attest to the capacity of industrious societies using simple machines and extensive human labor in this regard.

As urban places evolved in size and function, they gradually lost any vestiges of self-sufficiency. There came a need to establish linkages with the surrounding hinterland from which supplies, both foodstuffs and raw materials, could be obtained and with other urban places for the purpose of trade. Hence, a need arose for roads and highways. Borth (1969) recited a contention that "the first step in civilization is to make roads, the second is to make more roads, and the third to make more roads still."

Robinson (1971) and Borth (1969) traced the historical development of roads and highways. Some of the world's oldest roads were initially trade routes. Evidence of roads that served commerce in flint implements 6000 yr ago are found in Denmark. Later, and elsewhere in Europe, roads supported traffic in "freestone," salt, and amber.

The Great Royal Persian Road of King Darius I (c. 480 B.C.) is considered one of the first major highways, although it was not meant for vehicles, but travel by foot or beast. While trade may have provided the initial impetus for this route, it is believed that it was maintained for military purposes as well. A "pony express" system brought information to the central government from outlying regions.

Borth (1969) credits the Romans as the creators of the first highway system. The vastness of the network is revealed by the fact that the first-class roads alone extended for a distance equal to twice the earth's circumference. If all auxiliary roads are counted, the system could have encircled the earth ten times at the equator. Robinson (1971) noted that following the chaos of the Middle Ages in Europe, France was the first to achieve enough political cohesion and wealth to plan a national road system; it was largely complete by 1766.

Robinson (1971) also observed that except in the Aztec and Inca civilizations, road building in the Western Hemisphere lagged thousands of years behind. The Incas' royal roads were comparable to the Great Royal Persian Road in that their primary purpose was the rapid movement of the king's messengers. Here, however, runners were employed rather than horsemen. Legget (1973) reported that these roads were paved with bitumen for much of their 6500 km (4037 miles) length.

Clearly, road and highway construction has a long tradition worldwide and this alone confirms its significance to society. Greenwood and Edwards (1979) remarked that the automobile has become an economic mainstay; at least one out of every six businesses owes its existence to the use of private cars for transportation. Robinson (1971) asserted that as soon as a road attracts travelers, it goes beyond being a mere economic device and becomes a powerful social and cultural element.

From the foregoing, it seems appropriate to consider the construction of buildings and roads together. They have been historically connected in time and space. Gregory and Walling (1973) supported this proposition, commenting that road and highway construction is closely analogous to general building activity and can give rise to important contrasts in catchment response in otherwise rural areas. As we shall see, there are similarities in the nature of environmental disturbance caused by each.

The Construction Process

Most of us have observed the construction process for edifices and roads because they commonly take place near our homes or along the routes we travel. Basically, the procedure can be divided into these fundamental phases: (1) the surface upon which construction is to occur is prepared, (2) specified structures are emplaced, and (3) steps are taken to stabilize remaining surface areas not covered by structures or pavement. Surface preparation results from the need to change the topography in the area of development. First, the land is grubbed and cleared of any large obstacles to equipment movement. Topsoil may or may not be removed in the next separate step. Finally, earthmoving equipment is used to create the necessary surface configuration through a series of cuts and fills.

Following surface preparation, structures can be put into position. In the case of edifices, the buildings themselves, parking lots, sidewalks, drainage systems, and landscaping are emplaced. In the case of roads, the roadbed, curbs or shoulders, drainage systems, bridges, and surfacing materials are emplaced.

Lastly, some action is usually taken to stabilize the remaining surface area at the construction site. This is a form of disturbed land reclamation, and many procedures are similar to those used on surface mined lands. In contrast to mine land reclamation, however, the stabilization of construction sites frequently involves smaller land tracts and occasional maintenance is acceptable.

To be sure, there is a wide variety of procedures associated with edifice or road construction. Engineering design and implementation must accommodate diverse natural conditions. However, the three major steps cited above will likely be included in any project.

General Environmental Impacts

Construction activities produce both direct and indirect environmental impacts over both the short and long term. These may be broadly categorized as (1) noise pollution, (2) air pollution, and (3) water pollution. From the geomorphic perspective, we are principally concerned with the latter. The quantity and quality of surface water flow is dramatically altered by construction activities. Gregory and Walling (1973) commented that areas of building construction constitute a special, though transitory, zone of urban areas where removal of vegetation, earthmoving and excavation, and the operation of heavy machinery can completely disrupt the surface of a drainage basin. These conditions are associated with increased runoff and, more particularly, increased erosion and suspended-sediment transport. Holberger and Truett (1976) found that the major pollutant released from construction sites is sediment. White and Franks (1978) observed that poorly planned, constructed, and maintained developments can result in extremely severe erosion and sedimentation rates that, in turn, can have a massive impact upon surface water quality. This is exemplified by high levels of suspended sediment, turbidity, sediment deposition, and the destruction of aquatic life in streams whose watersheds contain such improper developments.

Similarly, the Highway Research Board (1973) remarked that it is almost axiomatic that highway construction causes soil erosion, erosion causes pollution, and pollution causes degradation of the environment. A highway project with an average right-of-way of 30 m (100 ft) could contribute up to 8500 kg/m (15,000 tons/mile) if no effort is taken to minimize the problem. Figure 1 shows the rilling of a road cut in Hawaii. Israelsen *et al.* (1980), in a Transportation Board report, commented that significant damage to the environment is brought about by uncontrolled water and wind erosion resulting from highway construction activities. The sediment that is produced pollutes lakes and reservoirs, restricts drainage, pollutes surface waters, damages adjacent property, and upsets the natural ecology of streams. Often construction costs are increased and delays and repairs result as a consequence of erosion.

In addition to the aforementioned short-term environmental impacts, there are other long-term consequences of interest to us. The land use changes brought about by construction activities are essentially permanent. While the technology

Figure 1. Rill development on a roadcut in Kauai, Hawaii.

exists to return urban land and roads back to agricultural or pseudonatural lands, the cost is substantial and there would usually be little incentive to do so. Of course, archeological discoveries of settlements beneath fields and pastures attest to nature's ability to claim the land over very long time periods. Nevertheless, alteration of geomorphic and hydrologic processes associated with changes in land use must be considered long term.

Various soil properties are modified through disturbance even when topsoil is stockpiled and replaced. Recreation of the profile may require several decades to several hundreds of years, depending upon pedogenic conditions. Particle size distributions can be permanently changed because soils may not be replaced from whence they came. Where the surface has been scraped to bedrock, the development of a mature soil may require hundreds of years.

Surface cover is altered throughout the impacted area. A part will be paved or roofed. Native or agricultural plant species are often replaced by other native or introduced species for purposes of ornamentation or erosion control. Even under favorable conditions, it would take several years for native vegetation to achieve densities and diversities characteristic of undisturbed sites.

Finally, manmade structures alter the geomorphic and hydrologic response of the surface. Dunne and Leopold (1978, pp. 275–276) nicely summarized some effects of urbanization:

> Gutters, drains and storm sewers are laid in the urbanized area to convey runoff rapidly to stream channels. Natural channels are often straightened, deepened, or lined with concrete to make them hydraulically smooth. Each of these changes increases the efficiency of the channel, so that it transmits the flood wave downstream more quickly and with less storage in the channel. Stormwaters can therefore accumulate downstream more quickly than in natural river systems and produce high flood peaks. Even if the total volume and peak rate of runoff from the land surface were not increased by urbanization (which of course they are), the increased velocities in the channels would still decrease the lag between rainfall and runoff.

Later, these authors noted that the increase of storm runoff has many costly consequences in urban areas. Frequent overbank flooding damages houses and gardens or disrupts traffic. The capacities of culverts and bridges are overtaxed. Channels become enlarged in response to larger floods, and residential lots suffer erosion and reduction of their value.

Regarding the effects of road and highway construction, Morgan (1980) remarked that the main problems associated with roads are posed by the excessive runoff generated from the road itself because water cannot penetrate its hard and compacted surface. Runoff is moved from the roadway through constructed channels and ditches designed to withstand expected velocities. However, this runoff must eventually leave the artificial system and enter the natural drainage system. Here and hereafter environmental damage may occur.

Debris from road embankment failure on the upslope side of the highway collects on the paved surface. While this is a traffic hazard and a maintenance problem, it does not usually result in off-site environmental damage. In contrast, failure on the downslope side may create the above problems, but in addition, often delivers colluvium directly into a stream or provides an unconsolidated mass of earth material that can be easily entrained and transported by runoff to nearby streams.

Purpose of Chapter

From the foregoing, it is evident that construction activities produce dramatic changes in landscapes and, hence, the magnitude and frequency of geomorphic processes in operation and, in turn, the rate and amount of geomorphic work accomplished. It is the aim of this chapter to demonstrate the relationship between the two selected construction activities and geomorphic principles. Again, this discussion centers around the following topics: (1) an assessment of land disturbance resulting from edifices and road or highway construction, (2) an examination of the specific nature of the disturbances to the environmental system, (3) some discussion of the legal constraints placed upon these activities, and (4) consideration of various techniques employed to mitigate the environmental impact and control geomorphic processes.

Assessment of Impact

We have previously presented our contention that a complete assessment of any environmental disturbance should be based upon consideration of three factors: (1) areal extent, (2) intensity or rate, and (3) duration of impact. This concept is now applied to disturbances caused by the construction of buildings and roads or highways.

Areal Extent of Disturbance

While there is a wealth of data compiled for edifice construction industries, areal extent is apparently not among them. Available estimates are based upon various definitions and assumptions. Thus, we are forced to use "urbanized area" as a surrogate measure, recognizing that this includes mostly areas disturbed by the building of edifices and related structures but also land disturbed by street construction.

Table 2 in Chapter 8 (on the surface mining of coal) indicates that there were about 27.8 million ha (68.7 million acres) of land within urban areas of the United States as of 1980. However, the data compiled by Frey (1983), and presented in Table 1 of this chapter, indicate only about 19.1 million ha (47.2 million acres) of urban land. Further, Frey (1983) stated that since 1960 there has been a trend toward the extension of official city boundaries to include territory essentially rural in character, and so the 19.1 million ha (47.2 million acres) urban total, the 0.51 million ha (1.3 million acre) annual increase of urban area, and, probably, the indicated increase in the annual increment from 0.37 to 0.51 million ha/yr (0.9 to 1.3 million acres/yr) all overstate the areal extent of urbanization. The discrepancy between the values provided by Johnson and Paone (1982) (see also Table 2 of Chapter 8, this volume) and Frey (1983) illustrate the problem of definition for urban or municipal lands and changes in the definitions through time. Unfortunately, more precise data are not available (H. T. Frey, personal communication 1985).

Areal extent can also be expressed in relative terms. For the sake of discussion, let us assume that the data provided by Frey (1983) overstated the areal extent of land subjected to urban construction activities by 20%; therefore, about 15.3 million ha (37.8 million acres) would have been disturbed as of 1980, approximately the same area as the State of Georgia. Comparing this value to the 919.2 million ha (2,271 million acres) total for the United States, we find that about 1.7% of the land area has been disturbed. Even the adjusted area (15.3 million ha) would cause urbanization to exceed the area devoted to all other land uses except agriculture, wildlife refuge, and national parks.

As with other types of land disturbance, it is somewhat unrealistic to compare the areal extent of urbanization to the entire surface area of the United States. Future population growth and urban development are likely to occur in selected areas. Hoy (1978) stated that most increases in population will occur through the expansion of existing urban areas and that the Piedmont province, the southern Ridge and Valley area, the Gulf Coast intermittently from southern Florida to Texas, and the far West and Southwest are urban-growth areas.

Roads and highways are also significant land uses in the United States.

Table 1 Area in Urban Areas in the United States[a]

Year	Urban area (million ha)	Change	
		Total (million ha)	Average annual (million ha)
1960	10.32	—	—
1970	14.00	3.68	0.37
1980	19.14	5.14	0.51

[a]Source: Frey (1983).

Mowbray (1969) remarked that there are 3,600,000 sq miles (9,324,000 km^2) of land in the United States and 3,600,000 miles (5,792,000 km) of roads and streets—one mile of road for every square mile of land. The land area covered by roads along with their rights-of-way is estimated to be about 24,000 sq miles (62,160 km^2), equal to the area of West Virginia. The author further noted that every mile of freeway in California consumes about 24 acres (9.7 ha) of land; every interchange occupies about 80 acres (32.4 ha). Finally, Mowbray (1969) stated that the 41,000-mile (65,969-km) interstate system, then under construction, would "gobble" up more land than the State of Rhode Island.

Wolman and Schick (1967, p. 456) provided somewhat different data concerning the amount of land disturbed by highway construction:

> Soil loss from roadside slopes may also be expressed in terms of sediment yield per linear mile of highway construction. The exposed area per linear mile of a divided highway ranges from 13 acres per mile on the Eastern shore to 26 acres per mile in central Maryland. For two-lane highways, the range is from 9 on the Eastern Shore to 16 acres per mile in western Maryland . . . . The lower unit areas are applicable to the low relief of the Coastal Plain of the Eastern Shore, whereas the higher figures apply to the rolling topography of the Piedmont and the high relief of the Appalachian Valley and Ridge Region. Richardson and Diseker (1961) imply somewhat larger area, about 30 acres per linear mile for major interstate highways. Where the rights-of-way on major highways have been designed to accommodate additional lanes in each direction, the cleared area per mile for a dual-lane highway in the Piedmont of Maryland would be increased by 12 to 15 percent to a value of about 30 acres per mile. For the major highway, these figures may be somewhat low, inasmuch as they do not include adjacent areas used for maneuvering equipment and for stockpiling soil.

Figure 2 shows the increase of highway length in the United States from 1955 to 1980. In the final year there are about 6.36 million km (3.95 million miles) of highways in various categories, from federal to county. According to Johnson and Paone (1982) (see also Table 2 of Chapter 8, this volume); there were about 8.7 million ha (21.5 million acres) of land disturbed by highways as of 1980. Given the relationships between linear and areal disturbance cited above, it seems likely that the areal estimate by Johnson and Paone (1982) is considerably low. Their data may refer solely to the area covered by the highway surface following construction rather than the total amount disturbed in the process.

If one were to use a conservative estimate of 15 acres/mile (3.8 ha/km) for areal disturbance during construction, then the 6.36 million km (3.95 million miles) of highways in the United States might have been responsible for roughly 24.2 million ha (59.3 million acres) of disturbance and could have been responsible for much more. If this estimate is accepted, then at least 2.6% of the land area of the United States has been subjected to modification through road construction activities. Again, however, this mode of land disturbance is not uniformly distributed across the countryside but tends to be concentrated in some regions.

It would appear logical that there is some upward limit to the building of road

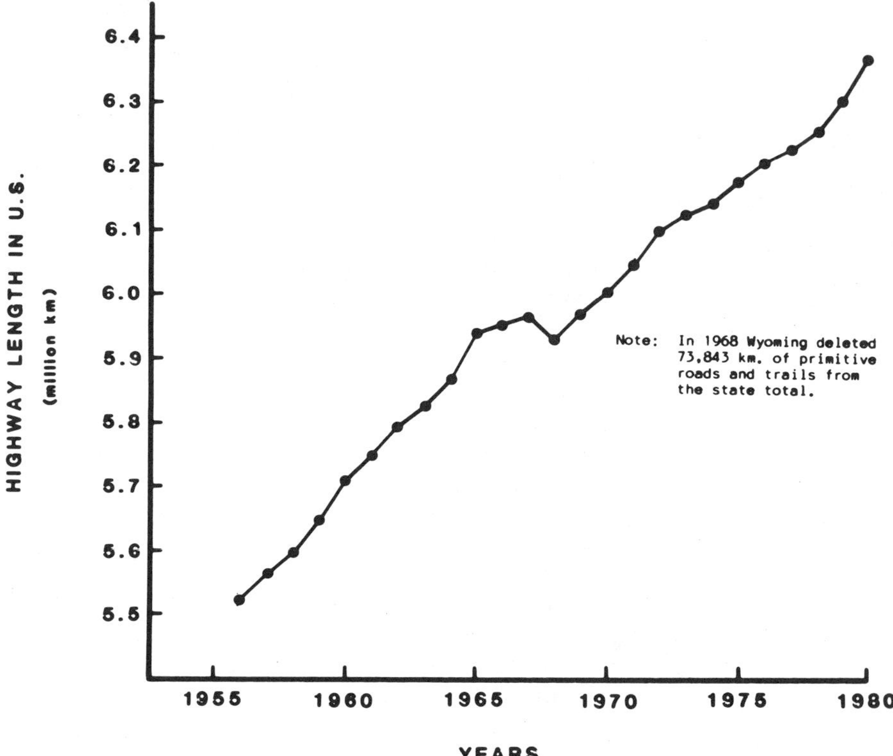

Figure 2. Increase in highway length through time for the United States (U.S. Dept. of Commerce, 1956–1980).

networks; if so, this condition has apparently not been reached because Figure 2 shows rather steady growth. It seems reasonable to suggest that future road construction will be closely associated with urban expansion and perhaps to some extent with reworking and enlargement of existing routes connecting urban places. Hence, we again confirm the relationship between the construction of edifices and roads or highways.

While the data concerning land disturbance by these two construction activities are not precise, it appears that only 4 to 5% of the total area of the United States has been impacted. This is substantially more land than affected by mining activities but much less than modified by agriculture. Although it could be argued that this surface area itself is sufficient cause for alarm, the following discussion points to an even better reason.

Intensity of Disturbance

The second component to an assessment of disturbance, intensity, is examined through consideration of changes in the rates of geomorphic processes. As for other types of land disturbance, we shall focus on erosion and mass movement processes. Holberger and Truett (1976) found a serious lack of field data pertaining to sediment loss at construction sites. Indeed, there are gaps in our knowledge concerning the consequences of construction on various soil types and in various climatic regimes, but there is adequate information upon which to judge the general intensity of impact.

Table 5 in Chapter 8 (on the surface mining of coal) provides one evaluation of soil erosion rates at construction sites. Herein, the U.S. Environmental Protection Agency (1973) suggested that a representative erosion rate for such locations is about 17,000 metric tons/km^2/yr (48,534 tons/sq mile/yr). This is about 2000 times the rate typical of a similar undisturbed area with a forest cover.

Table 2 in the present chapter presents a later evaluation by the U.S. Environmental Protection Agency (1976). According to this information, construction activities on sites formerly devoted to row crop agriculture increase surface erosion 10 times, on sites formerly under pasture by 200 times, and again on sites formerly forested by 2000 times.

Table 2 Some Reported Quantitative Effects of Man's Activities on Surface Erosion[a]

Initial status	Type of disturbance	Magnitude of impact by the specific disturbance[b]
Forestland	Planting row crops	100–1000
Grassland	Planting row crops	20–100
Forestland	Building logging roads	220
Forestland	Woodcutting and skidding	1.6
Forestland	Fire	7–1500
Forestland	Mining	1000
Row crop	Construction	10
Pastureland	Construction	200
Forestland	Construction	2000

[a]Source: U.S. Environmental Protection Agency (1976).

[b]Relative magnitude of surface erosion from disturbed surface, assuming "1" for the initial status. The first row of the table, for example, indicates that transforming a forestland into row crops may increase surface erosion 100 to 1,000 times.

Much of the data upon which these evaluations are based were collected or reviewed by Wolman (1964) and Wolman and Schick (1967). Table 3 contains a summary of this information. It is apparent that the sediment yields from the rural drainage basins are much lower than those disturbed by construction. It should be remembered that the forested watersheds probably provide the sediment yields closest to natural or undisturbed conditions while other rural areas possess various levels of disturbance, chiefly due to agriculture. Further, the group of disturbed watersheds possess various levels of surface modification.

There is a general inverse relationship between sediment yield per unit area per year ard drainage area. For large basins it is unlikely that the entire surface will

Table 3 Sediment Yield from Selected Areas[a]

Location	Sediment yield (metric ton/sq km/yr)	Surface condition	Reference
Rural drainage basins			
Broad Ford Run, Maryland	3.85	Entirely forested	Wolman, 1967
Fishing Creek, Maryland	1.75	Entirely forested	Wolman, 1967
Georges Creek at Franklin, Maryland	72.51	Rural, wooded	Wolman and Schick, 1967
Gunpowder Falls, Towson, Maryland	283.04	Rural: 1914–1943	Holeman and Geiger, 1965
Gunpowder Falls, Towson, Maryland	81.62	Rural: 1943–1961	Holeman and Geiger, 1965
Gunpowder Falls, Hereford, Maryland	319.82	Rural: 1933–1943	Holeman and Geiger, 1965
Gunpowder Falls, Hereford, Maryland	175.15	Rural: 1943–1961	Holeman and Geiger, 1965
Helton Branch, Somerset, Kentucky	5.25	Wooded	Collier *et al.*, 1962
Monocacy River, Frederick, Maryland	114.55	Rural	Wark and Keller, 1963
Northwest Branch, Anacostia River, Colesville, Maryland	164.64	Rural	Wark and Keller, 1963

Table 3 *(Continued)*

Location	Sediment yield (metric ton/sq km/yr)	Surface condition	Reference
Seneca Creek, Dawsonville, Maryland	112.10	Rural	Wark and Keller, 1963
Watts Branch, Rockville, Maryland	180.75	Rural	Wark and Keller, 1963
Drainage basins disturbed by construction			
Greenbelt Reservoir, Greenbelt, Maryland	1,961.68	Residential construction	Guy and Ferguson, 1962
Johns Hopkins Univ., Baltimore, Maryland	49,042.00	Construction site	Wolman, 1967; Wolman and Schick, 1967
Lake Barcroft, Fairfax, Virginia	11,384.75	Residential construction	Holeman and Geiger, 1959
Little Falls Branch, Bethesda, Maryland	812.70	Urban area plus construction area	Wark and Keller, 1963
Manor Run, Norbeck, Maryland	2,760.00	Construction area	Yorke and Davis, 1972
Northwest Branch of Anacostia River, Hyattsville, Maryland	648.06	Urban area plus construction area	Wark and Keller, 1963
Northeast Branch of Anacostia River, Riverdale, Maryland	371.32	Urban area plus construction area	Wark and Keller, 1963
Oregon Branch, Cockeysville, Maryland	25,221.60	Industrial park	Wolman and Schick, 1967
Rock Creek, Washington, D.C.	560.48	Urban area plus construction area	Wark and Keller, 1963
Stony Run, Maryland	18.92	Entirely urbanized area	Wolman, 1967
Tributary, Gwynn Falls, Maryland	3,958.39	Residential (stilling basin data)	Wolman and Schick, 1967
Tributary, Kensington, Maryland	8,407.20	Residential subdivision	Guy, 1963
Tributary, Minebank Run, Towson, Maryland	28,024.00	Commercial development	Wolman and Schick, 1967

[a]Compiled and adapted from Wolman (1967), Wolman and Schick (1967), and Chen (1974).

be disturbed all at once, and sediment storage is likely to increase as area increases. This association is illustrated in Figure 3. The term *dilution* is a rough separation based upon the percentage of the area under construction, according to Wolman and Schick (1967). Low dilution refers to sediments delivered directly to the stream channel from the construction site, and high dilution means that there is construction activity in the basin but also opportunity for sediment storage or dilution of sediment before reaching the point of measurement in the stream.

Keller (1985), using data assembled by Putnam (1972), presented the relationships between sediment discharge and drainage area for various land uses shown in Figure 4. Here, the total volume of sediment per year increases with the size of the disturbed basin and the severity of disturbance, ranging from completely exposed soil surfaces to undisturbed forested sites.

Although it is not possible, nor probably necessary, to fully discuss each study pertaining to erosion on construction sites or basins experiencing construction in various proportions, a few examples are useful.

Keller (1962; see also Wark and Keller, 1963) compares sediment rating curves for the Northwest Branch of the Anacostia River near Colesville, Mary-

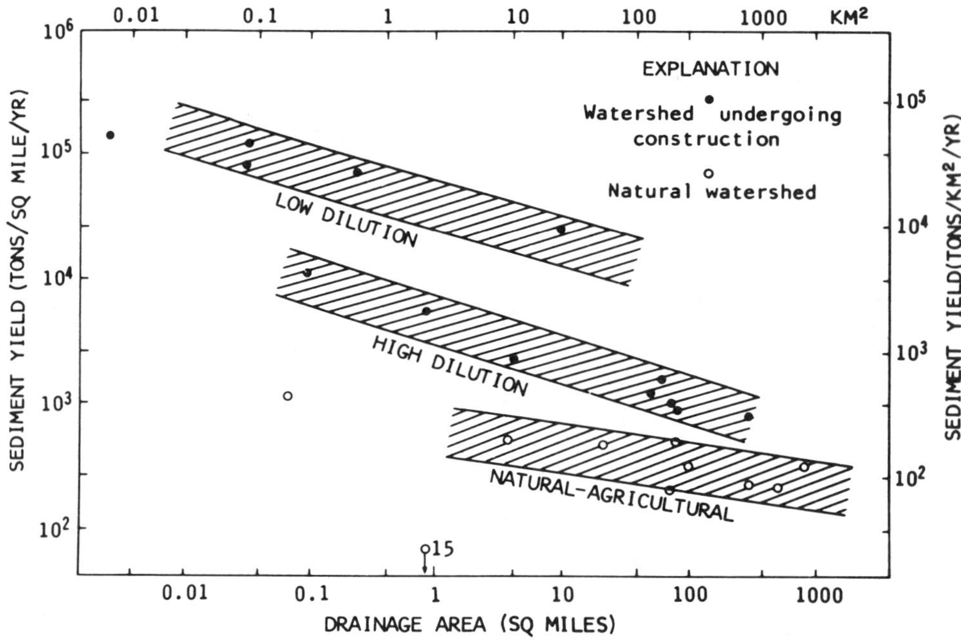

Figure 3. Relationship between sediment yield and drainage area for construction lands (Wolman and Schick, 1967).

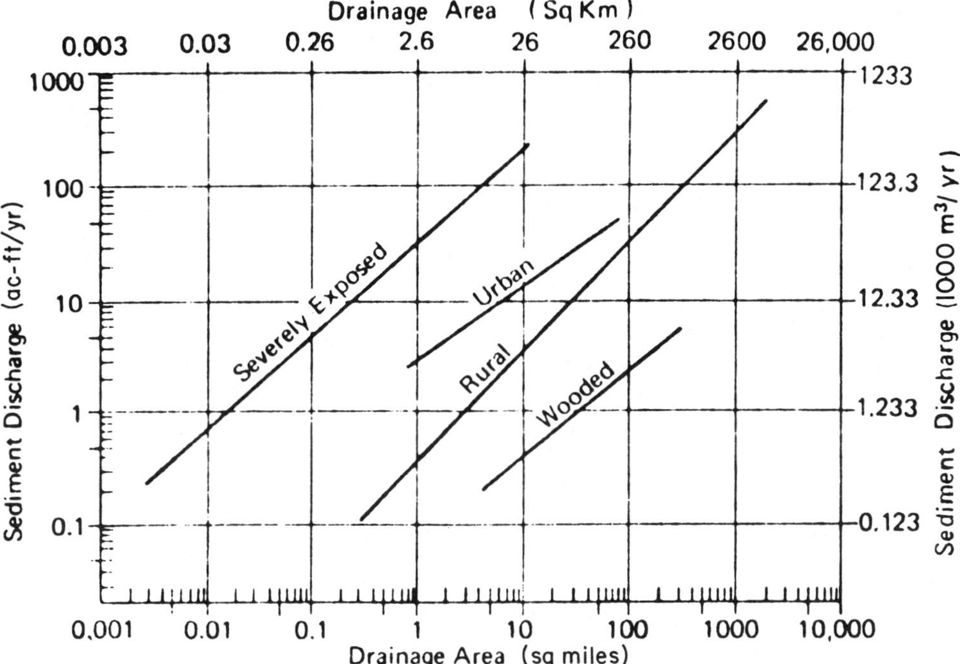

Figure 4. Relationship between sediment discharge and drainage area for various land uses (Keller, 1985; after Putnam, 1972).

land, located in a relatively rural watershed, with that for the Northwest Branch of the same river near Hyattsville, Maryland, situated in a partially urbanized watershed. The results indicate that sediment production is about four times greater in the watershed subjected to construction activities, as shown by the data in Table 3.

Yorke and Davis (1972) examined the Northwest Branch of the Anacostia River near Colesville, Maryland (1963–1967) following construction and found an average sediment yield of 211.8 metric tons/km^2/yr (0.9 tons/acre/yr) while the percentage of urbanized land increased from only 0.5 to 3.8. They also found that in Bel Pre Creek, storm runoff increased 30% and the suspended sediment load was 14 times greater while apartments and townhouses were being built on 15% of the basin than when the same area was covered by pasture and woodland.

Holeman and Geiger (1965) examined sediment yields for Holmes and Tripps Run, Lake Barcroff, Fairfax, Virginia, while in relatively undisturbed condition (1915–1945) and later (1948–1954) after some construction activity. The sediment yield increased from 126 metric tons/km^2/yr (0.6 tons/acre/yr) during the former period to 21,390 metric tons/km^2/yr (95.4 tons/acre/yr) during the latter (data from Chen, 1974).

Yorke and Herb (1976) examined a number of drainage basins in Montgomery County, Maryland, undergoing continuous change in land use from woodlots, pastures, and cultivated fields to houses, apartments, and shopping centers. Sediment yields ranged from 16 to 226 metric tons/ha/yr (7.2 to 101 tons/acre/yr) at construction sites, while cultivated fields produced 6.2 metric tons/ha/yr (2.8 tons/acre/yr) and established residential areas of 8.4 metric tons/ha/yr (3.7 tons/acre/yr).

Hilliard (1977) reported on the erosion and sediment inventory conducted in New York State. Here, soil loss from cropland ranged from 2.8 to 16.8 metric tons/ha/yr (1.25 to 7.5 tons/acre/yr), for pasture about 2.2 metric tons/ha/yr (1 tons/acre/yr), for established urban areas about 3.4 metric tons/ha/yr (1.5 tons/acre/yr), and for untreated construction sites about 69.5 metric tons/ha/yr (31 tons/acre/yr).

Changes in natural stream channels often accompany urbanization. These alterations may be either a part of the construction process or the result of natural processes responding to modifications of the hydrologic characteristics in the urbanized drainage basin. Emerson (1971) described channelization as the straightening of a stream course or the dredging of an entirely new channel to which the stream is diverted for the purpose of minimizing local flood hazard by shortening the distance traveled and thereby moving the floodwaters downstream more rapidly. However, changes in the gradient and geometry of these stream channels produces subsequent adjustments. For example, Emerson (1971) found that a channel of the Blackwater River in Johnson County, Missouri, increased from a cross-sectional area of 38 m^2 (408.9 sq ft) when newly dredged to an area ranging from 160 to 484 m^2 (1721.6 to 5207.8 sq ft), representing a maximum area increase of 1173% in 60 yr.

Hammer (1972) examined stream channel enlargement occurring in response to the change in streamflow regime accompanying urbanization for 48 small watersheds near Philadelphia, Pennsylvania. Major channel enlargement effects were the result of sewered streets and the area of major impervious parcels such as parking lots, while much smaller effects were observed for unsewered streets and impervious areas involving detached houses. Although the relative importance of various interactive factors is difficult to establish, the most critical determinant of the amount of channel enlargement resulting from a given level of urbanization appears to be basin slope.

From the foregoing, it is evident that construction activities associated with urbanization produce very large increases in erosion and sediment yield rates. It should be noted that the data presented above include some consequences of highway and road construction as well as the erection of edifices. Also, most of these studies are located in the Maryland region. Maps contained within Israelsen *et al.* (1980) showed this area to have "*R* factor" values (USLE), expressing the erodibility of rainfall, ranging from 150 to 175. Other locations in Texas, Louisiana, Mississippi, Alabama, and Georgia show values ranging from 400 to 800. Assuming the utilization of similar construction techniques in these areas, the erosion and sediment yields from disturbed and exposed soil surfaces could be many times greater than reflected in the reports mentioned above.

Several studies examine the acceleration of erosion or sediment yield rates due to road and highway construction alone. Some examples are contained in Table 4. Although these investigations utilize various research methods, there is ample evidence that this mode of disturbance again produces erosion and sediment yield rates substantially higher than those typically encountered on undisturbed areas, often many times higher. Vice *et al.* (1969) apparently discovered some of the greatest disparities, with sediment yields from a basin heavily impacted by highway construction amounting to about 2000 times that from an otherwise similar but forested watershed.

Reed (1978) examined stream discharge, sediment yields, and sediment control techniques in five adjacent drainage basins located about 16 km (10 miles) west of Harrisburg, Pennsylvania. On a monthly basis, the average streamflow in each of the five basins before highway construction was very close to the average flow during the 2 yr of construction. Further, he found that peak streamflow did not seem to be appreciably affected during highway construction. This is contrary to our usual expectations.

The average storm runoff, suspended sediment concentrations are summarized in Table 5 for periods before and during construction. The six highest daily mean suspended sediment concentrations are averaged for each month from October 1969 through September 1974. Most of the values represent periods when storms caused high suspended sediment concentrations. However, some values are included when operations in or near the streams caused increased concentrations. The variation of suspended sediment concentration during various construction phases is apparent.

During construction, baseflow suspended sediment concentrations in the streams increased from about 5 mg/liters to about 18 mg/liters, and during storm runoff periods suspended sediment concentrations increased from about 45 mg/liters to about 200 mg/liters.

The suspended sediment discharge for these basins is contained in Table 6. The totals are subdivided into the normal sediment discharge, that caused by highway construction, and that caused by Hurricane Agnes. The normal sediment discharge refers to the sediment discharge given no major changes in land use and no unusual storms. The normal sediment discharge for the 5-yr period is the total load minus the load transported during Hurricane Agnes (basin 1). Normal sediment discharge rates from the basins affected by highway construction are determined by comparing the amount of sediment discharged from basin

Table 4 Erosion Associated with Highway Construction

Location	Erosion rate	Relation to undisturbed area	Surface character	Reference
State Highway 92, Cartersville, Georgia	224 metric tons/ha/yr 312 metric tons/ha/yr 437 metric tons/ha/yr	Erosion rates more than 15 times those on steep agricultural lands nearby (1960 data)	Gently sloping plots Moderately sloping plots Steeply sloping plots	Diseker and Richardson, 1962
Scott Run basin, Virginia	1–10% of total area was under construction and contributed 85% of sediment; with normal precipitation 338 metric tons/ha/yr from highway construction areas	Sediment yield per unit area from this drainage basin was about 10 times greater than from cultivated lands, 200 times greater than from grasslands, 2000 times greater than from forest areas	11.8 km^2 drainage basin undergoing highway construction	Vice *et al.*, 1969
Deadwood River Road, central Idaho	1,241 metric tons/km^2/yr 62 metric tons/km^2/yr 695 metric tons/km^2/yr	No comparison given	Barren control plots on fill Mulched plots Tree plantings	Megahan, 1978
New York State	Average for state: 16.3 metric tons/bank km/yr	Untreated cropland, 16.8 metric tons/ha/yr; treated cropland, 2.8 metric tons/ha/yr; pasture, 2.2 metric tons/ha/yr	Roadbanks along sampled highways	Hilliard, 1977
Deer Creek Watershed, Maryland	Soil loss: 28.19 metric tons/km/yr along roadbanks	Cropland, 18.67 metric tons/ha/yr; pasture, 4.33 metric tons/ha/yr; woodland, 1.01 metric tons/ha/yr	Roadbanks along sampled highways	Stephens *et al.*, 1977
Cleveland County, Central Oklahoma	Mean annual ground retreat: 32.5 mm/yr; assuming a bulk density of 1.6 gm/cc soil loss becomes 520 metric tons/ha/yr	No comparison given	Roadbank incised into soil and bedrock, three profiles with erosion pins	Haigh, 1984
State Highway 209 near Port Carbon, Pennsylvania	Sediment yield: 332.89 metric tons/km^2/yr during construction; during construction about 50% of sediment load was produced by construction	Average annual sediment yield: 197.3–281.9 metric tons/km^2/yr	70.19 km^2 drainage basin undergoing highway construction	Helm, 1978

Table 5 Average Storm Runoff Suspended Sediment Concentration (in mg/liter) from October 1, 1969, to September 30, 1974[a]

		During construction			
Tributary	Prior to construction	Clearing, grubbing	Early earthwork	Late earthwork and final grading	Total construction period
1	45	45	45	45	45
2	45	200	200	200	200
2A	44	55	210	100	170
2B	68	225	820	120[b]	370
3	50	370	370	230	280

[a]Source: Reed, 1978.
[b]From data collected immediately below the onstream pond.

1 with with the amount discharged from other basins during the preconstruction phase of data collection. The effect of various sediment control facilities is apparent in this table.

Based upon all of the data collected during the 3-yr period before construction and the 2-yr period during construction, Reed (1978) found that the normal sediment discharge from these areas was about 59 metric tons (65 tons) per year. The passage of Hurricane Agnes in June 1972 caused 180 metric tons (200 tons)

Table 6 Suspended Sediment Discharge (in Metric Tons) from October 1, 1969, to September 30, 1974[a]

Tributary	Type of sediment controls after initial 3 yr	5 normal yr	Due to highway construction	Due to Hurricane Agnes	Total 5 yr
1	No construction	290.3	—	190.5	480.8
2	Construction—no sediment controls	317.5	392.8	187.8	1007.0[b]
2A	Construction—off-stream ponds	290.3	196.9	183.3	670.5
2B	Construction—on-stream ponds	344.7	614.2	229.5	1188.4
3	Construction—seeding, mulching, rock dams	235.9	377.4	88.9	702.2

[a]Source: Reed, 1978.
[b]Includes 108.9 metric tons that resulted from housing development in the basin in 1970.

of sediment discharge, while construction of the highway over a 2-yr period produced a sediment discharge that ranged from 197 to 614 metric tons (217 to 677 tons) in the four disturbed drainage basins.

Construction of the highway was divided into different phases to determine which phase contributed the largest increase in sediment discharge. Clearing and grubbing increased the sediment loads about 200%. Construction work on a bridge, that carried a two-lane roadway over the highway, resulted in an increased sediment load of about 100%. Early earthwork and culvert construction increased sediment loads about 700%.

The earthmoving was about 50% complete when construction was suspended for the winter. The average increase in sediment load during the winter and early spring period was about 200%. When construction resumed in the late spring, the average increase in sediment load was about 800%.

The highest increases in sediment discharge were observed during the period when areas were being fine graded and prepared for subbase placement. The increases averaged about 4000% for the 3-month period.

Before highway construction began, the particle size distribution of the suspended sediment averaged 7% sand, 51% silt, and 42% clay. Sediment discharged from the area of highway construction averaged 1% sand, 29% silt and 70% clay. There did not appear to be a very significant relationship between the size distribution of the soil used in the construction and the size distribution of the soil transported from the construction area as suspended sediment. This suggests selectivity in sediment transport.

A number of authors (Wolman and Schick, 1967; Diseker and Richardson, 1962; Megahan, 1978) described a distinct seasonality to erosion and sediment yield rates. This, of course, can be tied to seasonalities in the precipitation regime of an area, but could be related to the time of disturbance by construction activities. In some regions, frost action may also loosen soil for transportation by runoff.

Following construction, erosion from roadways often varies with the type of road and the level of maintenance afforded it. Turelle (1973) observed that although an excellent job is done to stabilize and beautify interstate and primary highways, secondary roads remain a problem. Briggs (1973) found that along the 139,983 km (87,000 miles) of Wisconsin roads, there are 21,000 sites that produce significant amounts of sediment. If these sites are placed in one continuous strip, it extends from Madison, Wisconsin, to New York City and then westward to Los Angeles, measuring 5971 km (3711 miles) long and 4.88 m (16 ft) wide. Further, his information shows that about 74% of the sediment-producing sites occurred along town roads, nearly 24% along county roads, and only about 3% on state highways. Similarly, Hafley (1975) commented that rural road systems deserve critical examination because they are generally accorded less sophisticated engineering and inadequate soil stabilization following disturbances by construction and periodic maintenance. Moreover, many kilometers of rural roads remain unsurfaced and, hence, vulnerable to accelerated erosion of the roadbed itself, as well as its banks and ditches.

In recent years, considerable research has been directed toward erosion from lands subjected to logging operations. This is of interest to us because, as stated by Packer (1967), observations and records indicate that most of the sediment from forest lands that reaches stream channels originates on logging roads. He further remarked that chief sources of sediment are roads that disrupt or infringe upon natural drainage channels. Roads that have steep gradients or that lack drainage facilities adequate to prevent swift concentrated overland flows of water from their surfaces are nearly as bad. Patric and Trimble (1972) asserted that it is almost universally recognized that poorly located and poorly managed logging roads are a major cause of forest soil erosion and sediment-laden streams. Finally, Berglund (1976, p. 3) observed:

> Forested areas contribute to geologic and accelerated erosion. The latter is primarily associated with existing roads and road construction activities. In general, timber harvesting is not directly the cause of erosion problems.

Table 2 provides some general information concerning the increase in erosion rates in disturbed forests. Woodcutting and skidding are estimated to increase surface erosion about 1.6 times, whereas the construction of logging roads is estimated to increase erosion about 220 times. Anderson (1975, p. 68) summarized research conducted in Oregon and concluded:

> If we assume that future logging will be as effective as past logging and that projected road development will follow past patterns and eventually occupy six pct [%] of each watershed, we may expect sediment to increase by a factor of four; eighty pct of the increase would be associated with road development; and 20 pct with logging independent of roads.

Reid and Dunne (1984) provided a detailed examination of sediment production from forest road surfaces in the Clearwater Basin, in the Olympic Mountains of Washington State. They found that erosion rates are very sensitive to traffic intensity. A heavily used road segment contributes 130 times as much sediment as an abandoned road. A paved road segment, along which cut slopes and ditches are the only source of sediment, yields less than 1% as much sediment as a heavily used road with a gravel surface. Table 7 contains a compilation of sediment yield data from that reference.

Megahan (1975) summarized the data from studies of logging activities in central Idaho in Table 8. Again, sediment production from roads greatly exceeds that from lands subjected to the felling and skidding of trees.

There are indications that construction activities alter rates of mass movement. Nelson and Martin (1981) commented that a reduction of hillslope stability may result from unloading the toe of the hillslope, loading the top of the hillslope, or the introduction of water onto the hillslope. Construction activities that would

Table 7 Calculated Sediment Yield per Kilometer of Forest Road[a] for Various Road Types and Use Levels[b]

Road type	Sediment yield for 1977 and 1978 (metric tons/km/yr)	Average sediment yield (metric tons/km/yr)
Heavy use	440	500
Temporary nonuse	58	66
Moderate use	36	42
Light use	3.4	3.8
Paved	1.9	2.0
Abandoned	0.43	0.51

[a]Roads are 4 m wide, have an average gradient of 10%, and are drained to an average of six culverts per kilometer. The 16% of the road surface that does not contribute runoff to the inboard ditch is not included in this tabulation.

[b]Source: Reid and Dunne, 1984.

produce such effects will decrease the factor of safety for hillslope movement. Gray and Leiser (1982) remarked that hillslope failures affect and disrupt transportation routes, urban development, mining, and timber harvesting operations; these activities are very often themselves a causative factor in mass movement of hillslopes. The authors also stated that individual hillslope failures are usually not as spectacular as certain other natural hazards such as earthquakes, major floods, and tornadoes; however, hillslope failures are collectively more widespread, and total financial losses from hillslope movements exceeds that of most other geologic hazards. Schuster (1978) estimated that the direct and indirect cost of hillslope failures or mass movements is in excess of 1 billion dollars per year

Table 8 Mean Sediment Production from a Logging Area in Central Idaho[a]

Type of disturbance	Sediment rate ($m^3/km^2/yr$)	Ratio to undisturbed land[b]
Undisturbed[c]	6.20	1.0
Disturbed[c]	942.49	155.2
Tree felling and log skidding[d]	9.74	1.6
Roads (surface erosion)[d]	1335.96	220.0
Roads (mass erosion)[d]	3340.19	550.0

[a]Source: Adapted from Megahan, 1975.

[b]Taken directly from source.

[c]Values expressed per unit area of watershed.

[d]Values expressed per unit area subjected to tree felling and log skidding or to road construction.

for the United States. Krohn and Slosson (1976) estimated the annual landslide damage to buildings and their adjacent property in the United States is about 400 million dollars (measured in 1971 dollars). Chassie and Goughnour (1976) concluded that 100 million dollars would be a conservative estimate of total annual cost of landslide damage to highways and roads in the United States. While these figures are impressive, they do not tell us the extent to which construction activities actually accelerate the magnitude or frequency of mass movement beyond the rates that would occur under natural conditions.

There are far fewer investigations of mass movement on lands disturbed by construction activities than concerning erosion. Cooke's (1984) assessment of geomorphic hazards in Los Angeles, California, demonstrated that at least the frequency of mass movement, if not the magnitude, has been accelerated by construction activities in this locale. A survey of over 3000 landslides in southern California by Leighton (1969, 1972) suggested that about 15% were the result of grading activities and 25–30% were caused by construction activities altogether.

In addition, there are several studies of mass movements associated with road construction in logging areas, and these provide some basis for inference. Berglund (1976) remarked that there are two basic soil loss problems commonly associated with forest roads in Oregon and the first is mass soil movement, landslides, and slump. Krammes and Burns (1973, pp. 5–6) observed for two watersheds in northern California:

> Winter rains in 1967–68 resulted in some mass movement within road prisms, particularly in the steeper cut slopes and from fill areas next to bridge sites. An estimated 500 cubic yards of soil and rock material from cut banks was deposited on the road in 19 separate slides along 2,000 feet of road. Sixteen slides along the fill slopes contributed an estimated 150 cubic yards directly to the stream.

Recall that in Table 8, Megahan (1975) reported mass erosion (mass movement) associated with roads built for logging in central Idaho produces 550 times as much sediment as undisturbed areas. Perhaps the best evidence of changes in mass movement rates due to construction is provided by Swanston and Swanson (1976). Examining data for several areas in the Northwest, they found that debris avalanche erosion associated with roads was 25 to 344 times greater than debris avalanche erosion in unroaded areas. Taken together, there seems to be sufficient evidence to suggest that construction increases mass movement rates considerably.

From all the foregoing, it is clear that construction activities cause dramatic increases in the intensity of geomorphic processes. Both erosion and mass movement rates are accelerated to an extent similar to those of surface mined lands. Further, it seems that the sediment produced often finds its way to nearby stream channels, whereas at mine sites these sediments are often trapped before leaving the area of disturbance due to the concomitant disruption of drainage networks and the installation of sediment ponds.

Duration of Disturbance

The third component in this assessment of disturbance, duration, concerns the length of time that process rates remain accelerated and impacted areas continue to manifest the effects of the disturbance. For construction lands this is generally a function of the length of time that the soil surface remains exposed during the building process and the length of time required for recovery by the natural system to restore a semblance of dynamic equilibrium under prevailing hydrologic and geomorphic conditions.

Numerous factors influence the length of time that a construction site remains "open." Wolman and Schick (1967, p. 460) provided some information in this regard:

> Approximately 50 percent of the sites were open for eight months, and 60 percent were open for nine. Read in reverse, the sample indicated that 40 percent of the sites were open for more than nine months, and 25 percent were open for more than one year. Reports from builders on 49 units constructed in the past five years, comprising about 1000 acres, showed a median completion time of ten months, 75 percent completed in less than twelve months, and two units of less than seven acres each completed in four months. However, field observations also demonstrated that many commercial, school, and industrial sites are open for periods of one, two, or more years.

Several writers comment on the apparent recovery time of systems disturbed by construction activities. Wolman (1967, p. 390) observed:

> Less well documented but suggestive is the evidence that sediment yield several years after completion of urban development is very low, perhaps as low as or lower than sediment yields from completely forested areas.

Wolman and Schick (1967) concluded that progressive urbanization causes an initial rise in the total sediment, soon followed by a steady decline. Guy and Ferguson (1970) stated that, unlike the continuing erosion on agricultural lands not checked by soil erosion practices, construction sites erode mainly during the brief periods between land clearing, shaping, and stabilization of the new surface. Guy (1970) compiled a table indicating that sediment yield is very heavy during urban construction, moderate during stabilization, and low to moderate following stabilization. Accordingly, stream channels rapidly aggrade with some bank erosion during construction, degrading with severe bank erosion during stabilization, and remain relatively stable following stabilization of the disturbed area. The U.S. Water Resources Council (1977, p. 37) stated:

> After the construction has been completed and lawns have been sodded, the overall sediment yield from the urban sites tends to drop below the prior condition. In terms of duration and character, large drainage areas undergoing urbanization tend to be affected by moderate to high levels of sheet erosion almost continuously over a long period of time. The smaller drainage areas—a single construction site, for example—may be severely affected for only a very short period, perhaps a few months.

Data from Stephens *et al.* (1977) showed that for Deer Creek watershed, Mary land, the soil loss from pastureland was 4.33 metric tons/ha/yr (1.93 tons/acre/yr), while that from urbanized areas was only 2.87 metric tons/ha/yr (1.28 tons/acre /yr). Brownlie and Taylor (1981) suggested that sand delivered to the sea by the San Gabriel and Los Angeles rivers in southern California may have been about 600,000 m^3/yr under natural conditions, but is now about 200,000 m^3/yr due to construction and river control. Wolman (1967) summarized the temporal pattern of sediment yield from construction lands in Figure 5.

Of course, the impact of disturbance by construction activities includes more than sediment yield. Guy and Ferguson (1970) noted that the sediment releases may have a lasting effect on channel shape as well as the aquatic environments downstream. Wolman and Schick (1967) cited evidence that highway and railroad construction had virtually eliminated trout from 126 km (78 miles) of stream channel in central Montana, owing to sedimentation associated with channel straightening, land clearing, and general construction. Similarly, on a tributary of Clark Fork of the Columbia River, successive studies of the trout population in a reach of river altered by highway construction show a 94% reduction in both number and weight of large size game fish in a period of 2 yr. But Wolman and Schick (1967) observed that the upper portion of Beaverdam Run, traversed by construction of a major interstate highway 7 yr prior to their study, was entirely free of sediment derived from the construction, indicating that a channel of this gradient and flow may be cleared of sediment in a period of 7 yr or less.

Other investigations have focused directly upon the duration of impacts caused by road and highway construction. Vice *et al.* (1969) found that for Scotts Run basin, Virginia, sediment concentrations were as high as 20,000 mg/liter during the period of active highway construction (1961–1962), but then decreased

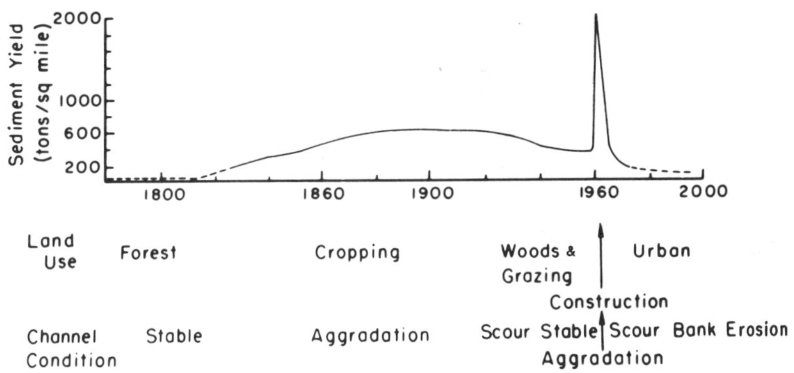

Figure 5. Changes in sediment yield through time resulting from construction activities (Wolman, 1967).

sharply, as depicted in Figure 6. Bethlahmy and Kidd (1966) reported erosion rates for steep fill hillslopes near Boise, Idaho, and concluded that most of the soil loss occurred during the first 3 months following construction and thereafter the rate of erosion declined markedly. At a roadbank in central Oklahoma, Haigh (1984) measured an average *decrease* in the depth of gully incision of about 17 mm/yr (0.67 in./yr) and noted that at this "rate of healing" it could be completely eliminated in approximately 25 yr or within 45 yr of its origin.

Additional information is available for road construction associated with logging activities in forested areas. Leaf (1974) reported that on Fool Creek, in the Fraser Experimental Forest of Colorado, road construction resulted in minimum erosion damage with apparently no reduction in water quality; the 5.2 km (3.3 miles) of main access roads are carefully located to avoid the stream channel and to minimize erosion. For the Caspar Creek watersheds in northern California, Krammes and Burns (1973, p. 1) stated:

> Generally, the immediate impact of right-of-way clearing, road building, and bridge construction was best reflected in the sediment yield. Turbidity levels were high during the road building period, but did not extend far downstream nor persist for extended periods during the summer. Suspended sediment yields the first winter were more than four times the preconstruction levels; subsequent winter levels, while still above preconstruction, were not excessive.

Patric (1977) commented that, as in the case of abandoned skid paths, with a lack of traffic on logging roads, erosion will often halt after 1 to 2 yr, even on those

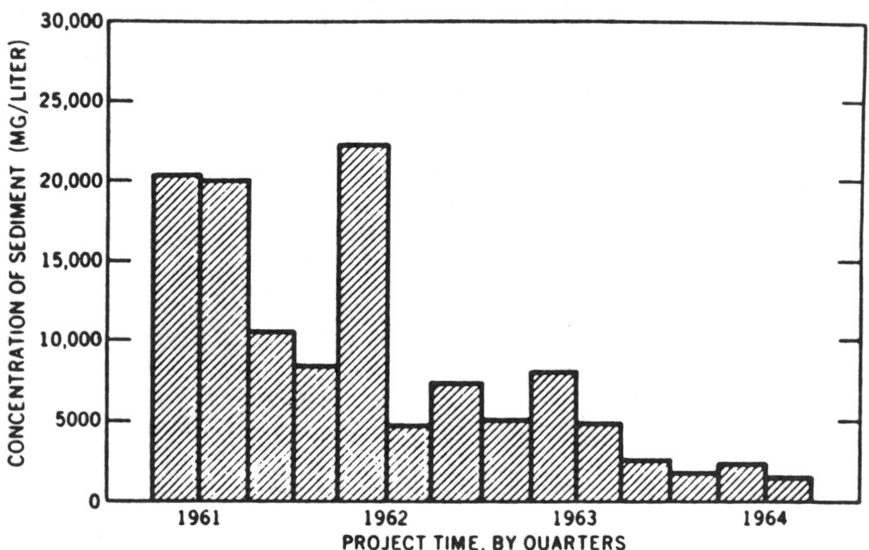

Figure 6. Variation in sediment concentration from storm runoff, 1961–1964 (Vice *et al.*, 1969).

that are poorly designed. In the humid eastern United States, natural revegetation by grass, weeds, shrubs, and trees begins soon after roads are abandoned and the litter cover is restored by the annual leaf fall. Elsewhere, Patric and Trimble (1972) remarked that the rate of return to pretreatment hydrologic conditions varies with the severity of treatment; less than 1 yr is needed after light cutting and careful road management. But forest revegetation is so prompt and vigorous in the central Appalachians that little more than 5 yr is needed for hydrologic recovery on even the most heavily cut and carelessly roaded watershed. Beschta (1978) described the effects of road construction and logging in two watersheds in Oregon. Sediment yields returned to pretreatment levels within 1 to 2 yr after mass failures so that these mass movements did not cause any long-term shifts in the sediment rating curves for Deer Creek. Increased sediment yields from Needle Branch returned to pretreatment levels by the 1972 water year or approximately 6 yr after road construction.

From the above, it appears that the duration of impact caused by construction activities lasts for about 1 to 7 yr, depending upon various site-specific factors including the length of time that the site is open, the effort expended in site stabilization, and the inherent characteristics of the hydrologic and geomorphic system operating in that particular locale. The land surface itself can probably be stabilized in timely fashion. Nearby streams, however, will likely experience a recovery lagtime, perhaps extending over several years, as the sediment produced by the construction activities is flushed from the system and the channel adjusts its gradient and geometry to accommodate the discharge and sediment regime of the new land surface. Frequently, the soil and vegetation properties of the disturbed land remain altered permanently or at least for a very long time.

The Nature of Disturbance

To fully understand the geomorphic consequences of construction activities, it is useful to elaborate on the processes by which the natural system is modified. Although the following is fairly typical of many construction projects, it must be recognized that each undertaking is somewhat unique in detail. The construction of edifices and roads or highways is considered concurrently because of the broad similarities in operations and impact.

Site Preparation

As mentioned earlier, there are three basic phases of most construction projects: (1) site preparation, (2) emplacement of structures, and (3) site stabilization. The first phase begins with clearing and grubbing of the surface. Any natural or manmade impediments to site access or traffic are removed, usually through the use of heavy equipment. The natural vegetation is disturbed and the underlying soil compacted during this process.

Next, topsoil may be stripped and stockpiled as required by contract. Because topsoil is a valuable and essentially nonrenewable resource, this step seems logical. However, it bears considerable cost and so is often by-passed.

The construction surface is now graded to the desired topographic configuration. A series of cuts and fills change the morphology of hillslopes and stream channels. Frequently, the objective is a relatively level building site or roadbed. Wolman and Schick (1967) found that when building sites are denuded for construction, the soil is piled nearby without cover or protection. Similarly, the Highway Research Board (1973) remarked that cuts and fills, with their exposure of bare ground, must be expected as part of normal highway construction. Machinery tracks up and down hillslopes of either unconsolidated or residual materials provide locations susceptible to rill and gully development.

It is also necessary many times to obtain additional earth materials from nearby "borrow" areas in order to create the specified surface shape. This, too, contributes to the total spatial extent of disturbance.

Stream channels may be diverted around the site, engineered through it, or placed in structures beneath it. Dunne and Leopold (1978, p. 694) stated:

> In urban development the smallest rills, swales, or incipient channels tend to be looked upon by the construction crew as being insignificant, unimportant, or a nuisance. Many a house is built over a buried pipe intended to carry runoff toward some obvious channel; so, in fact, first-order channels, ordinarily ephemeral, are either eliminated by grading or put in a pipe or conduit. The net result is the virtual expurgation of many first- or even second-order channels that under natural conditions played a role in keeping both sediment and runoff distributed or divided among many small channels, each of which played its part in delaying movement of flood peaks, providing channel storage and slowing the average speed at which water was delivered to the larger stream channels.

At least during this phase of construction, drainage density and stream frequency are reduced in the area of grading.

Thus, in addition to the piles of unconsolidated earth materials, the construction surface consists, in various proportions, of soil compacted by equipment traffic, illuviated soil horizons, partially weathered parent material of soils, or geologic strata laid bare by grading. In most cases, these materials possess an infiltration capacity substantially lower than the vegetation and topsoil cover that existed prior to disturbance.

Excavations into hillsides may intersect subsurface flows of water. Megahan (1972) commented that this is one of the more insidious effects of road construction because its occurrence often is not readily apparent. This water flows in road ditches, on road treads, on road cut and fill hillslopes, and on undisturbed slopes below a road and commonly causes considerable surface erosion. He noted that excess soil water raises pore pressures which reduce the shear strength of the soil and may contribute to mass movement above roadcuts, in road fills and underlying soils, and in soils below the road. Of course, these flows add to any surface runoff or streamflow that occurs.

Emplacement of Structures

The second phase of construction projects involves the emplacement of structures. For commercial or residential areas this includes the erection of buildings, parking lots, streets, sidewalks, storm sewers, drainage ditches, channels, and so forth. For roads or highways, this includes roadbeds, paved surfaces, bridges, culverts, and overpasses. Taken together, a considerable part of the construction site is rendered impervious by roofing, concrete, and asphalt.

In an analysis of the hydrologic effects of urbanization, Leopold (1968) observed an inverse relationship between residential lot size and the resulting percentage of the area made impervious. The current trend toward suburban development in the form of townhouses and condominiums, with the accompanying shopping mall, greatly reduces the surface area capable of natural infiltration rates. Hence, a given precipitation event generates much more runoff.

Toward the end of this construction phase, there exists a highly vulnerable, exposed soil surface together with large areas producing overland and channel flow. From the geomorphic perspective, resistance to erosion is low while erosive forces are relatively high. This explains the high rates of sediment yield previously discussed.

Site Stabilization

Under conditions described earlier, geomorphic processes are capable of substantial environmental degradation, both on- and off-site, and may even jeopardize the integrity and function of the structures erected. Thus, site stabilization becomes a necessary third phase and, as suggested earlier, this is regarded as a type of land reclamation. For large projects that progress and reach completion in stages, stabilization may accompany the second phase for parts of the construction area. Clearly, stabilization should be achieved in timely fashion.

Responsibility for this phase often depends upon the type of project. In the case of road or highway construction, the stabilization phase with detailed specifications can be included in the contract. In the case of residential construction, however, most of the landscaping is left to the eventual property owner. As a result, the exposed soil surface can persist for several months after the homes

Figure 7. Fine-graded residential lots in Greenriver, Wyoming. (Courtesy D. Longbrake.)

have been completed. Both economic and climate conditions influence the rate at which these surfaces are treated. Figure 7 shows a residential area in Wyoming where the land has been graded, but no lawns have been established as yet.

Land Disturbance and Geomorphic Principles

The foregoing demonstrates that significant land areas have been, are being, and will continue to be dramatically altered through various construction activities. The geomorphologist recognizes these impacted sites as parts of a nested hierarchy of drainage basins, composed of hillslopes and channels in orderly arrangement, that probably attain and maintain some semblance of an equilibrium prior to the disturbance.

The construction operation creates a disequilibrium among the characteristics or attributes of the natural system; then the system responds, through numerous geomorphic processes, to restore the balance. Viewed from this perspective, geomorphic principles apply to the area before, during, and after disturbance. Although the rates at which geomorphic processes function change through time, and with the three construction phases, they must still be controlled by the same physical and chemical laws manifested in established geomorphic principles.

Also from this vantage point, it is possible to explain and understand the results of these disturbances. But more importantly, we may be able to infer possible futures and design stabilization programs that facilitate the reestablishment of an equilibrium in a minimum amount of time. The response of hillslopes is considered first, followed by that of channels and of the entire drainage basin.

Hillslopes

The land surface of construction sites is composed of hillslopes with intervening channels. The hillslopes account for 90% or more of the surface area. While grading operations modify hillslope forms and positions, the regional relief remains similar to the preconstruction condition. So, if the objective of grading is to create relatively flat surfaces to facilitate construction in one part of this region, then it is likely to produce relatively steep hillslopes in other parts. These steepened hillslopes, frequently composed of unconsolidated material and lacking a vegetation cover, are potential locations for accelerated erosion and sediment production during rainfall events or windstorms. Additionally, the increase in hillslope gradient along with the removal of material from the toe of the hillslope through excavation may reduce the mass stability of the hillslope sufficiently to permit failure when triggered by precipitation or, perhaps, the blasting associated with construction activities. The diversion or placement of a stream or drainage ditch so that it actively erodes the base of the hillslope would, of course, have the same result as removal of the toe through excavation.

Manmade hillslopes are of two types: (1) cut hillslopes and (2) fill hillslopes. As the terms imply, cut hillslopes are created by excavation and often consist of exposed rock strata. Figure 8 shows a hillslope cut into competent rock along a highway in Wyoming. Fill hillslopes are created through the deposition of unconsolidated materials, frequently compacted to some specified density. Highway embankments are a type of fill hillslope; their function is to provide support for a pavement system above natural ground levels (Highway Research Board, 1971).

Erosion rates should be very different between cut and fill hillslopes in most environments. Erosion is likely to be "weather limited" on the cut hillslope and "transport limited" on the fill hillslope. Except in regions of extreme climatic conditions, the fill hillslope is likely to erode much faster than the cut hillslope and produce several times as much sediment. In very dry or cold regions, sediment production may be rather similar for both, although this remains supposition at present.

Nelson and Martin (1981, p. 134) examined hillslope stability as a part of stream-related hazards for highway bridges. They remarked in this report:

> It has been noted previously that reduction of slope stability may result from unloading the toe of the slope, loading the top of the slope, or the introduction of water into the slope. Construction activities that would produce such effects will decrease the factor of safety against slope movement. It is not intended to imply that any construction effort will cause failure, but if such

Figure 8. Nearly vertical roadcut in competent rock at Glendo, Wyoming.

effort is proposed, the slope must be analyzed to determine if the resulting factor of safety is adequate.

In this comprehensive treatise, the authors discussed the mechanics, types, causes, and consequences of hillslope failure, together with several actions that might be taken to reduce the probability of mass movement.

Megahan (1972) mentioned that excavations bringing groundwater flows to the surface, with subsequent infiltration of water elsewhere, could cause embankment failure. This is one of the problems encountered along the contact between cut, fill, and embankment hillslopes. The Highway Research Board (1971) commented that benching and adequate drainage provisions are the most widely accepted solutions for correcting stability problems on both sidehill fill and cut-to-fill transitions. At any construction site, seepage from water impoundments or drainage facilities increases the water content of underlying materials and affects the factor of safety accordingly. Here, one might speculate as to the consequences of lawn irrigation on hillslope stability.

Nelson and Martin (1981) observed that "industrialization and construction of structures on the tops of hillslopes will load the mass, thereby decreasing stability." The water contained within an impoundment or tower also constitutes a loading, of course.

There are numerous studies concerning the hillslope geomorphology of highway construction sites. The results of these investigations are likely transferable to similar situations at building sites. Wolman and Schick (1967) commented that soil loss per square mile from the roadside cuts in the Piedmont of Georgia (i.e., for roughly the same rock type as in Maryland but with slightly higher rainfall) is of the same order of magnitude as soil loss for the small catchments under construction, as included here in Table 3. The quantities of sediment derived from roadside areas are 17,514 metric tons/km^2/yr to 52,541 metric tons/km^2/yr (50,000 to 150,000 tons/sq mile/yr). Although much of the available data is summarized in Table 4, a few reports require further discussion. From the examination of roadbank erosion in central Oklahoma, Haigh (1984) found that total annual precipitation is a good predictor of average annual soil loss, according to the following equation:

$$E = 0.29(P - 90) \qquad (r = 0.98) \tag{1}$$

where E is the average annual soil loss (in millimeters) and P the total annual precipitation (also in millimeters). However, this equation apparently must be used with care in other areas because even modest annual precipitation totals result in high soil loss values.

Diseker and Richardson (1962) examined erosion rates on roadbanks in Georgia and found that soil loss is a function of rainfall, hillslope gradient, and aspect. They remarked that the factor having the most clear-cut effect on erosion is the aspect of the plot. The 3-yr average annual rate for the northwest-facing plots was about 459.5 metric tons/ha (205 tons/acre), whereas that for the southeast-facing plots was about 204.0 metric tons/ha (91 tons/acre). It was observed that banks facing northwest frequently froze, stayed frozen longer, and did not dry out as quickly as those facing southeast. When the northwest-facing banks did thaw, the wet soil rolled down the banks. In addition, the northwest-facing banks received precipitation more nearly at right angles than the southeast-facing banks because much of the rain comes from a westerly direction. Thus, raindrop impact is greater on the northwest-facing hillslopes. Interestingly, Haigh (1984) did not find the frequency of freeze–thaw to be significantly related to soil loss. The pattern of erosion described above is different than presented by Hadley (1962) for semiarid regions, where south-facing slopes experienced higher erosion rates.

Five years of data from six roadbank plots near Cartersville, Georgia, are used by Diseker and McGinnis (1967) to identify the factors that significantly influence sediment production and yield. The authors expressed general models thusly:

$$\hat{Y} = \gamma + A_i + B_j + C_{ij} + b_1EI + b_2T_i + \epsilon \qquad (R^2 = 0.861) \tag{2}$$

where: $\hat{Y}$ is the predicted annual sediment yield in tons per acre, γ the overall mean, A_i the constant for slope (gradient), B_j the constant for aspect, C_{ij} the constant for slope $\times$ aspect, EI the coefficient for annual rainfall erosion index as a covariant (b_1 = 0.0024), T_i the coefficient for annual average temperature below 32°F (b_2 = 0.8687), and ϵ the random error.

$$\hat{Y} = \gamma + A_i + B_j + C_r + D_{jr} + b_1EI + b_2M + \epsilon \qquad (R^2 = 0.579) \qquad (3)$$

where $\hat{Y}$ is the predicted monthly sediment yield for month A in tons per acre, γ the overall mean, A_i the constant for particular month, B_j the constant for slope (gradient), C_r the constant for aspect, D_{jr} the constant for slope × aspect, EI the coefficient for monthly rainfall erosion index (b = 0.0026), M the coefficient for soil moisture index (b = 2.3114), and ϵ the random error. The term EI is the Wischmeier and Smith (1958) rainfall erosion index and M is computed from the following relation:

$$M = \frac{P - r}{E_s} \qquad (4)$$

where M is the soil moisture, P the precipitation, r the runoff, E_s the potential evaporation.

In recognition of the fact that most sediment from forest lands that reaches stream channels originates on logging roads, Packer (1967) developed equations for estimating erosion on road surfaces and the distance that sediment moves downslope from those roads. Although full discussion of these models extends somewhat beyond our scope, both erosion and transport distance are functions of road and watershed characteristics, some of which were controllable through road design and others that could not be readily manipulated.

Finally, the form of hillslopes created by construction activities deserves comment because form affects rates of erosion. Meyer and Romkens (1976) asserted that a convex hillslope is more erodible than a uniform (straight) hillslope, and a uniform hillslope yields more sediment than a concave one. Further, a complex hillslope (convex–concave) generally yields less sediment than a uniform hillslope because the flat area at the base of the hillslope tends to reduce sediment yield. Often, space limitations resulting from property boundaries along the rights-of-way of roads dictate convex or straight hillslope profiles that can be expected to experience relatively high erosion rates. Maintenance of drainage ditches at the base of roadcuts often preserves the convex or straight profile. The question then becomes, Is the added cost of maintenance due to erosion over the years greater than the cost of acquiring sufficient land to construct concave or, at least, complex hillslope profiles?

Previously it was observed that valley head hillslopes (hollows) tend to concentrate flow at their base while spur end hillslopes (nose slopes) tend to disperse water. As a consequence, greater erosion, and perhaps even gully development, can be expected at the base of the valley head hillslope. Clearly, this general configuration is to be avoided if possible, although this may be difficult on the inside bend of a roadcut or the outside bend of a fill embankment. Here, the concave profile is especially desirable.

Channels

There is considerable research addressing the effects of construction on channels. Stated succinctly, there is a greater volume of runoff moving faster through the urban area than through adjacent, undisturbed areas. We also observe that erosion rates vary with the phases of construction activity. It is to these new and variable conditions of water and sediment discharge that the streams must adjust. Wolman (1967, pp. 255–256) described the applicable geomorphic theory thusly:

> If a set of river channels are in equilibrium with the prevailing conditions in a drainage basin, it follows that major disturbances on the drainage basin will result in changes in channel form and behavior. The process of urbanization of the landscape constitutes a major interruption of "prevailing" conditions on a watershed. If prior to urban development a kind of equilibrium prevailed in which channel gradient and form were related to water and sediment derived from the watershed, then the sequential changes which occur as urban development takes place on the watershed can be expected to alter markedly the equilibrium forms and may result in the eventual establishment of new conditions of equilibrium. At the present time the disturbance of equilibrium can be documented; but as the data here will show, it is currently difficult to determine whether a new equilibrium will be established or whether instead a condition of disequilibrium will persist.

Today, most geomorphologists probably agree that adjustment of the streams is toward the re-creation of an equilibrium, although the time required to achieve this goal is very difficult to estimate.

Leopold (1968) suggested that there are four interrelated but separable effects of land use changes on the hydrology of an area: (1) changes in total runoff, (2) changes in peak flow characteristics, (3) changes in water quality, and (4) changes in the hydrologic amenities. The hydrologic amenities, as defined by Leopold (1968), refer to the appearance or the impression that the river, its channel, and its valley leaves with the observer. He also commented that, of all land use changes affecting the hydrology of an area, urbanization is by far the most forceful. The first three of these listed effects comprise the general framework for the ensuing discussion.

The increase in total runoff or discharge caused by urbanization depends upon the infiltration capacity of the surface prior to disturbance and the proportion of the area made impervious following construction and stabilization of the land. Cohen *et al.* (1968) found that annual runoff volumes increase by up to 250% after residential development at East Meadow Brook on Long Island, New York. Yorke and Davis (1972) found that storm runoff from the Bel Pre Creek basin in Maryland is about 30% higher during a 30-month period of active construction than the expected amount based on the original grass and forest cover.

As a result of the increased efficiency of discharge transport through the urban drainage system, lag time for the flow decreases and the peak flow increases. As described by Dunne and Leopold (1978) and cited earlier, stormwaters can accumulate downstream more quickly than in natural river systems and produce higher flood peaks; even if the total volume and peak rate of runoff from the land surface are not increased by urbanization, the increased velocities in the channels would still decrease the lag between rainfall and runoff. Figure 9 shows the change in lag time for a standard storm producing 1 in. of runoff.

Numerous studies demonstrate the decrease in lag time that commonly accompanies urbanization. Research by Carter (1961) and Anderson (1970) showed that lag time is a function of basin properties (length and slope) and the extent of sewer installation. Figure 10 shows these relationships. Similarly, the reduction in lagtime is evident in the data of Cohen *et al.* (1968) for the Long Island study.

The increase in total runoff and the decrease in lagtime cause an increase in peak discharge. Leopold (1968) summarized the results of seven studies in Figure 11. The ratios of mean annual floods for urbanized to rural areas are

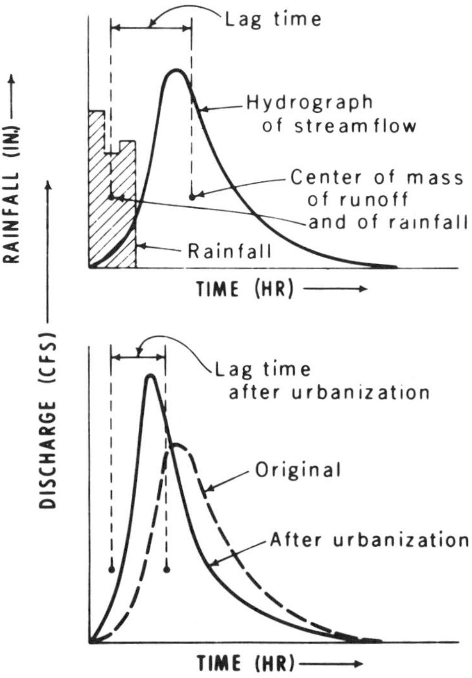

Figure 9. Hypothetical unit hydrograph showing effect of urbanization on discharge (Leopold, 1968).

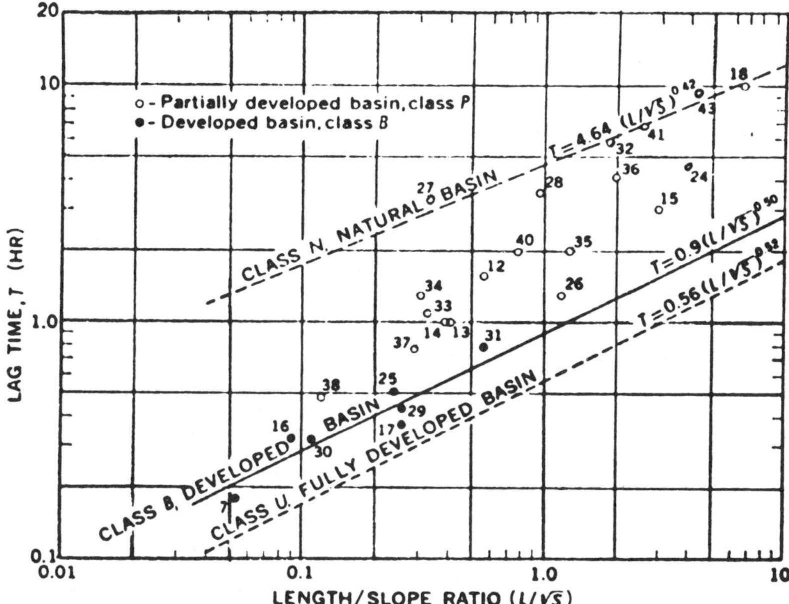

Figure 10. Relation between lag time and the slope and length of drainage basins for various levels of storm sewering (logarithmic scale) (Anderson, 1970).

presented for various percentages of sewered and impervious area. These curves show that for unsewered areas, peak discharge increases about 2.5 times as the impervious area changes from 0 to 100%. For areas that are 100% sewered, peak discharge increases from 1.7 to 8 times as the impervious area changes from 0 to 100%. Following this direction, Rantz (1971) provided a series of such curves for various other flood recurrence intervals.

From the work of Carter (1961) and Anderson (1970), it was possible to estimate discharge peaks based upon rainfall intensity and the extent of impervious area. The resulting equations are offered by Dunne and Leopold (1978):

$$Q_{pk} \text{ (after urbanization)} = I\left[\frac{0.30(100 - Z)}{100} + \frac{0.75(Z)}{100}\right] \quad (5)$$

which simplifies to

$$Q_{pk} = I(0.30 + 0.0045Z)$$

where Q_{pk} is the peak discharge, I the rainfall intensity of a storm, Z the percentage of impervious area in an urbanized basin. Similarly, the change of flood peak due to impervious area alone would be expressed by:

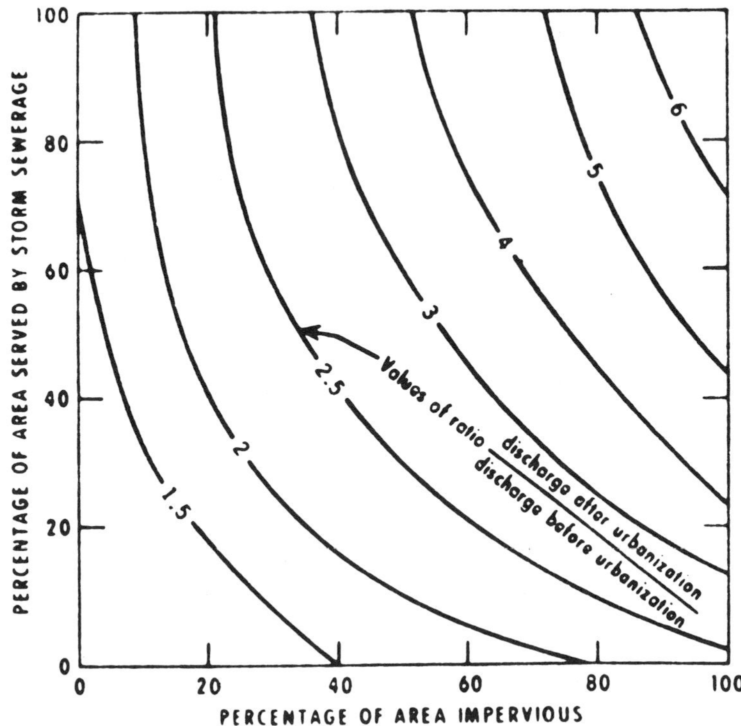

Figure 11. Effect of urbanization on the mean annual flood for a 1-sq-mile drainage basin (Leopold, 1968).

$$\frac{Q_{pk}\ \text{(after urbanization)}}{Q_{pk}\ \text{(before urbanization)}} = \frac{0.30 + 0.0045Z}{0.30} \qquad (6)$$

Anderson (1970) computed flood frequency curves for rural and completely impervious basins in the Washington, D.C., area and then estimated the curves for other proportions of impervious area. Figure 12 depicts the resulting suite of curves and again reveals the effect of impervious area on flood peaks. Here, the magnitude of a flood with a desired recurrence interval is determined as a multiple of the magnitude of the mean annual flood for any given percentage of impervious area in the drainage basin.

If urbanization changes the total discharge, lag time, and peak runoff from a basin, then it could be expected that it would also change the frequency of flooding. Precipitation events within the natural, undisturbed basins that produced low to moderate volumes of runoff, easily handled within the confines of stream banks, may generate sufficient flow following urbanization to exceed the capacity of the drainage network. A precipitation event previously harmless in magnitude is rendered hazardous due to the changes in the surface and drainage system of the basin.

Again using data from several sources, Leopold (1968) derived a relation between the extent of urbanization and the increase in the number of floods, as shown in Figure 13. For example, with an area 50% sewered and 50% impervious, the number of flows equal to or exceeding bankfull channel capacity would, over a period of years, increase nearly fourfold. Dunne and Leopold (1978) cited the case of Rock Creek, draining much of urbanized Washington, D.C., where the number of times each year the discharge reaches or exceeds bankfull has increased 10 to 20 times over its previous frequency.

Increases in runoff volumes and velocities, together with the exposure of unconsolidated soils and geologic materials, result in the extraordinary erosion and sediment yield rates mentioned earlier in this chapter. Walling and Gregory (1970) found up to a tenfold increase in suspended sediment concentrations during construction in a watershed near Exeter, England. Wolman (1967) cited research undertaken near Baltimore, Maryland, by Brosky (1966) wherein the suspended solids from a drainage basin without construction activity are almost exclusively granular and lack the clays characteristic of suspended sediments coming from areas undergoing construction. Hence, land disturbances alter the yield, concentration, and particle size distribution of sediments carried through

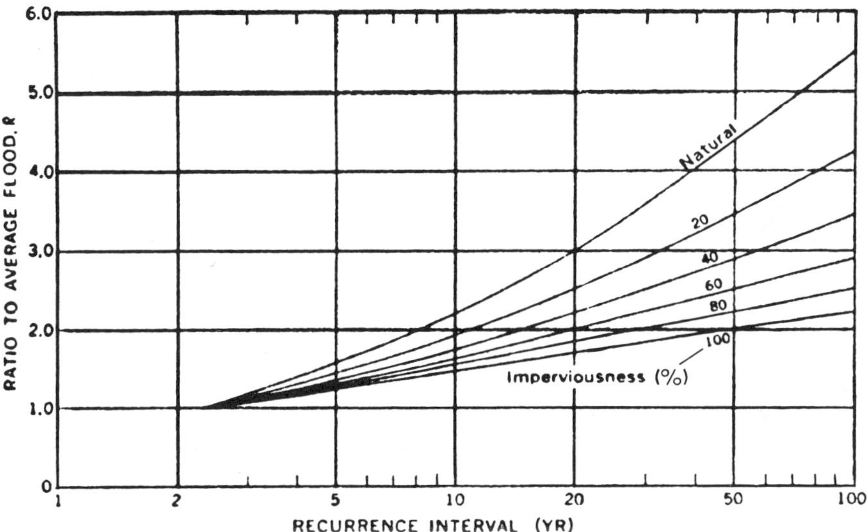

Figure 12. Regional flood frequency curves for drainage basins with various percentages of impervious area near Washington, D.C. (Anderson, 1970).

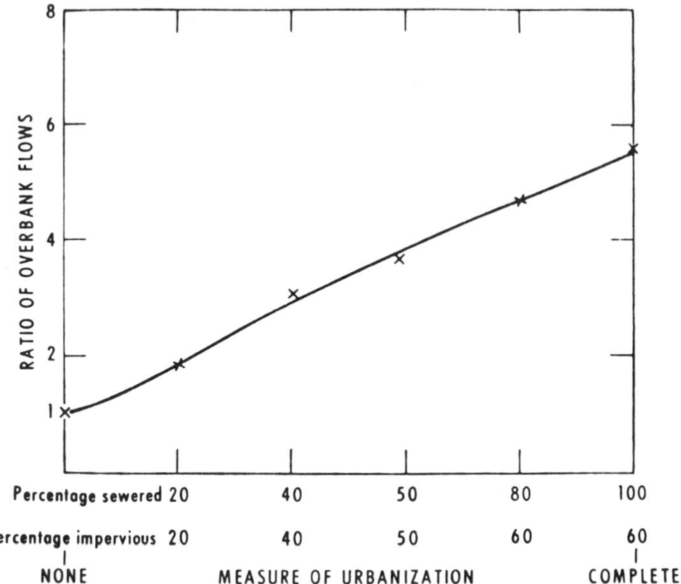

Figure 13. Increase in the number of flows per year equal to or exceeding original channel capacity as a ratio to the number of overbank flows before urbanization for different levels of urbanization (Leopold, 1968).

or deposited within the drainage system. It should be noted as well that the decrease in surface infiltration capacity that causes the increase in runoff and stream discharge also results in decreased groundwater recharge and consequently decreased baseflow of perennial streams. Low flows are reduced accordingly, and this may lead to a lowering of water quality during dry periods (Dunne and Leopold, 1978).

The physical properties of stream channels are largely a function of the water and sediment that pass through them. It follows that changes in these variables due to disturbance brings forth a change in channel morphology. The general nature of change, according to Schumm (1977), is presented in an earlier chapter.

The character of change also varies with the different phases of construction. During the active construction phase, the channel is encumbered with very large volumes of sediment; deposition within the channel is highly probable. Wolman and Schick (1967) stated that the imposition of large quantities of sediment in streams previously carrying relatively small quantities of primarily suspended material produces channel bars, erosion of channel banks as a result of deposition within the channel, obstruction of flow and increased flooding, shifting configuration of the channel bottom, blanketing of bottom-dwelling flora and fauna, alteration of the flora and fauna due to changes in light transmission and abrasive effects of sediment, and alteration of species of fish due to changes produced in the flora and fauna upon which fish depend. Graf (1976) observed an increase of floodplain width due to vertical accretion of alluvium during active suburbanization of basins near Denver, Colorado.

Following site stabilization, runoff rates remain high while sediment load generally decreases. Channel enlargement is often the result as described by Wolman (1967), Leopold (1968), Graf (1976), and Dunne and Leopold (1978). These latter authors comment that, although this enlargement may lag in time behind the changes of basin surface that cause it, the implication is that for small drainage areas the increase in flood peaks overcompensates for concurrent increases in sediment yield. This conclusion follows from the fact that enlargement means more channel erosion than deposition.

Leopold (1968) provided a series of computations suggesting that for a stream similar to Brandywine Creek in Pennsylvania, changes in flow regime could cause the channel to deepen by about 50% and increase in width to a little less than twice its original size. If such erosion takes place through at least 0.40 km (0.25 mile) of channel length in a drainage basin of 2.59 km^2 (1 sq mile), the amount of sediment produced by this enlargement would be 1416 m^3 (50,000 cu ft). At 1603 kg/m^3 (100 lb/cu ft), this amounts to 2270 metric tons (2,500 tons).

Not infrequently the hydrologic and geomorphic responses of channels to watershed urbanization pose significant threats to health, safety, and welfare of residents and damage to their property. Many times there ensues a sequence of steps to control the behavior of streams passing through populated areas. A dam may be constructed upstream to reduce flood hazards. The stream itself may be channelized (Emerson, 1971; Schumm *et al.*, 1984) or canalized (Wolman, 1967; Graf, 1976). Shen *et al.* (1981) and Schumm *et al.* (1984) provided extensive discussion of this subject. Drastic modification of the stream again changes the flow of water and sediment through the fluvial system and creates another round of readjustment in any reaches that are not completely enshrouded in erosion-resistant materials, such as riprap or concrete. Figure 14 shows a modified channel in Colorado. The urbanized stream channel becomes a battlefield between man and nature. Dunne and Leopold (1978, pp. 701–702) remarked:

> Another qualitative observation sufficiently general to be worthy of mention concerns the bank erosion immediately downstream from the end of a concrete urban channel. Where the trapezoidal or rectangular artificial section ends and a natural channel begins, the banks of the latter tend to erode back in bulbous, caving reentrants, making the channel exceptionally wide for a short distance. The danger in this local bank recession is that the lower end of the concreted section will be undermined laterally or at the toe. Of course, the original purpose of making a concrete channel is to decrease the flooding and bank erosion caused by urbanization, increasing the hydraulic efficiency by straightening and thus providing a smooth artificial channel. This speeds the runoff, but also increases the peak discharge, often necessitating further

Figure 14. Channelized stream in Denver, Colorado. (Courtesy J. Costa.)

downstream extension of the artificial channel section. And so each action creates a need for further construction and more concrete.

Drainage Basins

The relationship between geomorphology and construction activities in drainage basins is nicely described by Roberts (1972, pp. 389–390):

> The transition of a watershed from a rural to an urbanized one is accomplished by changes in land use and geomorphology, and from changes in these two parameters virtually all other watershed modifications can be traced. Geomorphic changes center around the alterations to the watershed geometry, surficial cover and drainage net by the urbanizing process. The geometric transformation of the watershed comes about by the grading practices which are used to develop roads and building sites. Slopes are changed both in magnitude and spatial position within the watershed with the result that patterns of overland flow are completely different after grading. First and second order stream basins are often completely removed by a combination of cuts and fills.

Roberts (1972) found that the disturbance of natural drainage basins does not always lower the drainage density of the urbanizing watershed because another network is formed by newly constructed streets and driveways. Using data from the Jackson Creek watershed near Bloomington, Indiana, he computed a drainage density of 4.8 before urbanization and 11.6 afterward. Dunne and Leopold (1978), however, suggested that in many cases there is a decrease in drainage density or number of actual channels to carry the water discharge and sediment load. The change in drainage density, then, is a function of the specific type of development, industrial or residential, and whether or not one is disposed to include both natural and manmade conduits in the computations.

Nevertheless, an alteration of drainage pattern is highly probable. The example provided by Roberts (1972) showed a dendritic pattern prior to urbanization and an anthropically produced rectangular pattern following the installation of storm sewers. In most areas, the use of a geographic coordinate system to arrange lots, streets, and driveways serves to reinforce the rectangularity of the postconstruction drainage patterns.

There are several models for estimating erosion and sediment yield at construction sites. Holberger and Truett (1976) modified the Universal Soil Loss Equation through the addition of two sediment-loading functions. One function uses the distance from the foot of the exposed area to the nearest perennial stream, while the second uses the percentage of the drainage basin undergoing construction. Comparison of predicted sediment yields with observed yields indicates that, for the first loading function, about 54% of the predictions fall within a range of ±50% of observed values. Approximately 30% of the predictions generated by the second function fall within ±50% of observed values. Despite these seemingly large errors, the authors concluded that loading functions provide the best available means for predicting sediment yield from a construction site, in the absence of specific data for that site.

Younkin (1973) examined erosion and sediment yields during construction of an interstate highway through White Deer Creek in central Pennsylvania. Based upon 86 sets of data, he found the following predictive equation:

$$Q_s = \frac{0.034R^{1.5}(\log A)^{2.45}(3.0)^{D}}{P^{0.72}} \tag{7}$$

where Q_s is the suspended sediment yield at a station (in tons), R the rainfall factor (after Wischmeier and Smith, 1958), A the area exposed to erosion by the clearing and grubbing phase of construction, D the depth (height) of cuts and fills, P the proximity factor, as the ratio of surface area between the upslope side of construction and the stream to the total area A. This equation has a standard error of 24%. Of the four independent variables included in the equation, the rainfall factor has the greatest effect on sediment yield. The three terms in the numerator provide an estimate of the quantity of sediment leaving the construction area. This quantity is then modified by the conditions that affect the overland transport process and are represented by the proximity variable in the denominator.

Extensive research provides general principles that permit anticipation of drainage basin, hillslope, and stream channel responses to the disturbances accompanying construction activities. Although the estimating equations or models usually require calibration for the region in which they are to be utilized, the appropriate logic and methodology are useful. As with other types of land disturbance, we expect subsequent, off-site transmission of impacts unless direct action is taken to control them.

Impact Mitigation and the Geomorphic Process–Response System

The goals and practices for reclamation of lands disturbed by various human activities are discussed elsewhere in this book. However, the term *reclamation* is less commonly used in relation to lands disturbed by construction enterprises, probably because the structures erected are intended to be permanent and, hence, returning the site to a condition similar to that which prevailed before construction is not desirable or feasible. Instead, attempts to control and mitigate the environmental impacts resulting from construction activities are referred to as "site stabilization" or more specifically "erosion and sedimentation control." Nevertheless, fundamental objectives remain to minimize the areal extent, intensity, and duration of impact.

There are numerous practices common to both reclamation and stabilization. Likewise, there are many stabilization procedures employed both by those engaged in the construction of edifices and roads or highways. From the geomorphic perspective, it is understood that the forces and resistances, together with the physical and chemical laws governing each, are virtually the same in these cases and so it is not surprising that the practices should be shared.

First, however, it should be recognized that contractors are usually required to function within the legal frameworks of federal, state, or local regulatory authorities. Thus, it is useful to briefly review some of the laws and regulations that affect construction industries.

Legal Controls for Lands Disturbed by Construction Activities

Concise and definitive discussion of this topic is difficult because there is no specific federal enactment that provides umbrella legislation, as the Surface Mining Control and Reclamation Act of 1977 (U.S. Congress, 1977) does for the coal industry. However, it should not be inferred that construction activities are entirely unregulated. Historically, soil erosion has been the primary concern, and like soil erosion itself, the nature of legal controls on construction lands is somewhat site specific.

Erosion control for construction lands is largely in the purview of agencies that issue permits and award construction contracts. They, in turn, receive their authority through one or more federal, state, or local statutes. It is desirable, then, to briefly review some of the legislation that may be brought to bear.

Cox and Walker (1977, p. 305) observed:

> The most visible form of law applicable to the erosion problem consists of various statutory enactments that place controls on erosion-producing activities, laws that are enforced by governmental agencies. Because erosion control traditionally has been viewed as a fundamental component of the conservation philosophy, government programs have existed for a considerable period of time. Increasing environmental awareness, particularly with regard to water quality, has resulted in recent expansion of these government efforts.
>
> A second less obvious, but nevertheless significant, form of applicable law consists of private constraints enforced by means of individual law suits. These judicially enforceable constraints arise from the rights of individual citizens to be free from certain adverse effects of artificially accelerated erosion that can arise from a wide range of land use activities. Mechanisms for enforcement of these rights include injunctions and awards of compensatory damages for injury already sustained. In addition to the direct constraints on a particular injury-producing activity, the existence of such private remedies serves as a deterrent to similar action by others.

First, let us examine federal legislation. Guy (1970) noted that sediment control was a result of Executive Order 11258 issued in 1966 through the authority of the Water Quality Act of 1965 (U.S. Congress, 1965). This order requires a review of all federal and federally aided operations where there is a significant potential for reduction of water quality by sediment. The reviewers may prescribe suitable remedial practices.

The National Environmental Policy Act of 1969 (NEPA) (U.S. Congress, 1969) requires that any federal agency (except the U.S. Environmental Protection Agency) prepare an environmental impact statement for any legislation it proposes or any project that it proposes, supports financially, or regulates if the proposed legislation or project can significantly affect environmental quality. The preceeding sections of this chapter have firmly established that construction activities can be detrimental to environmental quality. Further, many projects, both edifice and especially highway or road construction, utilize federal funds. Hence, this act can be regularly invoked to support erosion control requirements.

However, Greenwood and Edwards (1979) remarked that there is still some confusion as to whether state highway departments are subject to the National Environmental Policy Act. On the other hand, a member of a state highway department asserted that "around here, NEPA is the bible." According to a U.S. Environmental Protection Agency representative (Bruce Zander, personal communication, 1985), the EPA is usually not directly involved in the enforcement of erosion control requirements on construction lands. This is viewed as a matter of regional or local responsibility. The EPA provides technical support and often functions as an advisor to local governments.

There is one area of direct involvement by the EPA, according to the aforementioned individual. If erosion and sediment, or other nonpoint sources of pollution are channeled into a natural stream or reservoir (through a storm sewer system or artificial drainageway), they can be treated as a point source of pollution and a NPDES (National Pollution Discharge Elimination System) permit is necessary, as for a sewage treatment plant or industrial facility.

The Federal Water Pollution Control Act Amendments of 1972 (U.S. Congress, 1972) is perhaps the most commonly cited federal legislation pertaining to erosion and sedimentation control on construction lands. This act required each state to submit plans by November 1, 1978, for eliminating pollution from all point sources and reducing, to the extent feasible, that from nonpoint sources, such as farmland. Cox and Walker (1977) remark that this act contains explicit authority for land use control in a number of places.

Criteria for preparation of such plans place considerable emphasis on runoff pollution from nonpoint sources, and specific attention is given to erosion-related water quality problems. Such plans must include identification procedures and control measures for nonpoint sources of pollution related to agriculture, silviculture, mining, and construction activity. Cox and Walker (1977) also noted that such plans are subject to approval by the EPA, a requirement that establishes a degree of federal control over the planning process. The Colorado Department of Highway (State of Colorado, 1984) design manual (March, 1984) stated that the Water Pollution Control Act places strict regulations on the discharge of pollutants into the nation's waters. Special permits that regulate point source discharge of pollutants and fill encroachments into streams and wetlands are often required for highway construction.

Beyond the applicable federal legislation, regulations, and guidelines, individual states create further controls. For example, Miller (1979) observed that the 1970 Environmental Quality Act of California (State of California, 1970) goes farther than NEPA, requiring an environmental impact statement for any government or *private* project that can significantly affect environmental quality. Cox and Walker (1977) noted that the North Carolina Sedimentation Pollution Control Act of 1977 (State of North Carolina) encompasses land disturbing activities beyond agriculture, forestry, and mining operations; indeed, its primary focus is construction activities. In the Commonwealth of Virginia (1975), the Erosion and Sediment Control Law provides for the Virginia Soil and Water Conservation Commission to establish standards and criteria for control of erosion and sedimentation in connection with numerous land-disturbing activities. Actually, the State of Maryland (1970) was on the forefront in the enactment of erosion control legislation. Summaries of this legislation can be found in the Highway Research Board (1973) or Boysen (1977). Amimoto (1981) provides guidance for the development of state erosion control legislation.

The National Association of Conservation Districts (NACD) (Eugene Lamb, personal communication, 1985) has tabulated the components of erosion control legislation for 26 states. There is considerable variation in the contents of these laws and they change with some frequency. The NACD list is in the process of revision.

Occasionally, there exist local erosion and sedimentation controls. Boysen (1977) remarked that in Maryland, Montgomery County became the first jurisdiction in the state to pass an ordinance requiring sediment control for developing areas in 1965. This was 5-yr prior to the statewide legislation. The Denver Regional Council of Governments (DRCOG, 1980) specifies the planning procedures to be undertaken before commencement of construction activities, including consideration of erosion control.

Finally, research by Busby (1962, 1967) and summarized by Guy (1970, p. E5) leads to the general conclusion that several court cases make it "eminently clear that downstream owners can recover damages if changes (in erosion and sedimentation rates) and costs are well-documented."

Given some uncertainties regarding regulation of erosion and sedimentation from construction lands, an attempt was made to ascertain how the process actually works. Again, there is likely to be considerable variation from place to place, although the following is thought to be fairly representative. Clearly, anyone contemplating a construction project must become thoroughly familiar with the laws applicable at a particular construction site, or enlist legal advice.

First, we shall focus upon the construction of edifices in urban areas. Here, the local developer is probably already familiar with the erosion and sediment control requirements for a proposed project. Often, cities or counties provide published guidance. The preliminary plans are taken to the appropriate regulatory authority for review prior to application for a building permit. At this time further recommendations for erosion and sedimentation control, the identification of heretofore unforeseen problems, or the necessity of additional permits from other agencies can be conveyed to the developer. When the final plans are submitted to the regulatory authority for approval, they are again examined to ensure the inclusion of adequate erosion and sedimentation control measures. All of this should take place, together with the acquisition of required permits from other agencies, before any land is disturbed. The planning process is discussed in greater detail in a later section.

During and following construction, site inspections are made to determine compliance with the erosion and sedimentation control features of the approved plan. Where discharge and sediment load data are collected for urban streams, problems may appear in these data before or between inspections.

Highway construction commonly follows a different procedure. After extensive resource allocation review, priorities are established concerning future projects. Then, through a complicated planning and design process, a particular high-priority project is determined to be a federal or state endeavor and funding is appropriated accordingly. Whether federal or state, the state highway department commonly assumes responsibility for the project. After final plans are prepared, bids are accepted for the actual construction from independent contractors by the state highway department. A contract is then awarded. Next, a preconstruction meeting is held between the contractor and the state highway department, at which time the representatives of the company submit written schedules for the work, including temporary and permanent erosion and sedimentation control measures. No work is started until the erosion control schedules and methods have been accepted by the project engineer.

For highway construction projects, environmental safeguards, including the erosion and sedimentation control measures, are built into the initial plans and specified in the contract awarded. The Federal Highway Administration (1974, p. 1) (Federal Aid Highway Program Manual, Vol. 6, Chap. 7, Sect. 3, Subsect. 1) devotes several pages to the discussion of erosion and sediment control. Their general policy is stated thusly:

> It is the policy of the Federal Highway Administration (FHWA) that federal-aid highways and highways constructed under the direct supervision of the FHWA shall be located, designed, constructed, and operated according to standards that will minimize erosion and sediment damage to the highway and adjacent properties and prevent pollution of surface and groundwater resources.

Likewise, state highway departments usually provide specific design guidance that includes erosion and sedimentation control measures. The design manual of the Colorado Department of Highway states that soil erosion problems have been intensified in recent years due to wider highway templates and more uniform grades. Higher fills and deeper cuts now occur in comparison to the minimal earthwork on older highways.

Cox and Walker (1977) concluded that a substantial amount of legislation and common law has developed concerning soil erosion and related problems. Statutory law and the administrative programs created thereby have expanded from their original emphasis on agricultural land to encompass the total spectrum of land-disturbing activities. This increase in scope is accomplished both by enactment of additional legislation, specifically applicable to erosion producing activities, and by expansion of water quality legislation to include non-point sources of pollution. They contend, therefore, that it is likely that any land development activity will be subject to regulation with regard to the erosion aspects of such projects.

Finally, Israelsen *et al.* (1980, p. 3) offered an interesting perspective concerning the differences in the nature of erosion and sedimentation regulation from place to place:

> Current interest and activity in erosion control during highway construction vary greatly from state to state and seemingly depend to a great extent on the customs and values with which

people have grown up. If their streams have always run clear, they wish to keep them clear. If their streams have always carried a sediment load, they may be less concerned about a little more sediment as a result of highway erosion. These philosophies are reflected in present-day regulations and restrictions of the various states regarding requirements for controlling erosion from construction sites, including highways.

Heft (1977) discussed some of the political, social, and economic aspects of soil erosion and sedimentation control programs that may interest the reader.

Practices for Mitigation and Control of Impacts

In most cases today, construction contractors are required to mitigate and control the environmental impacts resulting from their operations. During the active construction phases, containment of the inevitable impacts is a principal objective. Permanent site stabilization becomes a subsequent goal.

Although construction activities produce a variety of environmental impacts, including noise and air pollution, alteration of vegetation communities, and reduction of wildlife habitat, the problems associated with erosion and sedimentation receive by far the greatest attention. Even mass movement phenomena are infrequently mentioned in the literature. The inference appears to be that standard engineering practice can effectively minimize the probability of hillslope failure in usual circumstances. Hence, we shall again limit our discussion to the mitigation and control of erosion and sedimentation.

The practices employed to manage these geomorphic processes are variously categorized. Based upon their expected effective longevity, actions may be regarded as temporary or permanent, as shown in Table 9. Based upon the materials involved, practices may be categorized as structural, vegetal, or a combination of both (biotechnical).

When properly designed and executed, erosion and sedimentation control practices are highly effective. White and Franks (1978) concluded on the basis of an examination of residential development at Lake Tahoe, California, that adequate technology exists to ensure that properly planned, constructed, and maintained developments will have a minimum, if not negligible, impact upon surface water quality from the standpoint of erosion and sediment loads. They also submit that the implementation of hillslope stabilization techniques to achieve control of erosion at a poorly designed development is expected to reduce the previously excessive sediment yield rates by about 80 to 90%.

Yorke and Herb (1976) documented similar success in controlling these geomorphic processes at construction sites located in the substantially different environment of Montgomery County, Maryland. As a result of improvements in grading practices and control measures, construction site sediment yields were reduced 60 to 80% between 1966 and 1974.

Table 9 Relating Sources or Causes of Accelerated Erosion to Erosion Control Practices[a,b]

Erosion and sediment control practices	Sources of sediment or causes of accelerated erosion			
	Loss of trees	Bare cut and fill slopes and graded areas	Unprotected road surfaces	Sensitive water courses
Protection of trees	P			
Vegetative measure		T, P	T	T, P
Protective covering of Mulch and other materials		T	T	T
Temporary diversion dike		T		
Permanent diversion dike		P		
Interceptor ditch		P		
Slope drain		T, P		
Diversion		T, P		
Interceptor dike			T	
Drainage dip			T	
Side ditch			T, P	
Open-top culvert			T, P	
Vegetative lining				P
Flexible lining				P
Rigid lining				P
Grade control structure				P
Channel realignment				P
Culvert				T, P
Paved ford				P
Bridge				T, P
Sediment trap				T, P
Sediment detention basin				T, P
Energy dissipator				T, P

[a]Source: Amimoto, 1981.

[b]T, Temporary erosion control practices that would prevent erosion during construction or before construction is completed; P, permanent erosion control practices that would permanently stay on the project area for erosion control.

Israelsen *et al.* (1980) contend that technology is available to control, within reasonable limits, the erosion and sedimentation that originates from highways both during and following construction. The extent of remedial action required on currently eroding highway roadcuts is variable. For example, from the inventory of roadside erosion in Wisconsin, Briggs (1973) found that 54% of the eroding sites can be controlled by fertilizers, seeding, and mulching and 37% required grading, fertilizing, seeding, and mulching. The remaining 9% need complete treatment, including structures, grading, fertilizing, seeding, and mulching.

Planning Process

The U.S. Soil Conservation Service (1978) remarked that resource plans should control and guide urban growth so that developments do not take place haphazardly and are not harmful or wasteful of resources; adequate planning reduces or eliminates many urban problems such as erosion, sediment, flooding, and land wetness. Highfill and Kimberlin (1977) observed that planning for erosion control at construction sites begins with the first thoughts in someone's mind about developing a tract of land. DRCOG (1980) assigned the responsibility for the direct effect of erosion and sediment on receiving water bodies and adjacent downslope areas receiving sediment to the developer, and therefore, erosion control measures should be incorporated into the initial planning phase of project development.

According to DRCOG (1980), there are two major elements in developing an erosion and sedimentation control plan: (1) an investigation and analysis of the natural characteristics of a site that help the developer anticipate where erosion problems might occur, and (2) the selection of effective erosion control practices. They suggested that these planning elements can be divided into five separate steps.

1. Gather information on topography, soils, drainage, vegetation, and other predominant site features.
2. Analyze the information in order to anticipate erosion and sedimentation problems.
3. Devise a plan that schedules construction activities and minimizes the amount of erosion created by development.
4. Develop an erosion control plan that specifies effective erosion and sedimentation control measures.
5. Follow the erosion control plan and revise it when necessary.

The remainder of that reference provides detailed descriptions and examples of the first four planning elements. Amimoto (1981) offered a rather comprehensive site evaluation checklist that is useful throughout the planning process.

The U.S. Soil Conservation Service (1975) stated that the sediment control plan consists of the best selection of erosion control practices and sediment trapping facilities in conjunction with an appropriate schedule in order to accomplish an adequate level of control. Particular attention must be given to concentrated flows of water, either to prevent their occurrence or to provide conveyance devices according to standards and specifications in order to prevent high rates of erosion. Sediment-trapping devices are usually required at all points of egress of sediment-laden water from the construction site.

The U.S. Soil Conservation Service (1978) offered the following planning considerations for urbanizing areas:

1. Select a suitable site.
   a. Avoid construction on areas subject to flooding.
   b. Consider erosion hazards and potential landslides, especially on steep slopes.
   c. Wet, shallow, and unstable soils are not desirable building sites. Recognize and overcome these soil limitations or develop these areas for parks, wildlife preserves or other such uses.
   d. Developers should be encouraged to contact the local Soil Conservation District regarding soil limitations for the intended landuses and for assistance in preparing conservation plans for developing areas.
2. Keep site modifications to a minimum.
   a. Avoid heavy grading that would alter drainage patterns or create very steep slopes.
   b. Use diversions to direct water away from erosive areas to points of safe disposal.
   c. Expose as small an area of soil for as short a time as possible.
   d. Stockpile topsoil removed by grading and redistribute over any disturbed areas.
   e. Use culverts when crossing streams with construction equipment.
   f. Protect bare areas with temporary vegetation or mulches during construction.
   g. Save and protect as much of the desirable native vegetation as possible. Mark or erect barriers to prevent damage from construction equipment.
3. Minimize pollution of air and water.
   a. Use debris basins to trap trash and sediment, preventing these materials from moving off-site.
   b. Keep dust levels low by sprinkling or mulching as necessary.
   c. Place construction facilities and store equipment away from drainageways.
4. Protect area after construction.
   a. Establish and maintain needed structural conservation practices.

b. Establish lawns or other suitable vegetation on all disturbed areas.
c. Maintain vegetated areas using proper horticultural practices.

Many of the concepts and principles discussed above in relation to the planning of developments in urban areas are equally applicable to highway construction. Israelsen *et al.* (1980) noted that, to be effective throughout the life of the project, erosion control measures should be carefully considered during the planning and design phases as well as during construction and operation of the highway. They propose a series of steps for the planning of erosion control along highways that are quite similar to the five steps offered above by DRCOG (1980) for the development of urban lands.

The Highway Research Board (1973, p. 18) observed that all new contracts involving federal highway funds now contain the following statement or variations thereof:

> At the preconstruction conference the contractor shall submit for acceptance his schedules for accomplishment of temporary and permanent erosion control work, as are applicable for clearing and grubbing; grading; bridge and other structures at watercourses; construction; and paving. He shall also submit for acceptance his proposed method for erosion control on haul roads and barrow pits and his plan for disposal of waste materials. No work shall be started until the erosion control schedules and methods of operations have been accepted by the engineer.

The Board also commented that erosion control planning should start with route selection. The route location choices that directly relate to the control of erosion on highway construction include: (1) proximity to critical areas or structures, (2) roadway grades, (3) side slopes, (4) depth of cut, (5) weight of fill, (6) streams and drainage channels, (7) channel changes, and (8) relationship of topography to alignment of the highway. However, the advantages and disadvantages of any particular route, based on soil erosion projections, are weighed against all of the other engineering, social, economic, and environmental considerations, such as right-of-way costs, safety, grades, drainage, access, noise, and alignment. Another comprehensive checklist is provided for the location, design, and construction of highways.

It is apparent that adequate erosion-control planning requires satisfactory estimates of potential erosion rates. Despite the oft-recited limitations, several authors (Megahan, 1978; Amimoto, 1981; Holberger and Truett, 1978; Wischmeier and Meyer, 1973; Israelsen *et al.*, 1980) have endorsed the use of the Universal Soil Loss Equation (USLE) for this purpose.

The manual prepared by Israelsen *et al.* (1980) provides maps, tables, and graphs and explains the use of the water and wind soil loss equations for calculating erosion potentials at construction sites and for evaluating the effectiveness of various erosion control measures. Results of research reported in this document indicate that the USLE is valid for use on steep slopes. Several maps of K-factors, R-factors, and seasonal distribution of rainfall energy are also included.

When applied to construction sites, the *C* and *P* factors of the USLE are replaced by the *VM* factor (vegetative and management measures). Israelsen *et al.* (1980) compiled a list of these factor values, shown in Table 10. Recall that all of the factors in this equation are multiplied together to arrive at an estimate of soil loss. So, the values in Table 10 are useful in determining the percentage increase or decrease in erosion due to particular practices.

Aside from the specific nature of construction and stabilization operations, the timing or scheduling of these activities must be carefully examined during the planning phase. Glover *et al.* (1978, p. 273) observe:

> Seasonal climatic variations play an important role in the scheduling of grading operations. The amount of rainfall and runoff during different periods of the year influences erosion. Because the soil is so vulnerable to erosion during the grading activities, those activities should be scheduled to coincide with periods of low precipitation. The spring and early summer months often have the highest precipitation rates. Therefore, the bulk of grading operations, especially in critical areas, should be scheduled for midsummer or fall.

For the Denver, Colorado, area DRCOG (1980) commented that erosion and runoff are greatest during the spring and early summer when the combination of snowmelt and seasonal intense rains are most common. If development is scheduled to take place during this time, then the erosion plan should reflect measures capable of handling the worst possible situation. Also, during the months of December through May, wind erosion potential is the greatest and soils should be protected from that process during this period.

Table 10 Typical *VM* Factor Values Reported in the Literature[a,b]

Condition	V_m factor
Bare soil conditions	
Freshly disked to 6–9 in.	1.00
After one rain	0.89
Loose to 12 in. smooth	0.90
Loose to 12 in. rough	0.80
Compacted bulldozer scraped up and down	1.30
Same except root raked	1.20
Compacted bulldozer scraped across slope	1.20
Same except root raked across	0.90
Rough irregular tracked all directions	0.90
Seed and fertilize, fresh	0.64
Same after 6 months	0.54
Seed, fertilizer, and 12 months chemical	0.38
Not tilled algae crusted	0.01
Tilled algae crusted	0.02

Table 10 *(Continued)*

Condition	V_m factor
Compacted fill	1.24–1.71
Undisturbed except scraped	0.66–1.30
Scarified only	0.76–1.31
Sawdust 2 in. deep, disked in	0.61
Asphalt emulsion on bare soil	
1250 gal/acre	0.02
1210 gal/acre	0.01–0.019
605 gal/acre	0.14–0.57
302 gal/acre	0.28–0.60
151 gal/acre	0.65–0.70
Dust binder	
605 gal/acre	1.05
1210 gal/acre	0.29–0.78
Other chemicals	
1000 lb fiberglass roving with 60–150 gal asphalt emulsion/acre	0.01–0.05
Aquatain	0.68
Aerospray 70, 10% cover	0.94
Curasol AE	0.30–0.48
Petroset SB	0.40–0.66
PVA	0.71–0.90
Terra-Tack	0.66
Wood fiber slurry[c], 1000 lb/acre fresh	0.05–0.73
Wood fiber slurry[c], 1400 lb/acre fresh	0.01–0.36
Wood fiber slurry[c], 3500 lb/acre fresh	0.009–0.10
Portland cement + Latex	
1000 lb/acre + 8 gal/acre	0.13
1500 lb/acre + 12 gal/acre	0.006
Seedings	
Temporary, 0 to 60 days	0.40
Temporary, after 60 days	0.05
Permanent, 0 to 60 days	0.40
Permanent, 2 to 12 months	0.05
Permanent, after 12 months	0.01
Brush	0.35
Excelsior blanket with plastic net	0.04–0.10

[a]Source: Israelsen *et al.*, 1980.

[b]Note the variation in values of *VM* factors reported by different researchers for the same measures. References containing details of research which produced these *VM* values are included in NCHRP Project 16-3 report, "Erosion Control During Highway Construction, Vol. III, Bibliography of Water and Wind Erosion Control References," Transportation Research Board, 2101 Constitution Avenue, Washington, D.C., 20418.

[c]This material is commonly referred to as hydromulch.

For road fill hillslopes in central Idaho, Megahan (1978) found that rainfall erosion potentials averaged about 7.5 times greater for summer–fall than for winter–spring according to a daily erodibility index. Hence, on the average, only about 20% of the total annual erosion occurred during the 232-day winter–spring period, extending from about October 16 to May 30. Deep snow covered the surface during much of this time.

For the Port Carbon, Pennsylvania, area Helm (1978) found that 70% of the total suspended sediment discharge and 18% of the total flow from an area disturbed by highway construction took place during eight storms that occupied only 4% of the study period from April 1975 to March 1977. In this case, the eight storms occurred in July, September, October, February, and March. Clearly, scheduling is difficult in this region if this storm distribution is characteristic of most years.

Lastly, for Fairfax County, Virginia, Vice *et al.* (1969) found that the average percentages of sediment discharge from areas disturbed by highway construction occurring in quarterly periods of each year were 30, 38, 11, and 21, which suggest that remedial measures to reduce erosion from construction areas have the greatest importance during the January to June period when about 68% of the annual sediment discharge occurs.

Scheduling of construction and stabilization operations is a complex task, subject to various inputs and uncertainties. The job is made even more difficult by the fact that, for many midlatitude locations, the construction season each year is substantially limited by severe weather conditions during a lengthy winter. In order to stay in business, the contractor must sometimes earn his living in a few months. Planning for erosion and sedimentation mitigation and control under conditions of the worst case scenario may be unavoidable.

Initial Water and Sediment Controls Practices

In many cases, erosion and sedimentation control begins even before site preparation. It is sometimes necessary to alter the flow of streams, redirect perennial and ephemeral channels, or construct diversions to protect the area of disturbance. Additionally, sediment-trapping structures should be emplaced in readiness for any runoff and sediment from the site once vegetation has been removed from the surface.

Gray and Leiser (1982) observed that the control of runoff and seepage water is required on many sites. Water flowing over cuts and fills is a common cause of erosion. Treatments, such as ditches, diversions, dike interceptors, or berms vary according to the source and quantity of water and are determined on a site-specific basis.

The U.S. Soil Conservation Service (1978) defines a diversion as a channel with a supporting ridge on the lower side constructed across a hillslope. They are either temporary or permanent. In the first case, they are installed as an interior measure to facilitate some phase of construction. These usually have a life expectancy of a year or less. Permanent diversions are installed as an integral part of an overall water disposal system and remain for the protection of property.

Diversions are used to reduce hillslope lengths, break up concentrations of runoff, and move water to stable outlets at nonerosive velocities. They serve to redirect water from (1) lower lying areas, (2) cut-or-fill hillslopes or steeply sloping land, (3) construction sites, (4) buildings (and residences), and (5) active gullies or other erodible areas. Construction of diversions and outlets must be in compliance with state drainage and water laws. The U.S. Soil Conservation Service *et al.* (1983) and Amimoto (1981) have provided standards and specifications for diversion design and construction.

Erosion Control Practices for Hillslopes: Temporary and Permanent

Of course, one of the best ways to minimize *off-site* impacts is to minimize *on-site* runoff and erosion and contain that which is produced on-site. Toward this end, the application of erosion and sedimentation control practices of a temporary nature should proceed concurrently with the actual construction operation whenever possible. Many of the permanent erosion and sedimentation control practices must await completion of construction phases, however.

It exceeds the scope of this chapter to present details concerning the construction or utilization of structural, vegetal, or combination techniques for erosion and sedimentation control. Such information can be found in Gray and Leiser (1982) or in a manual prepared by the U.S. Soil Conservation Service *et al.* (1983) entitled: "1983 Maryland Standards and Specifications for Soil Erosion and Sediment Control." Nonetheless, an overview of common practices illustrates the approaches taken in the effort to manipulate the frequency and magnitude of these geomorphic processes.

Prior to the erection of structures, the disturbed surface is graded to a desired configuration. For developments in urban areas, drainage basins will be created, intentionally or inadvertently, at this time. The U.S. Soil Conservation Service (1978) has provided guidelines for grading hillslopes at constitution sites; Table 11 contains the recommendations for the State of Colorado. These are slightly different from those for the State of Maryland (U.S. Soil Conservation Service *et al.*, 1983). Therefore, it is prudent to consult the local U.S. Soil Conservation Service office at a specific locale for the best advice.

There is considerable research concerning grading operations along highways. Wright *et al.* (1978, p. 555) offered a concise summary of recommended practices for cut hillslopes:

Table 11 Recommendations for the Grading of Urban Lands in Colorado[a]

1. Development and establishment of grading plan
   a. The cut face of earth excavation that is to be vegetated shall not be steeper than $3H{:}1V$. Cut slopes of materials not to be vegetated shall be at the safe angle of repose for the materials encountered. Unvegetated cut slopes shall be protected by mechanical treatment to protect them from erosion.
   b. The permanent exposed faces of fills shall be no steeper than $3H{:}1V$.
   c. Provisions are to be made to safely conduct surface water to storm drains or suitable natural water courses and to prevent surface runoff from damaging cut faces and fill slopes.
   d. Subsurface drainage is to be provided in areas having high water table or seepage conditions that would affect slope stability and building foundations or create undesirable wetness.
   e. Excavations shall not be made so close to property lines as to endanger adjoining property without supporting and protecting such property from erosion, sliding, settling, or cracking.
   f. No fill is to be placed where it will slide or wash up on the premises of another, or so placed adjacent to the bank of a channel as to create bank failure or reduce the natural capacity of the stream.
   g. Fills are to consist of materials from cut areas, borrow pits, or other approved sources.
2. Construction of Hillslopes
   a. Timber, logs, brush, rubbish, and vegetative matter that will interfere with the grading operation or affect the planned stability of fill areas shall be removed and disposed of according to the plan. Avoid unnecessary removal of trees and vegetation that could be left to enhance the attractiveness of the development.
   b. Topsoil is to be stripped and stockpiled in amounts necessary to complete finish grading of all exposed areas requiring topsoil for the establishment of vegetation.
   c. Fill material is to be free of brush, rubbish, rocks, logs, and stumps in amounts that will be detrimental to constructing stable fills.
   d. Cut slopes that are to be topsoiled will be scarified to a minimum depth of 3 in. prior to placement of topsoil.
   e. All fills intended to support buildings, structures, sewers, and conduits are to be compacted to a minimum of 90% of standard proctor with proper moisture control. Compaction of other fill will be as required to reduce slipping, erosion, or excess saturation.
   f. Frozen materials or soft, mucky, or easily compressible materials are not to be incorporated in fills intended to support buildings, parking lots, roads, structures, sewers, or conduits.
   g. Maximum thickness of layers of fills to be compacted are not to exceed 8 in.
   h. All areas are to be rough graded to within 0.2 ft of the planned elevation after allowance has been made for thickness of topsoil, paving, or other installations.
   i. All disturbed areas shall be left in a well drained, neat, and finished appearance.
   j. Construction operations shall be carried out in such a manner that erosion and air and water pollution will be minimized. State and local laws concerning pollution abatement shall be complied with.

[a]Source: U.S. Soil Conservation Service, 1978.

For long, sloping cuts, the grading operation should begin with establishing diversion ditches at the apex of cuts to impede and disburse water from the slopes above the area to be graded. Slopes should generally be no steeper than 33° (1.5:1), because the shallower slopes are less erosive and vegetation can more easily be established in them. The steepness of cut slopes should be determined by the length of grade, soil and rock materials, topography, width of highway corridor, and ease of establishing vegetation. For example, it is more difficult to establish vegetation on 'hot' sunny slopes than on 'cool' shaded slopes; hence, shallow slopes on sunny exposures would facilitate the establishment of vegetation (Wright *et al.*, 1975). Conversely, undecomposed rocky materials on cuts may be nearly vertical.

Slopes steeper than 18° (3:1) should be stair-step graded, left rough or grooved. Stair-step grading may be used on any materials soft enough to be ripped with a bulldozer. Slopes steeper than 27° (2:1) should be stair-step graded. The ratio of the vertical cut distance to the horizontal distance should be less than 1:1, and sloping toward the vertical wall to catch sloughing soil, increase infiltration, and reduce runoff. . . .

The surface of cut slopes less than 27° (2:1) should be 'rough' and undulating with stones left in place. High and low places giving variable micro-environments speed up the establishment of vegetative cover because of some coverage of seed and fertilizer and improved moisture.

The detrimental practice of constructing slopes with smooth, hard surfaces gives a false impression of 'finished grading' and a job well done, but vegetation often fails. Rough slope surfaces with rocks left in place give an 'ugly' appearance to the novice, but encourages water infiltration, speeds up establishment of vegetation, and decreases the rate of flow into drainage ways. . . .

Slopes along highways with smooth hard surfaces should be grooved or roughened to aid in establishing vegetation before the application of seed soil amendments, and mulch. . . . Such grooves collect sloughing soil, seed, and soil amendments, and enhance the rate of obtaining a protective vegetation cover. . . .

Of the four types of slope surfaces discussed, rough slope surfaces, or stair-step grading, is more desirable than smooth slope surfaces with or without lateral grooves for obtaining a vegetative cover quickly.

Wright *et al.* (1978, pp. 560–563) also discussed the grading of fill hillslopes:

It is generally easier to obtain vegetative covers for erosion control on fill slopes than on cut slopes because less compacted rock and soil materials encourage water infiltration and root growth. However, the common practice of blading and/or tracking fill slopes with a dozer is usually objectionable because compaction inhibits water infiltration and aeration, causing poor growth. Also, seeds and fertilizer on such hard surfaces are apt to wash away. Tracking clayery and silty soil materials, especially when wet, causes severe surface compaction which augments water runoff and erosion. . . . Tracking sandy materials may be desirable if the tracks leave indentations perpendicular to the slope. Tracking parallel to the slope is especially objectionable as the vertical rills, inadvertently occurring with such operations, leave channels for accelerated water movement downslopes.

When constructing fill slopes, berms and down troughs (rock, plastic, or other material) should be constructed on each side of the fill lift to direct water movement off the slope. As lifts of the fill slope are constructed, the soil and rock materials falling naturally onto the slope surface should not be removed. Variable steepness, looseness, and undulations within the fill slope create desirable mini-environments for establishing and maintaining vegetation. Also, by allowing materials to fall naturally, the variable contours inhibit water runoff. . . .

Smoothing of fill slopes may be justified when mowing is necessary. However, fill slopes steeper than 2:1 should never be mowed.

The configuration of hillslopes in a development or along the highway exerts substantial influence on erosion rates. The profile and plan shape, as well as the three-dimensional form, influence soil loss as previously discussed. The contractor should have access to sufficient land area to construct suitable hillslopes, rather than allow the available space to dictate landform characteristics.

Using intense simulated rainfall, Meyer *et al.* (1971) measured soil erosion and runoff from plots with six conditions, typical of denuded lands after they had been reshaped for developments or highway construction. They found that soil loss was greatest on the scalped-only and scarified treatments (about 49 metric tons, or 54 tons), less on compact fill (about 44 metric tons, or 48 tons), still less on loose fill and applied topsoil (about 28 metric tons, or 31 tons), and least on a straw mulch surface (about 9 metric tons, or 10 tons). The freshly graded surface was highly erosive, and the goal of erosion and sedimentation control required immediate, if temporary, action.

Some construction site specifications call for mulching. Blaser (1975) stated that mulching materials are used to moderate soil temperatures, improve soil moisture, and give a temporary protective cover as seedings become established. The U.S. Soil Conservation Service (1970) reported that the materials most widely used in mulching are small grain straw, hay, and certain processed materials.

A wide variety of mulches have been tested for their ability to control erosion and stimulate vegetation growth under assorted hillslope conditions and rates of application. Results of several tests are found in Diseker and Richardson (1962), McKee *et al.* (1964), Swanson *et al.* (1967), Meyer *et al.* (1971, 1972), Highway Research Board (1973), and Israelsen *et al.* (1980). Amimoto (1981) compiled the information included in Table 12.

Unfortunately, some results are contradictory. For example, Swanson *et al.* (1967) found that jute netting is among the materials that provide excellent protection against water erosion. Likewise, McKee *et al.* (1964) concluded that various nets give satisfactory to excellent results but that net mulches must be applied in such a fashion as to maintain firm contact with the ground in order to be effective. It is necessary to prepare a uniform soil surface to maintain good soil–net contact and this requires considerable hand labor and time on steep hillslope sites. Conversely, Wright *et al.* (1978) asserted that many experiments with various organic and inorganic nets on steep cut hillslopes in the Appalachian and Piedmont areas usually result in seeding failures. It is practically impossible to establish a continuous soil–net contact, and therefore water flows under nets causing serious erosion. It is thus imperative that one consult local sources of information in choosing the mulches best suited for a particular area and site conditions. General guidelines for mulch application are available through the U.S. Soil Conservation Service (1975, 1978), Amimoto (1981), Wright *et al.* (1978), and Tour (1985), among other sources.

The Highway Research Board (1973) asserted that the anchoring of the mulch

Table 12 Characteristics of Surface Covers and Mulches[a]

Type mulch	Effectiveness in immediate erosion protection	Effectiveness in establishing vegetation	Comment
Plastic sheet	High	—	Temporary use; mulch protection for shrubs
Straw alone	Medium	Medium	For flat slopes and nonwindy areas
Punched straw	Medium	High	Ideal for vegetating fill
Net-anchored straw	High	High	Retains straw on slope
Tackifiers with straw	High	Medium–high	Straw glued with tackifiers prevents its blowing away
Wood chip, sawdust	Medium	Medium	For flat slopes, stepped slopes, seed spot basins
Gravel, stones	High	Medium	Provides for mechanical protection and vegetative growth
Mulch blanket	Medium	Medium	Binder dissolved leaving fiber mat with net
Wood fiber	Medium	Medium	Applied on roughened surface
Washed dairy waste	Medium	Medium	Equivalent to wood fiber
Wood chips and asphalt	Medium–high	Medium	Good use of timber waste
Chemical mulch	Medium–high	None–low	Chemical film increases cohesion but reduces porosity
Wood fiber and chemical mulch	High	High	Improved retention of fiber on slope and improved germination
Wood excelsior mat	High	High	Covered by net and stapled in steep or windy areas
Fiberglass	Medium	Medium	Available in loose or matted form
Fiberglass and asphalt	High	High	May be difficult to mow
Jute	Medium–high	Medium–high	Slope must be smooth
Jute and straw	High	High	Slope may be rough
Sod	High	High	Some maintenance required
Building block	High	High	Stabilizes steep slopes and provides for vegetation

[a]Source: Amimoto, 1981

is of critical importance. The methods used include (1) pressing the mulch into the soil to a depth sufficient to anchor it but not bury it, (2) tacking it with various emulsified asphalts or other similar products, and (3) tying it down with netting or wire mesh. Some of the best results come from the use of organic mulches, such as straw, anchored with a binder, such as diluted asphalt or emulsified asphalt.

Chemical soil stabilizers, binders, or tackifiers are sometimes sprayed directly on soil surfaces. Here, however, test results seem consistent. Wright *et al.* (1978) found that of the approximately 15 chemical binders and soil stabilizers used in highway corridors on many soil and subsoil materials, all are unsatisfactory when used alone for erosion control and seedling establishment. Green *et al.* (1973) found that chemical stabilizers used alone are generally unsatisfactory; however, using them with organic mulches made the mulches more persistent and desirable for prolonged stabilization during difficult seeding seasons. Israelsen *et al.* (1980) observed that chemical erosion control methods and materials are fairly recent and new ones are being added regularly. These are used with or without vegetative measures and include such items as soil stabilizers, asphalt, chemical mulches, and soil sealants. They conclude that, generally speaking, chemicals have been less successful than other measures in controlling erosion from construction sites.

If chemical stabilizers have a place in revegetation and erosion control programs, the literature suggests that it may be to anchor and bind other mulches. In particularly troublesome areas, they may be used to reduce erodibility of the soil through direct application to the surface. This is a temporary and relatively expensive solution to an erosion problem in most cases.

The Highway Research Board (1973) comments that temporary seedings create an erosion protective cover until the proper season for permanent seeding. Here, the U.S. Soil Conservation Service *et al.* (1983) observed that rapidly growing plants, such as annual rye grass, small grain, sudangrass, and millet are the most commonly used temporary covers. Timely removal of seedheads to prevent voluntary reseeding is also important when this technique is employed so that temporary species do not compete with permanent species.

Permanent erosion control practices are utilized on hillslopes at the culmination of active construction. Again, these may be structural, vegetative, or some combination of both. Gray and Leiser (1982) provided a valuable discussion of structural approaches. They stated that properly designed structures help to stabilize the hillslope against mass movement and that they protect the toe or face of a hillslope against scour and erosion by running water. Structures are generally capable of resisting much higher lateral earth pressures and shear stresses than vegetation. Structures can be used to divert and convey running water away from critical areas or to dissipate the energy of flowing water within the structure.

Gray and Leiser (1982) reported that structures can be built from various materials, both natural and man-made. Natural materials include earth, stone, rock, and timber. These normally cost less, are environmentally more compatible, and are better suited to vegetative treatment or modification than man-made materials. Man-made materials include steel and concrete. Structures made from these are stronger and generally more durable than natural structures, but more energy and capital intensive.

Finally, Gray and Leiser (1982) described structures that are composed of both natural and man-made materials; examples include concrete crib walls, steel bin walls, gabion walls or revetments, welded wire walls, and reinforced earth. Steel and concrete provide the rigidity, strength, and reinforcement, while stone, rock, and soil provide the mass. These types of structures can often be revegetated on exposed earthen surfaces.

Gray and Leiser (1982) suggested that a retaining structure of some type will usually be required to protect and stabilize oversteepened hillslopes. A low toe wall or retaining structure at the foot of a hillslope permits oversteepening of the hillslope at its base and flattening above. The latter makes it possible to establish vegetation on the hillslope, and the former reduces the amount of clearance required between the base of the hillslope and the adjacent right-of-way or existing use.

There are a very large number of structural erosion control practices available to contractors. Israelsen *et al.* (1980) list more than a dozen common measures and the conditions of their use, along with pictures showing their construction. Aside from the Gray and Leiser (1982) reference, more detailed information is available from the U.S. Soil Conservation Service *et al.* (1983) and White and Franks (1978). Figures 15 and 16 show several common practices for cut and fill hillslopes along with their advantages and disadvantages.

For most of the surface area at construction sites, erosion control and stabilization involves the establishment of a permanent vegetation cover. Thus, the next step is often the spreading of topsoil, although the necessity of this has been challenged somewhat. Meyer *et al.* (1971) demonstrated that applied topsoil was by far the most successful of five surface treatments on which revegetation was tested. However, based upon research in Virginia, Wright *et al.* (1976) remarked that adequate soil amendments applied to a properly prepared seedbed along highways eliminates the need for topsoil, which is often of poor quality.

Wright *et al.* (1978) commented that topsoiling is expensive, costing \$5000 to \$10,000/per hectare, and it usually delays seeding operations, increasing the possibility of erosion and pollution. Further, most topsoil contains weed seeds that cause dense weed canopies that shade out desirable species unless the area is reseeded or herbicides are applied. They contend that experiments show rough-graded subsoil actually can be superior to topsoil for grass and legume establishment.

Wright *et al.* (1978) found that topsoiling is necessary for establishing vegetation in the following situations:

1. To cover xeric rocky environments
2. To cover and restrict root contact as with soil materials containing high amounts of pyrite (acid potential)
3. Where special ornamentals that demand especially good soil fertility and aeration are to be planted

Treatment Practice	Advantages	Problems
CUT SLOPES		
Berm at top of cut 	Diverts water from cut Collects water for slope drains/paved ditches May be constructed before grading is started	Access to top of cut Difficult to build on steep natural slope or rock surface Concentrates water and may require channel protection or energy dissipation devices Can cause water to enter ground, resulting in sloughing of the cut slope
Diversion Dike 	Collects and diverts water at a location selected to reduce erosion potential May be incorporated in the permanent project drainage	Access for construction May be continuing maintenance problem if not paved or protected Disturbed material or berm is easily eroded
Slope Benches 	Slows velocity of surface runoff Collects sediment Provides access to slope for seeding, mulching, and maintenance Collects water for slope drains or may divert water to natural ground	May cause sloughing of slopes if water infiltrates Requires additional ROW Not always possible due to rotten material, etc. Requires maintenance to be effective Increases excavation quantities
Slope Drains (pipe, paved, etc.) 	Prevents erosion on the slope Can be temporary or part of permanent construction Can be constructed or extended as grading progresses	Requires supporting effort to collect water Permanent construction is not always compatible with other project work Usually requires some type of energy dissipation
Seeding/Mulching 	The end objective is to have a completely grassed slope. Early placement is a step in this direction. The mulch provides temporary erosion protection until grass is rooted. Temporary or permanent seeding may be used. Mulch should be anchored. Larger slopes can be seeded and mulched with smaller equipment if stage techniques are used.	Difficult to schedule high production units for small increments Time of year may be less desirable May require supplemental water Contractor may perform this operation with untrained or inexperienced personnel and inadequate equipment if stage seeding is required
Sodding 	Provides immediate protection Can be used to protect adjacent property from sediment and turbidity	Difficult to place until cut is complete Sod not always available May be expensive
Slope Pavement, Riprap 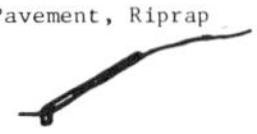	Provides immediate protection for high-risk areas and under structures May be cast in place or off site	Expensive Difficult to place on high slopes May be difficult to maintain
Temporary Cover	Plastics are available in wide rolls and large sheets that may be used to provide temporary protection for cut or fill slopes Easy to place and remove Useful to protect high-risk areas from temporary erosion	Provides only temporary protection Original surface usually requires additional treatment when plastic is removed Must be anchored to prevent wind damage
Serrated Slope 	Lowers velocity of surface runoff Collects sediment Holds moisture Minimizes amount of sediment reaching roadside ditch	May cause minor sloughing if water infiltration Construction compliance

Figure 15. Some stabilization practices for cut hillslopes (Highway Research Board, 1973).

Treatment Practice	Advantages	Problems
FILL SLOPES		
Berms at Top of Embankment 	Prevent runoff from embankment surface from flowing over face of fill Collect runoff for slope drains or protected ditch Can be placed as a part of the normal construction operation and incorporated into fill or shoulders	Cooperation of construction operators to place final lifts at edge for shaping into berm Failure to compact outside lift when work is resumed Sediment buildup and berm failure
Slope Drains 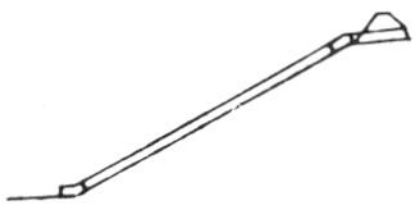	Prevent fill slope erosion caused by embankment surface runoff Can be constructed of full or half section pipe, bituminous, metal, concrete, plastic, or other waterproof material Can be extended as construction progresses May be either temporary or permanent	Permanent construction as needed may not be considered desirable by contractor Removal of temporary drains may disturb growing vegetation Energy dissipation devices are required at the outlets
Fill Berms or Benches 	Slows velocity of slope runoff Collects sediment Provides access for maintenance Collects water for slope drains May utilize waste	Requires additional fill material if waste is not available May cause sloughing Additional ROW may be needed
Seeding/Mulching	Timely application of mulch and seeding decreases the period a slope is subject to severe erosion Mulch that is cut in or otherwise anchored will collect sediment. The furrows made will also hold water and sediment	Seeding season may not be favorable Not 100 percent effective in preventing erision Watering may be necessary Steep slopes or locations with low velocities may require supplemental treatment

Figure 16. Some stabilization practices for fill hillslopes (Highway Research Board, 1973).

4. Where a quality turfgrass lawn is wanted, as at highway rest areas, and highway corridors adjacent to urban areas
5. In highway medians and shoulders, when good topsoil would otherwise be discarded

Item 4 in this list suggests topsoiling is desirable for most lands in residential developments. In any case, standards and specifications associated with the issuing of permits may render all arguments moot. Guidelines for topsoiling are available from the U.S. Soil Conservation Service (1975) and Amimoto (1981).

Seedbed preparation is the next component of effective revegetation, whether the graded surface is composed of topsoil or subsoil. Common agricultural techniques, such as plowing, disking, or raking, are utilized to relieve soil compaction due to equipment traffic and improve the physical condition of the seedbed. Where topsoil has been spread, this operation will also help to bond the topsoil and subsoil (Highway Research Board, 1973).

Usually, soil amendments are then added to the seedbed to enhance vegetal growth. These consist of various compounds that alter the acidity, alkalinity, sodicity, salinity, and plant nutrient levels of the growth medium. Chemical soil analyses are employed to determine the need for such soil amendments. Pot tests may be used to evaluate the response of selected plant species, raised in the available growth medium, to various combinations of fertilizers.

Wright *et al.* (1978) commented that introduced and improved grasses and legumes that develop protective vegetation covers quickly in highway construction projects generally require relatively high fertility and soil pH conditions. They noted that soil nitrogen is the primary limiting nutrient and the most costly for grass growth. Legume–grass associations eliminate the need for costly nitrogen fertilization to maintain sufficient vegetation cover to retard erosion. Finally, Wright *et al.* (1978) observed that, although applying certain elements may stimulate growth, the objective is to obtain a persistent protective cover of desirable botanical components rather than to produce high yields; excessive growth is objectionable because of diseases, rodents, increased mowing, and other management problems. Most state highway departments have information concerning specific amendments and application rates that prove successful in vegetation establishment under various conditions. In many areas, extensive research provides a basis for amending the growth medium. For example, studies in the State of Virginia have been conducted by Blaser *et al.* (1961), Green *et al.* (1973), and Wright *et al.* (1976); for the State of Maryland by the U.S. Soil Conservation Service (1975); and in the State of California by White and Franks (1978).

The Highway Research Board (1973) remarked that the basic principles for incorporating amendments thoroughly into the seedbed are understood, but are too often not written into the contracts or not enforced. Minimum incorporation should be 50 mm (2 in.) and 75 mm (3 in.) if mulch is to be anchored into the seedbed. Again, standard agricultural techniques are used to integrate the amendments into the soil.

Gray and Leiser (1982) asserted that the selection of plant species adapted to the climate and soils of the project area is one of the most important steps in achieving successful revegetation and erosion control. In a discussion of the factors involved in the selection process they considered the following: (1) desirable plant characteristics for erosion control; (2) native versus introduced species; (3) plant geography, variation, and adaptation; (4) plant succession; (5) nitrogen-fixing capabilities; (6) availability; and (7) use of mixtures composed of several species. Amimoto (1981) found that the types of plants used in establishing vegetative protection for erosion control should have the following characteristics: (1) must be self-sustaining, (2) require little or no maintenance, and (3) not increase the fire hazard. It is this latter attribute that is particularly interesting and often disregarded in dry climates where periodic fires are both natural and human-induced phenomena. Amimoto (1981) lists several fire hazardous plants including chamise (greasewood), sage (salvia), scrub oak, cheatgrass, and several sumac varieties.

It is generally accepted that monocultures are undesirable. Gray and Leiser (1982) observed that it is impossible to predict with certainty the success of any one species of plantings that are to receive little or no maintenance. A monoculture is always more susceptible to diseases or losses from abnormal weather conditions than a planting of a good species mix. By using an adequate mix, the chances of successful vegetation are greatly increased.

Vegetation covers can include grasses, forbs, shrubs, and trees, in various combinations, although grasses seem to be the most prevalent. Gray and Leiser (1982, Appendix II) provided an annotated list of references for the selection of plant species at construction sites. Lists of species commonly used in seed mixtures are also available in Wright *et al.* (1978), U.S. Soil Conservation Service (1975), Amimoto (1981), and White and Franks (1978), among others.

Finally, and of interest to geomorphologists, is the influence of microclimate on plant selection and revegetation success. McKee *et al.* (1965) measured microclimatic conditions on hillslopes of 1:1 gradient and facing as close to north and south as possible in the State of Virginia. Close relationships were found between soil surface temperature, light intensity, density of sod, clipping height, and hillslope aspect. On the basis of these results, they concluded that hillslope aspect should be considered in selecting plant species, mulches, and fertilizers for turf establishment. To take advantage of favorable temperature conditions, south-facing hillslopes might be seeded earlier in the spring and later in the fall than north-facing hillslopes. The U.S. Soil Conservation Service (1978) provided a table for the use of perennial grasses and legumes at critical areas receiving 12 to 15 in. of precipitation in Colorado. Herein, their recommendations take into account hillslope aspect (north and east versus south and west) when gradients exceed 6:1.

Transplanting and seeding methods are utilized to place live vegetal material on hillslope surfaces. Tree and shrub transplants, in the form of bareroot or container stock, are manually planted in most cases. Seeding may employ either drilling or broadcasting. Gray and Leiser (1982) indicated that drilling is the most satisfactory method of seeding herbaceous plants because the seeds are placed in the ground at the proper depth and covered; drilling is also more economical of seed and fertilizer than broadcast seeding. Likewise, Wright *et al.* (1978) remarked that seeding with a grain drill or other machines for seed coverage improves the moisture status in the soil–seed zone. Of course, seeds should be drilled along the contour wherever possible.

Gray and Leiser (1982) noted that broadcasting is used on most erosion control projects, especially for steep, rough, and otherwise inaccessible sites. Broadcasting of dry seed commonly employs a "breast seeder" or "cyclone seeder." Wright *et al.* (1978) suggested that hydromethods are the most practical for applying mulch, seed, and soil amendments on steep slopes. However, Gray and Leiser (1982) observed that these techniques are labor and equipment intensive and are much more expensive as a rule than other methods. Gray and Leiser (1982) and Amimoto (1981) provided additional discussion of planting procedures.

For rapid establishment of vegetation cover, grass sod is often used for lawns in residential developments and sometimes on critical or problem sites along highway corridors. This is also an expensive alternative and usually requires maintenance on a regular basis in order to sustain vigorous growth and aesthetic appeal. The U.S. Soil Conservation Service (1978) and U.S. Soil Conservation Service *et al.* (1983) offer guidance for this procedure.

Although a maintenance-free vegetation cover is a common revegetation objective, very often the species established will require periodic attention. Fertilizers and weed killers may be applied to the surface each year, and grasses generally require mowing, both to maintain and improve cover but also for safety reasons along roadways. The U.S. Soil Conservation Service *et al.* (1983) provided some examples of maintenance programs for various seed mixes and sod plantings.

While structural and vegetal practices for erosion and sediment control at construction sites are discussed separately in the foregoing sections, it is often possible to combine these techniques. Gray and Leiser (1982) referred to such combinations as "biotechnical slope protection and erosion control," and argued rather convincingly in their favor. Both biological and mechanical elements must function together in an integrated and complementary manner. Just as the principles of statics and mechanics are used to analyze and design biotechnical hillslope protection systems, so too must principles of horticulture and plant science be employed. The vegetative elements should not be regarded as a cosmetic adjunct to the structure.

In assessing the role of each element, Gray and Leiser observed that vegetation offers the best long-term protection against surface erosion on hillslopes and provides some degree of protection against shallow mass movements. Vegetation is self-regulating and self-repairing to a certain extent. Vegetative hillslope protection measures are also less costly than structural measures. Properly designed structures help to stabilize the hillslope against mass movement, and they protect the toe or face of a hillslope against scour and erosion by running water.

Beyond the fact that plants and structures can function together in mutually reinforcing or complementary roles, Gray and Leiser (1982) listed the following additional benefits of the biotechnical approach: (1) cost effectiveness, (2) environmental compatibility, (3) use of indigenous, natural materials, and (4) labor skill intensiveness. They remark that biotechnical hillslope protection methods are employed along both major highways and secondary road systems and that they should be considered for future use to protect hillslopes in housing developments and construction sites. Although it is not possible to fully describe the various combinations possible, Table 13 provides examples from the Gray and Leiser (1982) reference. Those engaged in the stabilization of hillslopes at construction sites are encouraged to peruse this entire text.

Erosion Control Practices for Channels: Temporary and Permanent

Natural channels are modified and new drainageways are created through the grading of construction lands. Hence, we shall consider two categories of conduits herein, disturbed natural streams and engineered conveyance channels. As with hillslopes, the practices for stabilization of each are temporary or permanent and structural or vegetal.

Streams and channels usually occur on a variety of scales at a construction site from the small side ditches along highways, to median drainages between adjacent lanes of large superhighways, to diversions on relatively flat ground, to ephemeral streams and perennial rivers. Nevertheless, the hydraulic and hydrologic principles of stabilization remain essentially the same. In broad terms, the objective is to maintain sufficient volume and velocity of flow to transport the sediment load, thus avoiding in-channel deposition and increased flooding potential, while preventing active erosion of the channel bed and banks. Selection, design, and implementation of stabilization practices is no mean challenge.

In forthcoming sections, we discuss techniques applicable to a variety of circumstances but perhaps more typical of development lands and large areas disturbed by highway construction. Those especially interested in techniques for roadway side ditches and median drainages are directed to Wright *et al.* (1978) and Amimoto (1981).

Boysen (1977) suggested that temporary structural practices can be divided into two groups: (1) water control (erosion control) and (2) sediment control. There are three basic types of water control techniques: (1) minidiversion terraces, or dikes; (2) miniwaterways, or swales, and (3) grade stabilization structures, both flume and pipe types. Sediment control practices, according to Boysen, consist of (1) sediment traps, for drainage areas up to 2 ha (5 acres), and (2) sediment basins, for drainage areas up to 40 ha (99 acres). Filter barriers are included within the category of sediment traps. The use of channels (dikes or swales) for water control was previously discussed in the section concerning the protection of hillslopes. Actually, grade stabilization structures may be either temporary or permanent. Often, they require extensive construction procedures and tend to be rather expensive. Hence, they are probably best regarded as permanent features and so will be discussed later.

The utilization of sediment control structures is akin to "closing the barn doors after the horses have run out." Erosion has already affected the surface of the

Table 13 Approaches to Hillslope Protection and Erosion Control[a]

Category	Examples	Appropriate uses	Stabilizing mechanism or role of vegetation
Live construction			
Conventional plantings	Grass seeding Transplants	Control of surficial rainfall and wind erosion To minimize frost effects	To bind and restrain soil particles To filter soil from runoff To intercept raindrops To maintain infiltration To change thermal character of ground surface
Mixed construction			
Woody plants used as reinforcement and as barriers to soil movement	Live staking Contour wattling Brush layering Reed trench terracing Brush mats	Control of surficial rainfall erosion (rilling and gullying) Control of shallow (translational) mass movement	Same as above, but also to reinforce soil and resist downslope movement of earth masses by buttressing and soil arching action
Woody plants grown in interstices of low, porous structures or benches of tiered structures	Vegetated revetments (riprap, grids, gabion mats, blocks)	Control of shallow mass movements and resistance to low–moderate earth forces	To reinforce and indurate soil or fill behind structure into monolithic mass.
	Vegetated retaining walls (open cribs, gabions, stepped back walls, and welded wire walls)	Improvement of appearance and performance of structures	To deplete and remove moisture from soil or fill behind structure.
Toe walls at foot of slope used in conjunction w/ plantings on the face	Low, breast walls (stone, masonry, etc.) with vegetated slope above (grasses and shrubs)	Control of erosion on cut and fill slopes subject to undermining at the toe	To stop or prevent erosion on slope face above retaining wall
Inert construction			
Conventional structures	Gravity walls Cantilever walls Pile walls Reinforced earth walls	Control of deep-seated mass-movement and restraint of high latitude earth forces Retention of toxic or aggressive fills and soil	Mainly decorative role

[a]Source: Gray and Leiser, 1982.

watershed. Topsoil has been removed and the hillslopes rilled or gullied. In the process, soil amendments, seeds, and mulches have been washed from the land. Nevertheless, some sediment production is likely to result from disturbed lands, and it is important to contain this material on or near site in order to minimize additional environmental impact.

Filter barriers (sediment barriers) are placed between the hillslope and channel or along the property boundary to prevent damage to the stream or adjacent land. These devices take several forms but all serve to slow and filter the runoff from an area. Filter berms are temporary ridges of porous materials such as gravel, crushed rock, or sand. Straw bale barriers are a type of filter berm. Filter fences consist of snow fence or other fencing material lined with a filter fabric. These are particularly useful where space is limited. Finally, vegetation filter strips are bands of natural or planted vegetation cover positioned across the flow of runoff and through which it must pass before entering a stream, storm sewer or conduit. Ohlander (1976) demonstrated a technique for estimating the sediment trapping characteristics of a vegetative filter strip, but additional research is required in this area.

Sediment traps are built through excavation or the construction of impervious berms. Here, runoff is temporarily impounded, allowing time for the sediment load to settle. Tryon *et al.* (1976) asserted that a decade of experience on large earth-moving jobs in the rugged Missouri Ozarks shows that excavated sediment traps are incomparably superior to small detention dams in terms of cost, industry acceptance, and sediment trap efficiency. Such structures can be a part of a permanent erosion and sedimentation control plan as well. All sediment traps including filters and structures require periodic maintenance in order to sustain their function. The U.S. Soil Conservation Service *et al.* (1983) and White and Franks (1978) presented detailed specifications for the design and construction of sediment traps.

Sediment basins (sediment retention basins, sediment detention basins, and debris basins) are small reservoirs that again hold flows allowing the sediment to settle. Boysen (1977) noted that they are highly effective and are the most complicated practice to design and construct. White and Franks (1978) suggested that, while they are frequently a part of a permanent erosion and sedimentation control plan, they should be used as the "last line of defense" against off-site sediment pollution and employed only if drainage control and stabilization of disturbed areas prove inadequate. The Highway Research Board (1973) observed that sediment basins are now included in the erosion control effort of most highway projects where rainfall is anticipated during the construction period. The Board stated that temporary basins must be maintained until permanent erosion control efforts are effective. Again, design and construction specifications are available through the U.S. Soil Conservation Service *et al.* (1983), Amimoto (1981), and U.S. Soil Conservation Service (1978).

Reed (1978, p. 1) examined several methods for controlling erosion and sediment transport during highway construction in four adjacent drainage basins of central Pennsylvania. From this research he concluded:

> The effectiveness of the various sediment-control measures were determined by comparing the sediment loads transported from the basins with sediment controls to those without controls. For most storms the offstream (sediment) ponds trapped about 60 percent of the sediment that reached them. The large onstream pond had a trap efficiency of about 80 percent; however, it remained turbid and kept the stream flow turbid for long periods following storm periods. Samples of runoff water from the construction area were collected above and below rock dams to determine the reduction in sediment as the flow passed through the device. Rock dams in streams had a trap efficiency of about 5 percent. Seeding and mulching may reduce sediment discharge by 20 percent during construction, and straw bales placed to trap runoff water may reduce sediment loads downstream by 5 percent.

The effectiveness of the practices investigated in this study seems low in comparison to others. Recall that Yorke and Herb (1976) found a 60–80% reduction in sediment yields from construction sites during 1966 to 1974 as a result of improvements in grading and erosion control measures.

The banks of streams and drainageways are temporarily stabilized using protective materials or structural devices. The application of coarse rock or other materials is known as riprapping and is commonly considered a permanent practice. The U.S. Soil Conservation Service (1978) described the use of willow poles driven or jetted into the eroding bank at or just below the normal waterline. These poles can be braced with wire after emplacement in the bank. Likewise, materials of various sorts can be used to protect or armor the bed of channels. Once it becomes established, vegetation planted in ephemeral water courses helps stabilize the channel. A potential problem here, however, is that a single flow prior to establishment can easily wash away amendments and seeds. At high velocities even sod covers are removed in a relatively short time.

Permanent channel stabilization practices also utilize structures and vegetation covers. Sometimes, culverts are used to transport flows. These are conduits that provide free passage of water under a highway, canal, or other embankment. Culverts are usually constructed of corrugated metal or reinforced concrete pipes. Amimoto (1981) provided information concerning their design and use.

Beside the aforementioned sediment traps and basins, other permanent structural erosion and sedimentation control measures include grade, bank, and bed stabilization devices. Amimoto (1981) described grade control measures as structures that reduce and maintain the channel gradients. The U.S. Soil Conservation Service (1978) expanded this definition, stating that grade stabilization structures are used to reduce or prevent excessive erosion by reduction in velocities in the watercourse *or* by providing channel linings or structures that can withstand high velocities. They also noted that these practices are applied to sites where the capability of earth and vegetative measures is exceeded in the safe handling of water at permissible velocities, where excessive grades or overfall conditions are encountered, or where water is to be lowered structurally from one elevation to another.

Numerous physical structures have been developed for grade stabilization and erosion control in channels. Amimoto (1981) discussed three categories: (1) check dams, (2) drop structures (spillways), and (3) erosion stops or erosion checks. Check dams are structures used to stabilize the grade and control head-cutting in natural or artificial channels. Drop structures are weirs in which waterflow passes through a weir opening, drops to an approximate level apron or stilling basin, then passes into the downstream channel. Erosion stops consist of porous, matlike material installed in a slit trench that is oriented perpendicular to the direction of flow in a ditch or swale. The mat can be composed of fiberglass, jute, plastic, or other flexible porous materials.

Channel linings of concrete, asphalt, or riprap are used where flow is of long duration, runoff velocities are high, gradients are steep, or vegetation cannot be established to successfully resist erosion (DRCOG, 1980). Such structures extending from the crest to the base of hillslopes are called "paved chutes" or "flumes." A revetment is a facing of stone or other material, either permanent or temporary, placed along the perimeter of a stream to stabilize the bank and to protect it from the erosive action of the stream (Amimoto, 1981). The U.S. Soil Conservation Service (1978) remarked that riprap is one of the most effective methods of streambank protection. The procedure involves placement of a bedding or filter blanket on the exposed streambank; then a layer of loose rock aggregate is spread upon this substratum. The toe of this structure must be firmly established. Linder (1976) described the design and performance of rock revetment toes, and the U.S. Soil Conservation Service (1975, 1978) also provided guidance for riprapping of streambanks. Several other references for riprap design are contained in Chapter 9 (on surface mining of uranium).

Gabions are also used to prevent streambank erosion. As described by Amimoto (1981), these are rock-filled, galvanized steel wire cages that when attached together form large, flexible, permeable protective blocks. Relatively small rock fragments or gravel 4–8 in. in diameter can be used to make a structure capable of streambank protection. The flexibility of gabions permits them to withstand differential settlement without fracturing. Further, the permeability of gabions prevents hydrostatic heads from developing behind the structure.

The bed of streams can be stabilized using lining materials or large rocks. This is a form of armoring that sometimes occurs naturally in streambeds. Of course, the materials employed must be able to resist the force of relatively high velocity flows.

Although the structures mentioned here are regarded as permanent, they nevertheless possess a finite design life. For the time frame entrenched in the minds

A

Treatment Practice	Advantages	Problems
ROADWAY DITCHES		
Check Dams	Maintain low velocities Catch sediment Can be constructed of logs, shot rock, lumber, masonry or concrete	Close spacing on steep grades Require clean-out Unless keyed at sides and bottom, erosion may occur
Sediment Traps/ Straw Bale Filters	Can be located as necessary to collect sediment during construction Clean-out often can be done with on-the-job equipment Simple to construct	Little direction on spacing and size Sediment disposal may be difficult Specification must include provisions for periodic clean-out May require seeding, sodding or pavement when removed during final cleanup
Sodding	Easy to place with a minimum of preparation Can be repaired during construction Immediate protection May be used on sides of paved ditches to provide increased capacity	Requires water during first few weeks Sod not always available Will not withstand high velocity or severe abrasion from sediment load
Seeding with Mulch and Matting	Usually least expensive Effective for ditches with low velocity Easily placed in small quantities with inexperienced personnel	Will not withstand medium to high velocity
Paving, Riprap, Rubble	Effective for high velocities May be part of the permanent erosion control effort	Cannot always be placed when needed because of construction traffic and final grading and dressing Initial cost is high

B

Treatment Practice	Advantages	Problems
PROTECTION OF STREAM		
Construction Dike	Permits work to continue during normal stream stages Controlled flooding can be accomplished during periods of inactivity	Usually requires pumping of work site water into sediment pond Subject to erosion from stream and from direct rainfall on dike
Cofferdam	Work can be continued during most anticipated stream conditions Clear water can be pumped directly back into stream No material deposited in stream	Expensive
Temporary Stream Channel Change	Prepared channel keeps normal flows away from construction	New channel usually will require protection Stream must be returned to old channel and temporary channel refilled
Riprap	Sacked sand with cement or stone easy to stockpile and place Can be installed in increments as needed	Expensive
Temporary Culverts for Haul Roads	Eliminate stream turbulence and turbidity Provide unobstructed passage for fish and other water life Capacity for normal flow can be provided with storm water flowing over the roadway	Space not always available without conflicting with permanent structure work May be expensive, especially for larger sizes of pipe Subject to washout
Rock-lined Low-Level Crossing	Minimizes stream turbidity Inexpensive May also serve as ditch check or sediment trap	May not be fordable during rainstorms During periods of low flow passage of fish may be blocked

C

Treatment Practice	Advantages	Problems
PROTECTION OF ADJACENT PROPERTY		
Energy Dissipators	Slow velocity to permit sediment collection and to minimize channel erosion off project	Collect debris and require cleaning Require special design and construction of large shot rock or other suitable material from project
Level Spreaders	Convert collected channel or pipe flow back to sheet flow Avoid channel easements and construction off project Simple to construct	Adequate spreader length may not be available Sodding of overflow berm is usually required Must be a part of the permanent erosion control effort Maintenance forces must maintain spreader until no longer required
Brush Barriers	Use slashing and logs from clearing operation Can be covered and seeded rather than removed Eliminates need for burning or disposal off ROW	May be considered unsightly in urban areas
Straw Bale Barriers	Straw is readily available in many areas When properly installed, they filter sediment and some turbidity from runoff	Require removal Subject to vandal damage Flow is slow through straw requiring considerable area
Sediment Traps	Collect much of the sediment spill from fill slopes and storm drain ditches Inexpensive Can be cleaned and expanded to meet need	Do not eliminate all sediment and turbidity Space is not always available Must be removed (usually)
Sediment Pools	Can be designed to handle large volumes of flow Both sediment and turbidity are removed May be incorporated into permanent erosion control plan	Require prior planning, additional ROW and/or flow easement If removal is necessary, can present a major effort during final construction stage Clean-out volumes can be large Access for clean-out not always convenient

Figure 17. A, Some stabilization practices for roadway ditches (Highway Research Board, 1973). B, Some stabilization practices for streams (Highway Research Board, 1973). C, Some stabilization practices for protection of adjacent lands (Highway Research Board, 1973).

of most earth scientists, namely, geologic time, they are decidedly temporary. Each structure requires periodic maintenance if it is to continue to function as intended over the long haul.

In addition to the foregoing, there are vegetative options for the stabilization of channels. Amimoto (1981) observed that vegetative linings in waterways reduce the erosion along the banks and provide for the filtration of sediment. Further, vegetative linings give an aesthetically pleasing appearance and improve wildlife habitat. Grasses, woody plant species, or a mixture of both compose an effective lining.

As with structural practices, the vegetative options are likely to require some maintenance. Should the vegetation cover be breeched by disease, fire, trampling, or the forces resulting from high flow velocities, barren areas must be quickly repaired. Erosion of the exposed surface material leads to the undermining and failure of progressively larger areas of cover during subsequent runoff events. The sediment produced can clog structures downstream and produce flooding.

Although no one has apparently developed the topic of structural and vegetative practices used in combination for channels, as Gray and Leiser (1982) have for hillslopes, several authors mention various possibilities. The U.S. Soil Conservation Service (1978) suggested that supplemental measures may be required for grassed waterways. These include such things as (1) grade control structures, (2) subsurface drainage to permit the growth of suitable vegetation and to eliminate wet spots that can be a public muisance, and (3) a paved channel bottom or buried storm drain to handle frequently occurring storm runoff or baseflow. Elsewhere, they discuss the construction of channels with stone centers and vegetated streambanks. Figure 17 illustrates several practices employed on roadway ditches, streams, or waterways and for the protection of adjacent land.

Lastly, it must be emphasized again that hillslopes and channels can be discussed separately, but in reality they function as component parts of an integrated system, the drainage basin. Failure to stabilize one component can eventually lead to the failure of the other. Accelerated hillslope erosion dramatically alters the sediment load of adjacent streams, and this produces a geomorphic response in channel gradient or pattern. Likewise, degrading channels can undercut hillslopes, turning a concave or convex–concave profile into a convex profile and generating greater erosion rates, rilling, or even gullying of these hillslopes. Hence, the effectiveness of construction site stabilization should be evaluated for the entire disturbed area, rather than for hillslopes and channels separately.

Geomorphology and Construction Site Stabilization

The objectives of construction site stabilization are actually quite simple, in comparison to those of surface mine reclamation. Perusal of the literature on this subject leaves the impression that mitigation and control of erosion and sedimentation are the overwhelming concerns. For many areas, the term erosion connotes entrainment and transportation of surface materials by water alone, although in some places, such as the State of Colorado, there is an effort to control wind erosion as well.

Further, the utilization of structural practices are allowed and even encouraged on occasion, despite the inevitable maintenance requirements and costs over extended time periods. The commitment of present and future funds for this purpose seems to be accepted as a cost associated with the particular land use. And it is recognized that highway construction or urbanization are, in all likelihood, permanent changes in land use. The metamorphosis of a natural landscape into an anthropic landscape is viewed as a cost of societal growth and development.

From the geomorphic perspective, construction operations constitute another disruption of the adjustment, or perhaps, equilibrium between landforms and geomorphic processes. When conditions of equilibrium prevail, it can be inferred that there is generally some balance between the forces associated with geomorphic processes and the resistances offered by surface materials, so that geomorphic work is usually performed at a relatively moderate rate. Under conditions of disequilibrium, forces often greatly exceed resistances and work is accomplished at high, or even extremely high, rates. At these times, on-site impacts can easily become serious and widespread off-site impacts.

The geomorphologist then, views stabilization as a process by which some semblance of equilibrium is reestablished at the disturbed site. Stability is a state in which slight perturbations of the variables defining an environmental system do not lead to a progression toward a new steady or equilibrium state. The concept of stability is discussed in greater detail in Chapter 9 (on the mining of uranium).

The geomorphologist is also keenly aware that many geomorphic processes operate simultaneously at a single site, although some may be imperceptible. The magnitude and frequency of each process is a function of the environmental setting, including such parameters as climate, geology, relief or topography, soils, and vegetation. Erosion and sedimentation are merely two of these processes, albeit commonly among the dominant ones.

And finally, the geomorphologist is atune to the operation of geomorphic processes throughout both short and long time frames. Stresses often tend to build and resistances often tend to diminish through time until a threshold is surpassed and work is performed. It is appropriate, from this perspective, that stabilization programs be founded upon geomorphic principles.

Initially it might seem that the stabilization process should attempt to recreate hillslopes and channels as they existed prior to disturbance. However, it is clear that changes in land use substantially alter many hydrologic and geomorphic characteristics of the construction site. Soil properties cannot be precisely reconstructed, even where topsoil is stripped in a separate step, stockpiled, and redistributed. Soil profiles and structures are lost, only to be regenerated by natural pedogenic processes over long durations. As a result, the quality of the soil as a growth medium and its erodibility are modified to some extent. In some cases, subsoil becomes the surface material where topsoil is not conserved. Both the disturbed topsoil and subsoil probably possess infiltration capacities different from those of natural soils.

Vegetation changes occur as a consequence of construction activities. Percentage of cover and root network density may develop toward values similar to those of predisturbance conditions with time. Depending upon revegetation practices, the species planted may not be those indigenous to the area. Cultural preferences often influence these choices. For example, urbanization in many semiarid regions frequently results in the replacement of "Buffalo grasses" with Kentucky bluegrasses that require extensive care.

The average and peak flow of streams passing through these areas increases markedly because a considerable proportion of the site is rendered impervious and because storm drainage systems convey surface flow with great efficiency. Thus, the streams and artificial waterways must contend not with discharges typical of the predisturbance period but with those higher flows of the postdisturbance period.

And so the environmental setting within which geomorphic processes operate are permanently altered by construction activities. Hence, the magnitude and frequency of these processes are expected to change. Therefore, any hillslope or channel forms that were in equilibrium with the environment prior to disturbance are unlikely to be in equilibrium with the "new" environment. What are the equilibrium landforms, then? At the present state of the art, geomorphologists can make reasonable estimates but cannot determine proper configurations with precision. Subsequent natural adjustment is inevitable following stabilization practices.

In the mitigation and control of erosion and sedimentation, the common goal is to reduce the rate of geomorphic work performed by these processes. This is accomplished by reducing the forces impinging upon the surface while concomitantly increasing the resistance of that surface. In simple terms, this means that the best strategy is to hold the soil on the hillslopes and prevent degradation of the stream channels.

Hillslopes

Many erosion control practices for hillslopes are well established as a result of research on agricultural lands. Zingg (1940) and Wischmeier and Smith (1965) documented the relationship between soil loss and the hillslope properties of gradient and length. As gradient decreases, the downslope component of gravitational force exerted on surface materials and the velocity of overland flow (and hence tractive force) decrease. Additionally, the normal component of gravitational force, which holds particles in place on the surface, increases. According to the Horton (1945) model, which appears acceptable for barren, disturbed surfaces, as hillslope length decreases, the depth and velocity of overland flow decrease. As a result, the tractive force produced by the flow also decreases. Meyer and Romkens (1976) discussed the relationship between erosion rates and profile shape. For convex hillslopes the steepest parts are near the toe where accumulated runoff is greatest, while for a concave hillslope the steepest parts occur where runoff is least.

Surface manipulations increase infiltration, increase surface roughness, and create surface depressions that serve as small water and sediment traps. The first two functions decrease the volume and velocity of runoff and, therefore the tractive force impinging upon surface materials.

The spreading of topsoil on the disturbed surface can reduce runoff volumes when properly handled because this material commonly has a higher infiltration capacity than subsoil.

Mulches and stabilizers increase the resistance of the surface to the forces of rain splash and overland flow. Additionally, they add another roughness factor to the surface and thereby reduce the velocity and tractive force exerted by runoff. Any reduction in flow velocity promotes deposition, and so a part of the sediment load carried by these flows from upslope may be trapped in the mulch along with seed and fertilizers.

Establishment of vegetation cover is critical to successful hillslope stabilization in almost every situation. The total force impinging upon the soil surface is greatly reduced because the energy of rainsplash is expended on the vegetal surfaces and because these same surfaces offer frictional resistance to overland flow with the effect of reducing its velocity. Further, vegetated surfaces possess higher infiltration capacities than their barren counterparts. And finally, root networks bind soil masses together to some extent and hence increase the resistance of individual particles to entrainment and transportation. Research summarized by Wischmeier and Smith (1965) attests to the value of vegetation cover in reducing soil loss. It might be appropriate to regard revegetation as "biological armoring" of the soil surface.

From the foregoing, it is apparent that stabilization practices operate in various ways to both reduce force and increase resistance. The rate of geomorphic work is correspondingly reduced. And, as a result, the amount of sediment arriving at the streams is decreased.

Channels

Sediment control measures near or in streams constitute a second and, perhaps, last line of defense in the avoidance of off-site impact. The facilities discussed usually operate on the same principle, namely, to reduce flow velocity so that the probability of bank and bed erosion is lessened and deposition takes place at predetermined locations rather than at various uncontrolled places along the watercourse. It is somewhat surprising to realize that these facilities actually can be too efficient. Should the water discharged from the basin contain a very small sediment load, it is possible that the flow will erode its channel downstream from the structure; this has occurred below some major dams.

Despite successes in controlling erosion and sedimentation, the problem of increased discharge from the disturbed drainage basin remains. Research by geomorphologists, such as Leopold and Maddock (1953) and Schumm (1977), tells us that the channel will increase in width and depth as flow velocities increase in order to accomodate this greater discharge. These changes are frequently incompatible with land uses adjacent to the stream or constitute a hazard to public safety in populated areas. There seems to be little practical alternative other than extensive and elaborate control of the channel through a series of engineering structures.

From the geomorphic perspective, the use of artificial structures for the stabilization of construction sites is somewhat troublesome. Although their necessity is often apparent, our understanding of weathering processes and the traditional long-term viewpoint forces acknowledgment of the fact that they are inevitably temporary. Some geomorphologists might regard such structures as merely "crutches" supporting the fascade of equilibrium for a time, a very brief time on the geologic scale.

However, geomorphologists must also accept the fact that land disturbance through construction is inevitable. Further, the state-of-the-art in geomorphology is presently incapable of providing nonstructural alternatives in many circumstances. Thus, structures become a matter of necessity rather than preference. The combination "biotechnical approach" advocated by Gray and Leiser (1982) seems to represent progress in the right direction.

Summary

The construction process generally consists of three phases. First, the construction site surface is prepared. Next, structures are built. Finally, steps are taken to stabilize the disturbed land. In assessing the impact of construction activities, the surface area involved is relatively small in comparison to the total area of the United States, but much greater than the amount of area affected by surface mining. There are considerable data showing that construction activities cause dramatic increases in the intensities or rates of geomorphic processes. The duration of impact depends upon the amount of time that the construction site remains "open" and the effort invested in subsequent stabilization. This can range from a few months to a few years.

Extensive project planning greatly reduces the extent of environmental impact. Some protective measures, such as water diversion around the construction site, are undertaken even before preparation of the surface. Other practices are implemented during the active construction phase to control erosion and sedimentation. These not only reduce on-site impacts but should virtually eliminate the off-site impacts if properly designed and implemented.

Although numerous geomorphic processes operate at a given site, stabilization programs strongly emphasize control of erosion and sedimentation. Stabilization practices are generally either structural or vegetative. Structures are designed according to accepted specifications and standards reflecting the state-of-the-art in engineering. Vegetative practices are commonly adopted from the techniques employed for soil conservation on agricultural lands and adapted to the specific situations encountered on construction sites. Gray and Leiser (1982) advocated utilization of various combinations of structural and vegetative practices and cited several advantages of this approach.

The effects of land disturbance by construction activities and the consequences of various stabilization measures can be examined in terms of force, resistance, and resultant geomorphic work. During site preparation and active construction phases, the forces impinging upon the surface could remain about the same in unique situations but in all likelihood will increase considerably. At the same time, there is a marked decrease in surface resistance. Geomorphic work, principally in the form of erosion, can progress at a substantially accelerated rate as a result. Left unchecked, these on-site impacts are transmitted off-site, greatly expanding the total area of disturbance.

From the geomorphic perspective, the objective of construction site stabilization is the reestablishment of some semblance of equilibrium between the landscape and the environment. More specifically, there should be an approximate balance between landform and process. This, in turn, connotes comparability in the forces and resistances operating on the disturbed surface. Under such conditions, geomorphic work is accomplished at a moderate rate (some might say a "normal" rate for a particular environmental setting) and the land is not degraded excessively on- site, nor are the consequences of the disturbance conveyed off-site.

Lastly, one additional point is noteworthy. Cooke (1984) commented that urban growth is directly responsible for transforming geomorphic processes into community hazards. Settlements have extended into areas where hillslope erosion, hillslope failure, and flooding have always occurred. Geomorphic processes are natural phenomena that have only become serious hazards because they have increasingly imposed themselves upon a vulnerable, often unsuspecting, and rapidly growing urban community.

References

Amimoto, P. Y., 1981, Erosion and sediment control handbook: State of California, Division of Mines and Geology, Department of Conservation for U. S. Environmental Protection Agency, Report No. EPA 4 40/3–78–003, 198 pp.

Anderson, D. G., 1970, Effects of urban development on floods in northern Virginia: U. S. Geological Survey, Water Supply Paper 2001-C, 22 pp.

Anderson, H. W., 1975, Relative contribution of sediment from source areas and transport processes: Present and prospective technology for predicting sediment yields and sources, U.S.D.A., Agricultural Research Service, ARS-S–40, pp. 66–73.

Berglund, E. R., 1976, Seeding to control erosion along forest roads: Oregon State University Extension Service, Extension Circular 885, 19 pp.

Beschta, R. L., 1978, Long-term patterns of sediment production following road construction and logging in the Oregon Coast Range: Water Resources Research, V. 14, No. 6, pp. 1011–1016.

Bethlahmy, N., and Kidd, W. J., 1966, Controlling soil movement from steep road fills: U.S.D.A., Forest Service, Intermountain Forest and Range Experiment Station, Research Note INT–45, pp. 1–4.

Blaser, R. E., 1975, Erosion control during road construction: Rural and Urban Roads, April 1975, pp. 38–40.

Blaser, R. E., Thomas, G. W., Brooks, C. R., Shoop, G. J., and Martin, J. B., 1961, Turf establishment and maintenance along highway cuts: National Academy of Science, Highway Research Board, Committee on Roadside Development, pp. 5–19.

Borth, C., 1969, Mankind on the move; the story of highways: Washington, D. C., Automotive Safety Foundation, 314 pp.

Boysen, S. M., 1977, Erosion and sediment control in urbanizing areas: Soil Erosion and Sedimentation, Proceedings of National Symposium on Soil Erosion and Sedimentation by Water, American Society of Agricultural Engineers, pp. 125–136.

Briggs, W. M., 1973, Inventory of roadside erosion in Wisconsin: Soil Erosion— Causes and Mechanisms; Prevention and Control: Proceedings of a Conference-Workshop, Highway Research Board, Special Report 135, pp. 77–81.

Brosky, D. L., 1966, Solids in a small urban watershed at extreme flows: unpublished manuscript, 29 pp.

Brownlie, W. R., and Taylor, B. D., 1981, Sediment management for southern California, coastal plains and shoreline: Part C, Coastal sediment delivery by major streams in southern California, California Institute of Technology, EQL Report 17-C.

Busby, C. E., 1962, Some legal aspects of sedimentation: American Society of Civil Engineers, Transactions, Vol. 127, Pt. I, pp. 1007–1044.

Busby, C. E., 1967, Aspects of American sedimentation law: Journal of Soil and Water Conservation, Vol. 22, No. 3, pp. 107–109.

Carter, R. W., 1961, Magnitude and frequency of floods in suburban areas: U. S. Geological Survey, Professional Paper 424-B, pp. B9-B11.

Chassie, R. G., and Goughnour, R. D., 1976, National highway landslide experience: Highway Focus, Vol. 8, No. 1, pp. 1–9.

Chen, C. N., 1974, Effect of land development on soil erosion and sediment concentration in an urbanizing watershed: Proceedings of the Paris Symposium, Effects of Man on the Interface of the Hydrological Cycle with the Physical Environment, IAHS Publication No. 113, pp. 150–157.

Cohen, P., Franke, O. L., and Foxworthy, B. L., 1968, An atlas of Long Island's water resources: New York Water Resources Committee Bulletin, Bulletin 62, 117 pp.

Collier, C. R., Pickering, R. J. and Musser, J. J., 1962, Influence of strip mining on the hydrologic environment of parts of Beaver Creek basin, Kentucky, 1955–1959: U. S. Geological Survey, Professional Paper 427-B, 83 pp.

Commonwealth of Virginia, 1975, Erosion and Sediment Control Law: Virginia Annotated Code.

Cooke, R. U., 1984, Geomorphological hazards in Los Angeles: London, George Allen and Unwin Ltd., 206 pp.

Cox, W. E., and Walker, W. R., 1977, Legal controls applicable to soil erosion: *in* Proceedings of a National Conference on Soil Erosion, Soil Erosion: Prediction and Control, Soil Conservation Society of America, pp. 305–313.

Denver Regional Council of Governments, 1980, Managing erosion and sedimentation from construction activities: Denver Regional Council of Governments, 101 pp.

Diseker, E. G., and McGinnis, J. T., 1967, Evaluation of climatic, slope, and site factors on erosion from unprotected roadbanks: Transactions of American Society of Agricultural Engineers, Vol. 10, pp. 9–14.

Diseker, E. G., and Richardson, E. C., 1962, Erosion rates and control methods on highway cuts: Transactions of American Society of Agricultural Engineers, Vol. 5, pp. 153–155.

Dunne, T., and Leopold, L. B., 1978, Water in environmental planning: San Francisco, W. H. Freeman and Co., 818 pp.

Emerson, J. W., 1971, Channelization—a case study: Science, Vol. 173, pp. 325–326.

Federal Highway Administration, 1974, Federal-Aid Highway Program Manual: U. S. Department of Transportation, Vol. 6, Chap. 7, Sec. 3, Subsec. 1, Erosion and Sediment Control on Highway Construction Projects, 5 pp. (plus attachments)

Frey, H. T., 1983, Expansion of urban area in the United States—1960–1980: U.S.D.A., Economic Research Service, Natural Resource Economic Division, ERS Staff Report No. AGES830615, 16 pp.

Glover, F., Augustine, M., and Clar, M., 1978, Grading and shaping for erosion control and rapid vegetative establishment in humid regions: *in* Schaller, F. W., and Sutton, P. (eds.), Reclamation of Drastically Disturbed Lands: Madison, Wisconsin, American Society of Agronomy, Crop Science Society of America, Soil Science Society of America, pp. 271–283.

Graf, W. L., 1976, Streams, slopes, and suburban development: Geographical Analysis, Vol. VIII, pp. 153–173.

Gray, D. H., and Leiser, A. T., 1982, Biotechnical slope protection and erosion control: New York, Van Nostrand Reinhold Company, 271 pp.

Green, J. T., Woodruff, J. M., and Blaser, R. E., 1973, Stabilizing disturbed areas during highway construciton for pollution control: Final Report for the Virginia Department of Highways, Virginia Highway Research Council, U. S. Department of Transportation, and the Federal Highway Administration, 68 pp.

Greenwood, N. J., and Edwards, J. M. B., 1979, Human Environments and Natural Systems: North Scituate, Massachusetts, Duxbury Press, 548 pp.

Gregory, K. J., and Walling, D. E., 1973, Drainage basin form and process: New York, John Wiley and Sons, Halsted Press, 456 pp.

Guy, H. P., 1963, Residential construction and sedimentation at Kensington, Maryland: presented at Federal Interagency Sedimentation Conference, Jackson, Mississippi, January 1963, 16 pp.

Guy, H. P., 1970, Sediment problems in urban areas: U. S. Geological Survey, Circular 601-E, 8 pp.

Guy, H. P., and Ferguson, G. E., 1962, Sediment in small reservoirs due to urbanization: Proceedings of American Society of Civil Engineers, Vol. 88, No. HY2, pp. 27–37.

Guy, H. P., and Ferguson, G. E., 1970, Stream sediment—an environmental problem: Journal of Soil and Water Conservation, Vol. 25, No. 6, pp. 217–221.

Hadley, R. F., 1962, Some effects of microclimate on slope morphology and drainage basin development: U. S. Geological Survey, Geological Survey Research, 1961, pp. B–32 to B–33.

Hafley, W. L., 1975, Rural road systems as a source of sediment pollution—a case study: Watershed Management Symposium, American Society of Civil Engineers, pp. 393–405.

Haigh, M. J., 1984, Microerosion processes and sediment mobilization in a road-bank gully catchment in central Oklahoma: *in* Burt, T. P., and Walling, D. E. (eds.), Catchment Experiments in Fluvial Geomorphology, Norwich, England, Geo Books, pp. 247–264.

Hammer, T. R., 1972, Stream channel enlargement due to urbanization: Water Resources Research, Vol. 8, No. 6, pp. 1530–1540.

Heft, F. E., 1977, Political, social, and economic aspects of soil erosion and sedimentation control programs: Soil Erosion and Sedimentation, Proceedings of National Symposium on Soil Erosion and Sedimentation, Proceedings of National Symposium on Soil Erosion and Sedimentation by Water, American Society of Agricultural Engineers, pp. 23–30.

Helm, R. E., 1978, Sediment discharge from highway construction near Port Carbon, Pennsylvania: U. S. Geological Survey, Water-Resources Investigations 78–35, 27 pp.

Highfill, R. E., and Kimberlin, L. W., 1977, Current soil erosion and sediment control technology for rural and urban lands: Soil Erosion and Sedimentation, Proceedings of National Symposium on Soil Erosion and Sedimentation by Water, American Society of Agricultural Engineers, pp. 14–22.

Highway Research Board, 1971, Construction of embankments: National Cooperative Highway Research Program, Synthesis of Highway Practice No. 8, 38 pp.

Highway Research Board, 1973, Erosion control on highway construction: National Cooperative Highway Research Program, Synthesis of Highway Practice No. 18, 52 pp.

Hilliard, R. L., 1977, The New York erosion and sediment inventory: Proceedings of a National Conference on Soil Erosion, Ankeny, Iowa, Soil Conservation Society of America, pp. 273–276.

Holberger, R. L., and Truett, J. B., 1976, Sediment yield from construction sites: Proceedings of the Third Interagency Sedimentation Conference, PB–245–100, Water Resources Council, Washington, D. C., 1–47 to 1–58.

Holeman, J. N., and Geiger, A. F., 1959, Sedimentation of Lake Barcroft, Fairfax County, Virginia: U.S.D.A., Soil Conservation Service, SCS-TP–136, 12 pp.

Holeman, J. N., and Geiger, A. F., 1965, Sedimentation of Loch Raven and Prettyboy reservoirs, Baltimore County, Maryland: U.S.D.A., Soil Conservation Service, SCS-TP–145, 17 pp.

Horton, R. E., 1945, Erosional development of streams and their drainage basins; hydrophysical approach to quantitative morphology: Bulletin, Geological Society of America, Vol. 56, pp. 275–370.

Hoy, D. R. (ed.), 1978, Geography and development: a world regional approach: New York, MacMillan Publishing Co., Inc., 728 pp.

Israelsen, C. E., Clyde, C. G., Fletcher, J. E., Israelsen, E. K., Haws, F. W., Packer, P. E., and Farmer, E. G., 1980, Erosion control during highway construction: Research Report and Manual on Principles and Practices, Transportation Research Board, National Cooperative Highway Research Program Report, Nos. 220 and 221, 30 and 23 pp.

Johnson, W., and Paone, J., 1982, Land utilization and reclamation in the mining industry, 1930–1980: U. S. Bureau of Mines, Information Circular 8862, 22 pp.

Jordon, T. G., and Rowntree, L., 1979, The Human Mosaic—A Thematic Introduction to Cultural Geography: (2nd ed.), New York, Harper and Row Publishers, 482 pp.

Keller, E. A., 1985, Environmental Geology: (4th ed.), Columbus, Charles E. Merrill Publishing Co., 480 pp.

Keller, F. J., 1962, Effect of urban growth on sediment discharge, Northwest Branch Anacostia River Basin, Maryland: U. S. Geological Survey, Professional Paper 450-C, pp. C129–131.

Krammes, J. S., and Burns, D. M., 1973, Road construction on Caspar Creek watersheds—10-year report on impact: U.S.D.A., Forest Service, Pacific Southwest Forest and Range Experiment Statio, Research Paper PSW–93, 10 pp.

Krohn, J. P., and Slosson, J. E., 1976, Landslide potential in the United States: California Geology, Vol. 29, No. 10, pp. 224–231.

Leaf, C. F., 1974, A model for predicting erosion and sediment yield from secondary forest road construction: U.S.D.A., Forest Service, Rocky Mountain Forest and Range Experiment Station, Research Note, RM–274, 4 pp.

Legget, R. F. 1973, Cities and Geology: New York, McGraw-Hill Book Company, 624 pp.

Leighton, F. B., 1969, Landslides: *in* Olson, R. A., and Wallace, M. M. (eds.), Proceedings of the Conference on Geologic Hazards and Public Problems: Santa Rosa, Office of Emergency Preparedness, Federal Regional Center 7, pp. 97–132.

Leighton, F. B., 1972, Origin and control of landslides in the urban environment of California: 24th International Geological Congress, Section 13, pp. 89–96.

Leopold, L. B., 1968, Hydrology for urban land planning—a guidebook for the hydrologic effects of urban land use: U. S. Geological Survey, Circular 554, 18 pp.

Leopold, L. B., and Maddock, T., 1953, The hydraulic geometry of stream channels and some physiographic implications: U. S. Geological Survey, Professional Paper 252, 57 pp.

Linder, W. M., 1976, Design and performance of revetment toes: Proceedings of the Third Inter-agency Sedimentation Conference, PB–245–100, Water Resources Council, Washington, D. C., 2–168 to 2–179.

McKee, W. H., Blaser, R. E., and Barkley, D. G., 1964, Mulches for steep cut slopes: Highway Research Record, Vol. 54, pp. 35–42.

McKee, W. H., Powell, A. J., Cooper, R. B., and Blaser, R. E., 1965, Microclimate conditions found on highway slope facings as related to adaptation of species: Highway Research Record, Vol. 93, pp. 38–43.

Megahan, W. F., 1972, Subsurface flow interception by a logging road in mountains of central Idaho: Proceedings of a Symposium on Watersheds in Transition, Urbana, Illinois, American Water Resrouces Association, pp. 350–356.

Megahan, W. F., 1975, Sedimentation in relation to logging activities in the mountains of central Idaho: Present and Prospective Technology for Predicting Sediment Yields and Sources, U.S.D.A., Agricultural Research Service, ARS-S–40, pp. 74–82.

Megahan, W. F., 1978, Erosion processes on steep granitic road fills in central Idaho: Soil Science Society of America Journal, Vol. 42, No. 2, pp. 350–357.

Meyer, L. D., and Romkens, M. J. M., 1976, Erosion and sediment control on reshaped land: Proceedings of the Third Inter-agency Sedimentation Conference, PB–245–100, Water Resources Council, Washington, D. C., pp. 2–65 to 2–76.

Meyer, L. D., Wischmeier, W. H., and Daniel, W. H., 1971, Erosion, runoff and revegetation of denuded construction sites: Transactions of American Society of Agricultural Engineers, Vol. 14, pp. 138–141.

Meyer, L. D., Johnson, C. B., and Foster, G. R., 1972, Stone and woodchip mulches for erosion control on construction sites: Journal of Soil and Water Conservation, Vol. 27, pp. 264–269.

Miller, G. T., 1979, Living in the Environment: (2nd ed.), Belmont, California, Wadsworth Publishing Company, 470 pp.

Morgan, R. P. C., 1980, Implications: *in* Kirkby, M. J., and Morgan, R. P. C. (eds.), Soil Erosion, New York, John Wiley and Sons, pp. 253–301.

Mowbray, A. Q., 1969, Road to Ruin: New York, J. B. Lippincott Company, 240 pp.

Nelson, J. D., and Martin, J. P., 1981, Slope stability as related to stream hazards to bridges: *in* Shen, H. W. *et al.*, Methods for Assessment of Stream-Related Hazards to Highways and Bridges, Federal Highway Administration, Offices of Research and Development, Environmental Division, Report No. FHWA/RD–80/160, pp. 87–143.

Ohlander, C. A., 1976, Defining the sediment trapping characteristics of a vegetative buffer—special case—road erosion: Proceedings of the Third Interagency Sedimentation Conference, PB–245–100, Water Resources Council, Washington, D. C., pp. 2–77 to 2–81.

Packer, P. E., 1967, Criteria for designing and locating logging roads to control sediment: Forest Science, Vol. 13, No. 1, pp. 2–18.

Patric, J. H., 1977, Soil erosion and its control in eastern woodlands: Northern Logger and Timber Processor, Vol. 25, No. 11, pp. 4, 5, 22, and 31.

Patric, J. H., and Trimble, G. R., 1972, Transition in research on small forested watersheds in West Virginia: Proceedings of a Symposium on Watersheds in Transitions, Urbana, Illinois, American Water Resources Association, pp. 272–275.

Putnam, A. L., 1972, The effects of urban development on floods in the Piedmont Province of North Carolina: U. S. Geological Survey, Open-File Report, 87 pp.

Rantz, S. E., 1971, Suggested criteria for hydrologic design of storm-drainage facilities in the San Francisco Bay region, California: U. S. Geological Survey, Open-File Report, Menlo park, California, 69 pp.

Reed, L. A., 1978, Effectiveness of sediment-control techniques used during highway construction in central Pennsylvania: U. S. Geological Survey, Water-Supply Paper 2054, 57 pp.

Reid, L. M., and Dunne, T., 1984, Sediment production from forest road surfaces: Water Resources Research, Vol. 20, No. 11, pp. 1753–1761.

Richardson, E. C., and Diseker, E. G., 1961, Roadside mulches: Crops and Soils, Vol. 13, No. 5, 1p.

Roberts, M. C., 1972, Watersheds in the rural-urban fringe: Proceedings of a Symposium on Watersheds in Transition, Urbana, Illinois, American Water Resources Association, pp. 388–393.

Robinson, J., 1971, Highways and our Environment: New York, McGraw-Hill Book Company, 340 pp.

Schumm, S. A., 1977, The Fluvial System: New York, John Wiley and Sons, 338 pp.

Schumm, S. A., Harvey, M. D., and Watson, C. C., 1984, Incised Channels—Morphology, Dynamics, and Control: Littleton, Colorado, Water Resources Publications, 200 pp.

Schuster, R. L., 1978, Introduction: *in* Schuster, R. L., and Krizek, R. J. (eds.), Landslides—Analysis and Control, Transportation Research Board, Special Report 176, pp. 1–9.

Shen, H. W., Schumm, S. A., Nelson, J. D., Doehring, D. O., Skinner, M. M., and Smith, G. L., 1981, Methods for Assessment of stream-related hazards to highways and bridges: Federal Highway Administration, Offices of Research and Development, Environmental Division, Report No. FHWA/RD–80/160, 241 pp.

State of California, 1970, California Environmental Quality Act of 1970.

State of Colorado, 1984, Design Manual: Colorado Department of Highways.

State of Maryland, 1970, Annotated Code, Natural Resources Article, Title 8, Subtitle 11.

State of North Carolina, 1977, Sedimentation Pollution Control Act of 1977: North Carolina, Gen. Stat.

Stephens, H. V., Scholl, H. E., and Gaffney, J. W., 1977, Use of the universal soil loss equation in wide-area soil loss: *in* the Proceedings of a National Conference on Soil Erosion, Soil Conservation Society of America, pp. 276–282.

Swanson, N. P., Dedrick, A. R., and Dudeck, A. E., 1967, Protecting steep construction slopes against water erosion: Highway Research Board, Highway Research Record, Vol. 206, pp. 46–52.

Swanston, D. N., and Swanson, F. J., 1976, Timber harvesting, mass erosion, and steepland forest geomorphology in the Pacific Northwest: *in* Coates, D. R. (ed.), Geomorphology and Engineering, Stroudsburg, Pennsylvania, Dowden, Hutchinson, and Ross, pp. 199–221.

Tour, J. W., 1985, Revegetating slopes with geotextiles and geogrid systems: U. S. Forest Service, Equipment Development Center, Project No. 1E12L56, 58 pp.

Tryon, C. P., Parsons, B. L., and Miller, M. R., 1976, Excavated sediment traps prove superior to dammed ones: Proceedings of the Third Interagency Sedimentation Conference, PB–245–100, Water Resources Council, Washington, D. C., pp. 2–42 to 2–46.

Turelle, J. W., 1973, Factors involved in the use of herbaceous plants for erosion control on roadways: Proceedings of a Conference-Workshop, Highway Research Board, Special Report 135, pp. 99–104.

U.S. Congress, 1965, Water Quality Act of 1965, Public Law 89–234.

U.S. Congress, 1969, National Environmental Policy Act of 1969: Public Law 91–190.

U.S. Congress, 1972, Federal Water Pollution Control Act Amendments of 1972, Public Law 92–500.

U.S. Congress, 1977, Surface Mining Control and Reclamation Act of 1977: Public Law 95–87.

U.S. Department of Commerce, 1956–80, Statistical Abstracts of the United States: Bureau of Census.

U.S. Environmental Protection Agency, 1973, Methods for identifying and evaluating the nature and extent of non-point sources of pollutants: GPA–4030/9–73–014, 261 pp.

U.S. Environmental Protection Agency, 1976, Loading functions for assessments of water pollution from non-point sources, Environmental Protection Technology Series.

U.S. Soil Conservation Service, 1970, Controlling erosion on construction sites: U.S.D.A., Soil Conservation Service, Agricultural Information Bulletin 347, 31 pp.

U.S. Soil Conservation Service, 1975, Standards and specifications for soil erosion and sediment control in developing areas: U.S.D.A., Soil Conservation Service, College Park, Maryland, 336 pp.

U.S. Soil Conservation Service, 1978, A guide for erosion and sediment control in urbanizing areas of Colorado (interim guide): U.S.D.A., Soil Conservation Service, 203 pp.

U.S. Soil Conservation Service, Water Resources Administration, and State Soil Conservation Committee, 1983, The 1983 Maryland Standards and Specifications for Soil Erosion and Sediment Control, abridged reprint, 85 pp.

U.S. Water Resources Council, 1977, The nation's water resources, the second national assessment of the U. S. Water Resources Council; Appendix, Erosion and sedimentation and related resource considerations: Water Resources Council, Washington, D. C., 111 pp.

Vice, R. B., Guy, H. B., and Ferguson, G. E., 1969, Sediment movement in an area of suburban highway construction, Scott Run Basin, Fairfax County, Virginia, 1961–1964: U. S. Geological Survey, Water-Supply Paper 1591-E, 41 pp.

Walling, D. E., and Gregory, K. J., 1970, The measurement of the effects of building construction on drainage basin dynamics: Journal of Hydrology, Vol. 11, pp. 129–144.

Wark, J. W., and Keller, F. J., 1963, Preliminary study of sediment sources and transport in the Potomac River Basin: Interstate Commission, Potomac River Basin, Technical Bulletin, 1963–11, 28 pp.

White, C. A., and Franks, A. L., 1978, Demonstration of erosion and sediment control technology: U. S. Environmental Protection Agency, Office of Research and Technology, EPA–600/2–78–208, 385 pp.

Wischmeier, W. H., and Meier, L. D., 1973, Soil erodibility on construction areas: Proceedings of a Conference-Workshop, Highway Research Board, Special Report 135, pp. 20–29.

Wischmeier, W. H., and Smith, D. D., 1958, Rainfall energy and its relationship to soil loss: Transactions of the American Geophysical Union, Vol. 39, No. 2, pp. 285–291.

Wischmeier, W. H., and Smith, D. D., 1965, Predicting rainfall-erosion losses from cropland east of the Rocky Mountains: U.S.D.A., Agricultural Handbook 282, 48 pp.

Wolman, M. G., 1964, Problems posed by sediment derived from construction activities in Maryland: Report to Maryland Water Pollution Control Commission, 125 pp.

Wolman, M. G., 1967, A cycle of sedimentation and erosion in urban river channels: Geografiska Annaler, Vol. 49A, pp. 385–395.

Wolman, M. G., and Schick, A. P., 1967, Effects of construction on fluvial sediment, urban and suburban areas of Maryland: Water Resources Research, Vol. 3, No. 2, pp. 451–464.

Wright, D. L., Perry, H. D., Green, J. T., and Blaser, R. E., 1975, Manual for establishing a vegetative cover in highway corridors of Virginia, Virginia Polytechnic Institute and State University, Blacksburg,

Wright, D. L., Perry H. D., and Blaser, R. E., 1976, Controlling erosion along highways with vegetation and other protective cover, Federal Highway Administration and Virginia Highway and Transportation Research Council, Report No. FHWA-VA–77-R17, 74 pp.

Wright, D. L., Perry, H. D., and Blaser, R. E., 1978, Persistent low maintenance vegetation for erosion control and aesthetics in highway corridors: *in* Schaller, F. W. and Sutton, P. (eds.), Reclamation of Drastically Disturbed Lands: Madison, Wisconsin, American Society of Agronomy, Crop Science Society of America, Soil Science Society of America, pp. 553–583.

Yorke, T. H., and Davis, W. J., 1972, Sediment yields of urban construction sources, Montgomery County, Maryland: U. S. Geological Survey, Open-File Report, 39 pp.

Yorke, T. H., and Herb, W. J., 1976, Urban-area sediment yield—effects of construction-site conditions and sediment-control methods: Proceedings of the Third Interagency Sedimentation Conference, PB–245–100, Water Resources Council, Washington, D. C., pp. 2–52 to 2–64.

Younkin, L. M., 1973, Effects of highway construction on sediment loads in streams: Proceedings of a Conference-Workshop, Highway Research Board, Special Report 135, pp. 82–93.

Zingg, A. W., 1940, Degree and length of land slope as it affects soil loss in runoff: Agricultural Engineering, Vol. 21, No. 2, pp. 59–64.

11

Geomorphic Perspectives on the Design and Management of Disturbed Lands

Introduction

Everyday human activities result in a wide variety of land disturbances. Some are virtually inevitable if a population is to secure the requirements for survival and enhancement of their standard of living. Other disturbances are merely a matter of convenience and lead to an unnecessary degradation of the environment. In essence, one challenge to our environmental custodians is to distinguish between "necessary" and "convenient" within a particular operational milieu, at a particular time and place.

In the previous chapters, several examples of land disturbance are examined in some detail. Often, the investigator is confronted with complex alterations of the natural landscape. The activities that cause the disturbance, such as surface mining or tailings disposal, are highly sophisticated operations. Usually, there are a host of interrelated environmental consequences resulting from the disturbance of an integrated natural system, including on-site, off-site, and cumulative impacts. The problem is complicated further by considering the legal and socioeconomic aspects of these activities.

However, we believe enough has been said and written regarding the *problems* of land disturbance. It is time to move on to the next generation of environmental concern, to the disentanglement of complexity and the development of solutions to these problems while laying a firm foundation for subsequent refinement of any techniques. Specifically, we believe that the science of geomorphology has much to offer in the formulation of these solutions and techniques. Indeed, this discipline is grounded in the character, evolution, and processes of landforms. The trend of geomorphology in recent years, process–response studies, is ideally suited to application in issues of land disturbance.

In this final chapter, we focus directly upon geomorphic perspectives concerning the design and management of disturbed lands. For the most part, we limit our discussion to surface and near-surface processes and relatively short time-scales, a few decades perhaps. Although these constraints cause us to bypass the valuable work of geomorphologists regarding the history of landforms, they do allow us to concentrate on the most immediate problems surrounding land disturbances. For most circumstances, this time frame is appropriate.

The exception, of course, is mill tailings disposal, where toxic or radiological hazards remain indefinitely. Here, time scales on the order of 10,000 to 100,000 yr or more are appropriate. As such, the historical approach is essential. Which processes can significantly affect the disposal site through very long spans of time? What would be the range of magnitudes and frequencies for these processes during this time framework? In what ways could the site respond to the forces generated and the work performed by these processes? Schumm and Chorley (1983) and Chorley *et al.* (1984) have provided comprehensive treatment of these issues, based upon current geomorphic theory.

Assessment of Disturbance

Throughout this book, it is proposed that land disturbance may be rationally evaluated in terms of three factors: (1) areal extent, (2) intensity, and (3) duration. Each can be assessed with some objectivity and even quantified in many cases. Ultimately, however, society must render a judgment concerning the perceived severity and acceptability of a given disturbance and its impact. This constitutes a collective decision regarding its necessity, if not acceptability.

Table 1 contains our assessment of the impacts caused by the land disturbances presented herein for the United States. Grazing affects large tracts of land, especially in the West. Erosion rates may be doubled as a result, but in comparison to changes in geomorphic processes that may result from other types of disturbance, this is a relatively small change. Ranching has existed in the United States for more than a century and will likely continue in about the same manner. On a geologic time scale, this a very short duration, but in terms of human occupation, it might be considered a relatively long time. While there have been numerous attempts to treat overgrazed pastures, there is no legislation requiring systematic rehabilitation or erosion control on these lands.

Recreational vehicles (ORV) and other recreational activities disturb a very small amount of land nationwide. However, their use often increases erosion rates many times. Coates (1984) indicated that the use of such equipment is a phenomenon dating from the end of World War II. The popularity of these forms of recreation and the development of a rather extensive supportive industry

Table 1 An Assessment of Environmental Impact Due to Various Land Disturbances

Type of disturbance	Areal extent[a]	Intensity[b]	Duration[c]
Grazing	Large	Low	Long
Recreational vehicles	Small	High	Moderate
Surface mining of coal	Small	High	Short
Surface mining of uranium	Small	High	Short
Construction	Moderate	High	Short

[a]Area extent: small, <100 km^2; moderate, 100–500 km^2; large, >500 km^2.

[b]Intensity: low, ≤ 2 times increase in process rates; moderate, 2–10 times increase; high, >10 times increase.

[c]Duration: short, few months to a few years; moderate, several decades; long, >100 yr.

allows the assumption that associated land disturbances are likely to continue for many years, unless significantly restricted by legislation or regulation. Additionally, there have been few attempts to rehabilitate lands subjected to recreational vehicle use, and apparently there is no requirement to do so. Perhaps it should be noted that changes in land use near cities occasionally result in a kind of rehabilitation as sites disturbed by vehicles are incorporated into developments and become construction lands. Taken together, we consider the duration to be moderate.

Construction affects much larger areas than recreational vehicles, but much less land than grazing. A moderate areal extent appears appropriate. Numerous studies document dramatic increases in erosion and sedimentation rates during construction activities and sometimes afterward; this warrants designation as a high-intensity disturbance. Several studies have also shown that stabilization of the disturbed land usually takes place within several months to a couple of years, and this may be regarded as an essentially temporary disturbance.

The surface mining of coal actually utilizes relatively small tracts of land in comparison to either grazing or construction activities. There is ample evidence, however, of manifold increases in erosion and sedimentation rates, and depending upon the terrain and type of mining, even the frequency of landsliding. Under current legislation, reclamation must proceed concurrent with the mining process. As a result, duration ranges from several months to a few years in most cases. Of course, some argue that changes in soil properties and groundwater hydrology are essentially permanent, but it has not been proven that these changes are detrimental to the environment for indefinite periods of time.

The extraction and processing of uranium ores affect even less land than the surface mining of coal. Although quantitative documentation is scarce, it seems permissible to infer that the changes in process rates are likely to be about the same as with the surface mining of coal. Again, regulations now require reclamation, but the nature of the open pit extraction method causes delays in commencement of these practices. Duration, then, may extend from a few years to perhaps a decade or more. We have chosen to call this a relatively short duration, although one could support a moderate duration as well. Some even argue in favor of a long duration on the basis of the "potential" hazard presented by toxic and radioactive tailings deposits. We feel, however, that the extensive technology available to stabilize these sites, together with essentially perpetual monitoring that permits early recognition of even slight impoundment failure and therefore timely maintenance, makes the assignment of a "long" duration unnecessary.

It is not our intent to entirely justify our assessment but rather to illustrate the utility of the approach. Surely, there is ample room for refinement. Nevertheless, if our evaluation presented in Table 1 is generally acceptable, then the results are rather interesting. It appears that public perception and concern focus upon the intensity of the disturbance resulting from various activities, while often ignoring the areal extent and duration. Certainly, the impact of grazing and other forms of agriculture has not produced the national outcry caused by surface mining.

Disturbance of the Vertical Column

Geologists and soil scientists commonly examine vertical columns of terrestrial materials (i.e., stratigraphic cross sections and soil profiles). Perhaps the same approach has value in assessing land disturbances. Some activities change only the characteristics of the vegetal cover; one type of vegetation is replaced by another or with a manmade product such as concrete or asphalt. Other activities affect both the vegetation cover and underlying soil properties. Grazing and recreational vehicle use fall into this category. Finally, other operations alter vegetation cover, underlying soils, and even the rock foundation of the landscape. Surface mining and site preparation for construction activities are included in this category.

The intensity of disturbance may be related to the depth of disturbance, although at present this approach is largely intuitive. Replacement or modification of vegetation cover alone substantially alters the functioning of hydrologic and geomorphic processes. The infiltration capacity of the surface is changed and, hence, the generation of runoff. The resistance provided by vegetation to the forces produced by rainsplash and overland flow is likewise changed.

It was also shown in the preceding chapters that disturbance of indigenous soils alters the hydrologic and geomorphic processes operating on a surface.

Even where topsoil is carefully stockpiled and eventually redistributed, the characteristics of profile and soil structure remain changed for some time. The benefits of topsoiling may be more biological than hydrologic or geomorphic, depending upon the disparity between the properties of the topsoil and the surficial materials that would otherwise be situated on the surface, such as subsoil or spoil.

Truncation or fragmentation of the rock foundation causes additional environmental impacts. It is not merely surface processes that are altered but subsurface processes as well. Groundwater systems are modified, at least for a time. Currently, there is considerable research in progress to evaluate the effect of surface mining on the quantity and quality of groundwater, both locally and regionally. The structural strength and bulk density of fragmented materials used as fill will be less than the original consolidated rock. This could result in subsidence of the disturbed area with time. Of course, this is sometimes a problem in areas of underground mining too.

Reclamation practices have generally proven quite successful in reestablishing vegetation cover on disturbed lands. However, the technology does not presently exist to restore physical properties of soil or geologic properties of bedrock with the same success. If this can be accomplished by natural processes, it will require extensive periods of time. Thus, both the intensity and duration of disturbance may be correlated with the depth of disturbance.

Geomorphic Analysis of Land Disturbance

For the geomorphologist, most of the physical consequences of land disturbance are far from mysterious; indeed, they are usually expectations. It is understood that a substantial proportion of environmental impacts is the direct result of geomorphic processes operating at accelerated rates. These geomorphic processes are the agents by which geomorphic work is performed, and this work occurs when force exceeds resistance. Further, these processes are governed by physical and chemical laws.

Conceptually, geomorphic analyses of land disturbances are best founded upon an examination of changes in force and resistance for a particular site. Overall, increases in force, decreases in resistance, or both occurring concurrently result in an increase in the rate and cumulative amount of geomorphic work accomplished. Climatic and gravitational sources of force impinging upon a surface usually remain constant, insofar as human activities on the land do not effectively alter the climatic regime of an area nor appreciably change the distance between a point on the earth's surface and the center of the spheroid.

However, other forces often increase as a result of land disturbance. Increases in hillslope gradient cause increases in the downslope component of gravitational force. Changes in the characteristics of surface materials often result in an increase in the rate and volume of runoff generation. This, in turn, causes an increase in the tractive forces generated by overland flow during precipitation events; increases in hillslope gradient and length exacerbate this process. Put simply, more runoff is moving faster at greater depth and generating greater force on surface materials. As a consequence, the discharge of water and sediment load in streams increases. The hydraulic geometry and gradient of the channel adjusts in order to accommodate these increases. The manner of change is complicated, as shown by Schumm (1977), depending upon the relative proportions of water and sediment delivered to the channel.

Except for paved areas, the surface resistance is generally reduced as a result of land disturbance. The protective vegetation cover, so important as a deterrent to erosion, as demonstrated by Wischmeier and Smith (1965) (among others), is inevitably altered. Soil properties associated with erodibility are frequently modified in such a way as to increase the susceptibility of materials to impinging forces. Resistance provided by geologic structure and lithology may likewise be reduced through fragmentation. Surely, there are other examples of changes in force and resistance in addition to the common scenarios cited above.

Although the foregoing provides a theoretical basis for understanding the effects of land disturbance on geomorphic processes, it does not really offer a practical analytical approach in most instances. It is very difficult to measure force, resistance, and their modification due to disturbance throughout a particular area because of natural variabilities in vegetation, soil, geology, landform, and perhaps even microclimate. Hence, recourse is given to the measurement of the work accomplished, the rate of erosion, the rate of surface creep, or the rate of water and sediment transport.

As referenced in Appendix A, various methodologies are available for measuring geomorphic work. A comparison of rates before and after disturbance provides documentation of the geomorphic consequences of the disturbance. An increase in work permits the inference of increase in force, decrease in resistance, or a combination of the two.

The fact that any work is performed indicates that the stresses caused by forces have surpassed the thresholds of strength for the materials involved. An allied concept is that of the "geomorphic threshold," defined by Schumm (1979), as a threshold of landform stability that is exceeded either by intrinsic change of the landform itself or by a change of an external variable. As we have seen, anthropic land disturbance is quite capable of rapidly modifying several external variables and so can easily cause the instability of landforms. On the other hand, inherent in the concept is a certain element of resiliency in the geomorphic system. A measure of change in some external variables can occur without causing a change in landform because the threshold is not reached. Further, landform instability and adjustment could be the result of intrinsic change, unrelated to the disturbance, although admittedly this is probably an unusual circumstance.

Additionally, it is noteworthy that the response to change in a geomorphic system is generally complex. Schumm (1977) and Chorley *et al.* (1984) discussed the concept of complex response in detail. In essence, one change begets another and another throughout the system. There may actually be several "waves" of adjustment and readjustment in the system.

Finally, the goal of these changes or adjustments is the reestablishment of a dynamic equilibrium or steady state within the system. Forms are in balance with processes, and processes are attuned among themselves. The time period required for a semblance of balance to occur is likely site specific but probably depends upon the magnitude of forces involved and the resistance encountered. This is to suggest that weak materials subjected to strong forces may produce a dynamic equilibrium or steady state sooner than strong materials subjected to weak forces. If so, then the increases in force and decreases in resistance accompanying land disturbance may ironically facilitate the achievement of a balanced state by allowing the rapid performance of geomorphic work. Of course, this work and its products constitute the environmental impacts of concern.

Taken together, these basic concepts provide a structure for the geomorphic analysis of land disturbance. As always, the problem is to apply the concepts and theories in the field. Issues and answers are rarely as simplistic as suggested above. However, it is possible to provide guidance for the design and management of disturbed lands.

Geomorphology and Disturbed Landscape Design

The design of a landscape for a disturbed area involves three phases: (1) analysis of existing and probable or possible environmental conditions, (2) formulation of a reclamation program, and (3) implementation of the program. In the first step, the historical approach mentioned earlier plays a significant role, depending upon the time frame adopted. In some cases, the magnitude and frequency of geomorphic processes likely to affect the site in the next few decades may be a satisfactory basis of design. In other circumstances, it may be necessary to consider, even speculate, as to the magnitude and frequency of any geomorphic processes that could occur in the disturbed area during the next several thousand years. Again, Schumm and Chorley (1983) address this prospect.

It is in the formulation of the reclamation program that fundamental principles of geomorphology find direct application. Of course, it is recognized that some of these precepts have been contributed through the research of soil scientists, agricultural engineers, hydrologists, and hydraulic engineers, among others. In conformance with the general order of this book, we shall divide the subsequent discussion into the topics of hillslopes, channels, and drainage basins.

Hillslope Design

The force impinging on the surface of a hillslope by running water can be controlled to a considerable extent through the manipulation of form. As discussed in the earlier chapter concerning hillslopes, the configuration of a valley head hillslope (hollow) is more likely to be problematic than the spur end hillslope (nose). In the former case, the flow lines converge toward the base of the hillslope with increases in the depth and velocity of overland flow. In the latter case, the flowlines diverge in the downslope direction. Hence, given a choice, the spur end hillslope design is preferable.

It has also been demonstrated that profile shape influences the magnitude of forces on the hillslope and the consequent work that takes place. Meyer and Romkens (1976) found that a convex hillslope is more erodible than a uniform (straight) hillslope and that the uniform hillslope, in turn, yields more sediment than a concave one. A complex hillslope (convex–concave) generally yields less sediment than a uniform hillslope. The convex hillslope is steepest where runoff is greatest; whereas the concave hillslope is steepest where the flow is least. Therefore, one can conclude that the concave profile is the most desirable design shape, with the complex, uniform, and convex forms as progressively less desirable.

Combining the above, it seems that a spur end hillslope plan, together with a concave or complex profile, would approach the optimum three-dimensional design. Of course, it would be rare that a reclaimed landscape could be constructed of a single hillslope form. Nevertheless, these general principles indicate where erosion problems are likely to be least, other things being equal, and where erosion control practices and supportive measures are likely to be most important.

Over the past four or five decades, there has been considerable research concerning the influence of hillslope angle and length on erosion rates. The findings are perhaps best summarized in the topographic factor (LS) of the Universal Soil Loss Equation (USLE) (Wischmeier and Smith, 1965). In the simplest terms, the downslope component of gravitational force increases as hillslope angle increases, accelerating overland flow. As hillslope length increases, this flow accumulates in the downslope direction, resulting in greater depth. In both cases, the tractive force produced by the runoff increases, again, other things being equal.

Research by Toy (1977) concerning hillslope form in various climatic regions of the United States indicates that hillslopes in arid regions tend to be steeper, shorter, and possess a smaller radius of curvature of their convex segments than counterparts in humid regions. This finding might be considered in the design of hillslopes on reclaimed lands. This supports the increase in drainage density with increasing aridity, as proposed by Melton (1957) and discussed later in this chapter. Further, this trend calls to question the use of nationwide, or possibly even statewide, regulations or standards concerning hillslope gradients.

Design hillslopes should be relatively gentle in gradient and short in length. Interestingly, there is a trade-off between these two features; as hillslopes are flattened, they become longer. Let us consider the simple case of uniform or straight hillslopes. Figure 1A shows the initial situation. The hillslope to be modified, which is presently undisturbed or composed of unconsolidated spoils material, is 100 ft (30 m) wide, 1600 sq ft (144 m^2) in cross section, 100% (45°) in gradient on each side (admittedly beyond the probable angle of repose for

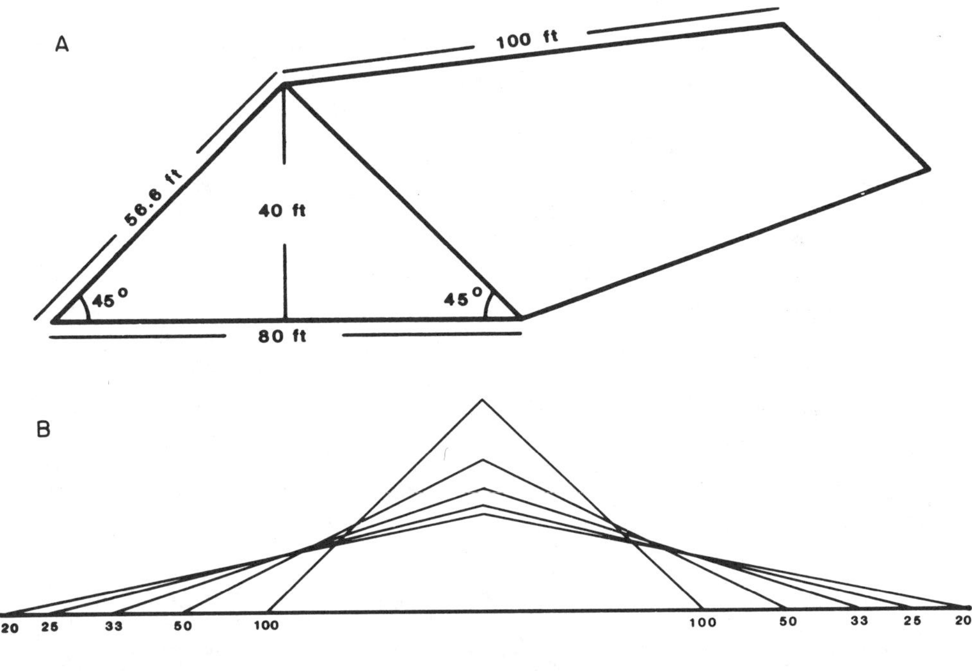

Figure 1. A, Initial characteristics of a hillslope to be graded. B, Effect of grading to achieve various gradients on hillslope cross section.

unconsolidated materials), and with hillslope lengths of 56.57 ft (16.9 m) on opposing sides for a total length of 113.1 ft (33.9 m). Figure 1B shows the effect of grading operations on the cross section to obtain 50%, 33%, 25%, and 20% gradients, respectively, while maintaining the 1600 sq ft (144 sq m) cross-sectional area, which together with the constant 100 ft (30 m) width retains a constant volume of material. We elect to utilize English units in this example because they are commonly employed by industry in the United States and, hence, render the example most meaningful for them in terms of the quantities of material to be moved.

Table 2 shows the change in hillslope characteristics as a result of grading. As hillslope gradient decreases, the base increases and the height decreases. This increase in base, of course, has definite implications concerning the surface area disturbed and the permit area or right-of-way required. Also, as hillslope gradient decreases, hillslope length increases. The relationship between hillslope gradient and base, height, and length is illustrated in Fig. 2.

The reduction of hillslope gradient is intended to reduce mass movement or soil erosion. In the latter case, the consequences of grading can be examined through consideration of the *LS* factor of the USLE (Wischmeier and Smith, 1965). Again, the estimates of this parameter contained in Table 2 show that the benefits derived from the reduction in gradient more than compensate for the accompanying increase in length. As the hillslope gradient is reduced, *LS* decreases sharply. It should be noted that the *LS* factor was not intended for use on very steep hillslopes, but the work of Israelsen *et al.* (1980), previously discussed in the chapter concerning construction lands, suggests that it is valid for the gradients used in this example.

From the foregoing, it is obvious that grading to reduce hillslope gradients is clearly a desirable practice from the geomorphic perspective. However, as the

Table 2 Hillslope Characteristics

Ratio	Percentage[a]	Degrees[a]	Base (ft)	Height (ft)	Length (ft)	*LS* factor[b]	Surface area[c] (sq ft)	Volume topsoil[d] (cu yd)
1:1	100	45°	80.0	40.0	113.1	31.8	11,313.7	1257.1
1:2	50	26°34′	113.1	28.3	126.5	14.0	12,648.3	1405.4
1:3	33.3	18°26′	138.6	23.1	146.1	8.0	14,607.6	1623.1
1:4	25	14°02′	160.0	20.0	165.0	5.4	16,495.8	1832.9
1:5	20	11°19′	178.8	17.9	182.4	3.9	18,235.1	2026.1

[a]Conversions from Young, 1972.
[b]Approximate values from nomograph and formula.
[c]Assumes hillslope width of 100 ft.
[d]To cover surface with 3 ft of topsoil.

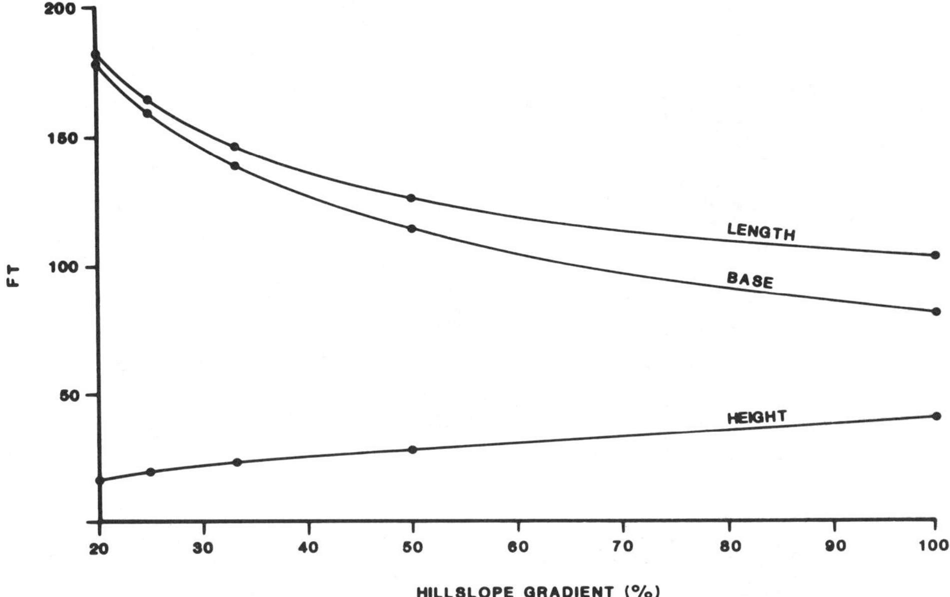

Figure 2. Effect of grading on the base, height, and length of hillslope.

gradients decrease and lengths increase, there is also an increase in the surface area upon which soil erosion may occur, as revealed in Table 2. Of course, at gentle gradients, this might be viewed as increased surface area for sediment storage.

From the industrial perspective, there are economic reasons that militate against gradient reduction beyond that which is necessary to control erosion. For example, in Fig. 1B, the area between cross sections, extended through a given hillslope width, comprises a volume of material to be moved; this quickly translates into the cost of machines and personnel. Further, it is sometimes necessary to cover the disturbed surface with stockpiled topsoil in order to achieve successful reclamation or comply with regulations. If a topsoil mantle of 3 ft is applied to the surface, then the values in Table 2 indicate the volume of topsoil to be transported and distributed. As the hillslope gradient decreases, the volume increases. At some point, the volume of topsoil saved may be insufficient to cover the surface to a required depth. Figure 3 shows the change in surface area and topsoil volume with changes in hillslope gradient.

Surface manipulations, such as terracing, dozer basins, listering, chiseling, or plowing, reduce the force of flowing water by (1) reducing the volume through improved infiltration, and (2) effectively shortening the hillslope length by divid-

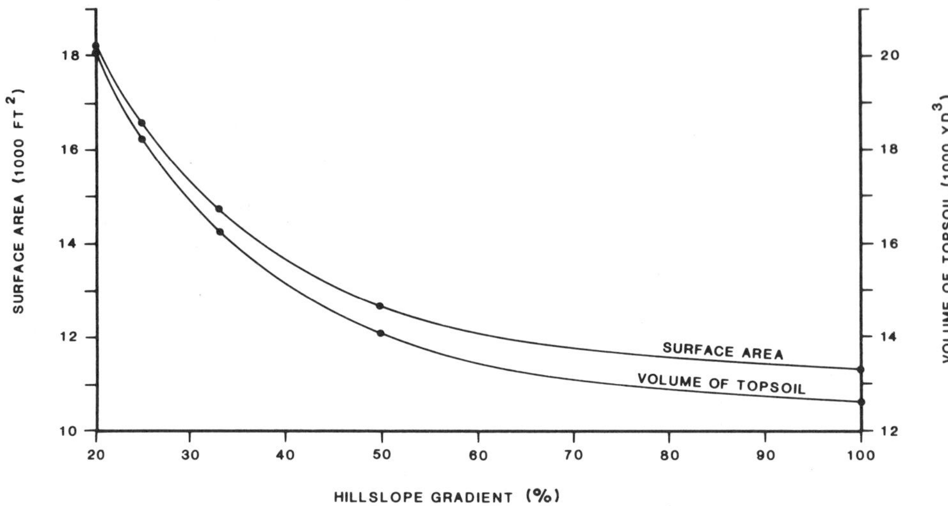

Figure 3. Effect of grading on the surface area and topsoil volume of hillslope.

ing it into segments with intervening sites for water and sediment storage. Concurrently, the resistance of the surface increases as a result of greater roughness that further reduces flow velocity and hence the tractive force. Such techniques are only effective for a few months to a few years, but this may be sufficient for the revegetation of the surface.

Surface amendments of geomorphic significance include mulches and soil tackifiers. Mulches operate much like permanent vegetation cover, albeit temporarily. Tackifiers increase surface resistance by binding soil particles together into larger and less transportable aggregates. Their effective life is often even shorter than surface manipulations, and they are commonly expensive to use. As a result, they are frequently reserved for particularly troublesome areas or special situations.

The reestablishment of a vegetation cover remains the most popular method of increasing surface resistance to the forces of soil particle entrainment and transportation. Vegetation functions in various ways to intercept, absorb, and reduce the kinetic energy of falling or flowing water. Observation of hillslope erosion on reclaimed sites at numerous surface mines throughout the western United States shows that rilling often begins in barren patches and extends into the areas of successful revegetation, sometimes undercutting the vegetal mat. The barren spot could be regarded as the so-called chink in the armor that leads to more serious failure of the reclaimed surface.

At this juncture, we could discuss the principles that can be incorporated into hillslope design to reduce the probability of mass movement or wind erosion. It must suffice, in the interest of brevity, to note that many of the concepts considered above with regard to erosion by water serve these purposes as well.

The reduction of hillslope gradients and heights increases mass stability, other things being equal. The reestablishment of a vegetation cover greatly reduces wind erosion. Interestingly, some practices have complex consequences. For example, techniques that increase infiltration increase pore pressures within unconsolidated masses and the propensity for mass movement; however, the increased soil moisture reduces wind erosion. It becomes evident that a complete understanding of geomorphic processes operating in a particular area is essential to truly effective hillslope design.

Channel Design

The most fundamental principle of channel design is that natural stream channels are formed by the waters that flow in them. Actually, these waters include two important components, the fluid itself and the sediment load in transit. Schumm (1977) noted that the channel-forming discharge is that which approaches bankfull. Other research concerning the recurrence intervals of various discharges suggests that such flows probably occur about once a year. Therefore, the magnitude of the mean annual flood has important design implications.

Additionally, Schumm (1977) observed that the proportion of bedload to total load has a major influence on the nature of alluvial channels. Those transporting large quantitites of suspended sediment are relatively sinuous and have a low width–depth ratio, whereas those in which the bedload sediment discharge is large tend to be relatively wide, shallow, and less sinuous. Both the total sediment load and its particle size distribution influence channel shape.

We submit that the design of stream channels on disturbed lands should rest upon the expected hydrologic properties of the disturbed surface rather than the properties of the natural, predisturbance surface. Replication of natural channel morphometry is necessarily based upon the assumption that the hydrologic characteristics of the drainage basin are the same after reclamation as they were before the disturbance. There is ample evidence that certain important characteristics, such as infiltration capacity and perhaps topography, are often substantially modified and are likely to remain so for a very long time. As a result, runoff and sediment loads can be quite different following disturbance, depending upon the specific nature of the human activities involved and the components of the reclamation program.

Shown *et al.* (1982) provide methodologies for estimating the discharge and sediment yields of streams crossing lands reclaimed following surface mining. Admittedly, these procedures rely upon numerous assumptions and include a measure of error. However, the approach seems more rational than merely reproducing the channel properties that prevailed prior to disturbance.

In the design of stable channels, two components of form should be given careful consideration: the channel geometry and the channel gradient. The direct relationship between discharge and the width and depth of channels has been well established, at least for perennial, alluvial streams. The problem, of course, is to calculate, estimate, or select a design discharge. Sometimes, a regulatory authority provides guidance in the determination of the recurrence interval for this discharge. Chorley *et al.* (1984) noted that there is usually a good statistical relationship between mean annual discharge $\bar{Q}$, bankfull discharge Q_b, and mean annual flood $Q_{2.33}$, so that almost any of these hydrologic variables can be used in an equation that relates channel dimensions to discharge. But how does one compute the discharge from a severely disturbed landscape? Presumably, the relief can be ascertained from the reclamation plan. A notion of precipitation regime can be obtained from the baseline data collected on-site before and during mining. Otherwise, reliance is given to the information available at the closest weather station, although it may be several miles away. Usually, this is the only source for long-term records. Estimation of infiltration rates offers a distinct challenge. Depending upon the depth of disturbance, the infiltration capacity of the reclaimed surface may be quite different from the predisturbance surface; vegetation cover, soil structure and texture, and possibly any near-surface geologic influences are all substantially altered. Further, the surface manipulations incorporated into the reclamation program add complexity to the estimation process. And, finally, the characteristics of the surface are in a transient state, in the short term, as vegetation develops and the effectiveness of surface manipulations diminishes, and in the long term, as soil properties regenerate.

Rainfall simulation or infiltrometers appear to provide a useful methodology for determining the infiltration capacity of the disturbed surface with some accuracy. Several representative sample sites would be chosen because surface conditions are commonly variable on disturbed sites. Nevertheless, at a given location, subjected to the same types of disturbance and reclamation practices, experience should allow reasonable approximation of infiltration rates within a fairly short time once the experimental data are available.

After the hydrologic properties of the reclaimed surface are determined, the probable discharge can be estimated for particular recurrence intervals. Thereafter, the appropriate channel geometry and channel gradient can be computed to accommodate that flow. If the material into which the channel is to be constructed is rather sandy, then a high width–depth ratio should be anticipated; conversely, high silt and clay contents in the bank material should dictate a low width–depth ratio.

Conservancy in assumptions and estimations seems warranted in light of numerous uncertainties. Additionally, erosion rates are likely to be highest immediately following reclamation because vegetation cover and root network development are relatively low. Hence, an overdesigned channel (reasonably overdesigned) serves as a sediment sink for a while, reducing accumulation in settling ponds or downstream impacts.

The gradient of the reclaimed channel also deserves specific attention. The fundamental principle in operation is that channels tend to adjust their gradients, or grade themselves, to the water and sediment load transported. As a result, reconstructed gradients that are either too steep or too gentle are modified by subsequent flows. This process may include aggradation, degradation, or changes in channel pattern. In the latter case, we cannot assume that a gently meandering channel is necessarily appropriate for the discharge and sediment load of the reclaimed surface. Because of the high initial rates of erosion, a braided pattern may emerge and transform itself into a meandering pattern with time as the surface of the watershed becomes stabilized by a vegetation cover. In most cases, however, we can be reasonably certain that a straight channel is inappropriate. Again, determination of channel gradient is encumbered by the same problems encountered in the design of the channel geometry.

From the geomorphic perspective, channel design requires answers to the following questions:

1. What recurrence interval of flow shall be used in the design?
2. What is the probable discharge of this event?
   a. What is the relief or average slope of the area?
   b. What is the precipitation regime?
   c. What is the infiltration capacity?
3. What is the sediment concentration of the discharge?
   a. What is the erosion rate for the hillslopes of the watershed?
   b. What is the sediment delivery ratio?
   c. What proportion of the sediment will be transported as bedload and as suspended load?
4. What is the appropriate channel geometry to accommodate the flow?
   a. What is the particle size distribution of the bank and bed material?
   b. What is an appropriate width-depth ratio for this material in the channel perimeter?
5. What is the appropriate channel gradient for the discharge and sediment load?
   a. Is there room within the channel or valley to accommodate changes in channel patterns?
   b. Is there a relationship between channel and hillslope gradients?

Clearly, the design of stable channels is a highly complicated undertaking. Even if accurate discharge and sediment load data are available for a selected recurrence interval of flow, the channel geometry and channel gradient cannot be determined precisely at the present state-of-the-art. However, reasonable estimates are entirely possible, and serious mistakes can be avoided in usual circumstances. Hopefully, settling ponds will contain the products of minor errors.

Boundaries between disturbed and undisturbed lands pose a potentially serious problem for channel design. In all likelihood, there are differences in the hydrologic properties and erosion rates of these two areas. As where natural streams cross geologic boundaries, the channel must adjust to these differences. A knickpoint produced by such adjustments can migrate upstream through the reconstructed channel causing an episode of rejuvenation.

The settling ponds are usually located near the boundary of disturbed and undisturbed lands, and they themselves are problematic. Settling ponds constitute a knickpoint in the longitudinal profile of the channel and may become the initial site of channel entrenchment unless properly removed and reclaimed. There probably will be an interval of several years between final reclamation of the disturbed surface and the removal of the settling ponds so that the disturbed area has an opportunity to approach a semblance of dynamic equilibrium and attain a measure of stability.

Given the tendency of channels to adjust to the discharge and sediment load impressed upon them, it is possible to create a scenario concerning off-site channels below a settling pond. We assume that they are in an equilibrium or a steady-state condition prior to the disturbance and installation of a settling pond. In accordance with regulations (described in Chapter 8), the sediment discharged from these settling ponds may be substantially lower than that characteristic of natural conditions. This release of relatively clean water may lead to channel degradation as frequently experienced below dams. It is possible that off-site environmental impacts, such as the draining of subirrigated meadows, are attributable to a regulation.

With the removal and reclamation of the settling pond, there is likely to be another period of adjustment in these off-site channels. The discharge of water and sediment will often be greater than experienced while the pond was in place. Depending upon the relative proportions of water and sediment, a period of aggradation or degradation may ensue, again causing some off-site environmental impact.

Although the above is likely to occur, it is necessary to put these events into perspective. The off-site and downstream system has the ability to dampen the effects of such adjustments with distance. Therefore, speculation of catastrophic consequences as a result of this scenario would not be supportable. That is to suggest that the additional area of impact is likely to be modest.

The above discussion applies most directly to drastically disturbed lands where it is necessary to design fully stable channels. It is sometimes permissible to construct engineered conveyance channels. Of course, these are not usually stable nor maintenance-free over long periods of time. From the geomorphic perspective, any apparent equilibrium or steady state is artificially created and perpetuated by manmade structures. Nevertheless, the same geomorphic principles are applicable and account for the periodic maintenance necessary to retain the channel in its design configuration.

Drainage Basin Design

The hillslopes and channels of the reclaimed surface are components of entire drainage basins. In a natural, undisturbed setting, there is an integration of these components, conducive to the efficient transport of water and sediment produced within the basin. The mutual adjustment of parts is revealed in the numerous statistical relationships discovered by various investigators and presented in an earlier chapter concerning drainage basins.

It may be necessary, at least in theory, to recreate all of the aforementioned relationships in order to design and construct truly stable drainage basins. Some would likely question whether or not this is even possible. If so, the next concern would be whether or not it is feasible, especially in economic terms. Therefore, we must generally be satisfied with a general approximation and allow natural processes to "fine tune" the system for us through time.

Drainage basins are composed of stream channels of various order numbers, from the small first-order channels, which lack tributaries, to the "trunk" stream of the basin, which is the largest stream therein, possesses the highest order number, and collects the discharge from all lower order streams in the drainage system. In his classic paper, Horton (1945) demonstrated that a close relationship exists between the order number of streams and other properties such as the number of stream segments of a particular order, the length of these stream segments, and the area they drain. Strahler (1952) and his colleagues simplified Horton's original work and modified it. However, most of Horton's laws of drainage composition exclude slope and may be functions of topological randomness. Abrahams (1980) summarized an opposing body of evidence and concluded that the factor of relief affects the planimetric geometry of drainage basins. In 1966, Shreve (1966) developed the random topology model. In contrast to Strahler (1952), Shreve defined the magnitude of a link to replace order; this represents the number of sources upstream from it. This concept may be more successful in explaining relationships in drainage basin morphometry, but many geomorphologists still use the Strahler modifications of Horton's laws. Again, to overdesign such that some channels disappear, shorten themselves, or enlarge their drainage areas is probably preferable to the other tendencies in adjustment because the former should result in sediment deposition while the latter could result in fairly rapid sediment production.

Additionally, research suggests that stream gradient is inversely related to order, whereby the lower order channels possess the steepest gradients and the highest order channel has the most gentle gradient. And, finally, we expect a direct relationship between stream discharge and order; here discharge increases with order. This is logical in view of the "collector" function of the trunk stream. As a consequence, the channel geometries of lower order streams should have smaller overall dimensions than those of higher order streams.

The incorporation of these principles into practical design is quite complicated, a matter for computer simulation. However, in most cases, the stream system of disturbed lands contains only a few orders, first through third or fourth perhaps. Another problem arises, however, when a high-order stream crosses the disturbed surface and we are obliged to mesh the reconstructed network with this channel.

Drainage basins also exhibit certain drainage patterns, or spatial arrangements of channels, that are largely controlled by the characteristics of the underlying geologic materials. The dendritic pattern is common to drainage systems found on homogeneous materials, where there is no influence of geologic structure or lithology. For drastically disturbed lands, such as those subjected to surface mining or possibly extensive construction activities, this pattern may be appropriate because the depth of disturbance has obliterated geologic controls.

Often the disturbance operation influences subsequent drainage patterns. For example, the area surface mining technique seems to lend a linear element to eventual drainage patterns due to the production of long, linear spoils banks in parallel series with valleys in between. Features of a trellis pattern are more appropriate to an area underlain by folded geologic strata than that developed upon graded spoils.

Chorley (1971) remarked that drainage density is one of the most sensitive and variable morphometric parameters of drainage basins, and one that controls the texture of landscape dissection and stream spacing. Drainage density is basically determined by the geology and climate of an area. Hadley and Schumm (1961) showed that there is an inverse relationship between the infiltration capacity of surface materials and drainage density. Melton (1957) showed that drainage density is inversely related to Thornthwaite's precipitation-effectiveness (P-E) index; as climate conditions become more arid, drainage density increases.

For design purposes, the above suggests that the drainage density of reclaimed surfaces should be higher than for the surrounding natural areas, if there is reason to believe that infiltration capacities are lower due to mechanical compaction by vehicle traffic or increases in the silt and clay contents of surface materials. Further, drainage density should generally increase with aridity, although this relationship has not been extensively tested in humid climate regions, so its validity cannot be assured in the eastern or central United States. It may, however, offer some guidance for regulatory authorities in states with considerable climatic diversity.

Other research documents direct relationships between drainage density and mean annual runoff and the magnitude of the mean annual flood. These imply that the width and depth dimensions of channels should increase as drainage density increases. As the amount of runoff generated within a basin increases, there is a need for more channels and greater landscape dissection to accommodate the flow. Greater discharge dictates proportionally larger channel cross sections. Abrahams and Ponczynski (1984) concluded that drainage density (D) is more closely related to precipitation intensity (P_I) than to mean annual precipitation (P_m). Also, contrary to expectation, the relationship between D and P_I is not direct but inverse, at least where P_I is greater than 50 mm/24 hr. This relationship, however, does not hold in desert climates.

The direct relationship between percentage of bare ground and drainage density is rather interesting. With greater barren surface, infiltration capacity is likely to be lower, other things being equal; runoff production is likely to be greater, and greater landscape dissection is the result. Noteworthy here is the prospect that drainage density should be higher in the early stages of reclamation and revegetation, with lower values appropriate after the vegetation cover is established. There may occur, then, a "healing" trend within drainage networks, presuming progressive vegetal development.

The length of overland flow is inversely related to drainage density. This means that as drainage density increases, the distance traveled by overland flow decreases. Based upon simple geometry, it could hardly be otherwise. We may infer, then, that as drainage density increases, hillslope length will decrease; and this is shown to be advantageous for erosion control.

Lastly, Melton (1958) found a strong direct relationship between stream frequency and drainage density, causing him to postulate that when the ratio of stream frequency to the square of drainage density equals 0.694, the system is in equilibrium. This, of course, is a bold assertion, but Melton bases this judgment on a rather large data base. Wells and Rose (1981) suggested that this relationship has utility in the design of reclaimed drainage basins. However, the concept probably requires additional verification and refinement, especially for various climatic regions beyond the southwestern United States field location where Melton's work was focused. Ritter (1978) reviewed the criticisms of the universal applicability of this relationship based upon the work of Abrahams (1972) and Wilcock (1975).

Drainage basin area is related to several channel, discharge, and sediment yield characteristics. The direct relationship between basin area and stream length should not surprise us; we expect large basins to contain more kilometers of channel than small basins, other things being equal. Hence, we expect to design and construct more kilometers of channel in larger reclaimed basins than in the smaller. But would we incorporate a proportion between basin area and channel length for the range of basin sizes found at our reclaimed site?

The direct relationship between drainage basin area and discharge appears almost intuitive. In most cases, greater surface area generates greater volumes of runoff and therefore greater stream discharge. Of course, this may not hold true for large basins in arid regions that experience fairly localized precipitation, covering only a part of the entire basin, and diminishing flow in the downstream direction due to channel infiltration and perhaps some evaporation of channel water.

Nevertheless, reclaimed basins are likely to be sufficiently small in size that precipitation events cover them completely. Thus, the direct relationship should be valid in usual circumstances. Again, this increase in discharge suggests corresponding adjustments in the geometry of channels. That is, the higher order streams in larger drainage basins should possess larger width and depth dimensions than the same order streams in smaller drainage basins.

Sediment yield per unit area is inversely related to drainage basin area because of greater storage opportunities in the larger basins, particularly in their downstream portions, usually composed of more gentle valley and hillslope gradients. As a result of this greater storage, sediment delivery ratios also decrease as drainage basin area increases. If reclaimed basins possess the topographic tendencies of natural basins and corresponding sediment storage opportunities, then we expect greater absorption of the erosion products within larger basins than within smaller ones. It must be emphasized here that this relationship specifically does not mean that total sediment yield is less from large basins than from small ones; the contrary is generally true. It is the sediment yield per unit area that decreases as basin area increases.

In most cases, total sediment yield increases with drainage basin area. This has important ramifications in the design of settling ponds positioned to control off-site release of sediment. Clearly, the ponds on streams draining larger areas should themselves be larger.

The average gradient of the drainage basin may be computed in a couple of ways, and one of the simplest is the relief ratio. From our previous discussion of hillslope erosion, we expect the rate of soil loss to increase with average basin gradient. Further, as hillslope gradients increase, corresponding increases in channel gradients are expected, and so we anticipate an increase in sediment transport as average basin gradient increases.

Sediment yield per unit area is directly related to the relief ratio of the basin. As the average gradient increases, the production and transport of sediment per unit area increases as well. This suggests that relief ratio of reclaimed drainage basins may be a useful variable in the design of settling ponds.

Abrahams (1972) discovered an inverse relationship between the basin relief and bifurcation ratio of headward-eroding basins. As the average gradient increases, the number of streams in a given order decreases in comparison to the number of the next higher order. It remains to be seen whether or not such a relationship has valuable design implications for reclaimed lands, but it could provide some guidance for planning the relationship between stream order and the number of stream segments.

Hadley and Schumm (1961) found an inverse relationship between the infiltration capacity of a basin surface and the sediment yield per unit area. As infiltration capacity increases, runoff decreases and also the generation and transportation of sediment. If the infiltration capacity of a reclaimed surface is less than that of adjacent undisturbed lands, or less than it was prior to disturbance, then sediment yield per unit area may be greater. Again, this relationship provides support for the formulation of settling pond design criteria.

Finally, numerous investigators have examined relationships between precipitation variables and sediment yield variables. However, the results vary considerably. Generally speaking, most support or at least utilize the Langbein and Schumm (1958) curve within the United States. Of course, any relationship best suits the conditions under which the data are collected, and so the relationships presented by others may be more appropriate for other parts of the world. Nevertheless, semiarid regions experience higher sediment yields than either very arid or humid regions in the United States. This has implications concerning not only settling pond design but also the sediment release regulations in various geographic regions. Regardless of which curve one endorses, a uniform regulation for the entire United States is inappropriate from the geomorphic perspective. Even a statewide regulation is ill-conceived in states experiencing a wide range of climatic conditions.

As discussed with regard to channels, the boundary problem may exist for entire drainage basins as well. Again, two basins can occur in juxtaposition with different geomorphic and hydrologic properties. One basin might have a lower baselevel than its neighbor or might erode more rapidly toward its divide. In either case, the conditions are present for stream capture and drainage diversion. This process results in the expansion of the drainage area by one stream at the expense of another as the capturing stream extends into the "territory" of another and diverts runoff from its course. Ritter (1978) remarked that captures are catastrophic events in basin development because they drastically alter basin properties, such as area, relief, and stream length, and that the internal adjustments to changes resulting from capture must take place rapidly. It should be noted that stream capture is a naturally occurring process in certain areas, such as piedmont regions. Other than controlling the rates of erosion and headward extension of drainage networks from or onto reclaimed lands, there is probably not much that can be done about this phenomenon over long time periods.

Geomorphic Principles and Design

Reflecting upon the numerous design considerations for hillslopes, channels, and complete drainage basins, one can hardly avoid a sense of awe. It becomes clear why the present state-of-the-art in reclamation cannot produce equilibrium landscapes. It is somewhat surprising that there are so many examples of quasi-equilibrium landscapes on disturbed and reclaimed lands. These are areas where geomorphic processes proceed at moderate rates, without rapid dissection or degradation of the landscape, and permit some economic activity such as grazing or agriculture or other land use such as wildlife habitat.

Despite the various relationships that should be constructed, current reclamation practices are usually rather successful. Nature adjusts form and process toward the equilibrium or steady state at rates generally acceptable to society in most cases. Nature, then, presumably re-creates the important relationships for us with time.

It should be equally apparent that there is ample room for improvement in the design and implementation of reclamation programs, and this may well involve the inclusion of geomorphic principles. Logic dictates that the closer the initial reclaimed landscape is to an equilibrium or steady state, the smaller the magnitude of natural adjustments and the less likelihood of land degradation with both on-site and off-site environmental impacts. A reclaimed landscape that possesses morphometric and process relationships, as discussed above, is likely to constitute a closer approximation of an equilibrium or steady state than one that does not.

Conversely, clear-cut cases of reclamation failure may be traceable to the violation of geomorphic principles. Simply, there is great disparity between force and resistance. The hillslope is too steep or too long to maintain low to moderate erosion rates, and the accelerated rate precludes the establishment of vegetation. The seed, fertilizer, and mulch all reside at the bottom of the hill after the first few precipitation events. Retreatment without regrading is futile.

Similarly, channel gradients can be too steep, drainage densities too low, or drainage basin areas too large. The ensuing channel entrenchment invades the entire basin, dissecting the reclaimed headlands and undercutting the hillslopes. Gully plugs or drop structures do not provide long-term solutions to such problems. The geomorphic disequilibrium pervades the entire system.

Standard engineering practice does not necessarily reestablish the sorts of geomorphic relationships described above. Therefore, some measure of disequilibrium exists in landforms and landscapes designed solely on the basis of these practices. Manmade structures of steel and concrete bear witness to this fact. Admittedly, there is little alternative on some occasions, although Gray and Leiser (1982) offered various compromises. We also submit that, from the geomorphic perspective, most engineering designs are relatively short-term and require periodic maintenance. Considerable attention is given today to relieving future generations of burdensome legacies caused by our actions or inactions. Incorporation of geomorphic principles into landform and landscape designs, whenever possible, appears to support that philosophy.

In certain circumstances, there is an approach to the design of reclaimed landscapes that greatly simplifies the process and indirectly incorporates many geomorphic relationships. For want of a better name, we call this procedure "infiltration matching." Let us suppose that within a region, there is general homogeneity of climatic characteristics, essentially no influence of geologic structure, but differences in the characteristics of surface materials. Under such conditions, topography, soils, and vegetation are likely to be largely reflective of the variabilities of these materials. We could measure the infiltration capacity of the reclaimed surface and compare it to the infiltration capacities of undisturbed surfaces on the various surface materials. If a reasonable match can be found, then the natural area can be examined in detail to ascertain selected drainage basin, hillslope, and channel properties. The replication of these attributes on the reclaimed land would likely provide an approximation of equilibrium or steady-state form.

To the best of our knowledge, this approach has never been attempted, but with some refinement, it might prove helpful where the conditions of its use can be fulfilled. Of course, the dynamic and evolutionary aspects of the initial reclaimed surface would not be taken into account, but the same is true for other design procedures.

Management of Disturbed Lands

Management refers to the control exercised over a parcel of land while disturbance is actually taking place, as in the case of grazing, or following the disturbance and reclamation of the area. In either case, a fundamental goal is to hold both on-site and off-site impacts to tolerable levels. This involves three components: (1) monitoring the impacts, (2) decisions and activities to reduce future impacts, and (3) remedial actions to repair existing impacts. The principles of geomorphology can play a role in each.

Monitoring the impacts often involves direct or indirect measurement of the work performed by geomorphic processes and erosion by water in particular. This work is an expression of the relationship between forces and resistances. Commonly, there is observational evidence of existing or potential problems. Deterioration of vegetation cover usually permits an increase in sheet erosion. Rilling or gullying indicates accelerated erosion by channelized flow. Tension cracks across hillslopes suggest mass instability.

Sometimes field data are collected using the techniques listed in Appendix A of this volume. Here, baseline data or information from nearby undisturbed areas becomes important for comparative purposes. This approach should be the most accurate but also the most costly.

When impacts become intolerable, according to some standard, it becomes necessary to select and implement procedures to mitigate further adverse consequences. If the impact is assessed in terms of areal extent, intensity, and duration, then each provides options for appropriate responses to such problems. Usually each specific type of disturbance has a number of possible approaches for reducing impacts. For example, the deterioration of vegetation in a pasture can be rectified by reducing the number of livestock on a given range or moving them to an entirely different location for a time. Likewise, it is suggested that a number of areas be designated for recreational vehicle use and the opening and closing of these sites be rotated to allow recovery of disturbed areas.

Sometimes considerable environmental damage is done before the problem is identifed or corrective action undertaken. Then it becomes necessary to employ remedial actions in order to halt further and progressive deterioration of the site. If the impact is the consequence of extreme events, such as intense precipitation and flooding, retreatment with past reclamation practices may be all that is necessary. However, if the cause cannot be traced to extraordinary circumstances, then there may be more serious problems creating a disequilibrium. It might be necessary to regrade hillslopes entirely and reconstruct channels in order to attain stability. Although this entails considerable expense, it may prove economical in the long run when compared to periodic retreatment with unsatisfactory results. Recall that in the case of surface mining of coal, each retreatment restarts the time clock of the liability period and delays bond release. In other cases, final payment for a project may be delayed until stability is demonstrated. The geomorphic principles addressed for the design of hillslopes, channels, and drainage basins are usually equally applicable in the selection and implementation of activities to reduce future impacts and remedial actions to correct existing impacts.

Throughout this text, we emphasize the notion that the earth's terrestrial surface is composed of drainage basins, each of which constitutes a process–response system. And each of these systems is, in turn, made up of numerous process–response subsystems in the form of hillslopes and channels. This concept is highly important in the management of disturbed lands because it reveals that all parts of the landscape are functionally linked to other parts. In other words, it is possible for an unstable landform to be linked to an otherwise stable landform, equilibria to disequilibria, or steady state to unsteady state. In the operation of the geomorphic system, it is entirely possible for the processes and products of the unstable part to disrupt the stability of the other parts, resulting in widespread failure throughout the landscape. For example, accelerated sheet erosion on hillslopes can lead to channel deposition that serves to steepen the gradient along a reach. This may trigger a period of channel entrenchment creating steep channel banks susceptible to mass failure or slumping. Lateral erosion of the stream against an adjacent hillslope steepens the hillslope gradient, removes the concave basal portion, and thereby increases surface erosion rates while reducing the opportunity for sediment storage at the bottom of the hillslope. Additionally, the removal of material from the toe of the hillslope could produce mass movement.

The principles underlying this scenario provide the rationale for attentive maintenance and timely remedial action. Indeed, the unstable site might evolve toward stability and repair itself in time. It could also remain unstable and cause additional instability in other areas, unless checked. The old adage, "a stitch in time saves nine," comes to mind.

Conclusion

From the foregoing, it is evident that human activities can be very effective in altering the relationship between landforms and geomorphic processes and, as a consequence, the very operation of the geomorphic system in a particular locale. The results of human interference in these natural systems are not mysterious; usually they can be anticipated to a considerable degree. Commonly, they can be explained in terms of changes in force and resistance. However, geomorphic systems possess two characteristics that complicate matters somewhat: (1) thresholds and (2) complex response mechanisms. These concepts help us to understand why some human activities seem to elicit minimal geomorphic response (the threshold has not been reached or exceeded), while other activities seem to cause the system to over-react; the whole landscape seems to come apart (the threshold has been surpassed and a chain reaction of complex responses ensue). Chorley *et al.* (1984) asserted that the geomorphologist's perspective of long time spans and large areas gives him or her an awareness of the causes of landscape form and of the complexity and potential for landscape change in time and space. Coates (1984, p. 97) stated:

> The geomorphologist is trained to recognize the equilibrium of landforms and the processes responsible for their formation. When the natural balance is distorted, he is best able to predict the type and amount of modification that societal activities will produce. By providing counsel to environmental planners, the geomorphologist can 1. suggest ways to minimize terrain degradation, 2. provide data to assist in maintaining the safety, health, and welfare of the public, and 3. act as a surrogate for nature when circumstances require such stability.

As amply demonstrated in this volume, the science of geomorphology offers powerful assistance in the design of reclamation programs and the understanding of past reclamation failures. Ideally, successful reclamation provides permanent and maintenance-free control of geomorphic processes. Under these conditions, the long-term stability of disturbed landscapes and landforms is a reasonable expectation.

It must be recognized, however, that the state-of-the-art in reclamation has not yet developed to the point that either permanence or freedom from maintenance can be guaranteed. Unforeseen conditions, such as extreme climatic events, disease, or fire, or perhaps even the transgression of a geomorphic threshold elsewhere in the undisturbed landscape, can cause the downfall of the best laid reclamation plans. To the geomorphologist, accustomed to dealing with events of various magnitudes and frequencies over extended periods of time, these occurrences are not especially unusual and certainly no cause for any allocation of blame for the failure.

Further, it is understood that some reclamation practices are inherently temporary from the geomorphic perspective. On hillslopes, surface manipulations such as terracing, ripping, and chiseling are only effective for a few years. Likewise, soil amendments such as mulches, fertilizers, neutralizers, and tackifiers are effective for short periods. Engineering structures such as retaining walls, gabions, and slope drains have finite design lives. In channels, rip-rap along the banks, revetments, and drop structures usually serve for several years or less. Eventually, the landscape or landform is on its own to remain stable or adjust to the prevailing environmental conditions and resultant geomorphic system. Unless the above devices receive periodic maintenance, the long-term stability of the landscape remains in doubt.

In recent times, several publications have appeared that focus upon "applied" geomorphology, which Jones (1980) defined as the application of geomorphic understanding to the analysis and solution of problems concerning land occupancy, resource exploitation, environmental management, and planning. Chorley *et al.* (1984) observed that if geomorphology has as its role the identification and description of landforms, the explanation of their origin and the prediction of future change, then applications of geomorphology should fall within these categories of description, prediction, and postdiction. In an interesting discussion of geomorphology and public policy, Coates (1984) suggested that there are five principal areas in which the environmental geomorphologist may serve and become involved with policy: publication of information, government work, industry, consulting work, and participation with special interest groups. Both Chorley *et al.* (1984) and Coates (1984) presented numerous case studies of applied geomorphology.

Although Chorley *et al.* (1984) acknowledged recent increasing interest among geomorphologists concerning the application of geomorphic principles to contemporary environmental problems, we are, nevertheless, still inclined to agree with Craig and Craft (1982, p. v), who contended that, "yet too much of our work remains obscure and poorly understood by those who are most likely to benefit from it." We believe that the issue of human land disturbance offers a valuable opportunity for geomorphologists to ply their trade in all of the ways listed above by Coates (1984). It is our fervent hope that this volume will serve as a catalyst for more of us to do so.

References

Abrahams, A. D., 1972, Environmental constraints on the substitution of space for time in the study of natural channel networks: Geological Society of America, Bulletin, Vol. 83, pp. 1523–30.

Abrahams, A. D., 1980, Channel link density and ground slope: Annals of Association of American Geographers, Vol. 70, No. 1, pp. 80–93.

Abrahams, A. D., and Ponczynski, 1984, Drainage density in relation to precipitation intensity in the USA: Journal of Hydrology, Vol. 75, pp. 383–388.

Chorley, R. J., 1971, The drainage basin as the fundamental geomorphic unit: *in* Chorley, R. J. (ed.), Introduction to physical hydrology: London, Methuen and Co., Ltd., pp. 37–59.

Chorley, R. J., Schumm, S. A., and Sugdin, D. E., 1984, Geomorphology: London, Methuen and Co., Ltd., 605 pp.

Coates, D. R., 1984, Geomorphology and public policy: *in* Costa, J. E., and Fleisher, P. J. (eds.), Developments and Applications of Geomorphology: New York, Springer-Verlag, pp. 97–132.

Craig, R. G., and Craft, J. L., 1982, Applied geomorphology: London, George Allen and Unwin, Ltd., 253 pp.

Gray, D. H., and Leiser, A. T., 1982, Biotechnical slope protection and erosion control: New York, Van Nostrand Reinhold Co., 271 pp.

Hadley, R. F., and Schumm, S. A., 1961, Sediment sources and drainage-basin characteristics in upper Cheyenne River basin: U. S. Geological Survey, Water Supply paper 1531-B, pp. 137–196.

Horton, R. E., 1945, Erosional development of streams and their drainage basins: hydrophysical approach to quantitative morphology: Geological Society of America, Bulletin, Vol. 56, pp. 295–370.

Israelsen, C. E., Clyde, C. G., Fletcher, J. E., Israelsen, E. K., Haws, F. W., Packer, P. E., and Farmer, E. E., 1980, Erosion control during highway construction: Research Report, National Cooperative Highway Research Program Report 220, Transportation Research Board, National Research Center, 30 pp.

Jones, D. K. C., 1980, British applied geomorphology: an appraisal: Zeitschrift für Geomorphology, Supp. 36, pp. 48–73.

Langbein, W. B., and Schumm, S. A., 1958, Yield of sediment in relation to mean annual precipitation: American Geophysical Union, Transactions, Vol. 30, No. 6, pp. 1076–1084.

Melton, M. A., 1957, An analysis of the relations among elements of climate, surface properties, and geomorphology: Office of Naval Research, Technical Report 11, Project NR 389–042, 102 pp.

Melton, M. A., 1958, Geometric properties of mature drainage systems and their representation in an E_4 phase space: Journal of Geology, Vol. 66, pp. 35–56.

Meyer, L. D., and Romkens, M. J. M., 1976, Erosion and sediment control on reshaped land: Proceedings of the Third Interagency Sedimentation Conference, PB–245–100, Water Resources Council, Washington, D.C., pp. 2–65 to 2–76.

Ritter, D. F., 1978, Process Geomorphology: Dubuque, Iowa, W. C. Brown Co., 603 pp.

Schumm, S. A., 1977, The fluvial system: New York, John Wiley and Sons, Inc., 338 pp.

Schumm, S. A., 1979, Geomorphic thresholds: the concept and its applications: Institute of British Geographers, Trans., Vol. 4, pp. 485–515.

Schumm, S. A., and Chorley, R. J., 1983, Geomorphic controls on the management of nuclear waste, U. S. Nuclear Regulatory Commission Report NUREG/CR–3276, 137 pp.

Shown, L. M., Frickel, D. G., Miller, R. F., and Branson, F. A., 1982, Methodology for hydrologic evaluation of a potential surface mine, Loblolly Branch Basin, Tuscaloosa County, Alabama: U. S. Geological Survey, Water Resources Investigation, Open File Report 82–50, 93 pp.

Shreve, R. L., 1966, Statistical law of stream numbers: Journal of Geology, Vol. 74, pp. 17–37.

Strahler, A. N., 1952, Hypsometric (area-altitude) analysis of erosional topography, Geological Society of America, Bulletin, Vol. 63, pp. 1117–1142.

Toy, T. J., 1977, Hillslope form and climate: Geological Society of America, Bulletin, Vol. 88, pp. 16–22.

Wells, S. G., and Rose, D. E., 1981, Applications of geomorphology to surface-coal mining reclamation, northwestern New Mexico: New Mexico Geological Survey, Special Publications, No. 10, pp. 69–83.

Wilcock, D. N., 1975, Relations between planimetric and hypsometric variables in third- and fourth-order drainage basins: Geological Society of America, Bulletin, Vol. 86, pp. 47–50.

Wischmeier, W. H., and Smith, D. D., 1965, Predicting rainfall-erosion losses from cropland east of the Rocky Mountains: U.S.D.A. Handbook 282, 44 pp.

Young, A., 1972, Slopes: London, Longman Group Ltd., 288 pp.

Appendix A

Methods of Geomorphic Analyses

Introduction

The science of geomorphology focuses upon the study of landforms, their origin, evolution, present configuration, and the processes that have, are, and may operate on them. Geomorphic processes are a major concern in the application of geomorphic principles to the understanding of the present and future conditions of disturbed lands. As described throughout this volume, it is clear that these processes are affected by many components of the physical environment in which the landforms of interest are situated. As a result, geomorphic analyses are necessarily multivariate in nature.

In gathering data, the geomorphologist commonly utilizes field and laboratory techniques developed by other scientific disciplines. Climatologists and meteorologists provide methods for compiling and assessing short-term and long-term weather data. Soil science and soils engineering provide procedures for evaluation of unconsolidated surface materials. Botanists and plant ecologists provide techniques for characterizing the nature of vegetal surface covers. Hydrologists provide a variety of ways for amassing and configuring surface and subsurface water information. In most cases, it is advisable to employ the standard procedures of these disciplines whenever possible.

In addition, geomorphologists themselves have generated numerous techniques for gathering data of particular interest to them. It is not feasible to fully describe all of the various procedures that have been used over the years. However, we offer direction for those interested in collecting their own information for special studies. The following is a selection of references that should be useful in this regard.

Topic	Source	Content
General geomorphology	King, C. A. M., 1967, Techniques in geomorphology: London, Edward Arnold Ltd., 342 pp.	Field techniques—form and process Experiment and theory models Cartographic and morphometric analysis Sediment analysis Statistical analysis
	Goudie, A., 1981, Geomorphological techniques: London, George Allen and Unwin Ltd., 395 pp.	Methods of investigation Morphometry—general, hillslopes, channels, basins Material properties—physical, chemical, strength Processes—weathering, hillslope, solutes, channel, glacial, aeolian, coastal Evolution—dating methods
	Gardiner, V., and Dackombe, R., 1983, Geomorphology field manual: London, George Allen and Unwin Ltd., 254 pp.	Topographic survey Geomorphological mapping Hillslope profiling Mapping of landscape materials Processes—fluvial, glacial, aeolian, coastal, hillslope Sampling
Statistical methods	Chorley, R. J. (ed.), 1972, Spatial analysis in geomorphology: New York, Harper and Row, 393 pp.	General Point systems Networks Continuous distributions Space partitioning Simulation
	Doornkamp, J. C., and King, C. A. M., 1971, Numerical analysis in geomorphology: London, Edward Arnold, Ltd., 372 pp.	Drainage basins Slopes Coastal forms Glacial forms
Drainage basin morphology	Gardiner, V., 1975, Drainage basin morphology: British Geomorphological Research Group, Technical Bulletin, No. 14, published by Geo Abstracts, Ltd., Norwich, England, 48 pp.	Data sources Stream-ordering measurements Morphometric variables Analysis techniques
	Chorley, R. J., 1971, The drainage basin as the fundamental geomorphic unit: in Chorley, R. J.	Morphometric units Linear aspects of basins Areal aspects of the basin

Topic	Source	Content
	(ed.), Introduction to Physical Hydrology, London, Methuen and Co., Ltd., pp. 37–59.	Relief aspects of the basin Consideration of scale
	Strahler, A. N., 1957, Quantitative Analysis of Watershed Geomorphology: Transactions, American Geophysical Union, Vol. 38, No. 6, pp. 913–920.	Dimensional analysis Order analysis Bifurcation ratio Frequency distribution of stream lengths Relationship of stream length to stream order Drainage basin area Drainage density and texture ratio Constant of channel maintenance Maximum valley slope Mean slope curve Slope maps Rapid slope sampling Relief ratio Hypsometric analyses
Drainage basin management	Food and Agriculture Organization of the United Nations, 1977, Guidelines for Watershed Management Rome FAO, 293 pp.	Land classification Results of conservation projects Evaluation of erosion Runoff plots Gully control Land management Impacts of landuse
Hillslope form	Leopold, L. B., and Dunne, T., 1971, Field methods for hillslope description: British Geomorphological Research Group, Technical Bulletin, No. 7, published by Geo Abstracts Ltd., Norwich, England, 24 pp.	Choice and location of profiles Survey of hillslope profiles Survey of cross profiles Description of vegetation Surficial material
	Young, A., 1974, Slope profile survey: British Geomorphological Research Group, Technical Bulletin, No. 11, published by Geo Abstracts Ltd., Norwich, England, 52 pp.	Definition of units Siting and alignment of profiles Profile survey Initial treatment of data Profile analysis
	Strahler, A. N., 1956, Quantitative slope analysis: Bulletin of Geological Society of America, Vol. 67, pp. 571–596.	Slope maps Frequency distribution of slope Forms of slope frequency distributions Simplified point-sampling methods
Hillslope processes	James, P. A., 1971, The measurement of soil frost-heave in the	Soil frost heave mechanisms Factors affecting soil frost heave

(*continued*)

Topic	Source	Content
Hillslope processes (*continued*)	field: British Geomorphological Research Group, Technical Bulletin, No. 8, published by Geo Abstracts, Norwich, England, 43 pp.	Technique for measurement of frost heave New design for the automatic frost heave recorder
	Anderson, E. W., and Finlayson, B., 1975, Instruments for measuring soil creep: British Geomorphological Research Group, Technical Bulletin, No. 16, published by Geo Abstracts Ltd., 51 pp.	Instruments that measure extremely accurately Instruments that measure in the long term Instruments that magnify and measure the changes
	Campbell, I. A., 1970, Microrelief measurements on unvegetated shale slopes: Professional Geographer, Vol. 22, No. 4, pp. 215–220.	Soil erosion
	Haigh, M. J., 1981/82, Microprofile measurement using the contour gauge: British Geomorphological Research Group, Technical Bulletin, No. 29, published by Geo Abstracts Ltd., Norwich, England, pp. 31–32.	Soil erosion
	Toy, T. J., 1983, A linear erosion/elevation measuring instrument (LEMI): Earth Surface Processes and Landforms, Vol. 8, pp. 313–322.	Soil erosion/heave (ground advance and retreat)
Miscellaneous hillslope properties	Finlayson et al., 1977, Shorter Technical Methods (II): British Geomorphological Research Group, Technical Bulletin, No. 18, 49 pp.	Measurement of soil creep at a stream bank Simple method for field measurement of slope profiles Field assessment of rock hardness Use of erosion pins
	Leopold, L. B. (ed.), 1968, Field method manual: Commission on Applied Geomorphology of International Geographical Union, Numero Special, pp. 147–188.	Slope profiles Soil investigations Slope processes and rates Vegetation Movement of blocks Overland flow traps Seismographic methods Photography Inclinometers Shift of rock fragments
Channel form	Richards, K. S., 1982, Rivers: Form and process in alluvial	Channel geometry relationships Sediment transport

Topic	Source	Content
	channels, London, Methuen and Co. Ltd., 358 pp.	Hydraulic geometry
	Schumm, S. A., Harvey, M. D., and Watson, C. C., 1984, Incised channels: Morphology, dynamics, and control: Littleton, Colorado, Water Resources Publications, 200 pp.	Channel morphology Erosion control
Channel processes	Leopold, L. B., 1968, as above listed under Miscellaneous Hillslope Properties	Mapping of river bank conditions Hydraulic measurements Crest stage Suspended sediment sampler Bedload trap Fluorescent sand
	Hadley, R. F., Frickel, D. G., Shown, L. M., and Miller, R. F., 1981, Methodology for hydrologic evaluation of a potential surface mine: East Trail Creek Basin, Big Horn County, Montana: U.S. Geological Survey, Water-Resources Investigations, Open-File Report 81-58, 73 pp.	Surface water estimating methods—peak discharges, runoff volumes, low flows Sediment yields
	Hedman, E. R., Moore, D. O., and Livingston, R. K., 1972, Selected streamflow characteristics as related to channel geometry of perennial streams in Colorado: U.S. Geological Survey, Water-Resources Division, Open-File Report (no number), 14 pp.	Estimating discharge from channel geometry and other variables
	McCain, J. F., and Jarrett, R. D., 1976, Manual for estimating flood characteristics of natural-flow streams in Colorado, Colorado Water Conservation Board, Colorado Department of Natural Resources, Technical Manual No. 1, 68 pp.	Estimating discharge of floods with various recurrence intervals from various channel and basin properties
Other field techniques	Toebes, C., and Ouryvaev, V. (eds.), 1970, Representative and experimental basins—an international guide for research and practice, Henkes-Holland, Haarlem, United Nations Educa-	Planning of observations according to research objectives Methods of observation and instrumentation for various climatic, hydrologic, and geomorphic variables

(continued)

Topic	Source	Content
Other field techniques (*continued*)	tional, Scientific and Cultural Organization, 348 pp.	Analytical techniques and interpretation of results
	Yatsu, E., Dahms, F. A., Falconer, A., Ward, A. J., and Wolfe, J. S., 1971, Research Methods in Geomorphology—First Guelph Symposium on Geomorphology, Don Mills, Ontario, Canada, Science Research Associates Ltd., 140 pp.	Research methods in karst geomorphology Research methods in glacial geomorphology (economic factors) Pedological methods in geomorphic research General discussion of research methods in geomorphology
	Miller, J. P., and Leopold, L. B., 1963, Simple measurements of morphological changes in river channels and hillslopes: United Nations Educational, Scientific, and Cultural Organization, Arid Zone Research, Vol. 20, pp. 421–427	Monumented cross sections of channels Pins for measuring bank recession Chains as indicators of bed scour Depth of flow Movement of individual rocks in streambeds Pins to measure hillslope erosion Stakes for measurement of soil creep or mass movement

Appendix B

Location of Available Data

The following are general repositories of information. Local sources, such as district or regional offices of the U.S. Geological Survey, U.S. Forest Service, U.S. Soil Conservation Service, U.S. Army Corps of Engineers, state geological surveys, or offices of the state climatologists may also possess useful data. Sometimes academic departments within the state land grant colleges and universities have research reports, Master's theses, doctoral dissertations, and extension service reports that are often very site specific.

Aerial Photography, Remote Sensing Data

Aerial Photography Field Office
USDA–ASCS
2222 West 2300 South
P.O. Box 30010
Salt Lake City, UT 84125

User Services
EROS Data Center
Tenth and Dakota Avenue
Sioux Falls, SD 57198

National Cartographic Information Center
U.S. Geological Survey
Room 1C107
507 National Center
Reston, VA 22092

For a detailed description of data available from these and other local services, see:

Smith, G. L., 1981. Documentation of national data sources relative to stream-related hazards and assessment of data for application to highway and bridge sites: *In* Shen, H. W. *et al.*, Methods for assessment of stream-related hazards to highways and bridges, Federal Highway Administration, FHWA Report No. FHWA/RD-80/160, Appendix C, pp. 183–214.

Geologic Maps

Distribution Section (continental United States)
U.S. Geological Survey
Federal Center
Box 25046
Denver, CO 80225

Distribution Section (Alaska)
U.S. Geological Survey
310 First Avenue
Fairbanks, AK 99701

Hydrologic Data

National Water Data Exchange (NAWDEX)

Program Office
National Water Data Exchange (NAWDEX)
U.S. Geological Survey
421 National Center
12201 Sunrise Valley Drive
Reston, VA 22092

National Water Data Storage and Retrieval System (WATSTORE)

Chief Hydrologist
U.S. Geological Survey
437 National Center
Reston, VA 22092

For a detailed description of data available from these sources, see:

Staubitz, W. W., and Sobashinski, J. R. 1983. Hydrology of Area 6, Eastern Coal Province, Maryland, West Virginia, and Pennsylvania: USGS, Water Resources Investigations, Open-File Report 83-33, pp. 63–67.

or any in series of Open-File Reports concerning Hydrology of Coal Provinces by USGS.

Land Use, Land Covering, and Associated Maps

Eastern Mapping Center
U.S. Geological Survey
536 National Center
Reston, VA 22092

Mid-Continent Mapping Center
U.S. Geological Survey
2400 Independence Road
Rolla, MO 65401

Rocky Mountain Mapping Center
National Mapping Division
U.S. Geological Survey
Box 25046, Mail Stop 510
Denver Federal Center
Denver, CO 80225

Western Mapping Center
U.S. Geological Survey
345 Middlefield Road
Menlo Park, CA 94025

Meteorologic, Weather, and Climate Data

National Oceanic and Atmospheric Administration
Environmental Data Service
National Climatic Center
U.S. Department of Commerce
Federal Building
Asheville, NC 28801

Soil Characteristics Data

U.S. Department of Agriculture
Soil Conservation Service
Washington, D.C. 20013

Soil maps are prepared on a county basis and contain a wealth of information but are not available for all counties. The Soil Conservation Service maintains an office in each state and usually several field offices as well. These can be located through telephone directories under the heading for U.S. Department of Agriculture.

Topographic Maps

See sources listed under *Geologic Maps*.

Appendix C

Unit Conversion Factors

Unit Conversion Factors[a]

To convert col. 1 into col. 2, multiply by:	Column 1 (metric)	Column 2 (English)	To convert col. 2 into col. 1, multiply by:
		Area	
0.386	kilometer^2 (km^2)	square mile	2.590
247.1	kilometer^2 (km^2)	acre	0.00405
2.471	hectare (ha)	acre	0.4047
		Length	
0.621	kilometer (km)	mile	1.609
1.094	meter (m)	yard (yd)	0.914
0.394	centimeter (cm)	inch (in.)	2.540
0.0394	millimeter (mm)	inch (in.)	25.40
		Mass	
1.102	metric ton	English ton (ton)	0.9072
2.205	kilogram (kg)	pound (lb)	0.454
0.035	gram (g)	ounce (oz)	28.35

(*continued*)

Unit Conversion Factors[a] *(Continued)*

To convert col. 1 into col. 2, multiply by:	Column 1 (metric)	Column 2 (English)	To convert col. 2 into col. 1, multiply by:
	Rates (yields)		
0.446	metric tons per hectare (metric tons/ha)	English tons per acre (tons/acre)	2.242
0.892	kilograms per hectare (kg/ha)	pounds per acre (lb/acre)	1.121
285.494	metric tons per hectare (metric tons/ha)	English tons per square mile (tons/sq mile)	0.0035
0.136	metric tons per hectare-meter (metric tons/ha-m)	English tons per acre-foot (tons/ac-ft)	7.357
0.0021	cubic meters per square kilometer (m^3/km^2)	acre-feet per square mile (ac-ft/sq mile)	476.23
35.311	cubic meters per second (m^3/sec)	cubic feet per second (cu ft/sec)	0.0283
	Volume		
35.311	cubic meters (m^3)	cubic feet (cu ft)	0.02832
8.108	hectare-meters (ha-m)	acre-feet (ac-ft)	0.1233
0.2642	liter	gallon (gal)	3.785
0.0353	liter	cubic feet (cu ft)	28.329

Other useful conversion factors

1 in./hr of runoff from 1 acre = 1 cu ft/sec or 0.0283 m^3/sec.
1 ac-ft = 325,851 gal or 1,233,346 liters.
1 cu ft/sec for 1 day = 1.98 ac-ft or 0.244 ha-m (hectare-meter).

[a]Note: Comprehensive lists of conversion factors can be found in Dietrich, R. V., Dutro, J. T., and Foose, R. M. (compilers), 1982, AGI data sheets for geology in the field, laboratory, and office: American Geological Institute, Falls Church, Virginia, pp. 37.1–39.2.

Index

D

G

H

M

N

O

P

R

S

T